Proceedings of the 7th International Conference on

CLUSTERING ASPECTS OF NUCLEAR STRUCTURE AND DYNAMICS

Proceedings of the 7th International Conference on

CLUSTERING ASPECTS OF NUCLEAR STRUCTURE AND DYNAMICS

Rab, Island of Rab, Croatia 14 – 19 June 1999

Editors

M. Korolija

Z. Basrak

R. Čaplar

Ruđer Bošković Institute, Zagreb, Croatia

World Scientific

Singapore • New Jersey • London • Hong Kong

Published by

World Scientific Publishing Co Pte Ltd
5 Toh Tuck Link, Singapore 596224
USA office: 27 Warren Street, Suite 401-402, Hackensack, NJ 07601
UK office: 57 Shelton Street, Covent Garden, London WC2H 9HE

British Library Cataloguing-in-Publication Data
A catalogue record for this book is available from the British Library.

CLUSTERING ASPECTS OF NUCLEAR STRUCTURE AND DYNAMICS

ISBN-13 978-981-02-4233-6
ISBN-10 981-02-4233-6

International Scientific Advisory Committee

Y. Abe (Kyoto)
D. Baye (Bruxelles)
R.R. Betts (Argonne/Chicago)
B. Borderie (Orsay)
P. Braun-Munziger (GSI)
R.A. Broglia (Milano)
E.B. Campbell (Gothenburg)
B.R. Fulton (Birmingham)
C.K. Gelbke (MSU)
W. Greiner (Frankfurt)
H. Horiuchi (Kyoto)
R.G. Lovas (Debrecen)
H.O. Lutz (Bielefeld)
W. Mittig (GANIL)
G. Münzenberg (GSI)
W. Nazarewicz (ORNL/Warsaw)
W. von Oertzen (HMI)
Y.Ts. Oganessian (Dubna)
D. Pelte (Heidelberg)
M. Petrovici (Bucharest)
A.V. Ramayya (Nashville)
R.A. Ricci (Legnaro)
G. Röpke (Rostock)
P. Schuck (Grenoble)
I. Tanihata (RIKEN)
D. Vinciguerra (Catania)

Organizing Board

Z. Basrak (Zagreb, Cochairman)
N. Cindro (Zagreb, Honorary Chairman)
R. Čaplar (Zagreb, Cochairman)
F. Haas (Strasbourg)
Đ. Miljanić (Zagreb)
V. Paar (Zagreb)
W. Scheid (Giessen)

Local Organizing Committee

Z. Basrak (Zagreb)
R. Čaplar (Zagreb)
M. Dželalija (Split)
M. Korolija (Zagreb)
M. Milin (Zagreb)
N. Pavin (Zagreb)
S. Szilner (Zagreb)

Sponsors

- International Union of Pure and Applied Physics (IUPAP)
- Ministry of Science and Technology of the Republic of Croatia
- Ruđer Bošković Institute
- Centre de Calcul de l'Institut National de Physique Nucléaire et de Physique des Particules
- Zagreb Fair

Supporters

- Pliva d.d.
- Osijek city printing-house
- Chromos catering Zagreb
- Zagrebačka banka d.d.
- City Council of Zagreb

This volume is dedicated to Professor Nikola Cindro for his endeavor to keep the subject of nuclear clusters - quasimolecular clusters in particular - alive.

Foreword

This volume is a collection of invited talks and contributions selected for oral presentation at the 7th *International Conference on Clustering Aspects of Nuclear Structure and Dynamics, CLUSTER '99*, held from June 14 to 19, 1999, at Rab, Island of Rab, Croatia. The Conference was organized by the *Laboratory for Heavy-Ion Physics* of the *Ruđer Bošković Institute*, Zagreb, Croatia.

The principal role of scientific meetings is to report on recent achievements in the field and to discuss possible developments. The Rab Conference had a particular feature, to be held 30 years after the first nuclear cluster conference, Bochum, Germany, 1969, and to be held at the turn of the century. Thus, a deeper retrospective look backward and an enhanced prospective look forward impose themselves naturally.

Initially, the Conference was intended to be about one third devoted to clustering phenomena in other branches of physics, mainly atomic physics, closely related to or issuing from nuclear concepts. However, an overwhelming number of suggestions for subjects on clustering aspects in nuclei was received from members of the *International Scientific Advisory Committee.* Thus, a decision was made to have a five-day *nuclear* conference and to add a sixth day for non-nuclear subjects. Owing to the political and military situation in the Croatian neighborhood during the months and weeks preceding the Conference, the organizers had to face an non-negligible number of cancellations of speakers. Therefore, the Program of the Conference does not fully reflect the chairmen wishes nor the Advisory Committee recommendations. The topics covering super- and hyperdeformed nuclei, astrophysics, and non-nuclear clusters were not fully covered.

Besides illuminating discussions on a number of questions raised at the *round table*, the scientific spirit of the Conference greatly benefited from numerous stimulating questions and comments on the presented talks. These are summarized in the Conference *summary talk*, emphasizing achievements and future developments.

A lively *poster session* inspiring discussions and exchange of ideas took place, creating a proper atmosphere for putting forward prospects for future experiments and collaborations between research groups from different countries. Summaries of most of the Conference talks as well as contributions and poster summaries are collected in the *Booklet of Summaries* distributed at the Conference. The complete *Booklet* is available in postscript format at the Conference WWW site (`http://icc99.irb.hr`). The interested reader may refer to the authors for additional information (see the attached *List of Participants*). Let me mention that one may also find a number of photos at the Conference WWW site (taken mostly during the Conference excursion).

The organizers are grateful to all participants for creating an ambience of friendship and well-being. I believe that this Conference has succeeded in furthering the communication between scientists working on various aspects of clustering phenomena. To my mind, this is especially valid for a relatively small community of *students* and *postdocs* attending the Conference. Let me anticipate that fruits of such contacts will be very beneficial to the cluster field of research as a whole. We hope that a number of these young people will meet again in Oxford and Nara.

Following the tradition, the *International Scientific Advisory Committee* held a meeting during the Conference and decided that this series of topical conferences should be continued. Two proposals were unanimously accepted: a Nuclear Cluster Workshop to be held in Oxford, United Kingdom, in 2001, and the 8th Nuclear Cluster Conference to be held at Nara, Japan, in 2003.

Finally, I wish to express my gratitude to the *International Scientific Advisory Committee*, the *Organizing Board*, and the *Local Organizing Committee* for their efforts which made the Conference possible. Without support of the Conference sponsors and supporters, the organizers' tasks would be much harder to fulfill. On behalf of all Committees let me express our gratitude to all of them.

To underline the role which *Professor Nikola Cindro* had in initiating the organization of the 7th international conference on nuclear clusters in Croatia, we have decided to *dedicate this volume* to him.

Zagreb and Nantes, October 1999 Z. Basrak

CONTENTS

STRUCTURE OF VACUUM AND ELEMENTARY MATTER

WALTER GREINER
Institut für Theoretische Physik, J.W. Goethe-Universität,
D-60054 Frankfurt, Germany

The extension of the periodic system into various new areas is investigated. Experiments for the synthesis of superheavy elements and the predictions of magic numbers are reviewed. Different ways of nuclear decay are discussed like cluster radioactivity, cold fission and cold multifragmentation, including the recent discovery of the tripple fission of ^{252}Cf. Furtheron, investigations on hypernuclei and the possible production of antimatter–clusters in heavy–ion collisions are reported. Various versions of the meson field theory serve as effective field theories at the basis of modern nuclear structure and suggest structure in the vacuum which might be important for the production of hyper– and antimatter. A perspective for future research is given.

There are fundamental questions in science, like e. g. "how did life emerge" or "how does our brain work" and others. However, the most fundamental of those questions is "how did the world originate ?". The material world has to exist before life and thinking can develop. Of particular importance are the substances themselves, i. e. the particles the elements are made of (baryons, mesons, quarks, gluons), i. e. elementary matter. The vacuum and its structure is closely related to that. On this I want to report today. I begin with the discussion of modern issues in nuclear physics.

The elements existing in nature are ordered according to their atomic (chemical) properties in the **periodic system** which was developped by Mendeleev and Lothar Meyer. The heaviest element of natural origin is Uranium. Its nucleus is composed of $Z = 92$ protons and a certain number of neutrons ($N = 128 - 150$). They are called the different Uranium isotopes. The transuranium elements reach from Neptunium ($Z = 93$) via Californium ($Z = 98$) and Fermium ($Z = 100$) up to Lawrencium ($Z = 103$). The heavier the elements are, the larger are their radii and their number of protons. Thus, the Coulomb repulsion in their interior increases, and they undergo fission. In other words: the transuranium elements become more instable as they get bigger.

In the late sixties the dream of the superheavy elements arose. Theoretical nuclear physicists around S.G. Nilsson (Lund) [1] and from the Frankfurt school [2,3,4] predicted that so-called closed proton and neutron shells should counteract the repelling Coulomb forces. Atomic nuclei with these special **"magic" proton and neutron numbers** and their neighbours could again

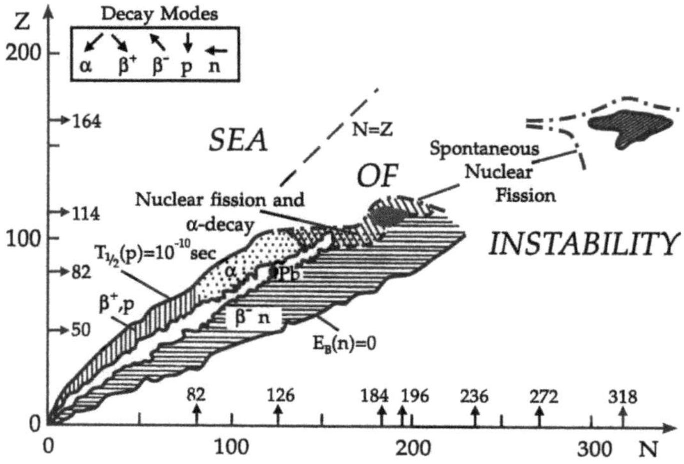

Figure 1. The periodic system of elements as conceived by the Frankfurt school in the late sixties. The islands of superheavy elements ($Z = 114$, $N = 184$, 196 and $Z = 164$, $N = 318$) are shown as dark hatched areas.

be rather stable. These magic proton (Z) and neutron (N) numbers were thought to be $Z = 114$ and $N = 184$ or 196. Typical predictions of their life times varied between seconds and many thousand years. Fig. 1 summarizes the expectations at the time. One can see the islands of superheavy elements around $Z = 114$, $N = 184$ and 196, respectively, and the one around $Z = 164$, $N = 318$.

The important question was how to produce these superheavy nuclei. There were many attempts, but only little progress was made. It was not until the middle of the seventies that the Frankfurt school of theoretical physics together with foreign guests (R.K. Gupta (India), A. Sandulescu (Romania)) [5] theoretically understood and substantiated the concept of bombarding of double magic lead nuclei with suitable projectiles, which had been proposed intuitively by the russian nuclear physicist Y. Oganessian [6]. The two-center shell model, which is essential for the description of fission, fusion and nuclear molecules, was developped in 1969-1972 together with my then students U. Mosel and J. Maruhn [7]. It showed that the shell structure of the two final fragments was visible far beyond the barrier into the fusioning nucleus. The collective potential energy surfaces of heavy nuclei, as they were calculated in the framework of the two-center shell model, exhibit pronounced valleys, such that these valleys provide promising doorways to the fusion of superheavy nu-

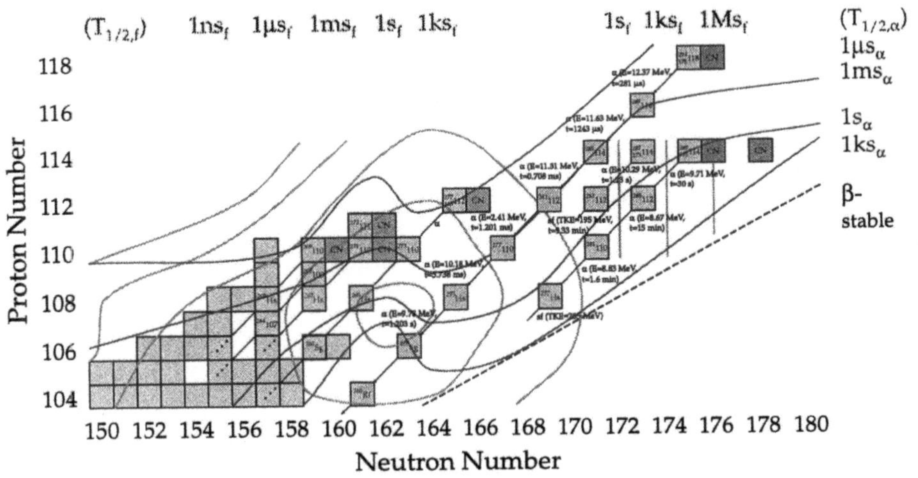

Figure 2. The $Z = 106 - 112$ isotopes were fused by the Hofmann–Münzenberg (GSI)–group. The two $Z = 114$ isotopes were produced by the Dubna–Livermore group. It is claimed that three neutrons are evaporated. Obviously the lifetimes of the various decay products are rather long (because they are closer to the stable valley), in crude agreement with early predictions [3,4] and in excellent agreement with the recent calculations of the Sobicevsky–group [11]. The recently fused $Z = 118$ isotope by V. Ninov et al. at Berkeley is the heaviest one so far.

clei for certain projectile-target combinations (Fig. 3). If projectile and target approach each other through those **"cold" valleys**, they get only minimally excited and the barrier which has to be overcome (fusion barrier) is lowest (as compared to neighbouring projectile-target combinations). In this way the correct projectile- and target-combinations for fusion were predicted. Indeed, Gottfried Münzenberg and Sigurd Hofmann and their group at GSI [8] have followed this approach. With the help of the SHIP mass-separator and the position sensitive detectors, which were especially developped by them, they produced the pre-superheavy elements $Z = 106, 107, \ldots 112$, each of them with the theoretically predicted projectile-target combinations, and only with these. Everything else failed. This is an impressing success, which crowned the laborious construction work of many years. The before last example of this success, the discovery of element 112 and its long α-decay chain, is shown in Fig. 4. Very recently the Dubna–Livermore–group produced two isotopes of $Z = 114$ element by bombarding ^{244}Pu with ^{48}Ca (Fig. 2). Also this is a cold–valley reaction (in this case due to the combination of a spherical and a deformed nucleus), as predicted by Gupta, Sandulescu and Greiner [9] in 1977.

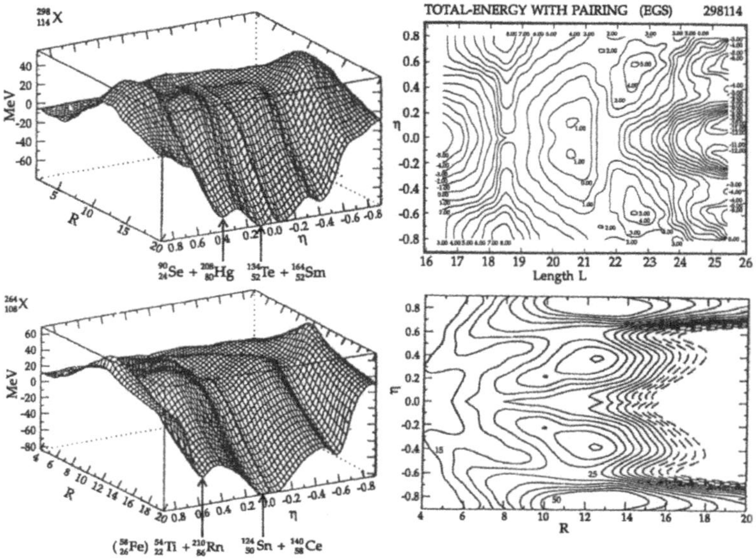

Figure 3. The collective potential energy surface of $^{264}108$ and $^{184}114$, calculated within the two center shell model by J. Maruhn et al., shows clearly the cold valleys which reach up to the barrier and beyond. Here R is the distance between the fragments and $\eta = \frac{A_1 - A_2}{A_1 + A_2}$ denotes the mass asymmetry: $\eta = 0$ corresponds to a symmetric, $\eta = \pm 1$ to an extremely asymmetric division of the nucleus into projectile and target. If projectile and target approach through a cold valley, they do not "constantly slide off" as it would be the case if they approach along the slopes at the sides of the valley. Constant sliding causes heating, so that the compound nucleus heats up and gets unstable. In the cold valley, on the other hand, the created heat is minimized. The colleagues from Freiburg should be familiar with that: they approach Titisee (in the Black Forest) most elegantly through the Höllental and not by climbing its slopes along the sides.

There exist also cold valleys for which both fragments are deformed [10], but these have yet not been verified experimentally. The very recently reported $Z = 118$ isotope fused with the cold valley reaction [12] ^{58}Kr + ^{208}Pb by Ninov et al. [13] yields the latest support of the cold valley idea.

Studies of the shell structure of superheavy elements in the framework of the meson field theory and the Skyrme-Hartree-Fock approach have recently shown that the magic shells in the superheavy region are very isotope dependent [14] (see Fig. 5). **According to these investigations $Z = 120$ being a magic proton number seems to be as probable as $Z = 114$.** Additionally, recent investigations in a chirally symmetric mean–field theory (see also

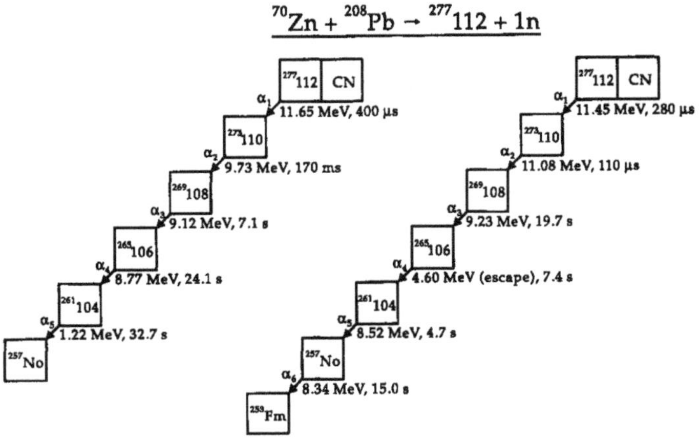

Figure 4. The fusion of element 112 with ^{70}Zn as projectile and ^{208}Pb as target nucleus has been accomplished for the first time in 1995/96 by S. Hofmann, G. Münzenberg and their collaborators. The colliding nuclei determine an entrance to a "cold valley" as predicted as early as 1976 by Gupta, Sandulescu and Greiner. The fused nucleus 112 decays successively via α emission until finally the quasi-stable nucleus ^{253}Fm is reached. The α particles as well as the final nucleus have been observed. Combined, this renders the definite proof of the existence of a $Z = 112$ nucleus.

below) result also in the prediction of these two magic numbers [38,40]. The corresponding magic neutron numbers are predicted to be $N = 172$ and - as it seems to a lesser extend - $N = 184$. Thus, this region provides an open field of research. R.A. Gherghescu et al. have calculated the potential energy surface of the $Z = 120$ nucleus. It utilizes interesting isomeric and valley structures (Fig. 6). The charge distribution of the $Z = 120, N = 184$ nucleus indicates a hollow inside. This leads us to suggest that it might be essentially a fullerene consisting of 60 α-particles and one binding neutron per alpha.

The determination of the chemistry of superheavy elements, i. e. the calculation of the atomic structure — which is in the case of element 112 the shell structure of 112 electrons due to the Coulomb interaction of the electrons and in particular the calculation of the orbitals of the outer (valence) electrons — has been carried out as early as 1970 by B. Fricke and W. Greiner [15]. Hartree-Fock-Dirac calculations yield rather precise results.

The potential energy surfaces, which are shown prototypically for $Z = 114$ in Fig 3, contain even more remarkable information that I want to mention cursorily: if a given nucleus, e. g. Uranium, undergoes fission, it moves in its potential mountains from the interior to the outside. Of course, this happens

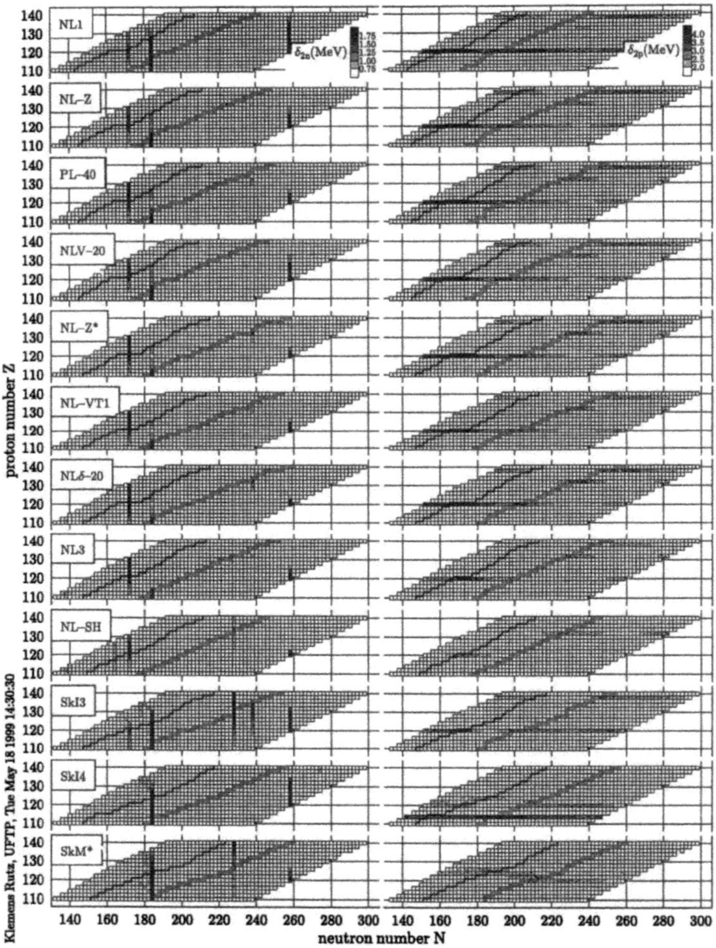

Figure 5. Grey scale plots of proton gaps (left column) and neutron gaps (right column) in the N-Z plane for spherical calculations with the forces as indicated. The assignment of scales differs for protons and neutrons, see the uppermost boxes where the scales are indicated in units of MeV. Nuclei that are stable with respect to β decay and the two-proton dripline are emphasized. The forces with parameter sets SkI4 and PL-40 reproduce the binding energy of $^{264}_{156}108$ (Hassium) best, i.e. $|\delta E/E| < 0.0024$. Thus one might assume that these parameter sets could give the best predictions for the superheavies. Nevertheless, it is noticed that PL-40 predicts only $Z = 120$ as a magic number while SkI4 predicts both $Z = 114$ and $Z = 120$ as magic numbers. The magicity depends — sometimes quite strongly — on the neutron number. These studies are due to Bender, Rutz, Bürvenich, Maruhn, P.G. Reinhard et al. [14].

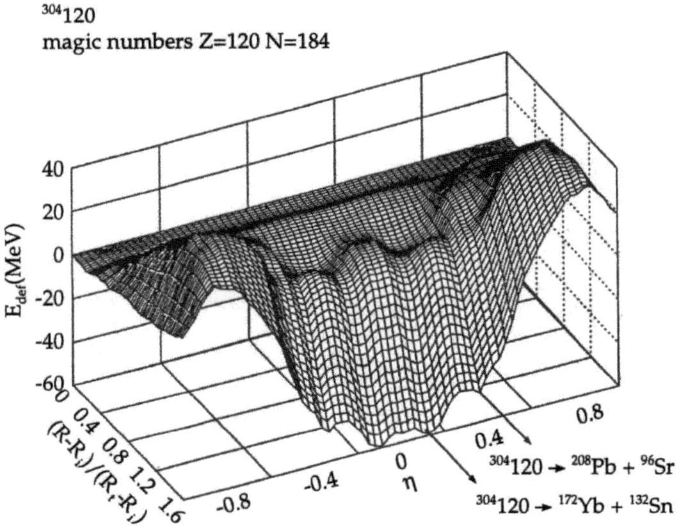

Figure 6. Potential energy surface as a function of reduced elongation $(R - R_i)/(R_t - R_i)$ and mass asymmetry η for the double magic nucleus $^{304}120$. $^{304}120_{184}$.

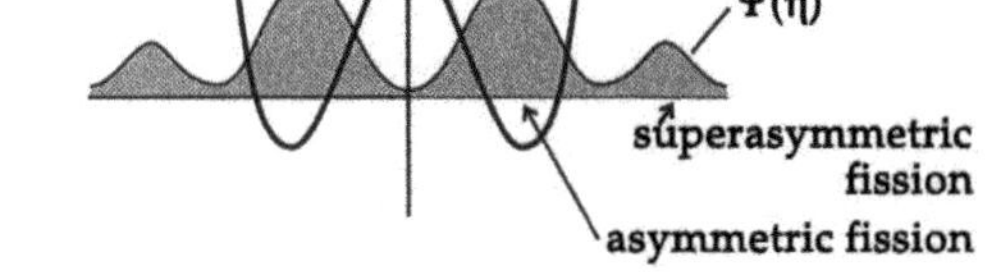

Figure 7. The collective potential as a function of the mass asymmetry $\eta = \dfrac{A_1 - A_2}{A_1 + A_2}$. A_i denotes the nucleon number in fragment i. This qualitative potential $V(R_{\text{fixed}}, \eta)$ corresponds to a cut through the potential landscape at $R = R_{\text{fixed}}$ close to the scission configuration. The wave function is drawn schematically. It has maxima where the potential is minimal and vice versa.

quantum mechanically. The wave function of such a nucleus, which decays by tunneling through the barrier, has maxima where the potential is minimal and minima where it has maxima. This is depicted in Fig. 7.

The probability for finding a certain mass asymmetry $\eta = \dfrac{A_1 - A_2}{A_1 + A_2}$ of the fission is proportional to $\psi^*(\eta)\psi(\eta)\mathrm{d}\eta$. Generally, this is complemented by a coordinate dependent scale factor for the volume element in this (curved) space, which I omit for the sake of clarity. Now it becomes clear how the so-called **asymmetric** and **superasymmetric** fission processes come into being. They result from the enhancement of the collective wave function in the cold valleys. And that is indeed, what one observes. Fig. 8 gives an impression of it.

For a large mass asymmetry ($\eta \approx 0.8,\ 0.9$) there exist very narrow valleys. They are not as clearly visible in Fig. 3, but they have interesting consequences. Through these narrow valleys nuclei can emit spontaneously not only α-particles (Helium nuclei) but also ^{14}C, ^{20}O, ^{24}Ne, ^{28}Mg, and other nuclei. Thus, we are lead to the **cluster radioactivity** (Poenaru, Sandulescu, Greiner [16]).

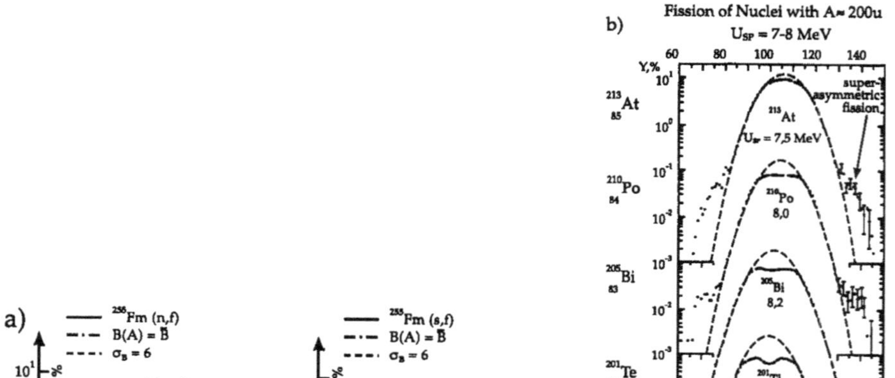

Figure 8. Asymmetric (a) and symmetric (b) fission. For the latter, also superasymmetric fission is recognizable, as it has been observed only a few years ago by the russian physicist Itkis — just as expected theoretically.

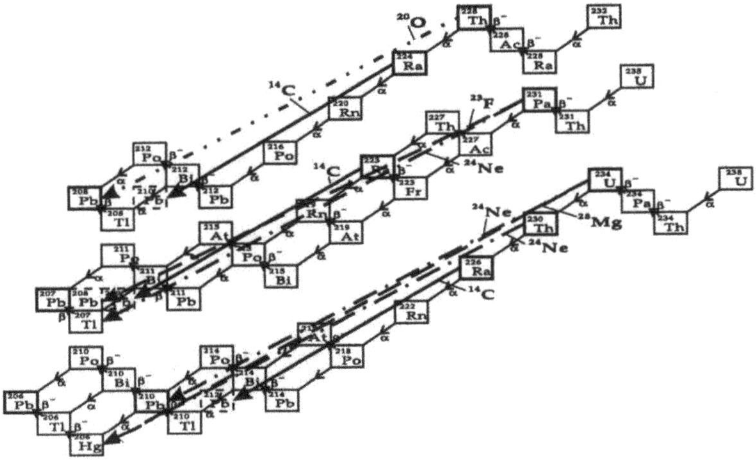

Figure 9. Cluster radioactivity of actinide nuclei. By emission of ^{14}C, ^{20}O,... "big leaps" in the periodic system can occur, just contrary to the known α, β, γ radioactivities, which are also partly shown in the figure.

By now this process has been verified experimentally by research groups in Oxford, Moscow, Berkeley, Milan and other places. Accordingly, one has to revise what is learned in school: there are not only 3 types of radioactivity (α-, β-, γ-radioactivity), but many more. Atomic nuclei can also decay through spontaneous cluster emission (that is the "spitting out" of smaller nuclei like carbon, oxygen,...). Fig. 9 depicts some nice examples of these processes.

The knowledge of the collective potential energy surface and the collective masses $B_{ij}(R,\eta)$, all calculated within the Two-Center-Shell-Modell (TCSM), allowed H. Klein, D. Schnabel and J. A. Maruhn to calculate lifetimes against fission in an "ab initio" way [17].
Utilizing a WKB-minimization for the penetrability integral

$$\mathcal{P} = e^{-\mathrm{I}}, \quad \mathrm{I} = \min_{\forall\,\mathrm{paths}} \tfrac{2}{\hbar} \int_S \sqrt{2m(V(R,\eta)-E)}\, ds$$

$$= \min_{\forall\,\mathrm{paths}} \tfrac{2}{\hbar} \int_0^1 \sqrt{\underbrace{2m g_{ij}}_{B_{ij}}(V(x_i(t)-E)\tfrac{dx_i}{dt}\tfrac{dx_j}{dt}}\, dt \qquad (1)$$

where $ds^2 = g_{ij}dx_idx_j$ and g_{ij} – the metric tensor – is in the well-known fashion related to the collective masses $B_{ij} = 2mg_{ij}$, one explores the minimal paths from the nuclear ground state configuration through the multidimen-

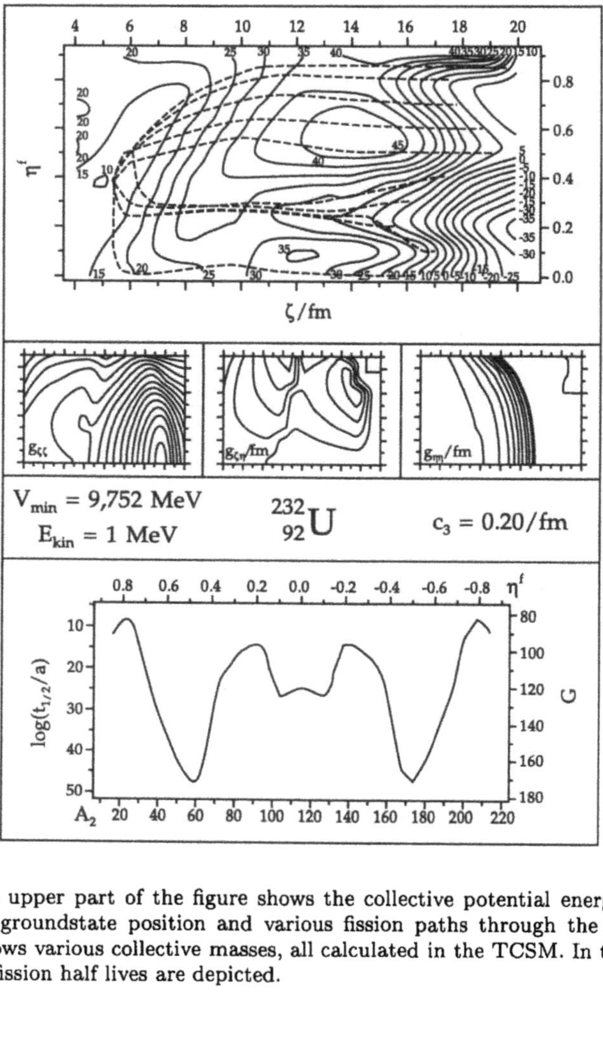

Figure 10. The upper part of the figure shows the collective potential energy surface for $^{232}_{92}U$ with the groundstate position and various fission paths through the barrier. The middle part shows various collective masses, all calculated in the TCSM. In the lower part the calculated fission half lives are depicted.

sional fission barrier (see Fig. 10).

The thus obtained fission half lives are depicted in the lower part of figure 10. Their distribution as a function of the fragment mass A_2 resembles quite well the asymmetric mass distribution. Cluster radioactive decays correspond to the broad peaks around A_2 = 20, 30 (200, 210). The confrontation of the calculated fission half lives with experiments is depicted in Fig. 11. One notices "nearly quantitative" agreement over 20 orders of magnitude, which

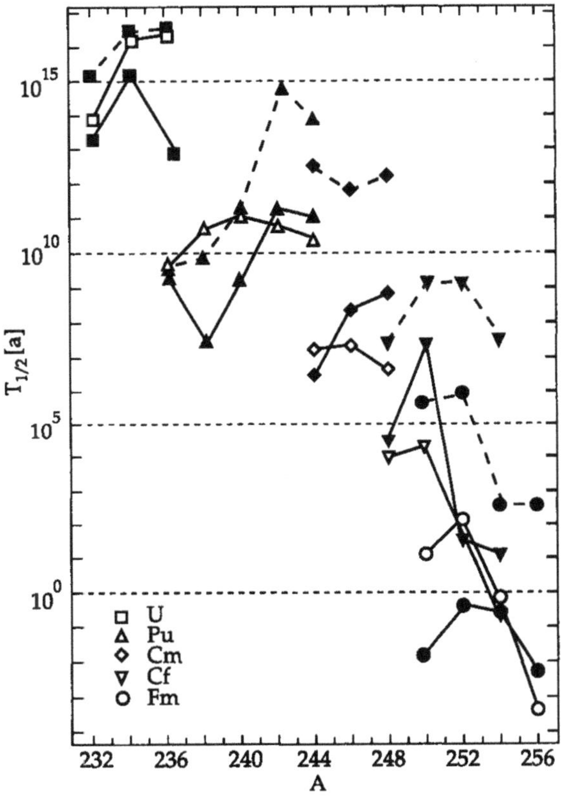

Figure 11. Fission half lives for various isotopes of $Z = 92$ (U), $Z = 94$ (Pu), $Z = 96$ (Cm), $Z = 98$ (Cf) and $Z = 100$ (Fm). The black curves represent the experimental values. The dashed and dotted calculations correspond to a different choice of the barrier parameter in the Two Center Shell Model ($c_3 \approx 0.2$ and 0.1, respectively).

is – for an ab-initio calculation – remarkable!

Finally, in Fig. 12, we compare the lifetime calculation discussed above with one based on the Preformation Cluster Model by D. Poenaru et al. [18] and recognize an amazing degree of similarity and agreement.

The systematics for the average total kinetic energy release for spontaneously fissioning isotopes of Cm and No is following the Viola trend, but ^{258}Fm and ^{259}Fm are clearly outside. The situation is similar also for ^{260}Md, where two components of fission products (one with lower and one with higher kinetic energy) were observed by Hulet et al. [19]. The explanation of these interesting observations lies in two different paths through the collective potential. One reaches the scission point in a stretched neck position (i.e. at a lower point of the Coulomb barrier - thus lower kinetic energy for the fragments) while the other one reaches the scission point practically

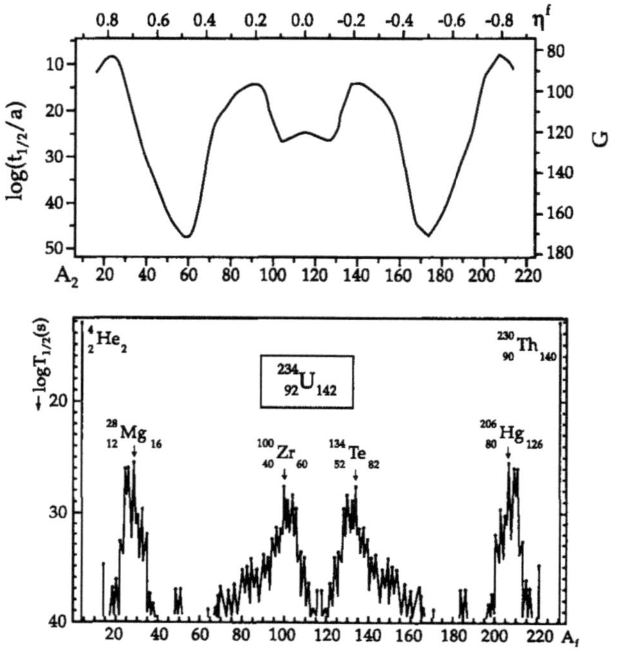

Figure 12. Comparison of the fission half lives calculated in the fission model (upper figure – see also Fig. 10) and in the Preformation Cluster Model [18]. In both models the deformation of the fission fragments is not included completely.

in a touching-spheres-position (i.e. higher up on the Coulomb barrier and therefore highly energetic fragments) [20]. The latter process is cold fission; i.e., the fission fragments are in or close to their ground state (cold fragments)and all the available energy is released as kinetic energy. Cold fission is, in fact, typically a cluster decay. The side-by-side occurence of cold and normal (hot) fission has been named bi-modal fission [20]. There has now been put forward a phantastic idea [21] in order to study cold fission (Cluster decays) and other exotic fission processes (ternary-, multiple fission in general) very elegantly: By measuring with e. g. the Gamma–sphere characteristic γ-transitions of individual fragments in coincidence, one can identify all these processes in a direct and simple way (Fig. 13). First confirmation of this method by J. Hamilton, V. Ramaya et al. worked out excellently [22]. This method has high potential for revolutionizing fission physics! With some physical intuition one can imagine that **triple** - and **quadriple** fission processes

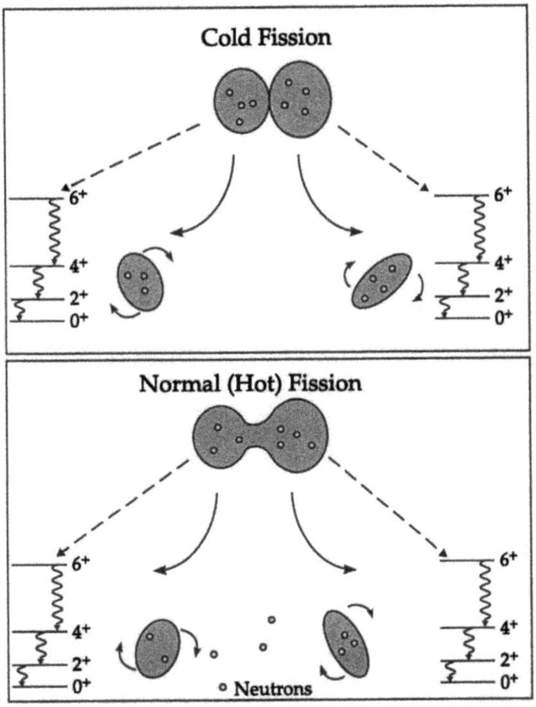

Figure 13. Illustration of cold and hot (normal) fission identification through multiple γ-coincidencs of photons from the fragments. The photons serve to identify the fragments.

and even the process of **cold multifragmentation** will be discovered - absolutely fascinating! We have thus seen that fission physics (cold fission, cluster radioactivity, ...) and fusion physics (especially the production of superheavy elements) are intimately connected.
Indeed, very recently, tripple fission of

$$\begin{aligned} {}^{252}\mathrm{Cf} &\rightarrow {}^{146}\mathrm{Ba} + {}^{96}\mathrm{Sr} \ + {}^{10}\mathrm{Be} \\ &\rightarrow {}^{112}\mathrm{Ry} + {}^{130}\mathrm{Sn} + {}^{10}\mathrm{Be} \\ &\rightarrow \ldots \end{aligned}$$

has been identified by measuring the various γ–transitions of these nuclei in coincidence (see Fig. 14). Even though the statistical evidence for the ^{10}Be line is small ($\approx$ 50 events) the various coincidences seem to proof that spon-

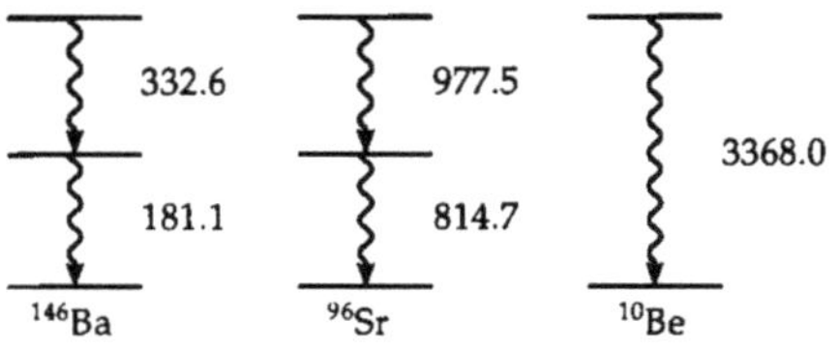

Figure 14. The γ–transitions of the three fission products of ^{252}Cf measured in coincidence. Various combinations of the coincidences were studied. The free 3368 keV line in ^{10}Be has recently been remeasured by Burggraf et al. [23], confirming the value of the transition energy within 100 eV.

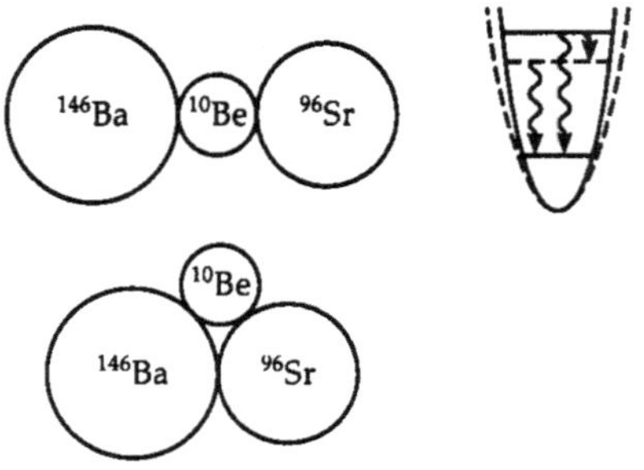

Figure 15. Typical linear cluster configuration leading to tripple fission of ^{252}Cf. The influence of both clusters leads to a softening of the ^{10}Be potential and thus to a somewhat smaller transition energy. Some theoretical investigations indicate that the axial symmetry of this configuration might be broken (lower lefthand figure).

taneous tripple fission out of the ground state of ^{252}Cf with the heavy cluster ^{10}Be as a third fragment exists. Also other tripple fragmentations can be expected. One of those is also denoted above. In fact, there are first indications, that this break–up is also observed. The most amazing observation is, however, the following: The cross coincidences seem to suggest that one deals with a simultaneous three–body breakup and not with a cascade process. For that one expects a configuration as shown in Fig. 15. Consequently the ^{10}Be will obtain kinetic energy while running down the combined Coulomb barrier of ^{146}Ba and ^{96}Sr and, therefore, the 3368 keV line of ^{10}Be should be Doppler-broadened. Amazingly, however, it is not and, moreover, it seems to be about 6 keV smaller than the free ^{10}Be γ–transition. If this turns out to be true, the only explanation will be that the Gamma is emitted while the nuclear molecule of the type shown in Fig. 15 holds. The molecule has to live longer than about 10^{-12} sec. The nuclear forces from the ^{146}Ba and ^{96}Sn cluster to the left and right from ^{10}Be lead to a softening of its potential and therefore to a somewhat smaller transition energy. Thus, if experimental results hold, one has discovered long living ($\approx 10^{-12}$ sec) complex nuclear molecules. This is phantastic! Of course, I do immediately wonder whether such configurations do also exist in e.g. U + Cm soft encounters directly at the Coulomb barrier. This would have tremendous importance for the observation of the spontaneous vacuum decay [26], for which "sticking giant molecules" with a lifetime

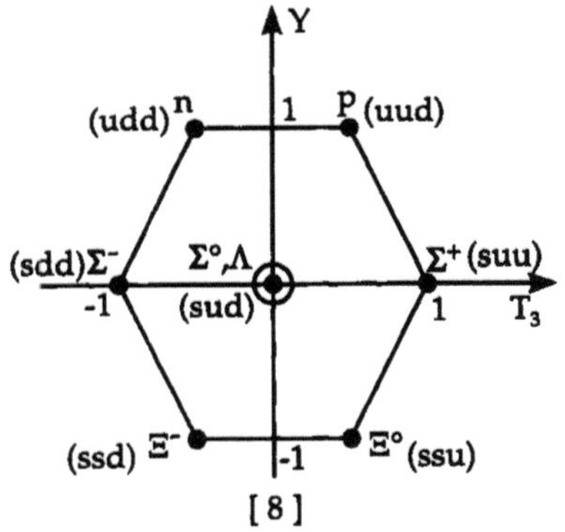

Figure 16. Important baryons are ordered in this octet. The quark content is depicted. Protons (p) and neutrons (n), most important for our known world, contain only u and d quarks. Hyperons contain also an s quark. The number of s quarks is a measure for the strangeness.

of the order of 10^{-19} sec are needed. The nuclear physics of such heavy ion collisions at the Coulomb barrier (giant nuclear molecules) should indeed be investigated!
As mentioned before there are other tri–molecular structures possible; some with ^{10}Be in the middle and both spherical or deformed clusters on both sides of ^{10}Be. The energy shift of the ^{10}Be–line should be smaller, if the outside clusters are deformed (smaller attraction ⇔ smaller softening of the potential) and bigger, if they are spherical . Also other than ^{10}Be–clusters are expected to be in the middle. One is lead to the molecular doorway picture. Finally, these tri–body nuclear molecules are expected to perform themselves rotational and vibrational (butterfly, whiggler, β–, γ–type) modes. The energies were estimated by P. Hess et al [24]; for example rotational energies typically of the order of a few keV (4 keV, 9 keV, ...). A new molecular spectroscopy seems possible!

The "cold valleys" in the collective potential energy surface are basic for understanding this exciting area of nuclear physics! It is a master example for understanding the **structure of elementary matter**, which is so important for other fields, especially astrophysics, but even more so for enriching our "Weltbild", i.e. the status of our understanding of the world around us.

Nuclei that are found in nature consist of nucleons (protons and neutrons) which themselves are made of u (up) and d (down) quarks. However, there also exist s (strange) quarks and even heavier flavors, called charm, bottom, top. The latter has just recently been discovered. Let us stick to the s quarks. They are found in the 'strange' relatives of the nucleons, the so-called hyperons (Λ, Σ, Ξ, Ω). The Λ-particle, e. g., consists of one u, d and s quark, the Ξ-particle even of an u and two s quarks, while the Ω (sss) contains strange quarks only. Fig. 16 gives an overview of the baryons, which are of interest

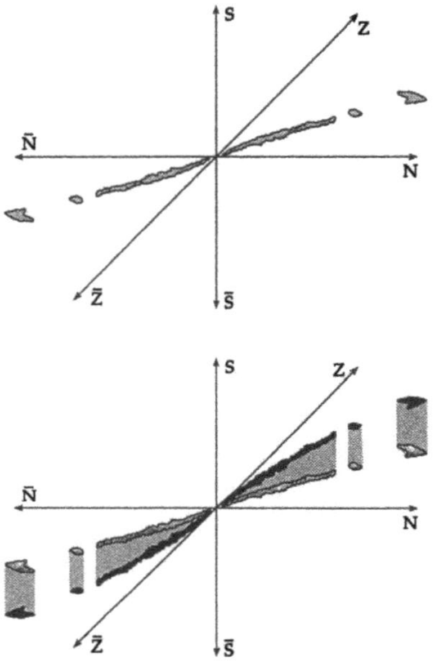

Figure 17. The extension of the periodic system into the sectors of strangeness (S, $\bar{S}$) and antimatter ($\bar{Z}$, $\bar{N}$). The stable valley winds out of the known proton (Z) and neutron (N) plane into the S and $\bar{S}$ sector, respectively. The same can be observed for the antimatter sector. In the upper part of the figure only the stable valley in the usual proton (Z) and neutron (N) plane is plotted, however, extended into the sector of antiprotons and antineutrons. In the second part of the figure it has been indicated, how the stable valley winds out of the Z-N-plane into the strangeness sector.

here, and their quark content.

If such a hyperon is taken up by a nucleus, a **hyper-nucleus** is created. Hyper-nuclei with one hyperon have been known for 20 years now, and were extensively studied by B. Povh (Heidelberg) [27]. Several years ago, Carsten Greiner, Jürgen Schaffner and Horst Stöcker [28] theoretically investigated nuclei with many hyperons, **hypermatter**, and found that the binding energy per baryon of strange matter is in many cases even higher than that of ordinary matter (composed only of u and d quarks). This leads to the idea of extending the periodic system of elements in the direction of strangeness.

One can also ask for the possibility of building atomic nuclei out of **antimatter**, that means searching e. g. for anti-helium, anti-carbon, anti-oxygen. Fig. 17 depicts this idea. Due to the charge conjugation symmetry antinuclei should have the same magic numbers and the same spectra as ordinary nuclei. However, as soon as they get in touch with ordinary matter, they annihilate with it and the system explodes.

Now the important question arises how these strange matter and antimatter clusters can be produced. First, one thinks of collisions of heavy nuclei, e. g. lead on lead, at high energies (energy per nucleon $\geq$ 200 GeV). Calculations with the URQMD-model of the Frankfurt school show that through **nuclear shock waves** [29,30,31] nuclear matter gets compressed to 5–10 times of its usual value, $\rho_0 \approx 0.17\ \text{fm}^3$, and heated up to temperatures of $kT \approx 200$ MeV. As a consequence about 10000 pions, 100 Λ's, 40 Σ's and Ξ's and about as many antiprotons and many other particles are created in a single collision. It seems conceivable that it is possible in such a scenario for some Λ's to get captured by a nuclear cluster. This happens indeed rather frequently for one or two Λ-particles; however, more of them get built into nuclei with rapidly decreasing probability only. This is due to the low probability for finding the right conditions for such a capture in the phase space of the particles: the numerous particles travel with every possible momenta (velocities) in all directions. The chances for hyperons and antibaryons to meet gets rapidly worse with increasing number. In order to produce multi-Λ-nuclei and antimatter nuclei, one has to look for a different source.

In the framework of meson field theory the energy spectrum of baryons has a peculiar structure, depicted in Fig. 18. It consists of an upper and a lower continuum, as it is known from the electrons (see e. g. [26]). Of special interest in the case of the baryon spectrum is the potential well, built of the scalar and the vector potential, which rises from the lower continuum. It is known since P.A.M. Dirac (1930) that the negative energy states of the lower continuum have to be occupied by particles (electrons or, in our case, baryons). Otherwise our world would be unstable, because the "ordinary" particles are found in the upper states which can decay through the emission of photons into lower lying states. However, if the "underworld" is occupied, the Pauli-principle will prevent this decay. Holes in the occupied "underworld" (Dirac sea) are antiparticles.

The occupied states of this underworld including up to 40000 occupied bound states of the lower potential well represent the **vacuum**. The peculiarity of this strongly correlated vacuum structure in the region of atomic nuclei is that — depending on the size of the nucleus — more than 20000 up to 40000 (occupied) bound nucleon states contribute to this polarization effect. Obviously, we are dealing here with a **highly correlated vacuum**. A pronounced shell structure can be recognized [32,33,34]. Holes in these states have to be interpreted as bound antinucleons (antiprotons, antineutrons). If the primary nuclear density rises due to compression, the lower well increases while the upper decreases and soon is converted into a repulsive barrier (Fig. 19). This compression of nuclear matter can only be carried out in relativistic nucleus-nucleus collision with the help of shock waves, which have been proposed by the Frankfurt school [29,30] and which have since then been confirmed extensively (for references see e. g. [35]). These **nuclear shock waves** are accompanied by heating of the nuclear matter. Indeed, density and temperature are intimately coupled in terms of the hydrodynamic Rankine-Hugoniot-

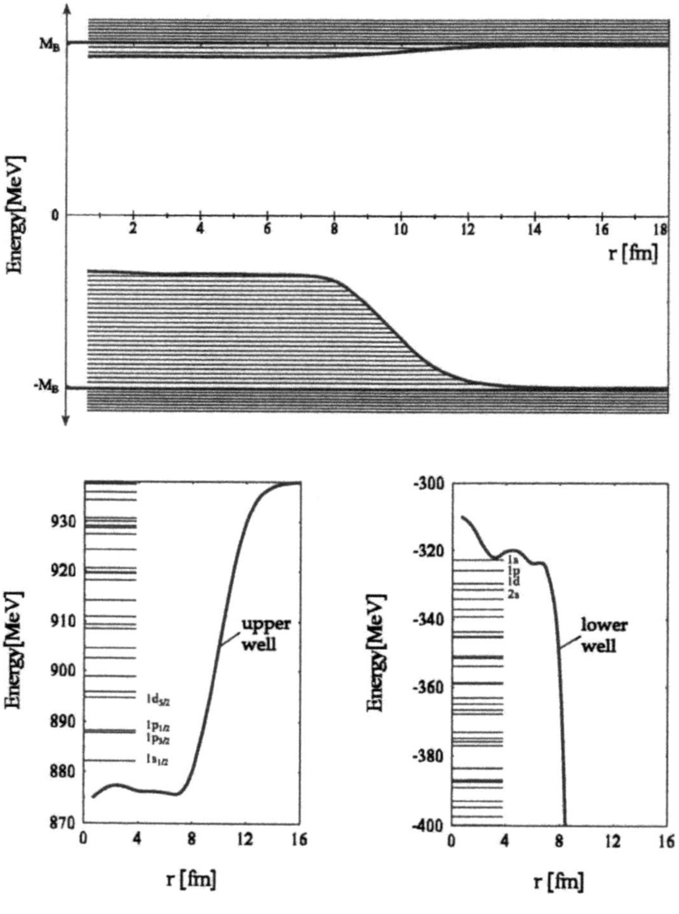

Figure 18. Baryon spectrum in a nucleus. Below the positive energy continuum exists the potential well of real nucleons. It has a depth of 50-60 MeV and shows the correct shell structure. The shell model of nuclei is realized here. However, from the negative continuum another potential well arises, in which about 40000 bound particles are found, belonging to the vacuum. A part of the shell structure of the upper well and the lower (vacuum) well is depicted in the lower figures.

equations. Heating as well as the violent dynamics cause the creation of many holes in the very deep (measured from $-M_B c^2$) vacuum well. These numerous bound holes resemble antimatter clusters which are bound in the medium; their wave functions have large overlap with antimatter clusters. When the

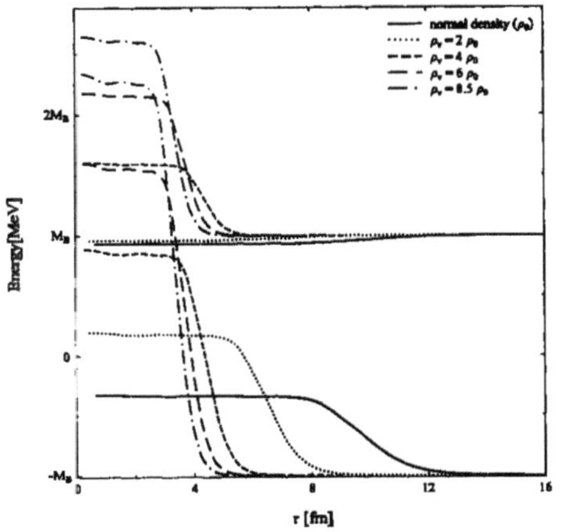

Figure 19. The lower well rises strongly with increasing primary nucleon density, and even gets supercritical (spontaneous nucleon emission and creation of bound antinucleons). Supercriticality denotes the situation, when the lower well enters the upper continuum.

primary matter density decreases during the expansion stage of the heavy ion collision, the potential wells, in particular the lower one, disappear.

The bound antinucleons are then pulled down into the (lower) continuum. In this way antimatter clusters may be set free. Of course, a large part of the antimatter will annihilate on ordinary matter present in the course of the expansion. However, it is important that this mechanism for the production of antimatter clusters out of the highly correlated vacuum does not proceed via the phase space. The required coalescence of many particles in phase space suppresses the production of clusters, while it is favoured by the direct production out of the highly correlated vacuum. In a certain sense, the highly correlated vacuum is a kind of cluster vacuum (vacuum with cluster structure). The shell structure of the vacuum levels (see Fig. 18) supports this latter suggestion. Fig. 20 illustrates this idea.

The mechanism is similar for the production of multi-hyper nuclei (Λ, Σ, Ξ, Ω). Meson field theory predicts also for the Λ energy spectrum at finite primary nucleon density the existence of upper and lower wells. The lower well belongs to the vacuum and is fully occupied by Λ's.

Dynamics and temperature then induce transitions ($\Lambda\bar{\Lambda}$ creation) and deposit many Λ's in the upper well. These numerous bound Λ's are sitting close to the primary baryons: in a certain sense a giant multi-Λ hypernucleus has been created. When the system disintegrates (expansion stage) the Λ's distribute over the nucleon clusters (which are most abundant in peripheral collisions). In this way multi-Λ hypernuclei can be formed.

Of course this vision has to be worked out and probably refined in many respects. This means much more and thorough investigation in the future. It

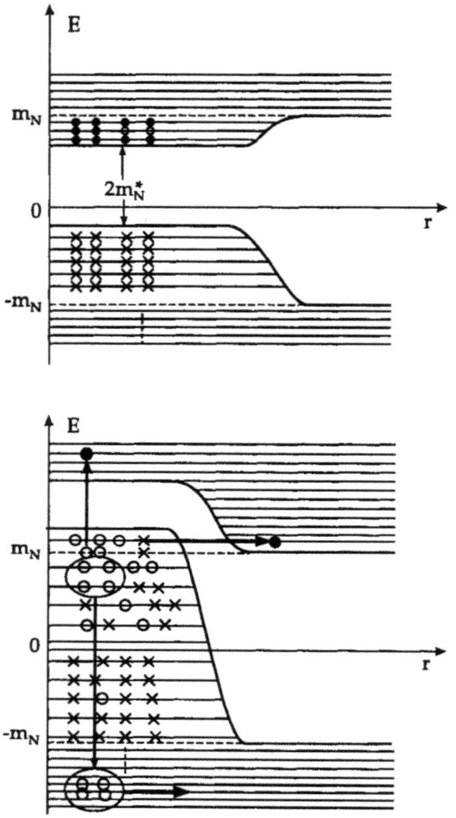

Figure 20. Due to the high temperature and the violent dynamics, many bound holes (antinucleon clusters) are created in the highly correlated vacuum, which can be set free during the expansion stage into the lower continuum. In this way, antimatter clusters can be produced directly from the vacuum. The horizontal arrow in the lower part of the figure denotes the spontaneous creation of baryon-antibaryon pairs, while the antibaryons occupy bound states in the lower potential well. Such a situation, where the lower potential well reaches into the upper continuum, is called supercritical. Four of the bound holes states (bound antinucleons) are encircled to illustrate a "quasi-antihelium" formed. It may be set free (driven into the lower continuum) by the violent nuclear dynamics.

is particularly important to gain more experimental information on the properties of the lower well by (e, e' p) or (e, e' p p') and also ($\bar{p}_c p_b$, $p_c \bar{p}_b$) reactions at high energy ($\bar{p}_c$ denotes an incident antiproton from the continuum, p_b is a proton in a bound state; for the reaction products the situation is just the opposite) [36]. Also the reaction (p, p' d), (p, p' ^{3}He), (p, p' ^{4}He) and others of similar type need to be investigated in this context. The systematic scattering of antiprotons on nuclei can contribute to clarify these questions. Problems of the meson field theory (e. g. Landau poles) can then be reconsidered. An effective meson field theory has to be constructed. Various effective theories, e. g. of Walecka-type on the one side and theories with chiral invariance on the other side, seem to give different strengths of the potential wells and also different dependence on the baryon density [37]. The Lagrangians of the Dürr-Teller-Walecka-type and of the chirally symmetric mean field theories look

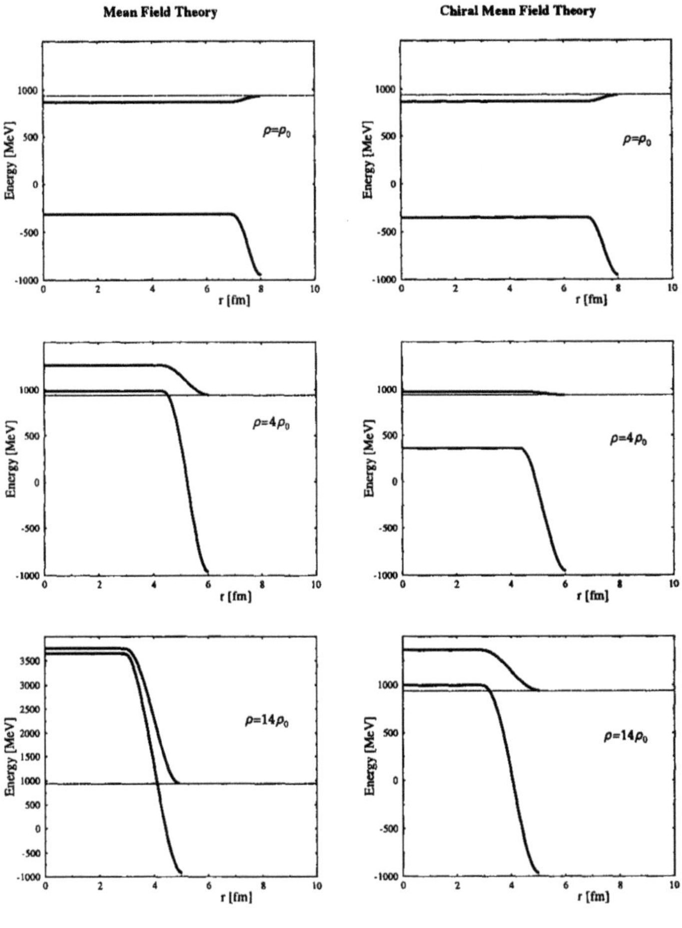

Figure 21. The potential structure of the shell model and the vacuum for various primary densities $\rho = \rho_0$, $4\rho_0$, $14\rho_0$. At left the predictions of ordinary Dürr-Teller-Walecka-type theories are shown; at right those for a chirally symmetric meson field theory as develloped by P. Papazoglu, S. Schramm et al. [37,38]. Note however, that this particular chiral mean–field theory does contain ω^4 terms. If introduced in both effective models, they seem to predict quantitatively similar results.

quite differently. We exhibit them — without further discussion — in the following equations:

$$\mathcal{L} = \mathcal{L}_{\rm kin} + \mathcal{L}_{\rm BM} + \mathcal{L}_{\rm vec} + \mathcal{L}_0 + \mathcal{L}_{\rm SB}$$

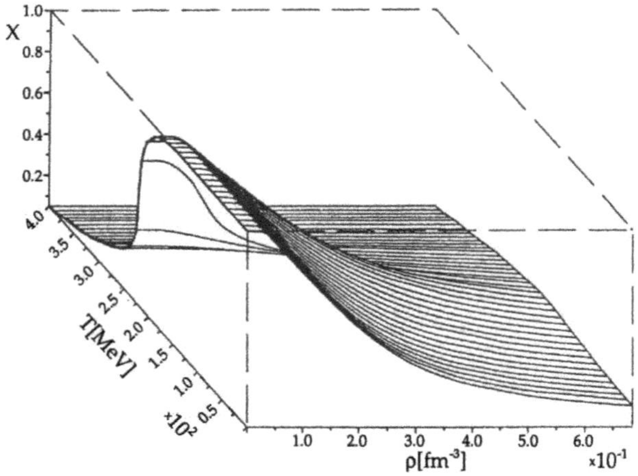

Figure 22. The strong phase transition inherent in Dürr-Teller-Walecka-type meson field theories, as predicted by J. Theis et al. [39]. Note that there is a first order transition along the ρ-axis (i.e. with density), but a simple transition along the temperature T-axis. Note also that this is very similar to the phase transition obtained recently from the Nambu-Jona-Lasinio-approximation of QCD [41].

Non-chiral Lagrangian:

$$
\begin{aligned}
\mathcal{L}_{\text{kin}} &= \frac{1}{2}\partial_\mu s \partial^\mu s + \frac{1}{2}\partial_\mu z \partial^\mu z - \frac{1}{4}B_{\mu\nu}B^{\mu\nu} - \frac{1}{4}G_{\mu\nu}G^{\mu\nu} - \frac{1}{4}F_{\mu\nu}F^{\mu\nu} \\
\mathcal{L}_{\text{BM}} &= \textstyle\sum_B \overline{\psi}_B \big[i\gamma_\mu \partial^\mu - g_{\omega B}\gamma_\mu \omega^\mu - g_{\phi B}\gamma_\mu \phi^\mu - g_{\rho B}\gamma_\mu \tau_B \rho^\mu \\
&\qquad - e\gamma_\mu \frac{1}{2}(1+\tau_B)A^\mu - m_B^* \big] \psi_B \\
\mathcal{L}_{\text{vec}} &= \frac{1}{2}m_\omega^2 \omega_\mu \omega^\mu + \frac{1}{2}m_\rho^2 \rho_\mu \rho^\mu + \frac{1}{2}m_\phi^2 \phi_\mu \phi^\mu \\
\mathcal{L}_0 &= -\frac{1}{2}m_s^2 s^2 - \frac{1}{2}m_z^2 z^2 - \frac{1}{3}bs^3 - \frac{1}{4}cs^4
\end{aligned}
$$

Chiral Lagrangian:

$$
\begin{aligned}
\mathcal{L}_{\text{kin}} &= \frac{1}{2}\partial_\mu \sigma \partial^\mu \sigma + \frac{1}{2}\partial_\mu \zeta \partial^\mu \zeta + \frac{1}{2}\partial_\mu \chi \partial^\mu \chi - \frac{1}{4}B_{\mu\nu}B^{\mu\nu} - \frac{1}{4}G_{\mu\nu}G^{\mu\nu} - \frac{1}{4}F_{\mu\nu}F^{\mu\nu} \\
\mathcal{L}_{\text{BM}} &= \textstyle\sum_B \overline{\psi}_B \big[i\gamma_\mu \partial^\mu - g_{\omega B}\gamma_\mu \omega^\mu - g_{\phi B}\gamma_\mu \phi^\mu - g_{\rho B}\gamma_\mu \tau_B \rho^\mu \\
&\qquad - e\gamma_\mu \frac{1}{2}(1+\tau_B)A^\mu - m_B^* \big] \psi_B
\end{aligned}
$$

$$\mathcal{L}_{\text{vec}} = \frac{1}{2}m_\omega^2\frac{\chi^2}{\chi_0^2}\omega_\mu\omega^\mu + \frac{1}{2}m_\rho^2\frac{\chi^2}{\chi_0^2}\rho_\mu\rho^\mu + \frac{1}{2}m_\phi^2\frac{\chi^2}{\chi_0^2}\phi_\mu\phi^\mu + g_4^4(\omega^4 + 6\omega^2\rho^2 + \rho^4)$$

$$\mathcal{L}_0 = -\frac{1}{2}k_0\chi^2(\sigma^2+\zeta^2) + k_1(\sigma^2+\zeta^2)^2 + k_2(\frac{\sigma^4}{2}+\zeta^4) + k_3\chi\sigma^2\zeta - k_4\chi^4 + \frac{1}{4}\chi^4\ln\frac{\chi^4}{\chi_0^4} + \frac{\delta}{3}\ln\frac{\sigma^2\zeta}{\sigma_0^2\zeta_0^2}$$

$$\mathcal{L}_{\text{SB}} = -\left(\frac{\chi}{\chi_0}\right)^2\left[m_\pi^2 f_\pi\sigma + \left(\sqrt{2}m_K^2 f_K - \frac{1}{\sqrt{2}}m_\pi^2 f_\pi\right)\zeta\right]$$

The non-chiral model contains the scalar-isoscalar field s and its strange counterpart z, the vector-isoscalar fields ω_μ and ϕ_μ, and the the ρ-meson ρ_μ as well as the photon A_μ. For more details see [37]. In contrast to the non-chiral model, the $SU(3)_L \times SU(3)_R$ Lagrangian contains the dilaton field χ introduced to mimic the trace anomaly of QCD in an effective Lagrangian at tree level (For an explanation of the chiral model see [37,38]).

The connection of the chiral Lagrangian with the Walecka-type can be established by the substitution $\sigma = \sigma_0 - s$ (and similarly for the strange condensate ζ). Then, e.g. the difference in the definition of the effective nucleon mass in both models (non-chiral: $m_N^* = m_N - g_s s$, chiral: $m_N^* = g_s\sigma$) can be removed, yielding:

$$m_N^* = g_s\sigma_0 - g_s s \equiv m_N - g_s s \tag{2}$$

for the nucleon mass in the chiral model.

Nevertheless, if the parameters in both cases are adjusted such that ordinary nuclei (binding energies, radii, shell structure,...) and properties of infinite nuclear matter (equilibrium density, compression constant K, binding energy) are well reproduced, the prediction of both effective Lagrangians for the dependence of the properties of the correlated vacuum on density and temperature is remarkably different. This is illustrated to some extend in Fig. 21. Accordingly, the chirally symmetric meson field theory predicts much higher primary densities (and temperatures) until the effects of the correlated vacuum are strong enough so that the mechanisms described here become effective. In other words, according to chirally symmetric meson field theories the antimatter-cluster-production and multi-hypermatter-cluster production out of the highly correlated vacuum takes place at considerably higher heavy ion energies as compared to the predictions of the Dürr-Teller-Walecka-type meson field theoories. This in itself is a most interesting, quasi-fundamental question to be clarified. Moreover, the question of the nucleonic substructure (form factors, quarks, gluons) and its influence on the highly correlated vacuum structure has to be studied. The nucleons are possibly strongly polarized in the correlated vacuum: the Δ resonance correlations in the vacuum are probably important. Is this highly correlated vacuum state, especially during the compression, a preliminary stage to the quark-gluon cluster plasma? To which extent is it similar or perhaps even identical with it? It is well known for more than 10 years that meson field theories predict a phase transition qualitatively and quantitatively similar to that of the quark-gluon plasma [39] — see Fig. 22.

The extension of the periodic system into the sectors hypermatter (strangeness) and antimatter is of general and astrophysical importance. Indeed, microseconds after the big bang the new dimensions of the periodic system, we have touched upon, certainly have been populated in the course of the baryo- and nucleo-genesis. Of course, for the creation of the universe, even higher dimensional extensions (charm, bottom, top) come into play, which we did not pursue here. It is an open question, how the depopulation (the decay) of these sectors influences the distribution of elements of our world today. Our conception of the world will certainly gain a lot through the clarification of these questions.

For the Gesellschaft für Schwerionenforschung (GSI), which I helped initiating in the sixties, the questions raised here could point to the way ahead. Working groups have been instructed by the board of directors of GSI, to think about the future of the laboratory. On that occasion, very concrete (almost too concrete) suggestions are discussed — as far as it has been presented to the public. What is missing, as it seems, is a **vision on a long term basis**. The ideas proposed here, the verification of which will need the **commitment for 2–4 decades of research**, could be such a vision with considerable attraction for the best young physicists. The new dimensions of the periodic system made of hyper- and antimatter cannot be examined in the "stand-by" mode at CERN (Geneva); a dedicated facility is necessary for this field of research, which can in future serve as a home for the universities. The GSI — which has unfortunately become much too self-sufficient — could be such a home for new generations of physicists, who are interested in the **structure of elementary matter**. GSI would then not develop just into a detector laboratory for CERN, and as such become obsolete. I can already see the enthusiasm in the eyes of young scientists, when I unfold these ideas to them — similarly as it was 30 years ago, when the nuclear physicists in the state of Hessen initiated the construction of GSI.

Acknowledgments

I am indebted to Thomas Bürvenich for helping technically preparing this manuscript and him as J. Maruhn and K. Rutz for many discussions.

References

1. S.G. Nilsson *et al*, *Phys. Lett.* B **28**, 458 (1969); S.G. Nilsson *et al*, *Nucl. Phys.* A **131**, 1 (1969); *ibid.* A **115**, 545 (1968).
2. U. Mosel, B. Fink and W. Greiner, *Contribution to "Memorandum Hessischer Kernphysiker"*, Darmstadt, Frankfurt, Marburg (1966).
3. U. Mosel and W. Greiner, *Z. Phys.* **217**, 256 (1968); *ibid.* **222**, 261 (1968).
4. a) J. Grumann, U. Mosel, B. Fink and W. Greiner, *Z. Phys.* **228**, 371 (1969); b) J. Grumann, Th. Morovic, W. Greiner, *Z. Naturforschung* **26a**, 643 (1971).
5. A. Sandulescu *et al*, *Phys. Lett.* B **60**, 225 (1976); R.K. Gupta *et al*, *Phys. Lett.* B **64**, 257 (1977); *ibid. Z. Phys.* A **283**, 217 (1977).
6. G.M. Ter-Akopian *et al*, *Nucl. Phys.* A **255**, 509 (1975); Yu.Ts. Oganessian *et al*, *ibid.* A **239**, 157 (1975), A **239**, 353 (1975).
7. D. Scharnweber, U. Mosel and W. Greiner, *Phys. Rev. Lett.* **24**, 601 (1970); U. Mosel *et al*, *Phys. Lett.* B **34**, 587 (1971).
8. G. Münzenberg *et al*, *Z. Phys.* A **309**, 89 (1992); S. Hofmann *et al*, *ibid.* A **350**, 277 (1995).
9. R.K. Gupta *et al*, *Z. Naturforschung* **32a**, 704 (1977).
10. A. Sandulescu and Walter Greiner, *Rep. Prog. Phys.* **55**, 1423 (1992); A. Sandulescu *et al*, *Int. J. Mod. Phys.* E **1**, 379 (1992).
11. A. Sobiczewski, *Phys. of Part. and Nucl.* **25**, 295 (1994).
12. R.K. Gupta, G. Münzenberg and W. Greiner, *J. Phys.* G **23**, L13 (1997).
13. V. Ninov *et al*, preprint
14. K. Rutz, M. Bender, T. Bürvenich, T. Schilling, P.-G. Reinhard, J.A. Maruhn, W. Greiner, *Phys. Rev.* C **56**, 238 (1997).
15. B. Fricke and W. Greiner, *Phys. Lett.* B **30**, 317 (1969); B. Fricke, W. Greiner, J.T. Waber, *Theor. Chim. Acta (Berlin)* **21**, 235 (1971).
16. A. Sandulescu *et al*, D.N. Poenaru, W. Greiner, *Sov. J. Part. Nucl.* **11(6)**, 528 (1980).
17. Harold Klein, *PhD thesis*, Inst. für Theoret. Physik, J.W. Goethe-Univ. Frankfurt a. M. (1992); Dietmar Schnabel, *PhD thesis*, Inst. für Theoret. Physik, J.W. Goethe-Univ. Frankfurt a.M. (1992).
18. D. Poenaru, J.A. Maruhn, W. Greiner, M. Ivascu, D. Mazilu and R.

Gherghescu, *Z. Phys.* A **328**, 309 (1987), *ibid.* A **332**, 291 (1989).
19. E.K. Hulet *et al*, *Phys. Rev. Lett.* **56**, 313 (1986).
20. K. Depta *et al*, *Int. J. Mod. Phys.* A **5**, 3901 (1990); K. Depta *et al*, in *Dynamics of Collective Phenomena*, ed. P. David, World Scientific, Singapore, 29 (1987); S. Cwiok *et al*, *Nucl. Phys.* A **491**, 281 (1989).
21. A. Sandulescu and W. Greiner in discussions at Frankfurt with J. Hamilton (1992/1993).
22. J.H. Hamilton, A.V. Ramaya *et al*, *J. Phys* G **20**, L85 (1994).
23. B. Burggraf *et al*, accepted for publication in *J. Phys* G.
24. P. Hess *et al*, in preparation.
25. E.K. Hulet *et al*, Phys. Rev. C **40**, 770 (1989).
26. W. Greiner, B. Müller, J. Rafelski, *QED of Strong Fields*, Springer Verlag, Heidelberg (1985). For a more recent review see W. Greiner, J. Reinhardt, *Supercritical Fields in Heavy-Ion Physics*, Proceedings of the 15th Advanced ICFA Beam Dynamics Workshop on Quantum Aspects of Beam Physics, World Scientific (1998).
27. B. Povh, *Rep. Progr. Phys.* **39**, 823 (1976); *Ann. Rev. Nucl. Part. Sci.* **28**, 1 (1978); *Nucl. Phys.* A **335**, 233 (1980); Progr. Part. Nucl. Phys. **5**, 245 (1981); Phys. Blätter **40**, 315 (1984).
28. J. Schaffner, C. Greiner and H. Stöcker *Phys. Rev.* C **45**, 322 (1992); *Nucl. Phys.* B **24**, 246 (1991); J. Schaffner, C.B. Dover, A. Gal, D.J. Millener, C. Greiner, H. Stöcker: *Ann. Phys.* **235**, 35 (1994).
29. W. Scheid and W. Greiner, *Ann. Phys.* **48**, 493 (1968); *Z. Phys.* **226**, 364 (1969).
30. W. Scheid, H. Müller and W. Greiner *Phys. Rev. Lett.* **13**, 741 (1974).
31. H. Stöcker, W. Greiner and W. Scheid *Z. Phys.* A **286**, 121 (1978).
32. I. Mishustin *et al*, *J. Phys.* G **19**, 1303 (1993).
33. P.K. Panda *et al*, *Int. J. Mod. Phys.* E **6**, 307 (1997).
34. N. Auerbach *et al*, *Phys. Lett.* B **182**, 221 (1986).
35. H. Stöcker and W. Greiner, *Phys. Rep.* **137**, 279 (1986).
36. J. Reinhardt and W. Greiner, to be published.
37. P. Papazoglou *et al*, *J. Phys.* G **23**, 2081 (1997); P. Papazoglou *et al*, *Phys. Rev.* C **57**, 2576 (1998).
38. P. Papazoglou *et al*, nucl-th/9806087, accepted for publication in *Phys. Rev.* C.
39. J. Theis, G. Graebner, G. Buchwald, J. Maruhn, W. Greiner, H. Stöcker and J. Polonyi, *Phys. Rev.* D **28**, 2286 (1983).
40. P. Papazoglou, *PhD thesis*, University of Frankfurt, 1998; C. Beckmann *et al*, in preparation.
41. S. Klimt, M. Lutz, W. Weise, *Phys. Lett.* B **249**, 386 (1990).

30 YEARS OF CLUSTERS IN NUCLEAR PHYSICS

W. VON OERTZEN

Hahn-Meitner-Institut, Glienicker Strasse 100, D-14109 Berlin, Germany, and Fachbereich Physik, Freie Universitaet Berlin

E-mail: oertzen@hmi.de

The concept of clusters in nuclei went through a fascinating evolution within the history of nuclear physics in the last three decades. In the first decade a fast development ocurred due to experiments based on the numerous Tandem accelerator laboratories. The emphasis has been on light nuclei, mostly with N = Z and the related Clusters built by alpha-particles. Experiments in nuclear physics started to move to higher incident energies. Cluster - cluster interaction has been studied in great detail, because new experimental techniques and models allowed a refined understanding of the two-center nuclear system. Cluster emission in heavy nuclei, directly or accompanied by fission has been discovered and studied systematically. The situation changed with the possibilities to study nuclei and nuclear dynamics under extreme conditions. Cluster emission from highly excited nuclear matter is studied. The new subjects of the last decade are based on results from the high energy accellerators and the radioactive beams and the new experimental detector arrays for gamma radiation and charged particles. Nuclei close to the neutron and proton drip lines show strong deviations from the mean field concept, namely clusterisations and molecular configurations with valence neutrons become a reality.

1 Introductory comments

Looking back into 30 years of studies of Clustering Phenomena in Nuclear Physics we notice strong oscillations of activities. The beginning was marked by the first conference in Bochum in 1969 (which I did not attend), and the conferences in Maryland (1975) and Winnipeg (1978), and for the future of the field we want to hope that not too many of the participants today of ICC99 have been able to attend those (a full list is given as the first 9 references). The earlier work was particularly focussed on light nuclei and their reactions, and with the advent of the Tandem van de Graaff accelerators on various aspects of heavy-ion reactions. This field has been the fore-front of the research, in particular the observation of the heavy-ion resonances. Many theoretical concepts have been clarified in these early days, specifically the equivalence of the shell model description with the cluster model approaches, once antisymmetrisation is considered in its full consequence. The shell model approaches made big progress because the microscopic models were able to include larger and larger model space with the complete antisymmetrisation. The equivalence of different cluster substructures in the same wavefunction has been realised and experiments with light nuclei were able to establish these concepts. The field

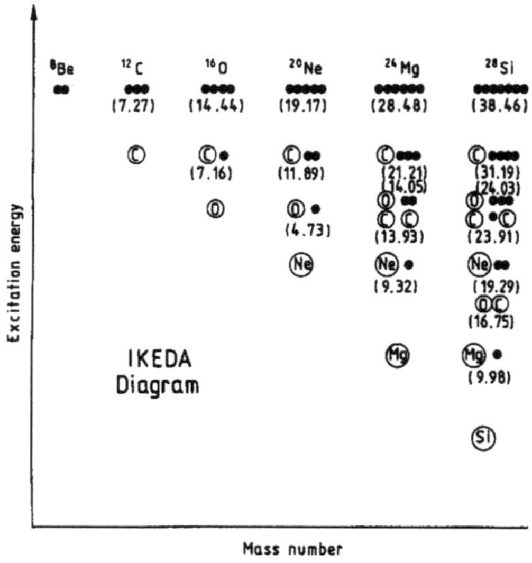

Figure 1. The IKEDA-Diagram for light α-cluster nuclei. The threshold energies for the illustrated configurations are indicated.

evolved faster in the last decades, important impacts came from new frontiers in nuclear physics, like high energy collisions, fragmentation and break-up reactions, and the reactions of weakly bound ("exotic") radioactive light nuclei close to the driplines. Mutual fertilisation occurred with atomic physics due to the emergence of the science of atomic clusters. The latter facts can be traced easily in the proceedings of the conferences in Chester (1984), Kyoto (1988), Strassbourg (1994), as well as in separate conferences dedicated to the common problems in Atomic and Nuclear Cluster Physics in Abo/Turku (1991), Santorini (1993), and Rauischholzhausen (1997).

The long standing problem how to define a cluster has been a line of thought all this time. One approach is perfectly illustrated by the diagram, [10,11] proposed by Ikeda in 1968. In Fig. 1 this diagram is shown. The concept suggests that states of a particular cluster-(sub)structure appear close to the threshold for the decay of the composite system into these "fragments". This can actually be observed with rising excitation energy from the view-point of bound states, as well as from the point of view of higher excitation energies, where condensation of nucleons into clusters sets in with lower temperature as we approach the binding energy limits of nucleons and α-particles. The main physical effect of the diagram resides in the properties of the nucleon-nucleon interaction, which gives rise to saturation and the stabilisation of clusters with particular "magic" numbers, and their mutually reduced interaction. This discussion need not to be repeated, however, the

Figure 2. THE WELL OF LIFE, sculpture by *Ivan Meštrović* in front of the National Theatre in Zagreb.

diagram of Fig. 1 can serve as a guideline through the present presentation. As a complementary illustration I want to show a rather different and unique example of clustering created by *Ivan Meštrović*, with the bronze sculpture named "The well of life" (shown in Fig. 2), to be found in front of the National Theater in Zagreb. There are particular correlations between members of this group, and there is a common "magic", which attracts them mutually to form a total cluster.

For my personal look back and across today's activities and into the future I chose a restricted selection of themes, which have marked the progress in the field. I will emphasise the enormous progress in experimental facilities, which have brought and will bring dramatic progress in the sensitivity of the experiments, compared to the situation 30 years back.

2 The highly efficient new experimental techniques

2.1 "4π"-detector systems

Nuclear physics has benefitted in the last decade from the development of new experimental techniques, which offer an increased sensitivity in a wide range of methods for nuclear structure studies. These advances refer to the operation of highly granular "4π"-detector systems for charged particles or for high resolution gamma-ray spectroscopy, or to the construction of complex systems comprising different detection methods, like magnets and particle detectors for high energy experiments. Several high energy accellerators use secondary radioactive beams with rather small intensities to study break-up processes and other secondary reactions of loosely bound nuclei. With these, e.g. the momentum distributions of the constituents of the ground states are determined; they are found to reflect indeed the low binding energy and the angular momentum content of the wave function. The experimental set-ups are quite large (see Fig. 3), they typically consist of a set of tracking devices for the incident beam, which can operate at the intensities of 10^4 to 10^6 particles/sec, a magnetic separator for momentum analysis of the charged particles, position sensitive charged particle detectors, as well as neutron walls, both used to detect the break-up fragments in coincidence. It is thus often possible to reconstruct the complete kinematics, and many experiments have been performed, where the excitation energy spectrum of the fragments is obtained by building the invariant mass spectrum. [12] Also at these high energies angular correlations can be obtained, which allow the determination of the relative angular momentum of the clusters. [13]

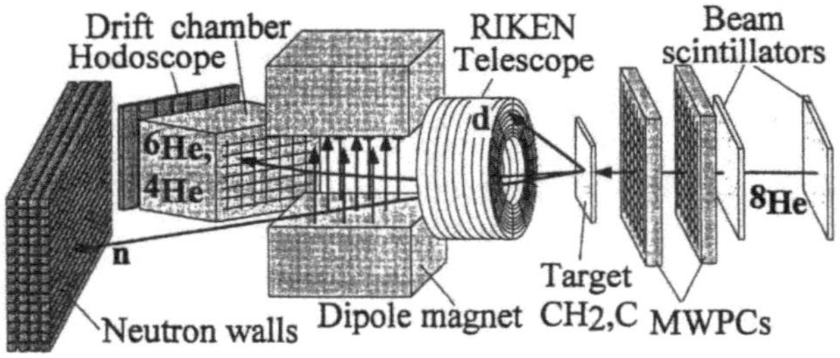

Figure 3. Layout of a complex system of detectors [12] used in research with radioactive secondary beams. The system consists of: tracking detectors for the beam (e.g.^{8}He), magnetic dipoles and telescopes for the separation of charged particles, and large area detectors for the position sensitive registration of charged particles and neutrons.

Complete coverage of final channels in three-body reactions and kinematic reconstruction of a specified reaction at lower energies for the study of cluster structure in light alpha-cluster nuclei has also become possible with silicon detector arrays, consisting of position sensitive silicon micro strip detectors, as demonstrated by Freer et al. [14] With such a system one can reconstruct the excitation energy spectrum of ^{24}Mg in the ^{12}C(^{16}O, ^{12}C + ^{12}C)α reaction. The two ^{12}C nuclei are registered in coincidence with the α-particle; the angular correlations of the ^{12}C nuclei for fixed total energy (excitation energy in ^{24}Mg) are shown as an example in Fig. 4. In this way it has become possible to determine rather precisely the spins of high lying resonances in ^{24}Mg.

In another contribution by Freer [15] to this conference, we see the application of this technique of angular correlations to the field of exotic nuclei; in

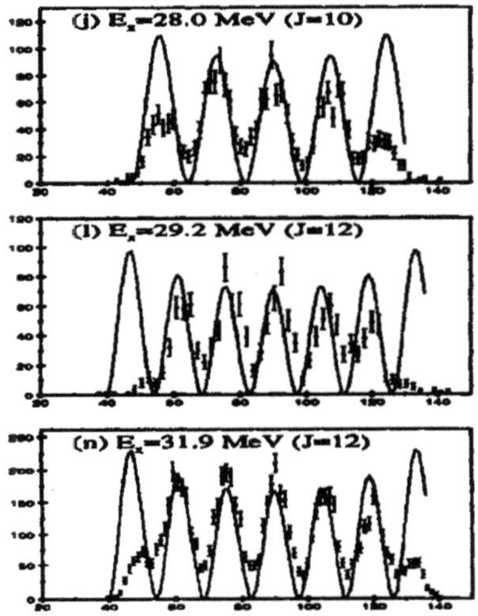

Figure 4. Angular correlations between the two ^{12}C nuclei emitted in the ^{12}C(^{16}O, ^{12}C + ^{12}C)α reaction for fixed excitation energies of the intermediate ^{24}Mg nucleus. The spins of the resonances can be determined unambiguously, from Freer et al. [14]

this work it is demonstrated that the resonant states at intermediate excitation energies in the ^{12}Be nucleus show the substructure of two ^{6}He clusters just above the threshold. The spins of these states have been determined by measuring the angular correlations of the emitted ^{6}He nuclei. It turns out that this result gives a confirmation of the "extended Ikeda diagram" proposed in section 5.3.

The measurement of angular correlations and angular distributions using two position sensitive detectors such as silicon strip detectors, or parallel plate counters, has produced a large body of very precise data for elastic scattering; at this conference results are shown for the ^{28}Si+ ^{28}Si system, and by Nicoli et al. [16] for the ^{16}O + ^{16}O scattering, the latter is discussed in sect. 3.2, where some angular distributions are shown in Fig. 6.

3 Clusters, molecular structure and nuclear dynamics

3.1 Heavy-ion and α-scattering potentials

The study of scattering of light nuclei, which has been started 30 years ago is the subject par excellence of cluster physics, because it involves the important aspect of nuclear dynamics, antisymmetrisation, inelastic couplings and the allways present problem of the *cluster-cluster potentials*. The effect of antisymmetrisation in heavy-ion scattering is consistently described by Y. Tang and D.R. Thompson in the Maryland Conference in 1975, and by Friedrich [17] in 1981. With the concept of phase equivalent potentials and the

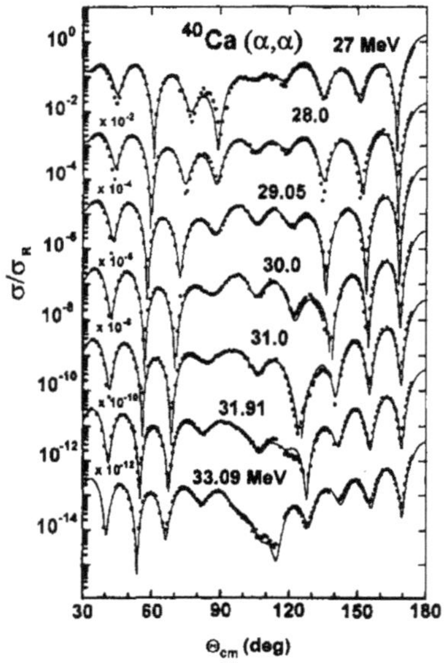

Figure 5. Angular distributions of α-particle scattering on ^{40}Ca nuclei at energies between 27 and 33 MeV with optical model fits. The potentials can be determined unambiguously because of the large angular range of the data, from Michel et al. [20]

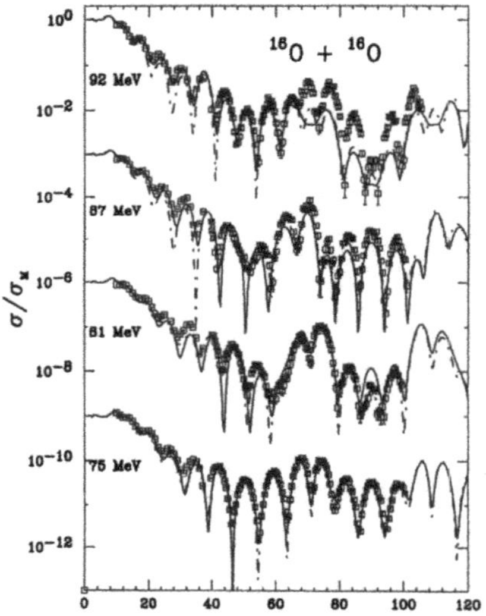

Figure 6. Angular distributions for the scattering of ^{16}O + ^{16}O as obtained by Nicoli [16] with position sensitive solid state detectors in kinematic coincidence. The observed structures around 70°–90° are due to Airy maxima and minima of 4th order and due to the Mott interference of the identical bosons.

supersymmetric transformation introduced in 1987 by Baye, [18] in which the bound states forbidden by the Pauli-Principle are removed, the discussion of the cluster-cluster potentials has now reached a clear foundation.

As an example the elastic scattering of α-particles on $N = Z$ nuclei has been an issue of intense studies over two decades. The work of Michel [19] reviewed in the Cluster conference in Kyoto in 1988, and more recently in a bigger compilation, [20] gives an impression of the quality of the fit to the data of α-scattering on ^{16}O and ^{40}Ca. The high precision of these angular distribution data spanning the angular range up to 180°, and the results at many energies pose a severe test to the theory; the result is shown in Fig. 5 and illustrates the maturity of the field. In conjunction with these potentials also the quasibound and bound states in ^{20}Ne, and other N=Z nuclei up to ^{44}Ti are correctly reproduced. [21] We can state that the description of this large body of precise data by the semimicroscopic theories for cluster potentials is one of the major successes in nuclear physics. The case shown in Fig. 5 gives an impression of this fact.

3.2 The deep ^{16}O + ^{16}O potentials

Probably the most conspicuous and controversial case of heavy-ion scattering [22] has been the case of ^{16}O + ^{16}O scattering, which is still the subject of intensive experimental and theoretical study. The most often shown result is the excitation function at 90° in the center-of-mass system with its pronounced structures covering a factor 100 between minima and maxima. After decades of theoretical analysis with shallow (10–30 MeV depth) potentials, finally the era of deep potentials started. It has been realised that at higher incident energies the potentials must be deep and the scattering process has refractive effects. [23] For ^{16}O + ^{16}O scattering finally ground was broken in 1989 with the analysis of Kondo et al. [24,25] of the low energy data, and with the new high precision data from the Hahn Meitner Institut at higher energies, first at $E_l = 250 - 480$ MeV, Refs. [26,27], as well as later studies at even higher and lower energies [28], and their analysis using the double folding model. [29,30]

The deep potentials for the ^{16}O + ^{16}O system have been a matter of debate already in the approach of the resonating group method summarised by Horiuchi [31] in 1991. The double-folding model [32] which had been applied to the analysis of heavy-ion scattering data for some time, was refined [29,33] and put on a semimicroscopic basis in which the Pauli-exchange effects are treated correctly *and* the nucleon-nucleon interaction in medium was changed so as to reproduce the saturation properties of nuclear matter; [28,29] this approach with a particular choice for the density dependence of the in-medium nucleon - nucleon interaction, opened actually a new way to determine the incompressibility of cold nuclear matter. [28,29,34,35] The potentials predicted in this method using a soft equation of state of cold nuclear matter give impressive descriptions of a wide variety of systems from alpha scattering to heavy ions. [28] One of the consequences of this analysis has been, as now widely accepted, that if parametrised directly, the potential shape should be chosen as (Wood-Saxon)-squared, in direct relation to the double folding of two densities.

We come back to the low energy scattering. High precision data in the range of $E_l = 100$ MeV to 125 MeV have been obtained by Sugiyama et al. [36] and more recently by Nicoli [16] for the range between 75 MeV and 124 MeV with a high density of points from 10° to 100° degrees. Using a shape for the optical potentials of squared Woods Saxon form with a depth which follows the systematics from the double folding model at higher energies, excellent fits of angular distributions (which show very detailed structures, see Fig. 6) are obtained. We can now state that the deep structures observed at 90° in the energy region of 70 to 90 MeV in the excitation function are due to the passage of the Airy minima of the 4th and 5th(!)-order of the nuclear rainbow produced by the strongly refractive potentials. The potentials of heavy-ion scattering remain deep down to energies of 4 MeV/nucleon, and we do observe the passage of the "Airy-elephants", as suggested by McVoy and Brandan. [37] The strong activity in this field has produced the remarkable result that heavy-ion elastic scattering is understood quantitatively today!

3.3 Inelastic couplings and the Band Crossing Models

The occurrence of intermediate and fine structure in resonances in the excitation functions of heavy-ion scattering has been the subject of controversy, because these structures can either be related to properties of the composite system (and thus to nuclear structure), or to dynamical features created typically by strong couplings, mostly strong inelastic transitions to low lying states. The classical example are the cases, which involve the ^{12}C with its well isolated 2^+ state. Actually detailed analysis over the last decades has given plenty of evidence that the coupling of the orbital and intrinsic angular momenta is the important feature. This has lead to the formulation of many models for nuclear resonances, reviewed by Abe [38] in 1994. The most basic structure could be formulated in a band crossing model. [39] A detailed discussion will not be persued here, I mention the fact that using particle - γ coincidences [40] for the system ^{28}Si + ^{28}Si the particular "butterfly mode" predicted by the inelastic coupling has been indeed observed for high lying resonances. The observation of resonances is restricted to certain systems, and we recognise constraints due to the concept of a molecular window, which is related to the matching of the elastic channel to reaction channels, and to the number of open channels as discussed in the next section 3.4.

3.4 Absorption and the molecular window

In the previous section we have discussed, that channel coupling plays a major role in the understanding of cluster-cluster interactions, depending on the strength of the couplings and their dynamical matching conditions. The general background of the many open channels which can carry away flux, although not directly, but connected to the resonant doorway states and determining their width, has been discussed since the beginning of the observation of the heavy-ion resonances. It became clear that the number of open channels (or those with positive Q-values) governs the "visibility" of the resonant structures. A modernised version [41] of such a diagram, which includes the old classical cases with ^{12}C and ^{16}O nuclei, but also possible new combinations with radioactive beams is shown in Fig. 7.

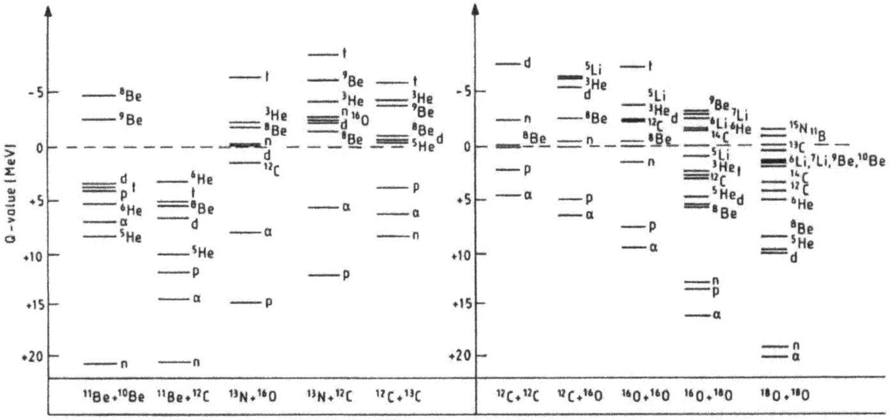

Figure 7. Q-values of important reaction channels relevant for the absorption in elastic scattering for a variety of systems with radioactive beams [41] and stable beams(right side).

Inspecting Fig. 7 we realise, as already 25 years ago, the uniqueness of the systems with closed shell clusters (and alpha-clusters), and that among the radioactive species the proton rich isotopes with masses close to a closed shell-cluster may become interesting candidates for the study of resonances. Neutron excess produces many open channels and immediately destroys the favourable conditions for the observation of resonances, as already noticed in the early stages of heavy-ion studies. [22]

However the description with the Q-values of the binary reaction channels is not complete, because absorption is strongly angular momentum dependent. The concept by Haas and Abe [42] has been to calculate the number of binary channels for a given flux with angular momentum L, and has been put onto a quantitative basis more recently in Ref. [43] for a larger number of heavy-ion systems. The result for the scattering combinations with ^{16}O is shown in Fig. 8. Again we see that the combinations with closed shell clusters give the most favourable conditions to observe resonances, in agreement with the experimental results accumulated in the last decades. Apart from the nuclear structure aspect of the compound system, the facts collected in this work, give the strongest constraints for the observation of energy dependent structures at higher excitation energies.

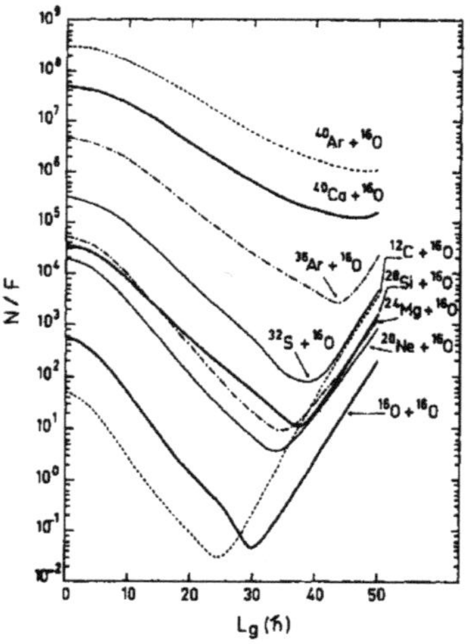

Figure 8. The number of open binary channel N, normalised to the incident flux for the indicated scattering systems as a function of angular momentum. [43]

Figure 9. The life time τ of a cluster in a medium will depend on the energetics (chemical potentials) for particle exchange or open binary channels. This may be influenced by the temperature of the clusters and of the "liquids".

The survival of a cluster, namely the lifetime of a cluster interacting with a medium is connected to the probability of nucleon exchange (and indirectly to antisymmetrisation represented by exchange terms). More generally we may draw a close analogy to physical chemistry, where we observe the different energetic conditions for the disolution of an atomic (molecular) cluster in a solvent (liquid); there it is related to the chemical potential of nucleons in the medium. The experiment can easily be done during the coffee-break as illustrated in Fig. 9. A strongly bound cold cluster has the maximum lifetime in a cold liquid (as for example elastic processes for α-particles in ^{16}O or in ^{40}Ca, and for $^{16}O + ^{16}O$ scattering).

4 Break-up, cluster decay and fission

4.1 Break-up reactions

Nuclei submitted to a short time pulse of the nuclear- or Coulomb-field will break into substructures, which are contained in the wave functions of their groundstates (with corresponding probabilities). For stable strongly bound nuclei the break-up process will be governed by a mixture of reaction mechanisms, where also multi-step interactions will contribute, mainly excitations and subsequent decay. For weakly bound ("exotic") nuclei the situation is changed because the ground state may be represented by one dominant cluster structure, the break-up process may proceed directly into the unbound continuum. For higher incident energies (~ 50 MeV/u and higher) the wave-functions of the ground states of many weakly bound isotopes have been probed. [44] Another important aspect of the direct break-up via the Coulomb field is the fact, that it is the time reversed radiative capture process, which is of paramount importance in stellar evolution, namely the radiative proton and α capture in explosive scenarios. The break-up reactions for astrophysically relevant processes are therefore studied with radioactive beams in many laboratories. [45]

4.2 Cluster radioactivity

The decay of a composite system is easily recognised as the time reversed scattering process in a fusion channel. The determination of scattering potentials helps to fix the potentials for the decay process. It is thus natural to persue an approach, which describes both, scattering of clusters and the cluster decay. This has been done by B.Buck and coworkers [46,47] over the last 10 years. The success of this approach is still not fully understood, but may be related to the problem discussed at the beginning, with the equivalent local potentials to be used in a one-dimensional dynamical theory. The other important part is the potential energy surfaces, which helped to predict the cluster radioactivity, see Ref. [48] for a complete coverage of the subject. For obvious reasons the potential energy landscape (the Q-values) determines that the emission is favoured for heavy neutron rich clusters like ^{14}C, ^{20}O up to ^{32}Si. We are now able to have experimental systematics as with alpha decay , this means a Geiger - Nutall plot for ^{14}C - decay over many decades of lifetimes has been established. [49]

4.3 Cold fission and cold reactions

Nuclear fission has been one of the first observations, which has been connected to large nuclear clusters, in particular the observation of nuclei with closed shells for neutrons or protons as fission fragments. In reference to cold fission we observe strong variations of the yields as function of fission fragment mass and total mass, which go beyond the simple statistical phase space approach. This is because the fission decays into different channels are connected through complicated rearrangements of nucleon configurations and coupling of the different pathways in the macroscopic nuclear landscape (shapes). Today it is established, that different fission modes (with different mass asymmetry) are connected to these cold rearrangement processes. The odd-even effect in the mass yields is another manifestation of this concept. The inverse cold fusion reactions give also clear indications that the minimum dissipation is observed if nuclei with closed shells are used, as discussed by Armbruster. [50] Similarly ternary fission is a manifestation of clustering, where very neutron rich species are emitted for the same reasons, as in cluster radioactivity discussed in the previous section. The asymptotic Q-values favour channels with strongly bound clusters in the final channel, irrespective of the fact that there are dynamical features to be considered in such a complicated decay process.

5 Nuclear structure and molecular structure

5.1 Cranked α-cluster models and super deformation

Since more than a decade systematic work, based on the cranked alpha-cluster model by Marsh and Rae and others, [51,52] has shown that in N=Z nuclei at higher excitations a class of states known (to "nuclear spectroscopists") as super- or hyper-deformed states occur, and that they are related to α-clustering. Experimentally a large variety of resonances [52–55] and shapes have been observed, which are partially found in the Ikeda-diagram. There exist now theoretical models, which allow the prediction of extreme shapes up to the α-chain states. A big step in the understanding of resonances in heavy-ion scattering came through the theoretical calculations of Leander and Larson. [57] These theoretical approaches are complementary to the cranked cluster model, they give very similar results. Further new developments came from the AMD (antisymmetrized molecular dynamics) calculations [57–59] and a consistent picture of strongly deformed shapes (shape isomers) up to extreme deformations like chain-states, can be given today. The ground state of ^{8}Be is actually the first nucleus with a deformation ratio of the two main axes of 2:1. The 3α-chain in ^{12}C is a much discussed structure, which is probably not stable against the bending mode, and it is located between 7–11 MeV excitation energy. Also, for these extreme structures we are able to establish the links between various nuclear structure models. As already mentioned in the introduction, many resonances observed in excitation functions are related to strongly deformed structures in the total nuclear systems, like in ^{12}C + ^{12}C (^{24}Mg), ^{16}O + ^{12}C (^{28}Si), ^{24}Mg + ^{24}Mg (^{48}Cr) and others. A nuclear spectroscopy of strongly deformed states in N=Z nuclei is still waiting for a detailed study with the new "4π"-detector systems for charged particles and gamma-rays. Strongly selective (due to the "Harvey-rules") binary reactions

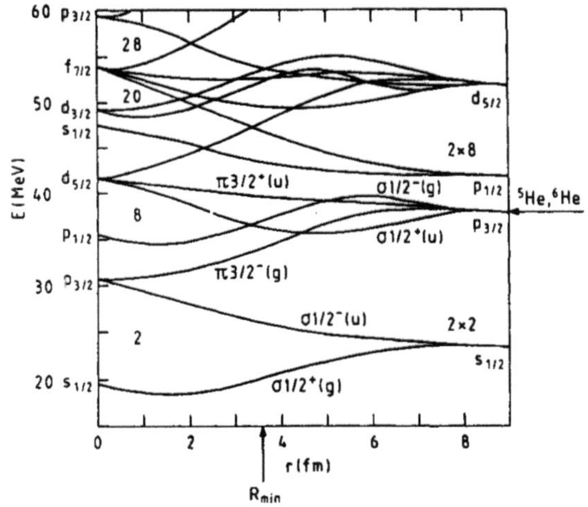

Figure 10. The correlation diagram of covalent molecular orbitals for the system with s1/2, p3/2and p1/2-configurations of the separated shells. [61] The quantum numbers used to specify molecular orbitals are indicated, K-quantum number, orbital angular momentum projection on molecular axis σ and π, gerade/ungerade property and parity. R_{min} is the distance of the α-α potential minimum.

are expected to play a major role in this venture.

5.2 Nuclear dimers and polymers

Based on the well known structure of the ^{8}Be nucleus with the 2:1 ratio of its deformation axes, the structure of excited states in heavier Beryllium isotopes has been reviewed by von Oertzen. [61] It has been found, [15,61,62] that there are excited bands in ^{10}Be, ^{11}Be and ^{12}Be, which have the same or even larger deformation as ^{9}Be . The discussion of the structure of these states can be based on a two-center correlation diagram, see Fig. 10, where the covalent molecular orbitals of neutrons in the two-α-cluster system determine the structure of these states, which we shall call Dimers. Chains bound by covalent neutrons based on longer α-chain states [63] we shall call Polymers (see also the contribution by Fulton in this conference).

The correlation diagram of Fig. 10 tells us that particularly stable isomers are expected with the population of the σ-orbits, which drive the system towards larger deformation; the optimum is reached at a distance between the α-particles, where the α-α potential has its very shallow minimum. We can collect all states in the Beryllium isotopes and plot their energies as for rotational bands of deformed states (see Fig. 11). The moments of inertia for the excited bands in ^{10}Be (K=1$^-$) and ^{11}Be (K=3/2$^-$), for example, correspond to configurations where the two α-particles are well separated.

Considerations of nuclear molecular orbitals have been persued over the last three decades and a large choice of literature exists;[63–68] in this work

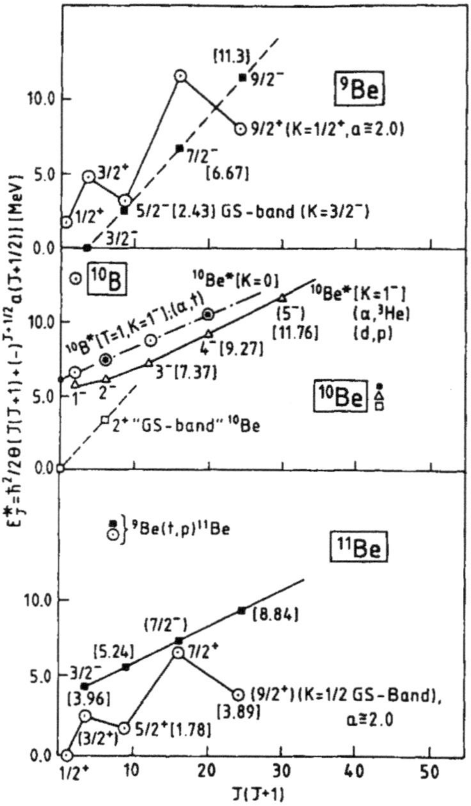

Figure 11. Excitation energies of states in the isotopes $^{9-11}$Be plotted as a function of angular momentum. [61] The K-quantum numbers are indicated; note the staggering due to Coriolis interaction in the (K=1/2)-bands in ^{9}Be and ^{11}Be.

mostly systems with heavier cores than α-particles have been considered, where no repulsion at small internuclear distances is observed, the total systems in these cases collapses to an excited compound nucleus. The observation of real shape-isomeric molecular states had to wait for the discussion of neutron rich light isotopes. [61,67]

Striking confirmation of the concepts of molecular orbitals in Be-isotopes [61] has come from the calculations based on the AMD-approach, [60] where from the slater determinants density distributions for particular states, for example the gground state 0^+, and the first 0^+ state in ^{10}Be can be deduced. Inspecting the diagram of Fig. 10, we expect that the ground state has a dominant $(\pi)^2$ and the excited 0^+ state a $(\sigma)^2$ squared configurations, which are very well confirmed in Fig. 12. Due to the strong binding energy of the alpha-particles, the low lying states are all determined by the properties of the (covalent) valence neutrons. Several recent studies of the structure of

^{11}Be and ^{12}Be have illustrated the potential of the future studies of extremely deformed shapes in neutron-rich exotic nuclei.

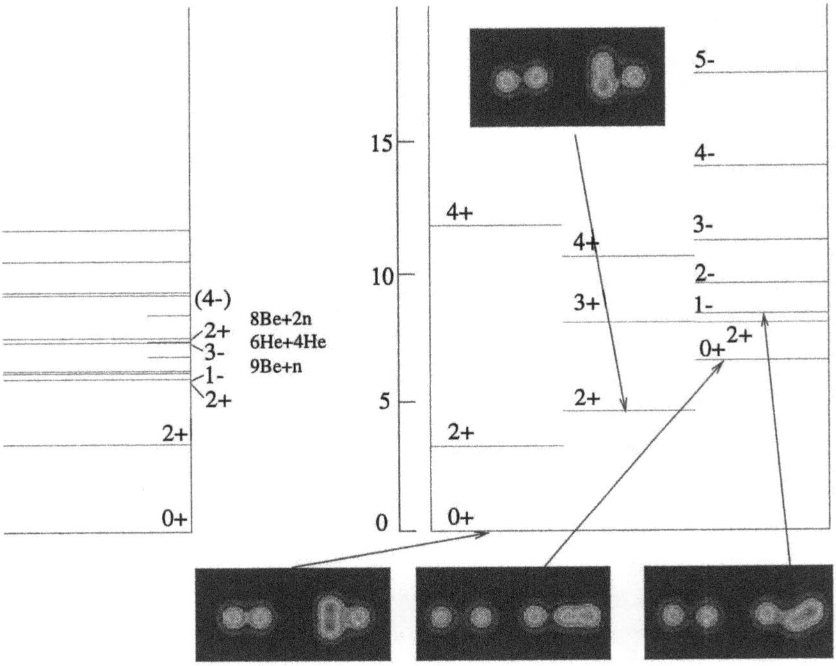

Figure 12. Density distributions of nucleons in ^{10}Be obtained by Kanada-En'yo [60] (left-hand side for protons, right-hand side for neutrons). The covalent extra neutrons can easily be distinguished: for the 0^+ ground state, the first excited 0^+ state, the second 2^+ and the band head of $K=1^-$. The first two states show the dominant π and σ configurations expected due to the correlation diagram of Fig. 10. Note the larger distance of the alpha-clusters (represented by protons) in the excited states. The left side shows the experimental level scheme.

5.3 Exotic shapes in very neutron-rich dripline nuclei

With the strong clustering phenomena in the excited states of "moderately" neutron-rich nuclei, we can speculate about the structure of very neutron-rich nuclei. This can actually be done by inspecting the calculations of Horiuchi et al., [58,60,70] who has studied the neutron and proton density distributions using the method of AMD already mentioned before. The observation from these model calculations is, that with the increase of the neutron number the clustering effects increase, so as to produce the maximum number of α-clusters with the neutrons filling the space in between these clusters in covalent orbitals. In close similarity to the linear molecular structures discussed in the previous section, these configurations offer for the neutrons states with the smallest possible momenta, and a maximum of interactions with the α-clusters. It is thus natural to expect that quasi-stationary stable molecules (which may not be the ground state of the nucleus) can be formed with covalent neutrons arranged in orbits corresponding to their eigenfunctions centered at individual clusters. I can formulate an *extended "Ikeda-diagram"* with the statement that: *covalently bound nuclear molecular states will occur in neutron rich nuclei at energies at or below the threshold for the decay into α-clusters (also ^{16}O) and neutrons.* Beams of very neutron rich projectiles will be needed to access these structures.

References

1. BOCHUM Conference, Int. Conf. on Clustering Phenomena in Nuclei (1969), Bochum (Germany), Proceed. IAEA, Vienna, 1969.
2. MARYLAND Conference, *2nd* Int. Conf. on Clustering Phenomena in Nuclei (1975), College Park (USA), Proceed., ed. D.A. Goldberg, USERDA/ORO-4856-26.
3. WINNIPEG Conference, Conf. on Clustering Aspects of Nuclear Structure (1978), Winnipeg (Can), Proceed., ed. W.T.H. van Oers, Amer. Inst. of Physics, New York, 1978.
4. CHESTER Conference, *4th* Int. Conf. on Clustering Aspects of Nuclear Structure and Reactions (1984), Chester (England), Proceed., ed. J.S. Lilley, D. Reidel (Dordrecht, 1985).
5. KYOTO Conference, *5th* Int. Conf. on Cluster. Aspects in Nuclear and Subnuclear Systems (1988), Kyoto (Japan), Proceed., ed. K. Ikeda, Phys. Society of Japan, 1989.
6. TURKU Conference, Int. Conf. on Nuclear and Atomic Clusters (1991), EPS Topical Conf., Turku (Finnland), Proceed., eds. M. Brenner, T. Loennroth, Springer (Berlin, Heidelberg) 1992.
7. SANTORINI Conference, Int. Conf. on Atomic and Nuclear Clusters (1993), Santorini (Greece), Proceed., eds. G.S. Anagnostatos, W. von Oertzen, Springer (Berlin, Heidelberg) 1995.
8. STRASBOURG Conference, *6th* Int. Conf. on Clusters in Nuclear Structure and Dynamics, Strasbourg (France), Proceed., ed. F. Haas, CRN Strasbourg.
9. RAUISCHHOLZHAUSEN Conference, *174th* WE-Heraeus-Seminar,

New Ideas on Clustering in Nuclear and Atomics Physics (1997), Rauischholzhausen (Germany), Proceed., eds. N. Cindro, W. Scheid, Phys. Society of Italy (Bologna) 1997.
10. K. Ikeda, N. Tagikawa, H. Horiuchi, *Prog. Theor. Phys. (Japan), Suppl.* 464 (1968).
11. H. Horiuchi, K. Ikeda, Y. Suzuki, *Prog. Theor. Phys. (Japan), Suppl.* **52**, Chapter 3 (1972).
12. A.A. Korsheninikov *et al*, *Phys. Rev. Lett.* **82**, 3581 (1999).
13. L.V. Chulkov, *Il Nuovo Cimento* A **111**, 791 (1998).
14. M. Freer *et al*, *Phys. Rev.* C **57**, 1277 (1998).
15. M. Freer *et al*, RAB conference 1999, and *Phys. Rev. Lett.* **82**, 1383 (1999).
16. M.P. Nicoli *et al*, RAB conference 1999.
17. H. Friedrich, *Phys. Rep.* **74**, 209 (1981).
18. D. Baye, *Phys. Rev. Lett.* **58**, 2738 (1987).
19. F. Michel, KYOTO Conference 1988, page 65.
20. F. Michel *et al*, *Prog. Theor. Phys. (Japan) Suppl.* **132**, 7 (1998).
21. T. Sakuda and S. Ohkubo, *Prog. Theor. Phys. (Japan) Suppl.* **132**, 103 (1998).
22. K.A. Erb and D.A. Bromley, Treatise in Heavy Ion Science, **Vol. 3** page 201, ed. D.A. Bromley, 1984.
23. K.W. McVoy and G.R. Satchler, *Nucl. Phys.* A **417**, 157 (1984).
24. Y. Kondo, B.A. Robson and R. Smith, KYOTO Conference 1988.
25. Y. Kondo *et al*, *Phys. Lett.* B **242**, 340 (1990).
26. E. Stiliaris *et al*, *Phys. Lett.* B **223**, 291 (1989).
27. H.G. Bohlen *et al*, *Z. Phys.* A **346**, 189 (1993).
28. Dao.T. Khoa *et al*, *Phys. Rev.* C **56**, 954 (1997).
29. Dao T. Khoa and W. von Oertzen, *Phys. Lett.* B **304**, 8 (1993).
30. Dao.T. Khoa *et al*, *Phys. Rev. Lett.* **74**, 34 (1995).
31. H. Horiuchi, Chapter 13 in TRENDS IN NUCLEAR PHYSICS Vol. 2. eds. P.J. Ellis and Y.C. Tang, (Addison Wesley Publ. Comp.).
32. G.R. Satchler and W.G. Love, *Phys. Rep.* C **55**, 183 (1979).
33. Dao.T. Khoa *et al*, *Phys. Rev.* C **49**, 1652 (1994).
34. W. von Oertzen, *Prog. Part. Nucl. Phys.* C **49**, 257 (1993).
35. Dao T. Khoa and W. von Oertzen, Int. School of Heavy Ion Physics, Erice, 1993, p 153; eds. R.A. Broglia, P. Kienle and P.F. Bortignon, World Scientfic, Singapure (1994).
36. Y. Sugiyama *et al*, *Phys. Lett.* B **304**, 8 (1993).
37. K.W. McVoy and M.E. Brandan, *Nucl. Phys.* A **542**, 295 (1992).
38. Y. Abe, STRASSBOURG Conference, 1994.

39. Y. Abe *et al*, *Prog. Theor. Phys. (Japan) Suppl.* **68**, (1980) chapter 4.
40. R. Nouicer *et al*, *Phys. Rev.* C **60**, 41303 (1999).
41. W. von Oertzen, to be published.
42. F. Haas and Y. Abe, *Phys. Rev. Lett.* **46**, 1667 (1981).
43. C. Beck, Y. Abe, N. Aissaoui *et al*, *Phys. Rev.* C **49**, 2618 (1994).
44. I. Tanihata, *J. Phys.* G **22**, 157 (1996).
45. Int. Conf. on Radioctive Nuclear Beams, *Nucl. Phys.* A **616**, 1-517c (1997).
46. B. Buck *et al*, STRASSBOURG Conference, 1994.
47. B. Buck *et al*, *Nucl. Phys.* A **644**, 306 (1998), and refs.
48. A. Sandulescu and W. Greiner, *Rep. Prog. Phys.* A **55**, 1423 (1992).
49. R. Bonetti and A. Guglielmetti, STRASSBOURG Conference, 1994.
50. P. Armbruster, STRASSBOURG Conference, 1994.
51. S. Marsh and W.D.M. Rae, *Phys. Lett.* B **180**, 185 (1986).
52. W.D.M. Rae, A.C. Merchant, STRASSBOURG Conference, 1994.
53. R. Betts *et al*, *Phys. Rev. Lett.* **47**, 23 (1981).
54. R.W. Zurmuehle *et al*, *Phys. Lett.* B **129**, 384 (1983).
55. R.W. Zurmuehle, KYOTO Conference 1988, page 37.
56. B.R. Fulton and W.D.M. Rae, *J. Phys.* G **16**, 333 (1990).
57. G. Leander and S.E. Larsson, *Nucl. Phys.* A **239**, 93 (1975).
58. H. Horiuchi *et al*, Proceedings, Int. Conf. Atomic and Nuclear Clustering, Santorini (1993), ed. W. von Oertzen and G.S. Anagnastatos, Springer (1995); and *Z. Phys.* A **349**, 142 (1994).
59. Y.K. Kanada-En'yo, H. Horiuchi, A. Ono, *Phys. Rev.* C **52**, (1995), also *Phys. Rev.* C **56**, 1844 (1997).
60. Y.K. Kanada-En'yo, H. Horiuchi and A. Dote, *J. Phys.* G **24**, 1499 (1998).
61. W. von Oertzen, *Z. Phys.* A **354**, 37 (1996), and *Z. Phys.* A **357**, 355 (1997), also *Il Nuovo Cimento* A **110**, 895 (1997).
62. H.G. Bohlen *et al*, *Prog. Part. Nucl. Phys.* A **42**, 17 (1999).
63. A.H. Wuosmaa, STRASSBOURG conference, 1994.
64. W. von Oertzen, W. Noerenberg, *Nucl. Phys.* A **207**, 113 (1973).
65. Y. Abe *et al*, *Prog. Theor. Phys.* **49**, 800 (1973).
66. J.Y. Park, W. Greiner, W. Scheid, *Phys. Rev.* C **21**, 958 (1980).
67. M. Seya, M. Kohno, S. Nagata, *Prog. Theor. Phys.* **65**, 204 (1981).
68. B. Imanishi, W. von Oertzen, *Phys. Rep.* **155**, 29 (1987).
69. W. von Oertzen, KYOTO Conference, 1988.
70. H. Horiuchi, Y.K. Kanada-En'yo, *Nucl. Phys.* A **616**, 394c (1997).

CLUSTERING AND THE PAULI PRINCIPLE

R. G. LOVAS

Institute of Nuclear Research of the Hungarian Academy of Sciences, Debrecen, P. O. Box 51, H-4001, Hungary

E-mail: rgl@atomki.hu

Some basic concepts of nuclear cluster physics will be reviewed. The Pauli principle, which is sometimes believed to destroy clusters, is a key factor in reconciling the apparent conflict of various cluster pictures with each other and with the shell model. The coexistence of the single-particle motion and various clusterings will be discussed, with examples from light to heavy nuclei. Guidelines will be given for the treatment of the Pauli principle in the analysis of clustering phenomena.

1 Introduction

Some nuclei decay into two fragments spontaneously. Some stable nuclei also tend to behave similarly: upon external perturbation, they prefer to break up into some particular fragments. This raises the question to what extent and in what sense composite fragments can be considered to exist in the nucleus as 'clusters' or as a 'cluster–core' pair before fragmentation. On the other hand, two colliding nuclei may stick together for a long time forming bands of 'quasimolecular' or binary cluster resonances. The lower parts of such bands often fall into the bound-state region; they may even include ground states (g.s.). Such states are described very well by cluster models [1], which are based on the assumption that the nucleus consists just of the clusters into which it tends to disintegrate. The cluster picture seems to defy the Pauli principle, which mixes up the nucleons irrespective of their cluster associations. It is in apparent conflict also with the whole conventional view of nuclei: the mean-field picture, the shell model as well as the usually assumed types of collective motion. And yet cluster models are the most satisfactory models of a range of light nuclei, including those built up from α-particles as well as a number of exotic nuclei. They are especially successful for states lying near the respective cluster breakup threshold; then the behaviour of the nucleus is mostly determined by the intercluster relative motion, and the cluster internal motion can be treated schematically, which is the underlying assumption of cluster models.

To dramatize the situation, in Sec. 2 I will give a few paradoxical examples, and resolve the paradoxes by qualitative arguments. In Sec. 3 we will define clustering quantitatively, and review some evidences for clustering and for the correctness of cluster models. In Sec. 4 we shall revisit the idea of intercluster relative motion, which is a key to phenomenological analyses on clustering.

2 Single-particle motion and clustering

As a simple example, let us describe an α-particle in a single-particle (s.p.) model. Its wave function is then $\mathcal{A}\{\phi\xi\} = \phi\mathcal{A}\{\xi\}$, where $\phi = \prod_{i=1}^{4} \varphi(\mathbf{r}_i)$ and $\xi = \prod_{i=1}^{4} \xi_i(i)$ are the spatial and spin-isospin factors, respectively. The antisymmetrizer $\mathcal{A} = n^{-1/2}\sum^n(-1)^P P$ takes care of spin-isospin coupling. Be the s.p. functions just 0s harmonic-oscillator (h.o.) states, $\varphi \equiv \varphi_{000}(\mathbf{r}) \sim e^{-\frac{1}{2}\beta\mathbf{r}^2}$. Let us also consider the following correlated motions: (i) each particle is strongly correlated with all others; (ii) particle 3 moves around the centre of mass (c.m.) of particles 1 and 2, and particle 4 around the c.m. of 1, 2 and 3; and (iii) the particles orbit around each other pairwise, with the two pair c.m. moving similarly around each other. These correlated motions are best represented, respectively, (i) by the six relative vectors $\mathbf{r}_i - \mathbf{r}_j$, (ii) by the Jacobi vectors

$$\mathbf{x}_j = \left(\tfrac{j}{j+1}\right)^{1/2} \left(\tfrac{1}{j}\textstyle\sum_{i=1}^{j} \mathbf{r}_i - \mathbf{r}_{j+1}\right) \quad (j = 1,\ 2,\ 3), \quad \mathbf{x}_4 = \tfrac{1}{2}\textstyle\sum_{i=1}^{4} \mathbf{r}_i \qquad (1)$$

and (iii) by the two-pair relative vectors

$$\begin{aligned} &\mathbf{x}'_1 = 2^{-1/2}(\mathbf{r}_1 - \mathbf{r}_2)\,, && \mathbf{x}'_2 = 2^{-1/2}(\mathbf{r}_3 - \mathbf{r}_4)\,, \\ &\mathbf{x}'_3 = 2^{-1}(\mathbf{r}_1 + \mathbf{r}_2 - \mathbf{r}_3 - \mathbf{r}_4)\,, && \mathbf{x}'_4 \equiv \mathbf{x}_4 = 2^{-1}(\mathbf{r}_1 + \mathbf{r}_2 + \mathbf{r}_3 + \mathbf{r}_4)\,. \end{aligned} \qquad (2)$$

Fig. 1 shows these vectors schematically, apart from the c.m. vector $\mathbf{x}_4$.

Let us build up states that show the above correlations. For the sake of comparability, let us assume that the correlations are described by Gaussians belonging to the same h.o. energy quantum $\hbar\omega$ as assumed for the s.p. state.

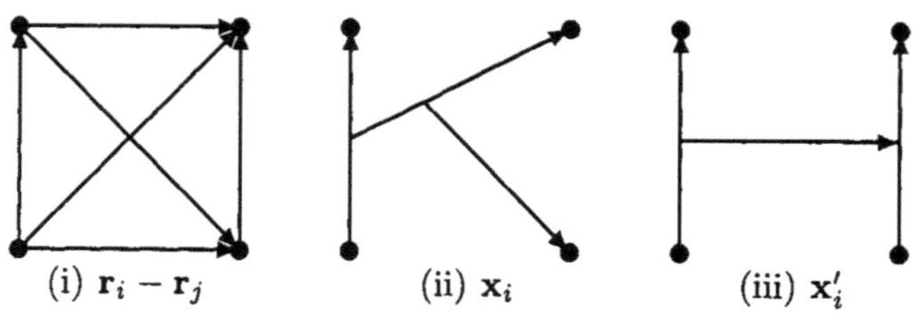

Figure 1. Two-particle relative vectors, Jacobi vectors and two-cluster relative vectors.

The result is paradoxical:

$$e^{-\frac{1}{2}\beta[\mathbf{x}_4^2+\frac{1}{4}\sum_{i<j}^{4}(\mathbf{r}_i-\mathbf{r}_j)^2]}=e^{-\frac{1}{2}\beta(\mathbf{x}_1^2+\mathbf{x}_2^2+\mathbf{x}_3^2+\mathbf{x}_4^2)}=e^{-\frac{1}{2}\beta(\mathbf{x}_1'^2+\mathbf{x}_2'^2+\mathbf{x}_3'^2+\mathbf{x}_4'^2)}=\phi. \quad (3)$$

We see that the s.p. state is identical to the most clear-cut pair-correlated state, which, for symmetry reasons, can be visualized as a regular tetrahedron, but at the same time it is equal to a stepwise correlated state, as well as to a state of a pair of two-nucleon clusters. This identity is a consequence of the Gaussian form and of the identity of the particles.

As another example, let us see the g.s. of ^{16}O in a similar s.p. model:

$$\Phi_{16}=\mathcal{A}\left\{\prod_{i=0}^{3}\varphi_{000}(\mathbf{r}_{i+1})\varphi_{01,-1}(\mathbf{r}_{i+5})\varphi_{010}(\mathbf{r}_{i+9})\varphi_{011}(\mathbf{r}_{i+13})\prod_{j=1}^{4}\xi_j(4i+j)\right\}. \quad (4)$$

(The antisymmetrizer prepares a state of spin zero out of the state in { }.) It can be shown that (4) can be identically rewritten, into a 4α state, in which all α-clusters are in intrinsic state ϕ. Their relative motion can be imagined as three $4\hbar\omega$ oscillations of an H or K shaped structure shown in Fig. 1, with $\hbar\omega$ equal to that involved in Eq. (4). Similarly, the configuration can also be visualized as a tetrahedron of α-clusters, with indefinite spatial orientation. It is to be noted that these identities are only valid when $\mathcal{A}$ is included.

As a third example, let us consider the cluster model of the ^{8}Be g.s.: $\Phi^{(8)}=\mathcal{A}\{\Phi^{(\alpha)}(1,2,3,4)\Phi^{(\alpha)}(5,6,7,8)\varphi(\mathbf{r}_{\alpha\alpha})\}$, where $\Phi^{(\alpha)}$ differs from Φ_α only in the c.m. factor: $\Phi_\alpha=e^{-\frac{1}{2}\beta\mathbf{x}_4^2}\Phi^{(\alpha)}$. When $\varphi(\mathbf{r}_{\alpha\alpha})=\varphi_{200}(\mathbf{r}_{\alpha\alpha})$, this is exactly equal to the shell-model state of $4\hbar\omega$ excitation, as the subspace defined by $L=S=0$ and energy $4\hbar\omega$ is one-dimensional. This is, of course, a spherically symmetric state, whose density involves an averaging over all $\alpha\alpha$ orientations. To unravel the imprint of clustering, one should omit this averaging, and then the clustering does appear even in the pure shell-model configuration [2].

The clue to these paradoxical examples lies in the symmetries of the states. They are invariant to rotation, reflection and permutation and form one-dimensional representations of the SU_3 symmetry group of the h.o. Any symmetry requirement is tantamount to a projection; e.g., $\mathcal{A}$ is proportional to a projector. What we see is that the projections of very different objects coincide, and that is quite natural. Cluster structure is often hidden by symmetry projections; e.g., for a system of two non-identical clusters, even parity-projecting is enough to conceal it. From the projection, the original object cannot be reconstructed unambiguously. Cluster models and s.p. models display two very different facets of a system, yet do not necessarily contradict each other. Representing each nucleon by a Gaussian wave packet pinned down in space, we obtain a cluster model (the antisymmetrized molecular dynamics [3]), which portrays some nuclei as finger-biscuit or comet-like objects (before projection), and yet we have no reason to say off-hand that such a model contradicts the mean-field picture.

The above examples may seem exceptional, but the message is general: the s.p. motion and various cluster-correlated motions do not exclude each other in a correctly symmetrized state space. The cluster models are not negations of the s.p. models; any particular shell-model state or cluster-model state may accommodate apparently different clusterings. To quantify this statement, one needs to define clustering more precisely.

3 Amount of clustering

Clustering for a nucleus means to be composed of clusters; the amount of clustering in a state Ψ is the probability of finding the system in a multi-cluster subspace [4]: $\mathcal{S} = \langle\Psi|P|\Psi\rangle$, where P is the corresponding projector. Each subspace may be subdivided according to orbital momenta and spins $[(l_1 l_2)...]L[(s_1 s_2)...]S$. A subspace may be spanned by "basis" states in which clusters are pinned down at definite relative locations. For an A+a two-cluster case, such a state is $\Phi_{\mathbf{r}} = \mathcal{A}\{\Phi^{(A)}\Phi^{(a)}\delta(\mathbf{r} - \mathbf{r}_{Aa})\}$. Observing the non-orthogonality, $\langle\Phi_{\mathbf{r}}|\Phi_{\mathbf{r}'}\rangle \equiv N(\mathbf{r},\mathbf{r}') \neq \delta(\mathbf{r}-\mathbf{r}')\}$, which is a consequence of antisymmetrization, P is to be constructed as $P = \int d\mathbf{r}|\Phi_{\mathbf{r}}\rangle\mathcal{N}^{-1}\langle\Phi_{\mathbf{r}}|$, where the norm operator $\mathcal{N}$ is the integral operator whose kernel is $N(\mathbf{r},\mathbf{r}')$. The inverse $\mathcal{N}^{-1}$ is defined so as to contain a projector Λ onto the null space of $\mathcal{N}$ (which is the strictly Pauli-forbidden part of the relative-motion space), i.e., $\mathcal{N}^{-1}\mathcal{N} = \mathcal{N}\mathcal{N}^{-1} = \Lambda$. Thus $\mathcal{S}$ compares with the spectroscopic factor s as follows:

$$s=\int d\mathbf{r}\langle\Psi|\Phi_{\mathbf{r}}\rangle\langle\Phi_{\mathbf{r}}|\Psi\rangle=\int d\mathbf{r}|g(\mathbf{r})|^2, \quad \mathcal{S}=\int d\mathbf{r}\langle\Psi|\Phi_{\mathbf{r}}\rangle\mathcal{N}^{-1}\langle\Phi_{\mathbf{r}}|\Psi\rangle=\int d\mathbf{r}|G(\mathbf{r})|^2, \quad (5)$$

where $g(\mathbf{r}) = \langle\Phi_{\mathbf{r}}|\Psi\rangle$ and $G(\mathbf{r}) = \mathcal{N}^{-1/2}g(\mathbf{r})$ are the spectroscopic and clustering amplitude, respectively.

A state can also be analysed in terms of shell-model subspaces. The weight of $Q\hbar\omega$ excitation is $\mathcal{S}_Q = \langle\Psi|P_Q|\Psi\rangle$, where P_Q projects onto the $Q\hbar\omega$ subspace. To have a feeling for these concepts, let us consider a realistic α+t+n model of ^{8}Li [6], in which the wave function is represented by a combination of non-orthogonal (αt)n,- (αn)t- and (tn)α sub-bases, each truncated such that the full relative-motion state space is still exact. In Table 1 we see that the nucleons are spread over many shells, but are clustered into very few cluster configurations. The amounts of clustering are given in terms of the subspaces that are used in the basis; the weights belonging to a particlar arrangement add up to less than unity just because of the sub-bases are not complete. With the amount of clustering we can quantify how much a nucleus is clustered, which is a measure of the performance of a cluster model. With two examples I will now advocate the perfection and generality of cluster models.

Some neutron-halo nuclei are supposed to be perfect core+n+n-type three-cluster systems. It was thus a cause for concern to find [7] that even ^{6}He falls short of this expectation. It was found that, added to α+n+n component, a t+t component deepens the binding by 0.65 MeV, which is comparable to the α+n+n binding energy (0.975 MeV). Meanwhile, the amount of t+t clustering, $\mathcal{S}_{tt}$, is increased from 0.49 to 0.54. This problem has been straightened out recently by Arai [8]. It turned out that $\mathcal{S}_{tt}$ can be further increased to 0.57 by using a {t+p,h+n}-type model for the α-cluster instead of the usual model of Eq. (3). But this result is nearly preserved when the t+t component is omitted from the basis. This implies that the main effect of the t+t component is to make the α-cluster more realistic, and the α+n+n picture remains valid. It is also instructive to see how the exact results [9] compare with the cluster model. There is a 3.5-MeV gain in the total binding but that goes into the binding of the α-cluster, and the α+n+n motion is nearly unaffected. Thus the exact calculation bears out the cluster ansatz: the intercluster motion can be separated from the internal motion very well.

Clustering in heavy nuclei is often thought to be insignificant since the shell-model α spectroscopic factors are low. It was pointed out by Fliessbach long ago [10] that $s \ll \mathcal{S}$ for clustering in heavy nuclei, thus s is not a good substitute for $\mathcal{S}$, and the shell-model predictions for the α-decay widths were still too low [11]. We have shown that this was caused by the technical limitations of these shell-model calculations, and we reproduced the correct absolute width of the ^{212}Po$\rightarrow$^{208}Pb+α decay in a combined shell and cluster model [12]. That the cluster-model ingredient does not bring in bias for clustering was subsequently confirmed by pure shell-model calculations in a

Table 1. Analysis of ^{8}Li in terms of the weights $\mathcal{S}_Q$ of Q h.o. excitation quanta [5] and in terms of the amount $\mathcal{S}_{(12)3}$ of (12)3 clustering [6] in an α+t+n cluster model

Q			0	2	4	6	8	10	12	14	16
$\mathcal{S}_Q$ (%)			61	18	11	4	2	1	<1	<1	<1
(12)3	(αt)n				(αn)t				(tn)α		
L, S	1,1	2,0	2,1		1,1	2,0	2,1		1,1	2,0	2,1
$\mathcal{S}_{(12)3}$ (%)	95	1	2		95	2	1		94	1	1

stochastic scheme [13]. The amount of α-clustering turned out to be 0.23–0.30. Is it then the closed-shell core in ^{212}Po that causes such an enormous clustering? No, the the reduced width as a function of A shows just a small jump at the shell closure. The α-decay systematics has been most successfully interpreted in the phenomenological cluster–core potential model of Buck *et al* [14]. This model involves a fully clustered ($\mathcal{S} = 1$) parent state, and the jump at the closed-shell core is well accounted for by a step in the node number of the relative motion. The assumption of $\mathcal{S} = 1$ is to be considered consistent with 0.3 since the phenomenological adjustments can easily smear out such differences. A similar systematics for heavy-cluster decay has also been well reproduced [15]. The potential model can furthermore describe excited states and γ-transitions between these states [16]. Moreover, a cluster–core potential model seems to work quite well for a range of heavy deformed nuclei without regard to whether cluster decay can be observed at all [17]. It seems that the exact way of cluster partitioning does not really matter. One is thus forced to conclude that clustering is a general mode of motion for any nuclear systems.

4 Intercluster relative motion

It is not trivial but in a microscopic scheme it is unnecessary to unravel intercluster motion from the dynamics of a multicluster problem. One should care about it only to interpret macroscopic approaches. The difficulties arise, again, from the Pauli principle, and I will present in a unified form what has been partially discussed in detail elsewhere [18,19,20]. Let us first consider an A+a two-cluster model, with wave function $\Phi = \int d\mathbf{r}\varphi(\mathbf{r})\Phi_{\mathbf{r}}$ and Hamiltonian

$$H = \sum_i T_i + \sum_{i<j} V_{ij} = H_A(\xi_A) + H_a(\xi_a) + T_{Aa}(\mathbf{r}_{Aa}) + V_{Aa}(\mathbf{r}_{Aa}, \xi_A, \xi_a), \quad (6)$$

where H_A and H_a are the cluster intrinsic Hamiltonians. Projected onto the subspace $\{\Phi_{\mathbf{r}}\}$, the Schrödinger equation $H\Phi = E\Phi$ reduces to

$$\mathcal{H}\varphi(\mathbf{r}) = E\mathcal{N}\varphi(\mathbf{r}), \quad (7)$$

where $\mathcal{H}$ is the integral operator whose kernel is $H(\mathbf{r}, \mathbf{r}') = \langle\Phi_{\mathbf{r}}|H|\Phi_{\mathbf{r}'}\rangle$. We need a new relative variable, $\mathbf{r}$, since the physical variables are mixed up by antisymmetrization. To get a Schrödinger-like equation $h\chi = E\chi$ with a Hermitean Hamiltonian for a properly normalized χ ($\int d\mathbf{r}|\chi|^2 = \langle\Phi|\Phi\rangle$), we have to multiply (7) by $\mathcal{N}^{-1/2}$ from the left and introduce χ and h as

$$\chi(\mathbf{r}) = \mathcal{N}^{1/2}\varphi(\mathbf{r}), \ h = \mathcal{N}^{-1/2}\mathcal{H}\mathcal{N}^{-1/2}; \text{ to end up with } h\chi(\mathbf{r}) = E\chi(\mathbf{r}). \quad (8)$$

To define an intercluster potential $\mathcal{U}$, one can substitute (6) into $h-E_A-E_a$:

$$\begin{aligned} h-E_A-E_a \equiv \mathcal{T}+\mathcal{U} &= \mathcal{N}^{-1/2}\mathcal{T}\mathcal{N}^{1/2}+\mathcal{N}^{-1/2}\mathcal{V}\mathcal{N}^{-1/2} \\ &= \mathcal{N}^{1/2}\mathcal{T}\mathcal{N}^{-1/2}+\mathcal{N}^{-1/2}\mathcal{V}^{\dagger}\mathcal{N}^{-1/2}, \end{aligned} \tag{9}$$

where $\mathcal{T} = T_{Aa}(\mathbf{r})$, and $\mathcal{V}$'s kernel is $V(\mathbf{r},\mathbf{r}') = \langle \mathcal{A}_{Aa}\{\Phi^{(A)}\Phi^{(a)}\delta(\mathbf{r}-\mathbf{r}_{Aa})V_{Aa}\}|\Phi_{\mathbf{r}'}\rangle$. With Eq. (9), the Schrödinger equation for the relative motion reads

$$(\mathcal{T}+\mathcal{U})\chi = \varepsilon\chi \qquad (\varepsilon = E - E_A - E_a). \tag{10}$$

The potential $\mathcal{U}$ is local and deep [21], apart from a trivial separable non-locality, which eliminates the Pauli-suppressed relative-motion components. It is this equation that any macroscopic relative-motion Schrödinger equation should be identified with.

A state Ψ for which the form $\int d\mathbf{r}\varphi(\mathbf{r})\Phi_{\mathbf{r}}$ is not assumed can be treated as follows. By projecting $H\Psi = E\Psi$ onto $\Phi_{\mathbf{r}}$, one obtains $(\mathcal{T}-\varepsilon)g(\mathbf{r})+w(\mathbf{r}) = 0$, where $w(\mathbf{r}) = \langle \mathcal{A}_{Aa}\{\Phi^{(A)}\Phi^{(a)}\delta(\mathbf{r}-\mathbf{r}_{Aa})V_{Aa}\}|\Psi\rangle$. One can then define $\mathcal{W}$ such that $w(\mathbf{r}) = \mathcal{W}g(\mathbf{r})$. With this, $(\mathcal{T}+\mathcal{W})g(\mathbf{r}) = \varepsilon g(\mathbf{r})$ follows. By acting with $\mathcal{N}^{-1/2}$ on this and introducing $\tilde{\mathcal{U}}$ as $\mathcal{T}+\tilde{\mathcal{U}} \equiv \mathcal{N}^{-1/2}(\mathcal{T}+\mathcal{W})\mathcal{N}^{1/2}$, one obtains

$$(\mathcal{T}+\tilde{\mathcal{U}})G(\mathbf{r}) = \varepsilon G(\mathbf{r}). \tag{11}$$

Note that $\tilde{\mathcal{U}}$ is non-Hermitean unless $\Psi \in \{\Phi_{\mathbf{r}}\}$.

It is Eqs. (10) and (11) that enter into the macroscopic descriptions of processes involving nuclear clusters. For the width of cluster decay, one has [20] $\Gamma = 2\pi|\langle\Psi|H-H_0|\Phi\rangle|^2$, where $H_0\Psi = E\Psi$ describes the parent state (in an inner region) and $H\Phi = E\Phi$, with $\Phi = \int d\mathbf{r}\varphi(\mathbf{r})\Phi_{\mathbf{r}}$, describes the decay channel (in an outer region), respectively. With the approximation $\langle\Psi|H_0|\Phi\rangle \approx E\langle\Psi|\Phi\rangle$, we have

$$\langle\Psi|H-H_0|\Phi\rangle = \int d\mathbf{r}g^*(\mathbf{r})(\mathcal{T}+\mathcal{W}^{\dagger}-\varepsilon)\varphi(\mathbf{r}) = \int d\mathbf{r}G^*(\mathbf{r})(\mathcal{T}+\tilde{\mathcal{U}}^{\dagger}-\varepsilon)\chi(\mathbf{r}). \tag{12}$$

We made the second step to introduce the macrosopically meaningful relative wave function χ. It is clear that, if one wants to do that, then the appearance of $G(\mathbf{r})$ is inevitable. This was shown by Fliessbach long ago [22] but it was not accepted generally because of the heuristic nature of his derivation. Fliessbach did not distinguish $\tilde{\mathcal{U}}$ from $\mathcal{U}$, and we see that if $\tilde{\mathcal{U}}^{\dagger}$ were replaced by $\mathcal{U}$, the width would be zero because of Eq. (10).

To describe a cluster transfer reaction A(a,b)B, where a=b+x, B=A+x, one has to re-derive (8)–(10) for a two-channel case. Following Refs. [18,19], let us start with the ansatz $\Phi = \int d\mathbf{r}[\Phi_{\mathbf{r}}^{(1)}\varphi^{(1)}(\mathbf{r}) + \Phi_{\mathbf{r}}^{(2)}\varphi^{(2)}(\mathbf{r})]$, where the channel indices 1 and 2 are used for Aa and Bb. By projecting onto $\Phi_{\mathbf{r}}^{(i)}$ and introducing matrix Hamiltonian and overlap operators, we obtain, by analogy with Eq. (7),

$$\begin{aligned}(\mathcal{H}_{11} - E\mathcal{N}_{11})\varphi_1(\mathbf{r}) + (\mathcal{H}_{12} - E\mathcal{N}_{12})\varphi_2(\mathbf{r}) = 0,\\ (\mathcal{H}_{21} - E\mathcal{N}_{21})\varphi_1(\mathbf{r}) + (\mathcal{H}_{22} - E\mathcal{N}_{22})\varphi_2(\mathbf{r}) = 0.\end{aligned} \tag{13}$$

Applying $\mathcal{N}_{ii}^{-1/2}$ and introducing $\chi_i = \mathcal{N}_{ii}^{1/2}\varphi_i$, $\mathcal{T}_i + \mathcal{U}_i$ and ε_i, we get

$$\begin{aligned}(\mathcal{T}_1 + \mathcal{U}_1 - \varepsilon_1)\chi_1 + \mathcal{N}_{11}^{-1/2}(\mathcal{H}_{12} - E\mathcal{N}_{12})\mathcal{N}_{22}^{-1/2}\chi_2 = 0,\\ \mathcal{N}_{22}^{-1/2}(\mathcal{H}_{21} - E\mathcal{N}_{21})\mathcal{N}_{11}^{-1/2}\chi_1 + (\mathcal{T}_2 + \mathcal{U}_2 - \varepsilon_2)\chi_2 = 0.\end{aligned} \tag{14}$$

Then, in the distorted-wave Born approximation the transition amplitude is

$$\begin{aligned}&T_{1\to 2} = \langle\chi_2^{(-)}|\mathcal{N}_{22}^{-1/2}(\mathcal{H}_{21} - E\mathcal{N}_{21})\mathcal{N}_{11}^{-1/2}|\chi_1^{(+)}\rangle\\ &= \binom{B+b}{b}^{1/2} \langle\chi_{Bb}^{(-)}\Phi_B\Phi_b|\mathcal{N}_{Bb}^{-1/2}(T_{Bb} + V_{Bb} - \varepsilon_{Bb})\mathcal{N}_{Aa}^{-1/2}|\mathcal{A}_{Aa}\{\Phi_A\Phi_a\chi_{Aa}^{(+)}\}\rangle\\ &\approx \left[\binom{a}{x}\binom{B}{x}\right]^{1/2} \langle\chi_{Bb}^{(-)}\Phi_B\Phi_b|\mathcal{N}_{Bb}^{-1/2}(T_{Bb} + V_{Bb} - \varepsilon_{Bb})\mathcal{N}_{Aa}^{-1/2}|\Phi_A\Phi_a\chi_{Aa}^{(+)}\rangle\\ &= [\ldots]^{1/2}\textstyle\sum_{b'x'}\int d\mathbf{r}\langle\chi_{Bb}^{(-)}\Phi_B\Phi_b|\mathcal{N}_{Bb}^{-1/2}(T_{Bb}+V_{Bb}-\varepsilon_{Bb})\mathcal{N}_{Aa}^{-1/2}|\Phi_{b'}\Phi_{x'}\delta(\mathbf{r}-\mathbf{r}_{bx})\rangle\\ &\times\langle\Phi_{b'}\Phi_{x'}\delta(\mathbf{r}-\mathbf{r}_{bx})|\Phi_A\Phi_a\chi_{Aa}^{(+)}\rangle\\ &\approx \langle\chi_{Bb}^{(-)}g_{B(Ax)}|\mathcal{N}_{Bb}^{-1/2}(T_{Bb} + V_{Bb} - \varepsilon_{Bb})\mathcal{N}_{Aa}^{-1/2}|g_{a(bx)}\chi_{Aa}^{(+)}\rangle\end{aligned} \tag{15}$$

Here $\chi_i^{(\pm)}$ are solutions to the single-channel approximants of Eqs. (14) with appropriate boundary conditions, and $\mathcal{N}_{Aa} \equiv \mathcal{N}_{11}$, $\mathcal{N}_{Bb} \equiv \mathcal{N}_{22}$. In the second line $\mathcal{A}^2 = n^{1/2}\mathcal{A}$ is used, and H is made to act to the left. The third line contains the "passive-core approximation": exchanges between A and b are neglected. In lines 4–5 a unity operator has been inserted, and in line 6 the "cluster transfer approximation" is applied, i.e., the summation is restricted to $x' = x$, furthermore, $V_{Bb}(\mathbf{r}_{Bb}, \xi_B, \xi_b) \approx V_{Bb}(\mathbf{r}_{Bb})$ is taken.

Fliessbach proposed [23] to supersede the amplitudes g by G in (15) like in Eq. (12). We see that this would imply assuming $\mathcal{N}_{Bb} \approx \mathcal{N}_{xb}$, $\mathcal{N}_{Aa} \approx \mathcal{N}_{Ax}$, which is consistent with the passive-core approximation, as well as taking $\mathcal{N}_{xb}^{-1/2}(T_{Bb}+V_{Bb}-\varepsilon_{Bb})\mathcal{N}_{Ax}^{-1/2} \approx \mathcal{N}_{Ax}^{-1/2}(T_{Bb}+V_{Bb}-\varepsilon_{Bb})\mathcal{N}_{xb}^{-1/2}$, which is, however, unjustifiable. Instead, one should either use high enough bombarding energies for the effect of $\mathcal{N}_{Aa}$ and $\mathcal{N}_{Bb}$ to be negligible or to use low enough bombarding energies for the Coulomb barrier to suppress the contribution of the regions where $\mathcal{N}_{Aa} \neq 1$ and $\mathcal{N}_{Bb} \neq 1$. In any case, one should accept that the amplitudes and weights involved in decay (G and S) and in rearrangement (g and s) are different. When I recognized this [18], I tried to contrive some way to bring G and S back. Since in a pure cluster model G is a relative wave function, G looked more fit to phenomenological analyses than g [10]. Later, however, we found just the contrary [20]. In Figure 2 we see that in the realistic

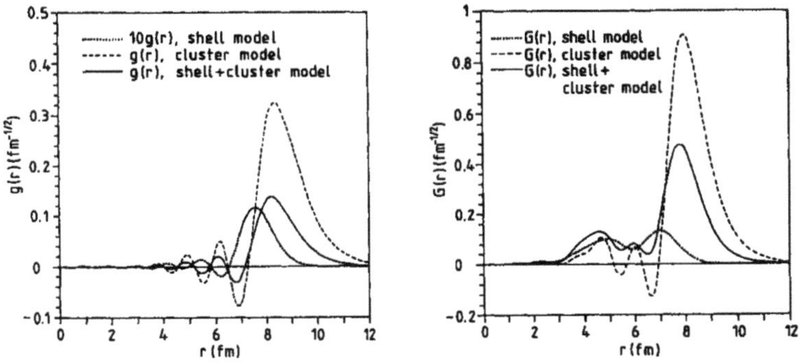

Figure 2. The $^{208}Pb+\alpha$ amplitudes g and G for the g.s. of ^{212}Po.

shell+cluster model g oscillates around zero as a wave function should, but G does not. The non-cluster channels render $\tilde{\mathcal{U}}$ strongly non-local and disfigure G. Thus we should not regret that G cannot be used in reaction analyses.

5 Conclusion

We have reviewed the paradoxical aspects of clustering and cluster models. These have been shown to arise from the fact that physical states must have correct symmetries, and the symmetry projections, especially antisymmetrization, make dissimilar objects similar. By definition, a state is said to show some clustering if it is not orthogonal to the subspace whose elements are, *before antisymmetrization*, products of cluster intrinsic and intercluster states. This idea is meaningful and useful because

(i) it says that clustering is what can be described by cluster models;
(ii) it allows to define intercluster relative motion despite antisymmetry;
(iii) there is a physical phenomenon that is proportional to the amount of clustering (cluster decay width).

Clustering phenomena and cluster models are thus intricately connected. The more widespread and clear-cut clustering phenomena are, the more powerful the cluster models will turn out to be. I quoted two subjects to emphasize this point. First, contrary to earlier indications, recent studies confirm that cluster models are excellent for light exotic nuclei. The counterindications have been explained as intracluster effects. Second, there is a mounting evidence that heavy cluster-decaying nuclei show a high amount of clustering. Cluster models can explain not only their decay but also the electromagnetic properties of a range of heavy nuclei. With this, it is also corroborated that an intercluster relative motion is meaningful.

We discussed the relative motion here to show how it can be used in phenomenological models of cluster decay and transfer reactions. This controversial issue is settled such that the amount of clustering is directly related to the decay width, but indirectly to the transfer cross section. To extract amounts of clustering from transfer reactions thus requires very careful analyses.

It would enhance the reputation of nuclear cluster physics if a state of concensus were reached on these basic issues.

References

1. K. Wildermuth and Y.C. Tang in *A Unified Theory of the Nucleus* (Vieweg, Braunschweig, 1977).
2. A. Săndulescu, E.W. Schmid and G. Spitz, *Few-Body Syst.* **5**, 107 (1988).
3. Y. Kanada-En'yo, H. Horiuchi and A. Ono, *Phys. Rev.* C **52**, 628 (1995).
4. R. Beck, F. Dickmann and R.G. Lovas, *Ann. Phys. (N.Y.)* **137**, 1 1987.
5. Y. Suzuki *et al*, *Phys. Rev.* C **54**, 2073 (1996).
6. K. Varga, Y. Suzuki and I. Tanihata, *Phys. Rev.* C **52**, 3013 (1995).
7. A. Csótó, *Phys. Rev.* C **48**, 165 (1993).
8. K. Arai *et al*, *Phys. Rev.* C **59**, 1432 (1999); this conference.
9. K. Varga and Y. Suzuki, *Phys. Rev.* C **52**, 2885 (1995).
10. T. Fliessbach, *Z. Phys.* A **272**, 39 (1975).
11. I. Tonozuka and A. Arima, *Nucl. Phys.* A **323**, 45 (1979).
12. K. Varga, R.G. Lovas and R.J. Liotta, *Nucl. Phys.* A **550**, 421 (1992).
13. K. Varga and R.J. Liotta, *Phys. Rev.* C **50**, R1292 (1994).
14. B. Buck *et al*, *Phys. Rev. Lett.* **65**, 2975 (1990).
15. B. Buck, A.C. Merchant and S.M. Perez, *J. Phys.* G **20**, 351 (1994).
16. B. Buck *et al*, *Phys. Rev. Lett.* **72**, 1326 (1994); S. Ohkubo, *Phys. Rev. Lett.* **74**, 2176 (1995).
17. B. Buck, A.C. Merchant and S.M. Perez, *Phys. Rev.* C **59**, 750 (1999).
18. R.G. Lovas, *Z. Phys.* A **322**, 589 (1985).
19. A. Arima and A. Weiguny, *Z. Phys.* A **331**, 299 (1988).
20. R.G. Lovas *et al*, *Phys. Rep.* **294**, 265 (1998).
21. H. Friedrich, *Phys. Rep.* **74**, 209 (1981); E.W. Schmid, *Nucl. Phys.* A **416**, 379c (1984); H. Horiuchi in *Trends in Nuclear Physics*, Vol. 2, ed. P.J. Ellis and Y.C. Tang (Addison-Wesley, New York, 1991) p. 277.
22. T. Fliessbach and H.J. Mang, *Nucl. Phys.* A **263**, 75 (1976).
23. T. Fliessbach, *Z. Phys.* A **278**, 353 (1976).

PAIR FLUCTUATIONS IN FERMI SYSTEMS

P. SCHUCK

Institut des Sciences Nucleaires, Universite Joseph Fourier, CNRS-IN2P3
53, Avenue des Martyrs, F-38026 Grenoble Cedex, France

J. DUKELSKY

Instituto de Estructura de la Materia, C.S.I.C., Madrid, Spain

G. RÖPKE

FB Physik, Universität Rostock, D-18051 Rostock, Germany

A. SCHNELL

Institute for Nuclear Theory, University of Washington, Box 351550
Seattle, WA 98195-1550, U.S.A.

Strong pair fluctuations are considered in two systems: i) A multi-level pairing Hamiltonian describing superconductivity in ultrasmall metallic grains, but also superfluid deformed nuclei, is studied in the region close to the phase transition. ii)Hot nuclear matter at low density is investigated for temperatures close to T_c. The Cooper pole in the two body T-matrix translates in the single particle density of states to the formation of a pseudo gap.

In Fermi systems such as finite nuclei or ultra small metallic grains but also in infinite systems close to a phase transition quantum fluctuations can become very important and even dominate over the mean field. In such cases it is very important to have a proper theory to treat the fluctuations. In this contribution we will deal with pair fluctuations of systems which are normal fluid but close to the point of turning into the superfluid state. Such systems are on the one hand metallic grains which are so small that their individual electron levels become discrete [1]. We will treat this system in the so-called picket fence model which has become by now quite popular in this field. It is a multi level pairing model with all states only two fold degenerate. This model is also relevant for deformed nuclei and in fact it is there where it has been invented and studied first [2]. The second system is hot nuclear matter in the close vicinity of the superfluid phase transition. The quantity of interest will be the single particle level density aroud the Fermi energy in order to see what happens when one is approaching the critical point. For this system we will bear on a recent publication [3].

Let us start with the picket fence model. It consists of an equidistant multilevel pairing Hamiltonian with each level two fold degenerate, *i.e.* only spin up/down fermions of one kind (the case considered here) can occupy one level. The corresponding Hamiltonian is therefore given by

$$H = \sum_{i=1}^{\Omega} (\varepsilon_i - \lambda) N_i - G \sum_{i,j=1}^{\Omega} P_i^{\dagger} P_j$$

where

$$P_i^\dagger = c_i^\dagger c_{\bar{i}}^\dagger \quad , \quad N_i = c_i^\dagger c_i + c_{\bar{i}}^\dagger c_{\bar{i}} \quad , \quad P_i = \left(P_i^\dagger\right)^\dagger \tag{2}$$

and single particle energies are $\varepsilon_i = i\,\varepsilon$ with ε being the level spacing chosen to be $\varepsilon = 1$ and $\bar{i}$ means the time reversed of i . The level density of the model is thus a constant and therefore imitates a 1 D situation. However, our treatment below can be extended to higher dimensional cases and actually we think that our approach will even work better in $D > 1$ cases. The chemical potential λ in (1) determines the origin of energy and chosen as $\lambda = \frac{1}{2}(\Omega + 1 - G)$ the model becomes particle-hole symmetric for the case considered here where there are as many particles as levels (half filling). In the non interacting case ($G = 0$) the lowest $\Omega/2$ levels are occupied with two particles in each level. We will classify the states into particle states (p) for those levels above the Fermi energy and hole states (h) for those below the Fermi energy. The hamiltonian can then be written as

$$\begin{aligned} H = &\sum_p (\varepsilon_p - \lambda) N_p + \sum_h (\varepsilon_h - \lambda) N_h - G\sum_{pp'} P_p^\dagger P_{p'} - G\sum_{hh'} P_h^\dagger P_{h'} \\ &-G\sum_{ph}\left(P_h^\dagger P_p + P_p^\dagger P_h\right) \end{aligned} \tag{3}$$

As mentioned already the above model Hamiltonian can for example schematically describe pair correlations in deformed nuclei and it has recently extensively been used to mimic the physics of ultrasmall superconducting grains [1]. The picket fence model has been solved exactly with a Bethe ansatz type of solution by Richardson [2] and more recently with a density matrix renormalization group (DMRG) procedure for very large number of levels [4]. The aim here is to test the SCRPA approach, introduced in nuclear physics long time ago [5] and further developed by the present authors in a series of papers [6,7,8,9] against the exact results. The SCRPA equations have been derived in a number of preceding papers [5,6,7,8,9,10] and consequently we will be very short. In the particle-particle RPA (ppRPA) we define pair addition and removal operators [11]

$$A_\mu^\dagger = \sum_p X_p^\mu\, \overline{P}_p^\dagger - \sum_h Y_h^\mu\, \overline{P}_h^\dagger \quad , \quad R_\lambda^\dagger = -\sum_p Y_p^\lambda \overline{P}_p + \sum_h X_h^\lambda \overline{P}_h \tag{4}$$

where $\overline{P}_p = P_p/\sqrt{1-\langle N_p\rangle}$ and $\overline{P}_h = P_h/\sqrt{\langle N_h\rangle - 1}$. The amplitudes obey the normalization condition

$$\sum_p X_p^\mu X_p^{\mu'} - \sum_h Y_h^\mu Y_h^{\mu'} = \delta_{\mu\mu'} \quad , \quad \sum_h X_h^\lambda X_h^{\lambda'} - \sum_p Y_p^\lambda Y_p^{\lambda'} = \delta_{\lambda\lambda'} \tag{5}$$

and the closure relations

$$\begin{aligned} \sum_\mu X_p^\mu X_{p'}^\mu - \sum_\lambda Y_p^\lambda Y_{p'}^\lambda = \delta_{pp'} \quad , \quad &\sum_\lambda X_h^\lambda X_{h'}^\lambda - \sum_\mu Y_h^\mu Y_{h'}^\mu = \delta_{hh'} \quad , \\ &\sum_\lambda X_h^\lambda Y_p^\lambda - \sum_\mu X_p^\mu Y_h^\mu = 0 \end{aligned} \tag{6}$$

With (5) and (6) one can invert (4) to give

$$P_p^\dagger = \sqrt{1-\langle N_p\rangle}\left[\sum_\mu X_p^\mu A_\mu^\dagger + \sum_\lambda Y_p^\lambda R_\lambda\right] \quad ,$$

$$P_h^\dagger = \sqrt{\langle N_h\rangle - 1}\left[\sum_\lambda X_h^\lambda R_\lambda + \sum_\mu Y_h^\mu A_\mu^\dagger\right] \tag{7}$$

With the usual RPA vacuum condition [5,6] $A_\mu |RPA\rangle = 0$ and $R_\lambda |RPA\rangle = 0$ it is then easy to express the expectation value of the two pair operators that enter in the pairing interaction in terms of the RPA amplitudes and the occupation numbers

$$\langle P_p^\dagger P_{p'}\rangle = \sqrt{(1-\langle N_p\rangle)(1-\langle N_{p'}\rangle)}\sum_\lambda Y_p^\lambda Y_{p'}^\lambda \tag{8}$$

and analogous for $\langle P_p^\dagger P_h\rangle$ and $\left\langle P_h^\dagger P_{h'}\right\rangle$.

The SCRPA for e.g. the addition mode is written as

$$\begin{pmatrix} \overline{A} & \overline{B} \\ -\overline{B} & \overline{C} \end{pmatrix}\begin{pmatrix} X \\ Y \end{pmatrix} = E\begin{pmatrix} X \\ Y \end{pmatrix} \tag{9}$$

where with $\overline{A}_{pp'} = \left\langle \left[\overline{P}_p\left[H, \overline{P}_{p'}^\dagger\right]\right]\right\rangle$ we have

$$\begin{aligned} \overline{A}_{pp'} &= 2\delta_{pp'}(\varepsilon_{p'} - \lambda) + 2G\delta_{pp'}\frac{\sum_{p''}\left\langle P_{p''}^\dagger P_p\right\rangle + \sum_h \left\langle P_h^\dagger P_p\right\rangle}{(1-\langle N_p\rangle)} \\ &-G\frac{\langle 1 - N_p - N_{p'} + N_p N_{p'}\rangle}{\sqrt{(1-\langle N_p\rangle)(1-\langle N_{p'}\rangle)}} \end{aligned} \tag{10}$$

and similar for $\overline{B}$ and $\overline{C}$. The pair removal mode can be treated analogously. Replacing in (10) the SCRPA correlated expectation values by the uncorrelated ones yields the standard pp-RPA equation for this model [12].

As usual in the SCRPA the occupation numbers in $\langle N_i\rangle$ give some problems since they are not directly expressed by the RPA operators (4) and consequently cannot be given as a function of the RPA amplitudes X, Y using the vacuum condition. However, in the present model we can very easily obtain the occupation numbers from the exact identity

$$N_i = 2P_i^\dagger P_i \tag{11}$$

which follows from the relation of the SU(2) Casimir operator(see also below). From (11) we directly obtain with (4)

$$\langle N_p\rangle = 1 - \frac{1}{1+2\sum_\mu (Y_p^\mu)^2} \quad ; \quad \langle N_h\rangle = 1 + \frac{1}{1+2\sum_\mu (Y_h^\mu)^2} \tag{12}$$

One should also realize that (12) can be obtained from the lowest order scheme advocated by Catara [13] that becomes an exact relation in our model. The standard RPA expressions for the occupation numbers are recovered from (12) to lowest order in Y^2.

Since the levels are only twofold degenerate, the Pauli principle is of great importance here and it is very helpful that (12) corresponds to the exact relation (11).

On the other hand in (10) we also have to know correlation functions of the form $\langle N_i N_j \rangle$. It was shown in [14] that the simple factorization

$$\langle N_i N_j \rangle \simeq \langle N_i \rangle \langle N_j \rangle \tag{13}$$

leads to satisfying results. With (13) and (12) the matrix elements (10) can now entirely be expressed by the RPA amplitudes X, Y and the SCRPA equations are ready for a numerical solution. It should, however, be realized that above approximation which we used here for reasons of simplicity can be avoided. Indeed in inserting into $\langle N_i N_j \rangle$ expression (11) for N_i and using again the inversion of the RPA operators one can arrive at a linear system of equations for those correlation functions. The use of the approximation instead of this exact relation may be the reason for the slight overbinding of the energy which we will see below. It is also interesting to point out that the relation (11) can be generalised for cases with levels of higher degeneracy $\Omega_j = j + 1/2$. This can be seen most easily in using the Holstein Primakoff boson image of the Fermion pair operators [12]:

$$P_i^\dagger = \sqrt{\Omega_i} B_i^\dagger \sqrt{1 - B_i^\dagger B_i / \Omega_i} \tag{14}$$

$$N_i = 2 B_i^\dagger B_i \tag{15}$$

From these relations one easily constructs the following implicit equation for N_i:

$$N_i = \frac{2}{\Omega_i} P_i^\dagger [1 - \frac{N_i}{2\Omega_i}]^{-1} P_i \tag{16}$$

Above equation for N_i can be solved iteratively. To lowest order we find the exact relation (11) for $\Omega_i = 1$. For $\Omega_i = 2$ we have to expand to second order:

$$N_i = \frac{2}{\Omega_i} P_i^\dagger P_i + \frac{2}{\Omega_i^3} P_i^\dagger P_i^\dagger P_i P_i \quad , \quad \Omega_i = 2 \tag{17}$$

The expectation value of this equation and evaluating again the rhs with the inversion of the RPA operators gives the exact expression of $\langle N_i \rangle$ for the case $\Omega_i = 2$. For higher degeneracies one could in principle develope to respective higher orders but this becomes quickly very combersome. On the other hand,since the series seems to converge very rapidly,we propose to take the relation developed to second order as an approximate expression for any value of $\Omega_i > 1$. We developed this here for the case of particle-particle correlations but it is evident that analogous developments can be performed for the case of ph-correlations.

In Fig. 1 we show the result for the correlation energy $E_{corr} = \langle H \rangle + \varepsilon \frac{\Omega^2}{4}$ for the case of ten levels as a function of the interaction strength G. One sees that the SCRPA solution stays very close to the exact results even beyond the phase transition point $G \simeq 0.33$ where standard RPA breaks down. We notice the usual feature that standard RPA due to the inherent violation of the Pauli principle strongly overbinds. It is remarkable how well apparently SCRPA restores the Pauli principle. This can also be seen in Fig. 2 where we show for the given interaction strength $G = 0.21$ the correlation energy as a function of the number of levels.

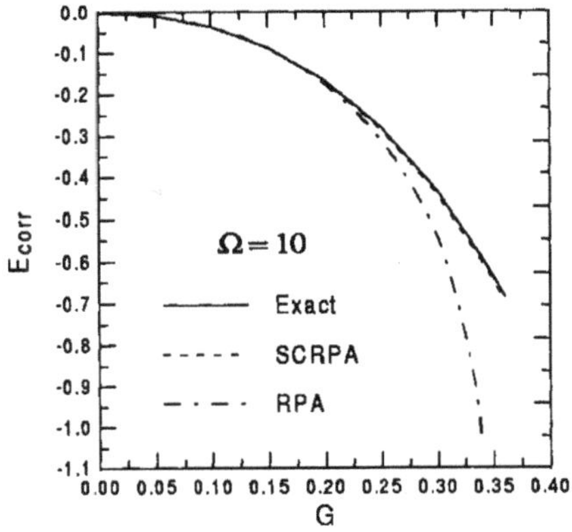

Figure 1. Groundstate correlation energies of the system with $\Omega = 10$ as a function of the pairing strength G.

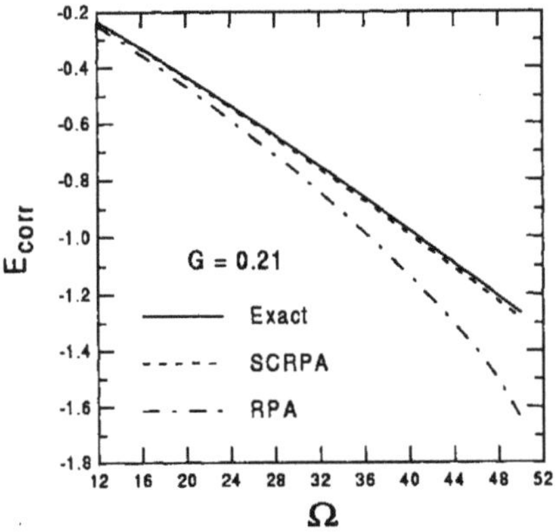

Figure 2. Groundstate correlation energies as a function of the number of levels Ω for $G = 0.21$.

The exact results have been obtained using the DMRG method of Ref. [4]. Again SCRPA stays very close to the exact result up to the critical point of $\Omega \simeq 50$ where the standard RPA breaks down. In Figs. 3, 4 we show the groundstate energy and the first excited state energy of the system with $\Omega = 10$ and with $N = \Omega + 2$ relative to the groundstate energy of the system with the same Ω and $N = \Omega$ as a function of G, which are given by the first and second RPA eigenvalues respectively. The first eigenvalue of the SCRPA fails at the phase transition point by $\sim 10\%$ while the RPA eigenvalue collapses. The second SCRPA eigenvalue performs much better. Considering these numbers one should, however, realize that the screening of the interaction inherent in SCRPA has accomplished a tremendous performance. Indeed the self-consistency overscreened the interaction turning it from attraction into repulsion in agreement with the trend of the exact solution! One also sees from the SCRPA solution that due to the self-consistent treatment of quantum fluctuations the mean field phase transition point inherent to standard RPA is, in agreement with the exact solution, completely washed out without any remnence at all.

The interesting point from these studies is indeed the fact that the phase transition to superfluidity is completely smooth with no sign of a critical behavior. Yet we know that in the limit of very high level densities, i.e. in the limit where e.g. all levels collaps into one (the seniority model) the system is superfluid. This is also confirmed from the exact solution of the model [4]. For a correct description it was important to treat the quantal pair fluctuations self-consistently in properly defining the meanfield created by the fluctuations.

Let us now come to a second example where pair fluctuations are extremely important. We consider low density nuclear matter and approach the critical temperature from above. What will happen to the level density? We will study this problem also self-consistently in using in the particle-particle T-matrix single

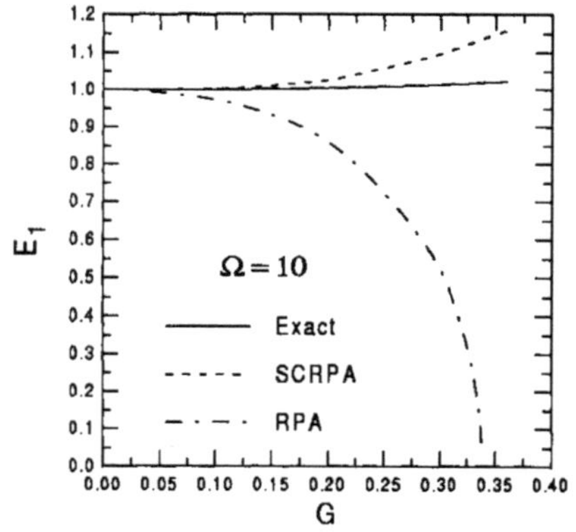

Figure 3. Groundstate of the system with $\Omega = 10$ and $N = 12$ particles relative to the groundstate of the system with $\Omega = N = 10$.

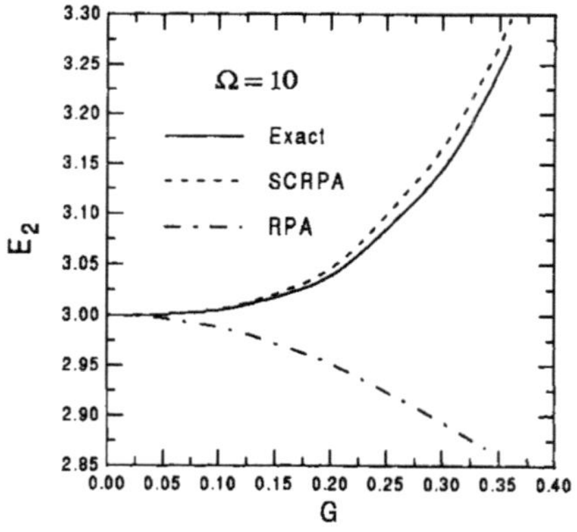

Figure 4. First excited state energy of the system with $\Omega = 10$ and $N = 12$ particles relative to the groundstate of the system with $\Omega = N = 10$.

particle propagators which are obtained with a mass operator in T-matrix approximation. This is similar to Brueckner-Hartree-Fock theory besides the fact that we use for the T-matrix the Feynman-Galitskii form (equivalent to pp-RPA) with hh contributions absent in Brueckner theory. The latter are, however, crucial for a correct description of pairing phenomena.

We start with the single-particle propagator $G_1(11', z)$ obeying the Dyson equation [15] whith the self-energy $\Sigma(1, z)$. The spectral function $A(1, \omega)$ follows as

$$A(1,\omega) = \frac{2\,\mathrm{Im}\;\Sigma(1,\omega+i0)}{\left[\omega - \frac{p_1^2}{2m} - \mathrm{Re}\;\Sigma(1,\omega)\right]^2 + \left[\mathrm{Im}\;\Sigma(1,\omega+i0)\right]^2}\,. \tag{18}$$

The notation '1' stands for momentum $\vec{p}_1$, spin σ_1, and isospin τ_1 quantum numbers. Based on the spectral function the level density reads (Ω – normalization volume)

$$N(\omega) = \frac{1}{\Omega}\sum_p A(p,\omega)\,. \tag{19}$$

For the determination of the single-particle spectral function the single-particle self-energy Σ can be calculated within a cluster decomposition [6,8] from the two-particle T matrix $T_2(12, 1'2', z)$ taking only two-particle correlations into account. The imaginary part of Σ is given by

$$\mathrm{Im}\,\Sigma(1,\omega) = \sum_2 \mathrm{Im}\,T_2(12,12,\omega+\epsilon_2+i0) \times\{f(\epsilon_2) + g[\omega+\epsilon_2]\}\,. \tag{20}$$

with $f(E) = [\exp\{(E-\mu)/T\}+1]^{-1}$ being the Fermi and $g(E) = [\exp\{(E-2\mu)/T\}-1]^{-1}$ the Bose distribution function. The real part is obtained from the imaginary part according to the Kramers-Kronig relation. The two-particle ladder T matrix is given by the Galitskii-Feynman equation [15]

$$T_2(12,34,z) = V(12,34) + \sum_{1'2'} V(12,1'2') \times G_2^0(1'2',z)\, T_2(1'2',34,z)\,. \tag{21}$$

The uncorrelated two-particle Green function G_2^0 is given as a product of two single-particle propagators via the one-particle spectral function

$$G_2^0(12,z) = \int \frac{\mathrm{d}E}{2\pi}\frac{\mathrm{d}E'}{2\pi}\frac{1-f(E)-f(E')}{z-E-E'} \times A(1,E)A(2,E') \tag{22}$$

Equations (18) and (20) together with (21) and (22) form a self-consistent set of equations which can be solved iteratively. Using in a first step the quasi-particle approximation for the intermediate states one obtains

$$G_2^0(12,z) = \frac{1-f(1)-f(2)}{z-\epsilon(p_1)-\epsilon(p_2)}\,. \tag{23}$$

The usual first-order Brueckner-Hartree-Fock approximation [16] can be recovered replacing in (23) the expression $1-f(1)-f(2)$ by $(1-f(1))(1-f(2))$, i.e., neglecting the "backward going" part. For the determination of the quasi-particle energies $\epsilon_1 = p_1^2/(2m) + U(p_1), U(p_1) = \mathrm{Re}\,\Sigma(1,\epsilon_1)$ we perform a self-consistent Galitskii-Feynman Hartree-Fock calculation (cf. [17]) to evaluate the real part of the on-shell self-energy $\mathrm{Re}\,\Sigma = \mathrm{Re}\,\Sigma^{(1)} + \mathrm{Re}\,\Sigma^{(2)}$ as a function of the momentum from

$$\mathrm{Re}\,\Sigma^{(1)}(1,\omega) = \sum_2 \mathrm{Re}\,T_2(12,12,\omega+\epsilon_2)f(\epsilon_2)\,, \tag{24}$$

$$\mathrm{Re}\,\Sigma^{(2)}(1,\omega) = \sum_2 \mathcal{P}\!\int \frac{\mathrm{d}\omega'}{\pi}\frac{\mathrm{Im}\,T_2(12,12,\omega'+i0)}{\omega'-\omega-\epsilon_2}g(\omega')\,. \tag{25}$$

For an exploratory calculation we use a separable representation of the nucleon-nucleon potential according to Yamaguchi [18] $V_\alpha(p,p') = -\lambda_\alpha/\Omega\, v(p)v(p'), v(p) = 1/(p^2+\beta^2)$ with the relative momenta p and p', α denotes the interaction channel, and λ_α the coupling strength of the respective channel. The parameters given in [18] are chosen to reproduce the low-energy scattering phase shifts as well as the deuteron binding energy. This potential yields somewhat too strong pairing properties as compared to those calculated, e.g. with the Paris potential [19], but this will not invalidate the important conclusions which we will draw from our study below. Using a separable potential a solution of (21) is particularly simple. Transforming to relative $\vec{p} = (\vec{p_1} - \vec{p_2})/2$ and total momentum $\vec{P} = \vec{p_1} + \vec{p_2}$ one obtains

$$T_\alpha(pp'P,z) = \frac{V_\alpha(p,p')}{1-J_\alpha(P,z)}\,, \tag{26}$$

$$J_\alpha(P,z) = \frac{-\lambda_\alpha}{2\pi^2}\int \mathrm{d}p\, p^2 v^2(p) G_2^0(p,P,z)\,. \tag{27}$$

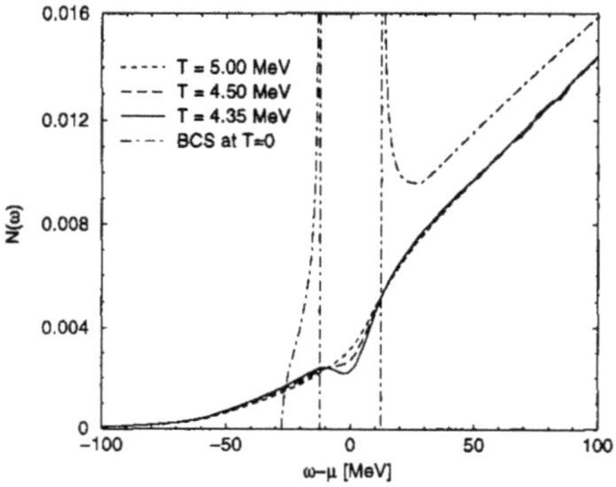

Figure 5. Nucleon level density (19) as a function of the energy $\omega-\mu$, relative to the chemical potential μ, given for the density $\rho=\rho_0/3$ and three values of the temperature. For comparison, the BCS result at zero temperature is also shown.

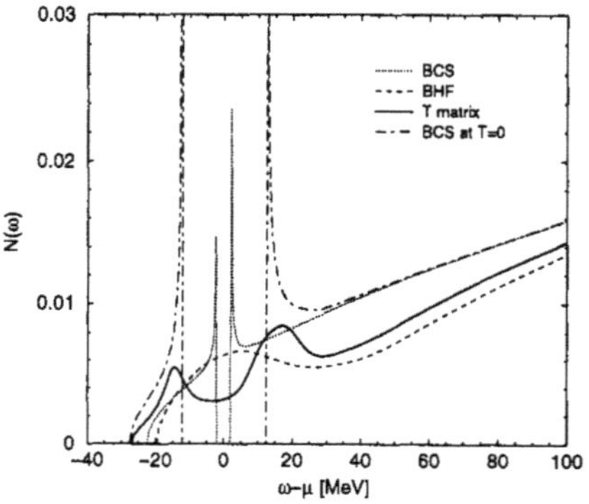

Figure 6. Nucleon level density in quasi-particle approximation (28) at $\rho=\rho_0/3$ and $T=5\,$MeV. Results are shown for Brueckner-Hartree-Fock (BHF, dashed), BCS (dotted), and T matrix approximation. The dot-dashed curve is the $T=0$ calculation in BCS approximation.

In the following we are going to present results of a calculation for symmetric nuclear matter at the density $\rho=\rho_0/3$. At this density, using the self-consistent procedure for the real part of the on-shell self-energy, one obtains from the Thouless criterion of the T matrix [20] a critical temperature $T_c=4.34$ MeV for pairing in the $^3S_1$-channel.

The density of states is shown in Fig. 5 for three different temperatures ($T=5\,$MeV, $T=4.5\,$MeV, and $T=4.35\,$MeV) as a function of the energy $\omega-\mu$ relative to the chemical potential μ. At $\omega\approx\mu$ one observes a depletion of the level density when the temperature reaches the critical value of the phase transition. This depletion, the so-called pseudogap (see below), is a precursor signal of the onset of superfluidity which happens at $T_c=4.34$ MeV. For comparison, in Fig. 5 also the level density in BCS approximation at zero temperature is given (dashed-dotted line, see also below). It exhibits a gap of twice the pairing gap around $\omega-\mu$ where $N(\omega)=0$ is flanked by sharp peaks.

As for a more detailed discussion of the behavior of $N(\omega)$ we are looking at the quasi-particle approximation for the level density, i.e., the limit of vanishing imaginary part of the self-energy in (18) by using $A(1,\omega)=2\pi\delta(\omega-\epsilon_1)$ in (19). Then, the level density reads

$$N(\omega)=\left[\frac{1}{p^2}\frac{\partial\epsilon(p)}{\partial p}\right]^{-1}\;;\qquad \omega=\epsilon(p)\,. \tag{28}$$

For comparison the quasi-particle energy ϵ_1 is self-consistently determined in three different types of approximation. The corresponding results for the level density are given in Fig. 6 as a function of the energy $\omega-\mu$ at the temperature $T=5\,$MeV. In Brueckner-Hartree-Fock (BHF) approximation [16] (dashed curve) the hole-hole contribution to the effective interaction which is necessary for the description of pairing is neglected. Interestingly, the BHF calculation yields an opposite behavior compared to the T matrix approach which is known as the enhancement of the effective mass at the Fermi surface [16]. The mean-field BCS approach at $T = 5\,\text{MeV}$ (dotted curve) which is closely below the critical temperature of $T_c^{\text{BCS}} = 5.08\,\text{MeV}$ exhibits a tiny gap around the chemical potential which becomes considerably larger at zero temperature (dot-dashed). The self-consistent determination of ϵ_1 according to (24,25) leads to the solid curve. Compared with the calculation via the full spectral function (Fig. 5), the structures in $N(\omega)$ obtained from the quasi-particle approximation are more pronounced and show a striking similarity with the BCS results at zero temperature. The scenario of the phase transition to superfluidity in this strongly correlated quantum system is characterized by pre-critical fluctuations forming the pseudogap, in contrast to BCS where a gap opens spontaneously at T_c^{BCS}. It would be interesting to study the single particle level density around the Fermi surface also for the picket fence model from above, in order to see whether in this exactly solvable model also a depression of the density of states is realised.

The fact that due to pair fluctuations a depression of the level density around the Fermi energy occurs,is a novel feature in nuclear physics. In finite nuclei,due to the strong collectivity of ph correlations(surface vibrations)the density of states is rather enhanced than depressed. However,also in nuclei this final enhancement certainly results from the battle between the ever opposing effects:ph- and pp-correlations. Finally we want to mention that a pseudo gap has been found experimentally in high T_c superconductors (see, e.g., Ref. [21]). It is believed by some people that this pseudo gap is indeed due to pair fluctuations.

In conclusion we presented in this contribution two systems, small metallic grains and hot nuclear matter, both close to the phase transition, where the influence of pair fluctuations is very important. In such cases a self-consistent treatment of the fluctuations creating their own mean field is crucial to obtain qualitative and quantitative correct results.

We acknowledge useful discussions with M. Baldo, W.H. Dickhoff, J. Hirsch, U. Lombardo, P. Nozières, and N. Vinh Mau.

References

1. F. Braun and J. von Delft *Phys. Rev. Lett.* **81**, 4712 (1998).
2. R.W. Richardson, *Phys. Rev.* **141**, 949 (1966); J. Bang and J. Krumlinde, *Nucl. Phys.* **141**, 18 (1970).
3. A. Schnell, G. Röpke, and P. Schuck, *Phys. Rev. Lett.* **83**, 1926 (1999).
4. J. Dukelsky and G. Sierra, cond-mat/9906166 and *Phys. Rev. Lett.* **83**, 172 (1999).
5. D.J. Rowe, *Phys. Rev.* **175**, 1283 (1968); *Rev. Mod. Phys.* **40**, 153 (1968); P. Schuck, *Z. Phys.* **241**, 395 (1971); P. Schuck and S. Ethofer, *Nucl. Phys.* A **212**, 269 (1973).
6. J. Dukelsky and P. Schuck, *Nucl. Phys.* A **512**, 466 (1990); J. Dukelsky, G. Röpke, and P. Schuck, *Nucl. Phys.* A **628**, 17 (1998).
7. P. Krueger and P. Schuck, *Europhys. Lett.* **72**, 395 (1994); F. Catara, G. Piccitto, M. Sambataro, and N. Van Giai, *Phys. Rev.* B **54**, 17536 (1996); S. Schaefer

and P. Schuck, *Phys. Rev.* B **59**, 1712 (1999).
8. G. Röpke, T. Seifert, H. Stolz, and R. Zimmermann, *Phys. Stat. Sol.* (b) **100**, 215 (1980); G. Röpke, *Ann. Physik (Leipzig)* **3**, 145 (1994); G. Röpke, *Z. Physik* B **99**, 83 (1995).
9. J. Dukelsky and P. Schuck, *Phys. Lett.* B **387**, 233 (1996).
10. D. Karadjov, V.V. Voronov and F. Catara, *Phys. Lett.* B **306**, 197 (1993); A.N. Storozhenko, D.S. Kosov and A.I. Vdovin, *Yad. Fiz.* (1998); ibid., to be published in *Nucl. Phys.* A; J. Toivanen and J. Suhonen, *Phys. Rev. Lett.* **75**, 410 (1995); J. Hirsch, P.O. Hess, and O. Civitarese, *Phys. Lett.* B **390**, 36 (1997); F. Krmpotic, T.T.S. Kuo, A. Mariano, E. J. V. Passos, and A.F.R. de Toledo Piza, *Nucl. Phys.* A **612**, 223 (1997).
11. J. Dukelsky and P. Schuck, *Phys. Lett.* B, in print.
12. P. Ring and P. Schuck, *The Nuclear Many Body Problem*, (Springer, New York, 1980).
13. F. Catara, N. Dinh Dang, and M. Sambataro, *Nucl. Phys.* A **579**, 1 (1994).
14. J. Dukelsky, J.G. Hirsch, and P. Schuck, nucl-th/9904083 (to be published in *EPJ* A).
15. A. L. Fetter and J. D. Walecka, *Quantum Theory of Many-Particle Systems* (McGraw-Hill, 1971).
16. A. Lejeune, P. Grange, M. Martzloff, and J. Cugnon, *Nucl. Phys.* A **453**, 189 (1986); H.-J. Schulze, J. Cugnon, A. Lejeune, M. Baldo, and U. Lombardo, *Phys. Rev.* C **52**, 2785 (1995).
17. B.E. Vonderfecht, W.H. Dickhoff, A. Polls, and A. Ramos, *Nucl. Phys.* A **555**, 1 (1993); C. C. Gearhart, W. H. Dickhoff, A. Polls, and A. Ramos, *J. Modern Phys.* E **5**, 461 (1996).
18. Y. Yamaguchi, *Phys. Rev.* **95**, 1628 (1954).
19. M. Baldo, J. Cugnon, A. Lejeune, and U. Lombardo, *Nucl. Phys.* A **515**, 409 (1990); M. Baldo, U. Lombardo, and P. Schuck, *Phys. Rev.* C **52**, 975 (1995).
20. M. Schmidt, G. Röpke, and H. Schulz, *Ann. Phys.* (N.Y.) **202**, 57 (1990).
21. H. Ding *et al.*, *Nature* **382**, 51 (1996); A. G. Loeser *et al.*, *Science* **273**, 325 (1996).

CLUSTER MEAN-FIELD APPROACH FOR CORRELATIONS IN DENSE MATTER

G. RÖPKE

FB Physik, Universität Rostock, D-18051 Rostock, Germany

A. SCHNELL

Institute for Nuclear Theory, University of Washington, Box 351550 Seattle, WA 98195-1550, U.S.A.

P. SCHUCK

Institut des Sciences Nucleaires, Universite Joseph Fourier, CNRS-IN2P3 53, Avenue des Martyrs, F-38026 Grenoble Cedex, France

The formation of correlations is considered in strongly coupled quantum liquids, such as dense plasmas and nuclear matter. A Green function approach is presented, which allows for a self-consistent treatment of a few-body system (cluster) in a correlated medium. As examples, calculations for two- and three-particle clusters in nucler matter are given. The formation of quantum condensates is considered, with special attention to the occurrence of higher order (α-particle like) correlations. Medium dependent reaction rates are derived to describe the formation of clusters in expanding hot dense matter.

1 CORRELATIONS IN CORRELATED MATTER

Starting from a microscopic picture the many-particle system is considered to consist on elementary particles of mass m, interacting by the potential V. In second quantization it shall be described by the Hamiltonian

$$H = \sum_1 E(1)\, a_1^\dagger a_1 + \frac{1}{2} \sum_{121'2'} V(121'2')\, a_1^\dagger a_2^\dagger a_{2'} a_{1'} \tag{1}$$

with $1 = \{\vec{p}_1, \alpha_1\}$; $\vec{p}_1$ denotes the momentum, α_1 spin and further internal quantum numbers. The creation operator $a^\dagger$ and the annihilation operator a for the elementary particles obey fermion or boson commutation relations; $E(1) = p_1^2/(2m_1)$ is the kinetic energy.

As a consequence of the interaction V, correlations will occur in the equilibrium as well as in the non-equilibrium state of the system. If the interaction is weak, we can use the concept of a nearly free system of quasi-particles where the energies are shifted by mean-field terms (Hartree-Fock approximation).

In strongly coupled fermion systems with attractive interaction, according to the Schrödinger equation for an A-particle system the formation of **bound**

states may be possible. This is an important quantum effect which modifies macroscopic properties such as the thermodynamic equation of state or transport properties. In the low-density limit, we can apply the chemical picture where the system is treated as a nearly ideal mixture of different components, and reactions due to collisions may occur. At low temperatures and densities, bound states may determine the properties of the considered system in an essential manner, and any reasonable approximation in the quantum statistical treatment should take the formation of bound states into account.

There arises the following interesting aspect in statistical physics: From a more fundamental point of view, the so-called "elementary" particles may prove to be composites, consisting of more fundamental "elementary" particles which form bound states. For instance, molecules are consisting of atoms, atoms are consisting of electrons and nuclei, nuclei are consisting of protons and neutrons, etc. It is an important task of statistical physics to find the appropriate degrees of freedom within a given range of excitation energy. For instance, a description of a system as consisting of atoms interacting via an effective atom-atom interaction is not possible at high excitation energies or densities, where a description by electrons and ions interacting via the Coulomb force may be more appropriate. Thus, an important criterion for systematic many-particle approximations is whether bound states (**clusters**) are treated on the same footing as the "elementary" particles.

As an example, the well-known Hartree-Fock approximation mentioned above is based on the concept of the mean field on the level of the elementary particles. This concept has to be extended in two directions: not only the single-particle properties, but also the properties of the clusters are modified by the mean field. Furthermore, the mean field is produced not only from single particle states, but also from clusters. The self-consistent approach where the modifications of a cluster by a medium composed of modified clusters are calculated in first order with respect to the interaction potential is denoted as **Cluster Mean-Field Approximation** [1]. It is given by the following wave equation for an A-particle cluster

$$[E(1) + \ldots + E(A) - E_{A\alpha p}]\psi_{A\alpha p}(1\ldots A) + \sum_{1'\ldots A'}\sum_{i<j}^{A} V_{ij}^{A}(1\ldots A, 1'\ldots A')$$
$$\times\psi_{A\alpha p}(1'\ldots A') + \sum_{1'\ldots A'} V_{\text{matter}}^{A,\text{HF}}(1\ldots A, 1'\ldots A')\psi_{A\alpha p}(1'\ldots A') = 0\ , \quad (1)$$

with $V_{12}^{A}(1\ldots A, 1'\ldots A') = V(12, 1'2')\delta_{33'}\ldots\delta_{AA'}$. The effective potential $V_{\text{matter}}^{A,\text{HF}}(1\ldots A, 1'\ldots A')$ accounts for the influence of the medium on the cluster bound states and takes the form

$$V^{A,\mathrm{HF}}_{\mathrm{matter}}(1\dots A,1'\dots A') = \sum_i \Delta(i)\delta_{11'}\dots\delta_{AA'} + \sum_{i,j}{}' \Delta V^A_{ij}(1\dots A,1'\dots A') \,, \tag{2}$$

with

$$\Delta(1) = \sum_2 V(12,12)_{\mathrm{ex}}\tilde{f}(2) - \sum_{B=2}^{\infty}\sum_{\alpha p}\sum_{2\dots B}\sum_{1'\dots B'} f_B(E_{B\alpha p})$$
$$\times \sum_j^B V^B_{1j}(1\dots B,1'\dots B')\psi_{B\alpha p}(1\dots B)\psi^*_{B\alpha p}(1'\dots B') \,,$$

$$\Delta V^A_{12} = -\left\{\frac{1}{2}\left[\tilde{f}(1)+\tilde{f}(1')\right]V(12,1'2') + \sum_{B=2}^{\infty}\sum_{\alpha p}\sum_{\bar{2}\dots\bar{B}}\sum_{\bar{2}'\dots\bar{B}'} f_B(E_{B\alpha p})\right.$$
$$\left.\times \sum_j^B V^B_{1j}(1\bar{2}'\dots\bar{B}',1'\bar{2}\dots\bar{B})\psi^*_{B\alpha p}(2\bar{2}\dots\bar{B})\psi_{B\alpha p}(2'\bar{2}'\dots\bar{B}')\right\}\delta_{33'}\dots\delta_{AA'},$$

$$\tilde{f}(1) = f_1(1) + \sum_{B=2}^{\infty}\sum_{\alpha p}\sum_{2\dots B} f_B(E_{B\alpha p})|\psi_{B\alpha p}(1\dots B)|^2, \tag{3}$$

where $f_A(E_{A\alpha p})$ denotes the (in general time dependent) distribution function for the A-particle cluster; in equilibrium we have

$$f_A(E) = (\exp\beta(E - A\mu) + (-1)^{A+1})^{-1} \,. \tag{4}$$

Note that as in the ordinary HF approximation the effective potential $V^{A,\mathrm{HF}}_{\mathrm{matter}}$ remains energy independent, i.e. instantaneous. The quantity $\tilde{f}(1)$ describes the effective occupation of the state (1) due to free and bound states, whereas the exchange terms obtained from the additional terms in ΔV^A_{ij} and $\Delta(i)$ thus realizing the **Pauli exclusion principle**.

An interesting problem is to which extent the cluster-mean field approximation is complete, e.g. respects the usual **conservation laws**. It has been shown [2] that balance equations are fulfilled where similar equations were derived using the equation of motion for the Green functions. In general, the Ward-Takahashi identity determines a conserving approximation [3].

2 FEW-PARTICLE STATES IN AN UNCORRELATED MEDIUM

The self-consistent solution of the Cluster Mean-Field Approximation is highly sophisticated. At present, calculations can be performed where the medium is considered as **uncorrelated**, i.e. consisting of quasiparticles with distribution function reproducing the correct particle density. Such an approximation may be valid in a region where the contribution of correlations to the total density is small.

Restricting us to small clusters with $A \leq 4$, we have to consider the following set equations for the A-particle Green functions ($\bar{f} = 1 - f$)

$$G_1(1,1',z) = \frac{1}{z - E^{\rm HF}(1)}\delta_{11'}\,, \tag{5}$$

$$G_2(12,1'2',z) = \frac{\bar{f}_1\bar{f}_2 - f_1 f_2}{z - E^{\rm HF}(1) - E^{\rm HF}(2)}\delta_{11'}\delta_{22'} \tag{6}$$

$$+\ \frac{1 - f_1 - f_2}{z - E^{\rm HF}(1) - E^{\rm HF}(2)}V(12,\bar{1}\bar{2})G_2(\bar{1}\bar{2},1'2',z)\,, \tag{7}$$

$$G_3(123,1'2'3',z) = \frac{\bar{f}_1\bar{f}_2\bar{f}_3 - f_1 f_2 f_3}{z - E^{\rm HF}(1) - E^{\rm HF}(2) - E^{\rm HF}(3)}\delta_{11'}\delta_{22'}\delta_{33'}$$

$$+\frac{(1 - f_1 - f_2)V(12,\bar{1}\bar{2})\delta_{3\bar{3}}G_3(\bar{1}\bar{2}\bar{3},1'2'3',z)}{z - E^{\rm HF}(1) - E^{\rm HF}(2) - E^{\rm HF}(3)} + \text{perm.}\,, \tag{8}$$

$$G_4(1234,1'2'3'4',z) = \frac{(\bar{f}_1\bar{f}_2\bar{f}_3\bar{f}_4 - f_1 f_2 f_3 f_4)\delta_{11'}\delta_{22'}\delta_{33'}\delta_{44'}}{z - E^{\rm HF}(1) - E^{\rm HF}(2) - E^{\rm HF}(3) - E^{\rm HF}(4)}$$

$$+\frac{(1 - f_1 - f_2)V(12,\bar{1}\bar{2})\delta_{3\bar{3}}\delta_{4\bar{4}}G_4(\bar{1}\bar{2}\bar{3}\bar{4},1'2'3'4',z)}{z - E^{\rm HF}(1) - E^{\rm HF}(2) - E^{\rm HF}(3) - E^{\rm HF}(4)} + \text{perm.} \tag{9}$$

The effects of the medium are contained in the Hartree-Fock shifts of the single-particle energies, $E^{\rm HF}(1) = E(1) + \sum_2 V(12,12)_{\rm ex} f(2)$, and in the blocking factors $(1 - f_1 - f_2)$ in front of the interaction potential V.

These additional density corrections arise from the fact that the constituents of the composite are fermions and have to obey the Pauli exclusion principle also with respect to the medium (Pauli blocking). As a consequence, bound states in a dense medium are dissolved (**Mott effect**) because the phase space occupied by fermions is no longer available to form a bound state wave function, see [4,5,6] for further details.

Using a separable representation of the two-particle interaction, the in-medium two-particle problem can be solved immediately. Depending on density, temperature and total momentum, the shift of the two-particle binding energy as well as the modification of the bound state wave function and scattering phase shifts were calculated and broadely discussed, see [5,6]. Recently, it was possible to treat the three-body problem in a medium on the basis of a Faddeev-type equation where the medium is included in mean-field approximation [7]. Up to now, the four-body problem in a medium has been solved only within variational or perturbative approaches [1,8].

3 QUANTUM CONDENSATES

A further aspect of interaction is the formation of quantum condensates. The standard BCS theory of pair condensation is based on the mean-field concept. The inclusion of clusters and correlated states is a difficult task in many-particle theory [9,10].

The pairing transition is obtained from the solution of the effective wave equation corresponding to the two-particle Bethe-Goldstone equation, see Eq. (6), if the energy z coincides with 2μ. In symmetric nuclear matter, pairing is strong in the isospin-singlet (deuteron) channel, where a **cross-over from the Bose-Einstein condensation of deuterons at low densities to *pn* Cooper pairing at high densities** is obtained [11].

A quantum condensate can arise also from the solution of the effective wave equation corresponding to the four-particle Bethe-Goldstone equation, see Eq. (8), if the energy z coincides with 4μ [8]. This quartetting transition describes the Bose-Einstein condensation of α particles in the low-density limit. As shown in Fig. 1, in symmetric nuclear matter it beats the pairing

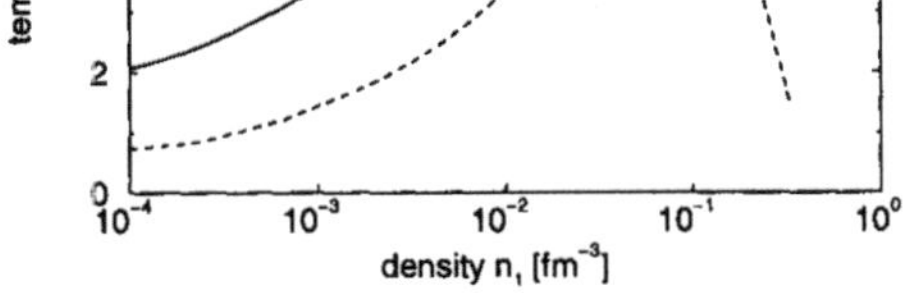

Figure 1. Critical temperatures for the onset of two-particle pairing (T_t^c) and four-particle condensation (T_4^c) in symmetric nuclear matter as a function of the uncorrelated density, model calculation.

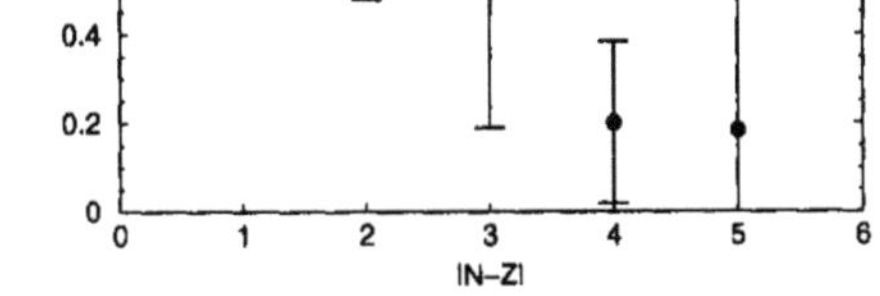

Figure 2. Ratio $b_{N-Z}(Z)/b_0(Z)$ averaged over Z as a function of the absolute value of the difference between proton and neutron numbers. The additional binding energy $\Delta B(Z,N) = b_{N-Z}(Z)$, Eq(10), contains the contributions of isospin singlet (pn) pairing and quartetting.

transition (deuteron channel) if the density is smaller than $0.03\,\mathrm{fm}^{-3}$. Such situations may occur at the surface of heavier nuclei or in heavy ion collisions, e.g. in the neck region during multifragmentation.

Isospin singlet ($S = 1, T = 0$) *pn* **pairing** as well as **quartetting** should occur in nuclei near $Z = N$. In a recent paper [12], an analysis of binding energies of nuclei near $Z = N$ has been performed using different filters such as $h(Z,N) = 2B(Z,N) - B(Z,N-2) - B(Z,N+2) - 2B(Z-2,N) + B(Z-2,N-2) + B(Z-2,N+2)$. The empirical binding energies of nuclei are well approximated by the Bethe-Weizsäcker formula, containing the contributions of bulk, surface, Coulomb and asymmetry energy, pairing in the isospin triplet channel (*pp*, *nn*) and shell structure effects.

Decomposing the binding energy $B(Z,N)$ into different contributions as

$$\begin{aligned} B(Z,N) = {} & B_{\mathrm{bulk}}(Z,N) + B_{\mathrm{surf}}(Z,N) + B_{\mathrm{Coul}}(Z,N) + B_{\mathrm{asymm}}(Z,N) \\ & + B_{\mathrm{shell}}(Z,N) + B_{\mathrm{pair}}(Z,N) + \Delta B(Z,N) , \end{aligned} \tag{10}$$

it has been found that the **additional binding energy** $\Delta B(Z,N) = b_{N-Z}(Z)$ containing the effects isospin singlet (*pn*) pairing and quartetting arises if $|N - Z| \leq 4$, see Fig. 2. It can be obtained from $h(Z,N)$. Using a Local Density Approximation, reasonable agreement between the gain in binding energy due to isospin singlet pairing in asymmetric nuclear matter and the observed values has been obtained. Considering excited states, a reduction of the pairing gap in agreement with the corresponding temperature was obtained.

4 CORRELATED MEDIUM IN NON-EQUILIBRIUM STATE

The cluster-mean field approximation allows to construct a basis in the Fock space consisting of stationary solutions. Damping effects will occur if the frequency-dependent (retardation) contributions to the effective Hamiltonian are included, what can be done within a perturbation theory [9].

To construct the nonequilibrium statistical operator $\rho(t)$, we start with the A-particle distributions at time $t_0 \leq t$:

$$n_1(11', t_0) = \langle a_1^\dagger a_{1'} \rangle^{t_0}, \qquad n_2(12, 1'2', t_0) = \langle a_1^\dagger a_2^\dagger a_{2'} a_{1'} \rangle^{t_0}, \tag{11}$$

etc. It is possible to introduce another orthonormal basis $|\nu_1\rangle$ of single-particle states with $\psi_{\nu_1}(k) = \langle k|\nu_1\rangle$, of two-particle states $\psi_{\nu_2}(k_1 k_2) = \langle k_1 k_2|\nu_2\rangle$, etc., so that

$$n_1(\nu_1;\nu_1';t) = \sum_{kk'} \psi^*_{\nu_1}(k) n_1(k;k';t) \psi_{\nu_1'}(k') ,$$

$$n_2(\nu_2;\nu_2';t) = \sum_{k_1k_2k_1'k_2'} \psi^*_{\nu_2}(k_1k_2)n_2(k_1k_2;k_1'k_2';t)\psi_{\nu_2'}(k_1'k_2') \ , \quad (12)$$

etc.

In particular, it is possible to introduce a basis system of quantum states which are optimally stationary, i.e. for which the damping is small. We propose [1] to use the **Cluster Mean-Field Basis** $\psi_{A\alpha p}(1,\ldots,A,t_0)$ introduced above as such an optimized basis system.

In the homogeneous case the relevant statistical operator can be found from the maximum of information entropy [13]

$$\rho_{\rm rel}^{\rm corr}(t_0) = Z_{\rm rel}^{-1}(t_0)\exp[-\sum_{A\alpha p} s_{A\alpha p}(t_0)c^\dagger_{A\alpha p}c_{A\alpha p}] \ , \quad (13)$$

where we used the abbreviation $c_{A\alpha p} = \sum_{1\ldots A}\psi^{\rm corr}_{A\alpha p}(1\ldots A)a_A\ldots a_1$. The thermodynamic parameters $s_{A\alpha p}(t_0)$ have to be determined from the given mean values

$$f_{A\alpha}(p,t_0) = \sum_{1\ldots A,1'\ldots A'} \psi^{\rm corr}_{A\alpha p}(1\ldots A)^* n_A^{\rm corr}(1\ldots A,1'\ldots A',t_0)\psi^{\rm corr}_{A\alpha p}(1'\ldots A'), \quad (14)$$

with the correlated part of the A–particle distribution such as

$$n_2^{\rm corr}(12,1'2') = n_2(12,1'2') - n_1(11')n_1(22')_{\rm ex} \ . \quad (15)$$

With the relevant operator $\rho^{\rm corr}_{\rm rel}(t')$ we can construct the non–equilibrium statistical operator in the usual way [13] and derive equations for the time dependence of the thermodynamic parameters $s_{A\alpha p}(t)$. With the time evolution operator $U(t,t')$, obtained from $i\hbar\partial U(t,t')/\partial t = H(t)U(t,t')$ with $U(t,t)=1$, we find [13] ($\lim\epsilon\to 0$)

$$\rho(t) = \epsilon\int_{-\infty}^{t} e^{\epsilon(t'-t)}U(t,t')\rho_{\rm rel}^{\rm corr}(t')U^\dagger(t,t')dt' \ . \quad (16)$$

The derivation of **kinetic equations** within that approach is given in [9,13,14]. In particular, it is shown in [14] that the Vlasov term of the kinetic equation for the A-particle cluster contains the cluster mean-field.

Based on the solution of the in-medium three-body problem, it was possible to evaluate formation and break-up reaction rates for deuterons in dense nuclear matter [7,15]. A considerable enhancement of the reaction rates was observed which has also consequences for the time evolution of composition of expanding nuclear matter, e.g. in heavy-ion collisions. The implementation of medium dependent reaction rates into nuclear matter transport codes is in progress.

References

1. G. Röpke, T. Seifert, H. Stolz, and R. Zimmermann, Phys. Stat. Sol. (b) **100**, 215 (1980); G. Röpke, M. Schmidt, L. Münchow, and H. Schulz, Nucl. Phys. **A 399**, 587 (1983); J. Schmelzer, G. Röpke, and R. Mahnke, *Aggregation Phenomena in Complex Systems* (Wiley-VCH Publishers, Weinheim 1999).
2. J. Dukelsky, G. Röpke, and P. Schuck, Nucl. Phys. **A 628**, 17 (1998); see also P. Schuck, Z. Phys. **241**, 395 (1971); P. Schuck and S. Ethofer, Nucl. Phys. **A 212**, 269 (1973); J. Dukelsky and P. Schuck, Nucl. Phys. **A 512**, 466 (1990).
3. T. Tayoda, Ann. Phys. (N.Y.) **173**, 226 (1987).
4. W. D. Kraeft, D. Kremp, W. Ebeling, and G. Röpke, *Quantum Statistics of Charged Particle Systems* (Plenum, New York, 1986).
5. G. Röpke, L. Münchow, and H. Schulz, Nucl. Phys. **A 379**, 536 (1982); Phys. Lett. **B 110**, 21 (1982).
6. G. Röpke, A. Schnell, and P. Schuck, in *Condensed Matter Theories*, Vol. 12, edited by J. W. Clark and P. V. Panat (Nova Science Publishers, Commack, NY, 1997), p.237 ff.
7. M. Beyer, G. Röpke, and A. Sedrakian, Phys. Lett. **B 376**, 7 (1996); M. Beyer and G. Röpke, Phys. Rev. **C 56**, 2636 (1997);
8. G. Röpke, A. Schnell, P. Schuck, and P. Nozières, Phys. Rev. Lett. **80**, 3177 (1998).
9. G. Röpke, Ann. Physik (Leipzig) **3**, 145 (1994); Z. Phys. **B 99**, 83 (1995).
10. R. Haussmann, Z. Phys. **B 91**, 291 (1993); Phys. Rev. **B 49**, 12975 (1994).
11. T. Alm, G. Röpke, M. Schmidt, Z. Phys. **A 337**, 355 (1990); T. Alm, B. L. Friman, G. Röpke, and H. Schulz, Nucl. Phys. **A 551**, 45 (1993); B. E. Vonderfecht, C. C. Gearhart, W. H. Dickhoff, A. Polls, and A. Ramos, Phys. Lett. **B 253**, 1 (1992); M. Baldo, I. Bombaci, and U. Lombardo, Phys. Lett. **B 283**, 8 (1993).
12. G. Röpke, A. Schnell, P. Schuck, and U. Lombardo, Phys. Rev. **C**, in print.
13. D. N. Zubarev, V. Morozov, and G. Röpke, *Statistical Mechanics of Nonequilibrium Processes* (Wiley-VCH Publ., Weinheim, 1998).
14. G. Röpke and H. Schulz, Nucl. Phys. **A 477**, 472 (1988).
15. M. Beyer, W. Schadow, C. Kuhrts and G. Röpke, Phys. Rev. **C**, in print; C. Kuhrts, M. Beyer and G. Röpke, nucl-th/9908032, Nucl. Phys. **A**, in print; M. Beyer, Few Body Syst. **99**, 1 (1998).

DISCRETE AMBIGUITIES IN THE CLUSTER-CLUSTER INTERACTION AND SUPERSYMMETRY

D. BAYE AND J.-M. SPARENBERG

Physique Nucléaire Théorique et Physique Mathématique, C.P. 229, Université Libre de Bruxelles, B 1050 Brussels, Belgium

E-mail: dbaye@ulb.ac.be, jmspar@ulb.ac.be

Supersymmetric transformations allow eliminating bound states from the spectrum of a potential without affecting its phase shifts. Phase-equivalent potentials possess different off-shell properties which may lead to different behaviours for a three-body system. Several applications are considered: the $\alpha + \alpha$ bremsstrahlung, the ^{16}O + ^{17}O collision and a three-body model of the ^6He, ^{11}Li and ^{14}Be halo nuclei.

1 Introduction

The deep or shallow nature of the nucleus-nucleus potentials is a long-standing problem. The interactions extracted from experimental data are affected by the discrete ambiguity related to their number of bound states. This number is not clearly constrained by physical reasons. Microscopic models suggest that realistic potentials should involve a simulation of the Pauli principle between the interacting clusters through the introduction of a specific number of forbidden bound states. Elastic scattering alone is not able to settle this discrete-ambiguity problem since different potentials may be *exactly* phase equivalent, i.e. provide *identical* phase shifts and cross sections. However, equivalent potentials usually have significantly different angular-momentum dependences.

Equivalent potentials may lead to different properties when they are used in a three-cluster system. Such studies should in principle allow determining the best potential in an equivalent family. Supersymmetric transformations[1] allow eliminating bound states from the spectrum of a potential for a given partial wave without affecting its phase shifts.[2,3,4,5,6] The resulting potential has an r^{-2} singularity at the origin and depends on the relative orbital momentum. This technique can be used for real,[2,3,4,6] complex[7,8] and coupled-channel potentials.[9] It can even afford a linear energy dependence.[10]

The principle of supersymmetric transformations is presented in Sec. 2. Applications are described in Sec. 3. Off-shell properties of different types of three-body systems are discussed in Sec. 4. Concluding remarks are presented in Sec. 5.

2 Derivation of Phase-Equivalent Potentials

For $\hbar^2/2\mu = 1$ (where μ is the reduced mass of the system), the radial Schrödinger equation at fixed orbital momentum l reads

$$\left(-\frac{d^2}{dr^2} + V_0(r)\right)\psi(E,r) = E\psi(E,r), \tag{1}$$

where the radial wave function $\psi(E,r)$ at energy E vanishes at the origin. The potential V is bounded except possibly near the origin where it may present a centrifugal barrier of the form $l(l+1)/r^2$.

Let $\psi_0(E_0,r)$ be a normalized bound-state wave function at energy E_0. The Hamiltonian H_0 in (1) can be factorized as

$$H_0 = A_0^+ A_0^- + E_0 \tag{2}$$

with

$$A_0^\pm = \pm\frac{d}{dr} + \frac{\psi_0'(E_0,r)}{\psi_0(E_0,r)}. \tag{3}$$

This factorization is easily verified by a direct calculation in which the fact that $\psi_0(E_0,r)$ satisfies the Schrödinger equation (1) is used.

The supersymmetric partner of H_0 is defined as[1]

$$H_1 = A_0^- A_0^+ + E_0 = H_0 - 2\frac{d}{dr}\left(\frac{\psi_0'(E_0,r)}{\psi_0(E_0,r)}\right). \tag{4}$$

If $\psi_0(E,r)$ is an eigenfunction of H_0 at a positive or negative energy E, an eigenfunction of H_1 at the same energy is given by

$$\psi_1(E,r) = (E-E_0)^{-1/2} A_0^- \psi_0(E,r) \tag{5}$$

for $E \neq E_0$. The function $\psi_1(E,r)$ can also be written as[3,4]

$$\psi_1(E,r) = (E-E_0)^{1/2}\,\frac{\int_0^r \psi_0(E_0,t)\psi_0(E,t)dt}{\psi_0(E_0,r)}. \tag{6}$$

As shown by (5), the function $A_0^-\psi_0(E_0,r)$ vanishes and energy E_0 does not correspond any more to a normalizable solution for H_1.

In a second step, H_1 is factorized as[2,3]

$$H_1 = A_1^+ A_1^- + E_0 \tag{7}$$

with

$$A_1^\pm = \pm\frac{d}{dr} - \frac{\psi_0'(E_0,r)}{\psi_0(E_0,r)} + \frac{\psi_0^2(E_0,r)}{\int_0^r \psi_0^2(E_0,t)dt}. \tag{8}$$

The supersymmetric partner of H_1 is therefore

$$H_2 = A_1^- A_1^+ + E_0 \tag{9}$$

which corresponds to a potential

$$V_2(r) = V_0(r) - 2\frac{d}{dr}\left(\frac{\psi_0^2(E_0,r)}{\int_0^r \psi_0^2(E_0,t)dt}\right). \tag{10}$$

The denominator does not vanish, even if $\psi_0(E_0,r)$ has nodes.[5,6] Potential V_2 is thus bounded, except near the origin where it behaves like $(l+2)(l+3)/r^2$. The transformation increases the singularity at the origin.[2]

Bounded eigensolutions of Hamiltonian H_2 at discrete bound-state energies ($\neq E_0$) or at positive energies read

$$\psi_2(E,r) = \psi_0(E,r) - \psi_0(E_0,r)\,\frac{\int_0^r \psi_0(E_0,t)\psi_0(E,t)\,dt}{\int_0^r \psi_0^2(E_0,t)dt}. \tag{11}$$

Since $\psi_0(E_0,r)$ decreases exponentially, the asymptotic behaviour of $\psi_2(E,r)$ in (11) is identical to the asymptotic behaviour of $\psi_0(E,r)$. Therefore, the potential $V_2(r)$ is phase equivalent to $V_0(r)$. Relation (10) can be iterated. The potential V_{2n} is obtained when n bound states are removed.[4]

3 Applications

Supersymmetric transformations have been applied to different systems such as α + n,[11] $\alpha + \alpha$,[2,3] α + ^{16}O[7], and ^{16}O + ^{16}O.[8] In Fig. 1, we show the effect of removing the two forbidden states at −72.6 MeV and −25.8 MeV in the s wave of the deep α + α potential of Ref. [12]. Potentials V_2 and V_4 are phase equivalent to V_0 and have respectively one and two bound states less. Potential V_4 is similar to the shallow phenomenological potential of Ref. [13]. The intermediate potentials V_1 and V_3 have the same number of bound states as V_2 and V_4 but are not phase equivalent to V_0. Each transformation increases the r^{-2} repulsive singularity at the origin.[2] With (10), one can successively obtain V_2 and V_4 in a simple and direct way and avoid calculating V_1 and V_3.

For the ^{16}O + ^{16}O elastic scattering,[8] deep and shallow optical potentials provide essentially similar fits of excitation functions over the 10 to 35 MeV range of centre-of-mass energies. After removing four complex normalizable solutions from the deep optical potential of Ref. [14] for $l = 18$ (V_0 in Fig. 2, thick lines), the real and imaginary parts of the resulting potential V_8 resemble those of the shallow potential of Ref. [15] (Fig. 2, dotted lines).

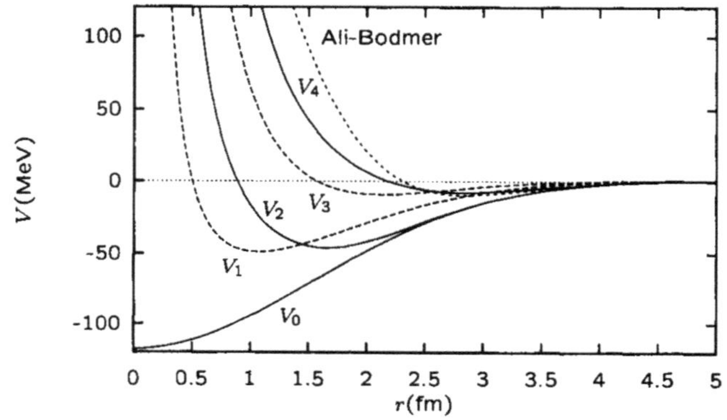

Figure 1. Deep $\alpha + \alpha$ potential of Ref. [12] for the s wave (V_0) and its successive supersymmetric partners ($V_1, \ldots, V_4$). The potential of Ref. [13] (dotted line) is shown for comparison.

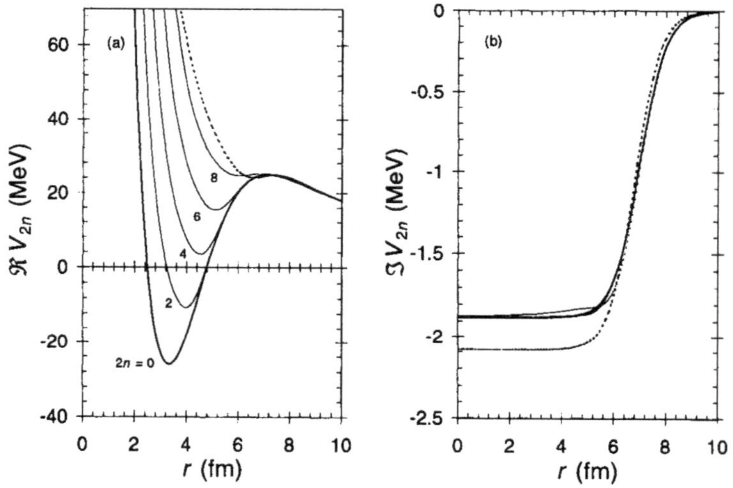

Figure 2. Deep $^{16}O + {}^{16}O$ potential of Ref. [14] for the $l = 18$ wave ($n = 0$) and its phase-equivalent supersymmetric partners with n removed forbidden bound states, compared with the shallow potential of Ref. [15] (dotted lines).

4 Off-shell effects in three-body systems

4.1 $\alpha + \alpha$ bremsstrahlung

Nucleus-nucleus bremsstrahlung corresponds to a transition between continuum states, the two-particle relative motion being slowed down by the emission of a photon. Model calculations involve an electromagnetic transition between initial and final scattering waves with different energies. The α + α bremsstrahlung has been studied by the resonating-group microscopic model[16,17] and by the potential model[18] with both the deep potential (DP) of Ref. [12] and the phase-equivalent shallow potential (SP) (see V_0 and V_4 in Fig. 1 for the s waves). For the d waves, only one forbidden state has to be removed from the deep potential, while for larger l values there is no forbidden bound state.[12,2] At the long-wavelength approximation, the E2 contribution dominates the α + α bremsstrahlung.

In available experimental data,[19] the α particles have an initial centre-of-mass energy E_i below 10 MeV (see Fig. 3). After the collision, they are detected in the coplanar directions $\theta_1 = \theta_2 = 35°$ (Harvard geometry) while the photon emitted with energy $0.51\,E_i$ remains undetected. All theoretical calculations reproduce reasonably well the data. However, the shallow-potential cross section is smaller beyond 13 MeV than the deep-potential one by a factor of two. Below 13 MeV, the microscopic calculation (dashed line) fairly agrees with the deep-potential result. Additional data with improved statistics would allow discriminating between the models.

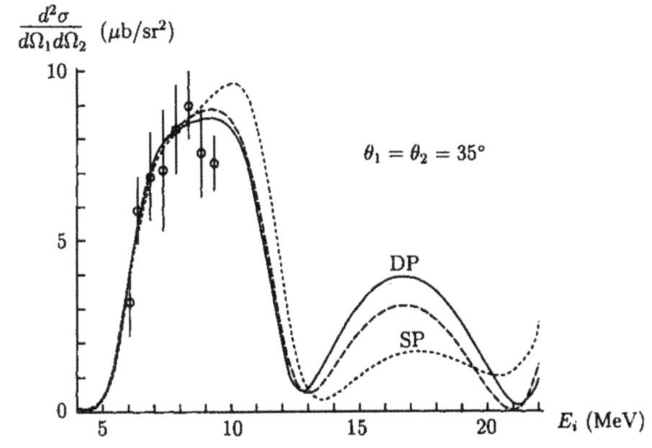

Figure 3. $\alpha(\alpha,\alpha\gamma)\alpha$ experimental cross sections (circles), compared with microscopic (dashed line), deep-potential (DP, full line) and shallow-potential (SP, dotted line) calculations.

4.2 $^{16}O + {}^{17}O$ *collision*

This collision is studied with the coupled-reaction-channel model[20] (see also Ref. [21]). The ^{17}O nucleus is described as a pointlike ^{16}O cluster plus a neutron occupying different single-particle orbitals. The three-body calculation involves $^{16}O + {}^{16}O$ and ^{16}O + neutron interactions. Figure 4 shows that calculations with the deep potential of Ref. [14] and with the shallow potential of Ref. [15] provide very different results above 30 MeV. In order to test whether

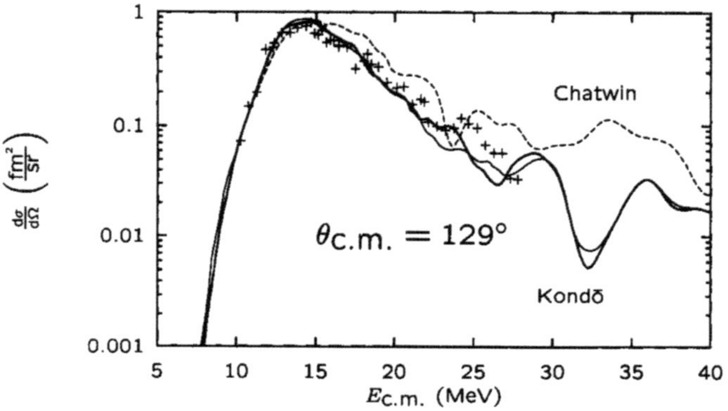

Figure 4. $^{17}O(^{16}O,^{16}O)^{17}O^{*}$ (0.871 MeV, $\frac{1}{2}^{+}$) excitation function at $\theta_{c.m.} = 129°$. Experimental data of Ref. [22] are compared with a calculation using the deep potential of Ref. [14] (thick line), its supersymmetric partner (thin line) and the shallow potential of Ref. [15] (dotted line).

this sensitivity is related to the number of forbidden bound states of these potentials, the calculation is also performed with the supersymmetric partner of the deep potential (see for example V_8 in Fig. 2 for $l = 18$). No significant difference between these exactly phase-equivalent potentials is found neither for elastic nor for inelastic cross sections at $E_{c.m.} < 40$ MeV, except at extreme backward angles. The large difference between the potentials of Refs. [14] and [15] is thus due to other properties than the discrete ambiguity.

4.3 ^{6}He, ^{11}Li and ^{14}Be halo nuclei

Light nuclei with a neutron excess and a small binding energy possess much larger radii than other nuclei. The neutron density distribution displays an extended shape known as a halo. Two-neutron halo nuclei such as ^{6}He, ^{11}Li and ^{14}Be are well described by a three-body model involving cluster-neutron and neutron-neutron interactions. Occupied neutron orbitals of the core cluster are forbidden to the halo neutrons in the deep core-neutron potential.

For the three-body system, the non-physical two-body bound states lead to unrealistic binding energies. The forbidden states must be removed. They are usually pushed to high energies with a projection technique.[23] Here, we compare that approach with the supersymmetric removal method presented above.

Table 1. Binding energy B in MeV and rms matter radius r_m in fm of the ^{6}He, ^{11}Li and ^{14}Be nuclei. See text for the definitions of calculations (1), (2) and (3).

	^{6}He			^{11}Li			^{14}Be		
	(1)	(2)	(3)	(1)	(2)	(3)	(1)	(2)	(3)
B	0.97	1.18	0.97	0.33	0.71	0.33	1.18	1.76	1.16
r_m	2.38	2.33	2.37	3.21	2.95	3.20	3.17	2.97	3.11

In Table 1 are compared calculations for the three halo nuclei using (1) the deep cluster-neutron potential and the projection technique with an interaction scaled to reproduce the experimental binding energy, (2) the shallow potential equivalent to the scaled deep interaction, and (3) a shallow potential fitting the experimental binding energy cooresponding to a slightly rescaled deep potential (see Ref. [11] for the conditions of the calculation). The phase-equivalent deep and shallow potentials [cases (1) and (2)] lead to slightly different binding energies. The larger binding energy obtained with the shallow potential induces a smaller radius. However, when the binding energy is refitted [case (3)], the original radius is essentially recovered. The energies and radii obtained in this model do therefore not allow to physically discriminate between the off-shell properties of the different potentials. In the case of ^{14}Be however, the wave functions obtained with the deep and shallow potentials are significantly different (see Fig. 2 in Ref. [11]) and should be discriminated by experiments.

5 Conclusion

Pairs of supersymmetric transformations provide a useful tool for studying the influence of discrete ambiguities on off-shell properties of potentials. This technique is now mastered for real, complex, coupled-channel and linearly-energy-dependent potentials. Different applications are found to be weakly sensitive to the off-shell behaviour. The most sensitive one is nucleus-nucleus bremsstrahlung for which new experimental studies would be welcome.

Acknowledgments

We would like to thank P. Descouvemont, M. Hesse, B. Imanishi, M. Kruglanski, G. Lévai and H. Leeb for fruitful collaborations. J.-M.S. acknowledges the support of the National Fund for Scientific Research (FNRS). This text presents research results of the Belgian program P4/18 on interuniversity attraction poles initiated by the Belgian-state Federal Services for Scientific, Technical and Cultural Affairs.

References

1. C.V. Sukumar, *J. Phys.* A **18**, 2917 (1985).
2. D. Baye, *Phys. Rev. Lett.* **58**, 2738 (1987).
3. D. Baye, *J. Phys.* A **20**, 5529 (1987).
4. L.U. Ancarani and D. Baye, *Phys. Rev.* A **46**, 206 (1992).
5. D. Baye, *Phys. Rev.* A **48**, 2040 (1992).
6. D. Baye and J.-M. Sparenberg, *Phys. Rev. Lett.* **73**, 2789 (1994).
7. D. Baye, G. Lévai and J.-M. Sparenberg, *Nucl. Phys.* A **599**, 435 (1996).
8. J.-M. Sparenberg and D. Baye, *Phys. Rev.* C **54**, 1309 (1996).
9. J.-M. Sparenberg and D. Baye, *Phys. Rev. Lett.* **79**, 3802 (1997).
10. J.-M. Sparenberg, D. Baye and H. Leeb, to be published.
11. M. Hesse, D. Baye and J.-M. Sparenberg, *Phys. Lett.* B **455**, 1 (1999).
12. B. Buck, H. Friedrich and C. Wheatley, *Nucl. Phys.* A **275**, 246 (1977).
13. S. Ali and A.R. Bodmer, *Nucl. Phys.* **80**, 99 (1966).
14. Y. Kondō, B.A. Robson and R. Smith, *Phys. Lett.* B **227**, 310 (1989).
15. R.A. Chatwin, J.S. Eck, D. Robson and A. Richter, *Phys. Rev.* C **1**, 795 (1970).
16. D. Baye, P. Descouvemont and M. Kruglanski, *Nucl. Phys.* A **550**, 250 (1992).
17. Q.K.K. Liu, *Nucl. Phys.* A **550**, 263 (1992).
18. D. Baye, C. Sauwens, P. Descouvemont and S. Keller, *Nucl. Phys.* A **529**, 467 (1991).
19. U. Peyer *et al.*, *Phys. Lett.* B **41**, 151 (1972).
20. B. Imanishi and W. von Oertzen, *Phys. Rep.* **155**, 29 (1987).
21. J.-M. Sparenberg, D. Baye and B. Imanishi, contribution to this conference.
22. C.K. Gelbke *et al.*, *Nucl. Phys.* A **219**, 253 (1974).
23. V.M. Krasnopol'skii and V.I. Kukulin, *Sov. J. Nucl. Phys.* **20**, 470 (1975).

MICROSCOPIC ASPECTS OF α-CLUSTERING IN THE NUCLEONIC SEA
-NEW APPROACH TREATING THE AGGREGATION OF NUCLEONS-

AKIHIRO TOHSAKI
Department of Fine Materials Engineering, Shinshu University, Ueda, 386-8567

Mathematical formulism is given a full microscopic approach treating the aggregation of nucleons free from periodical characters. The energetic properties of the surface and deep-sea nucleons are investigated simultaneously. In particular, light is shed on α-clustering process in rarefied density of nucleons. At first, α-clusters might come out on the surface region of sea.

1 Introduction

We have been fascinated by investigating the meeting and parting of nucleons in their aggregation from a microscopic point of view. Only geometrical description for the configuration of nucleons allows us to complete this aim by using the generator coordinate space which can be exactly transformed to the real space.

Equation of motion for a representative nucleon or a cluster with a few nucleons in the aggregation can be solved on the basis of the eigenequation with several eigenstates in the GC space. In other words, the motion of a nucleon or a cluster in a cavity, which is surrounded by the other nucleons, is considered from microscopic aspects. The diagonal elements before diagonalization can be regarded as the binding energy in a static treatment which includes a spurious energy concerning the c.m. coordinate.

In this talk, we give how to treat the aggregation of nucleons within a microscopic framework where the Pauli exclusion principles among all the nucleons and internucleon force are completely taken into account following ref.(1). After that, we show a certain critical value with respect to α-clustering in nucleonic sea.

2 Formalism

Imagine nucleonic sea in the GC space in which treatment is equivalent to that of resonating group method [2,3]. As shown in Fig.1. we cut out a certain domain Γ centered at a nucleon or a cluster in nucleonic sea where $4N$ nucleons exist with four kinds of spin-isospin states. It is possible to contain the surface of sea with or without waves by this cutting which allows us to estimate surface effects of semi-infinite aggregation of nucleons. In Fig.1(a), different sorts of nucleons are distinguished symbolically. The total wave function is written by

$$\Psi = \frac{1}{\sqrt{(4N)!}} \mathcal{A}[\prod_{p=1}^{4} \{\phi_1^{(p)}, \cdots, \phi_N^{(p)}\}\chi_p]. \tag{1}$$

The spatial wave function, $\phi_i^{(p)}$, of the i-th nucleon is expressed relative to the dimensionless spatial parameter $\mathbf{S}_i^{(p)}$ by

$$\phi_i^{(p)} = (\pi)^{-3/4} \exp\{-\frac{1}{2}(\mathbf{q} - \mathbf{S}_i^{(p)})^2\}, \tag{2}$$

where the realistic variable, $\mathbf{q}$, is integrated over infinite space. The parameter $\mathbf{S}_i^{(p)}$ is directly related to the spatial generator coordinate, $\mathbf{R}_i^{(p)} = b\mathbf{S}$, where b is the size parameter of $(0s)$ wave function. When the labels, p, differ from each other, the GCs are also different due to the spatial distribution of nucleons although they have the same label, i. In Fig.1(b), the nucleons with labels, i_1, i_2, i_3 and i_4 are placed on the same position, then four nucleons make an α-cluster in nucleonic sea with random arrangement of nucleons.

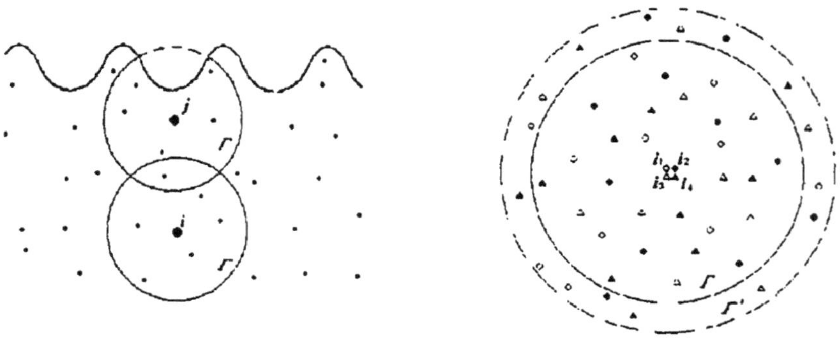

Fig.1(a), Image of nucleonic sea with surface. Figl(b), Schematic model of microscopic calculation.

After inserting a microscopic Hamiltonian, $\mathcal{H}$, the energy overlapping is written by

$$\begin{aligned} < \Psi|\mathcal{H}|\Psi > &= < \Psi|\Psi > [T_r(TB^{-1}) + T_r\{(V_d^{(2)} - V_e^{(2)} + V_d^{(c)} - V_d^{(c)})(B^{-1} \otimes B^{-1})\} \\ &\quad + T_r\{(V_1^{(3)} - V_2^{(3)} - V_3^{(3)} - V_4^{(3)} + V_5^{(3)} + V_6^{(3)})(B^{-1} \otimes B^{-1} \otimes B^{-1})\}] \\ &= < \Psi|\Psi > \sum_i E_i, \end{aligned} \tag{3}$$

where the matrices are defined as

$$B = \left(< \phi_i^{(p)} \chi_p | \phi_j^{(q)} \chi_q > \right), \tag{4}$$

$$T = \left(< \phi_i^{(p)} \chi_p | -\frac{\hbar^2}{2M} \nabla_j^2 | \phi_j^{(q)} \chi_q > \right), \tag{5}$$

$$V_d^{(2)} = \frac{1}{2} \left(< \phi_i^{(p)} \chi_p \phi_j^{(q)} \chi_q | v_{kl}^{(2)} | \phi_k^{(r)} \chi_r \phi_l^{(s)} \chi_s > \right), \tag{6}$$

$$V_d^{(c)} = \frac{1}{2} \left(< \phi_i^{(p)} \chi_p \phi_j^{(q)} \chi_q | \frac{e^2}{q_{kl}} | \phi_k^{(r)} \chi_r \phi_l^{(s)} \chi_s > \right), \tag{7}$$

$$V_1^{(3)} = \frac{1}{6} \left(< \phi_i^{(p)} \chi_p \phi_j^{(q)} \chi_q \phi_k^{(r)} \chi_r | v_{lmn}^{(3)} | \phi_l^{(s)} \chi_s \phi_m^{(t)} \chi_t \phi_n^{(u)} \chi_u > \right), \tag{8}$$

and the other ones are attributed to those with exchange characters. The matrices for internucleon force have a tensor representation in eqs.(6)~(8), and the symbol, $\otimes$, means the tensor product. In the static model the respective binding energies of nucleons are straightforwardly regarded as corresponding elements of traces in eq.(3). The domain Γ' is given by being slightly expanded from Γ as shown in Fig.1(a). If the binding energy of a nucleon in the center of Γ' is almost unchanged from that in Γ, we can take the convergence of the value as its binding energy. However the Coulomb interaction inevitably leads to the logarithmic divergence which is eliminated by subtracting the Coulomb energy without the Pauli exchange effects. This term and the background coming from negative charges are correctly canceled out each other. The remaining value is just regarded as the Coulomb exchange term which rapidly converges into a definite value as expanding the domain Γ.

Now, we come to the basic difficulty on considering an infinite number of nucleons, which inevitably include the spurious energy originating from the impossibility of dynamical treatment of system. Therefore, we propose an approximate method to remove the spurious energy. We assume that the α-cluster whose labels $i_1 \sim i_4$ are on the same one i (or the i-th nucleon) in a cavity, which is surrounded by the remaining nucleons, moves freely. Then we solve the equation of motion for α-cluster (or the i-th nucleon) in the cavity in Γ using the Hill-Wheeler variational method as

$$\int < \Psi(\mathbf{S}_i) | \mathcal{H} - \lambda | \Psi(\mathbf{S}_i') > f(\mathbf{S}_i') d\mathbf{S}_i' = 0. \tag{9}$$

The deviation of $\sum_i E_i$ and λ_{min} may correspond to the spurious energy for the centered α-cluster (or the i-th nucleon), thus the spurious free energy is easily evaluated. The weight function, f(S), expresses characteristics of motion for the representative cluster or nucleon. This method can exactly remove the spurious energy in the rarefied limit of nucleon gas. One of the developments of this treatment is to enable us to investigate intercluster potential in a medium within a microscopic framework. The procedure mentioned here can be applicable for infinite number of nucleons even with non periodical property.

3 Validity of model

Taking Fibonacci arrangement of nucleons on a line as a non-periodical example as shown in Fig.2, we verify the convergence of kinetic energy and two-body interaction energies with direct and exchange terms in Fig.3 where h_k, h_d and h_e correspond to above ones. Here, we assume an appropriate two-body operator, and take a unit into the nearest dimensionless distance, d, in the GC space. The radii for Γ and Γ' are 6 and 7, respectively. The remarkable coincidence is given for every term as shown in the comparison in the both sides in Fig.3. The domain, Γ, here contains 4 nucleons on which the Pauli principle works. Of course, as the dimensionless distance, d, becomes shorter, the antisymmetrized effects work for the more number of nucleons. Since the interior nucleon is influenced from the both side nucleons, the Γ contains $7 \sim 9$ nucleons. Anyhow, even the domain in the three dimensional space contains at most 1000 nucleons for the dense region. We can obtain almost the exact result for the normal density in nucleons calculating only 100 nucleons which exert on the representative nucleon placed on the center in the Γ.

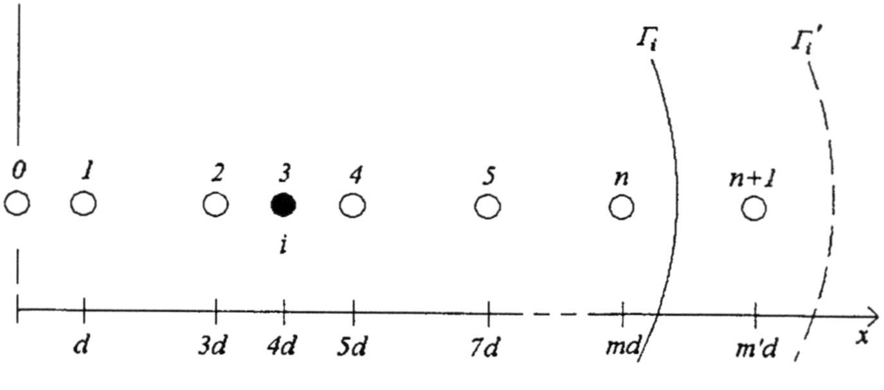

Fig.2, The domaing Γ_i and Γ'_i corresponding to the nucleon with label $i = 3$.

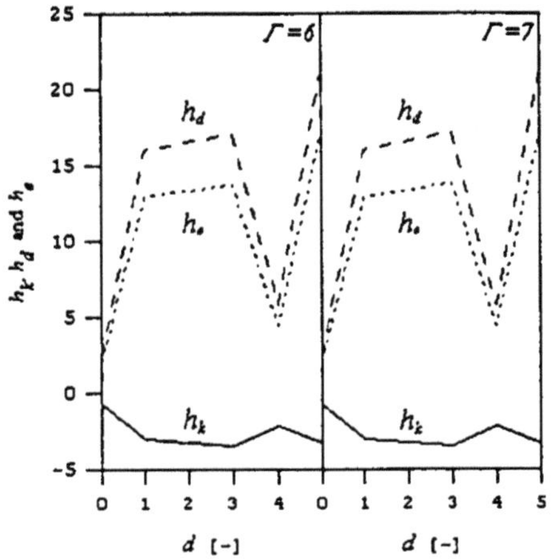

Fig.3, The binding energies, h_k,h_d and h_e for Γ and Γ' which show remarkable coincidence each other.

4 Simple lattice of nucleons

As a special case a static model being comprised of the nucleons on the meshes on the Cartesian coordinate, in short,the static model with the nucleons on the simple lattice can be attributed to one-dimensional problem in order to see the energetic features of the aggregation which is not restricted to an infinite lattice. We have mentioned the details of mathematical foundation in ref.(4). It is also easy to develop this thought to the different densities between protons and neutrons because this model is available for every lattice with same spin-isospin state. Furthermore, an imaginative lattice in the higher dimension is also reduced to linear lattice. Here, we show only the examples for the infinite lattice which is derived from an analytical use of a computer memory.

First of all, we demonstrate the relationship between overlapping matrix B_{ij} and its inverse B_{ij}^{-1}, which can be given only by the relative distance in a line with respect to $|i-j|$. As decreasing the distance, d, the shape of $B_{ij} = f(|i-j|)$ becomes broader, and the inverse, $f(|i-j|)^{-1}$, approaches the δ-function as seen in Fig.4. As increasing, d, the matrix element, $f(|i-j|)$, can fulfill almost the orthogonal condition.

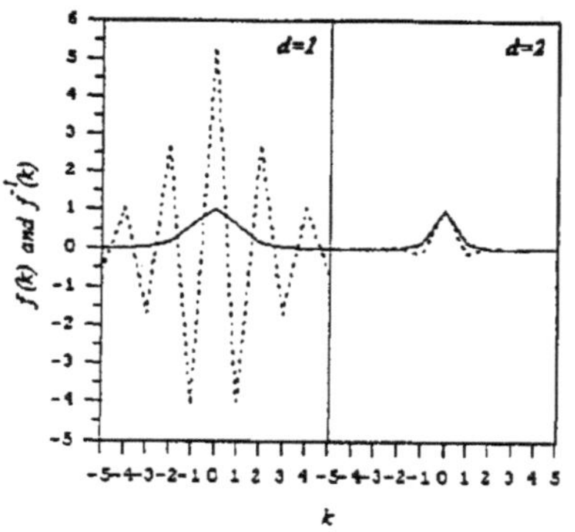

Fig.4, B and B^{-1} matrices for $d = 1$ and $d = 2$.

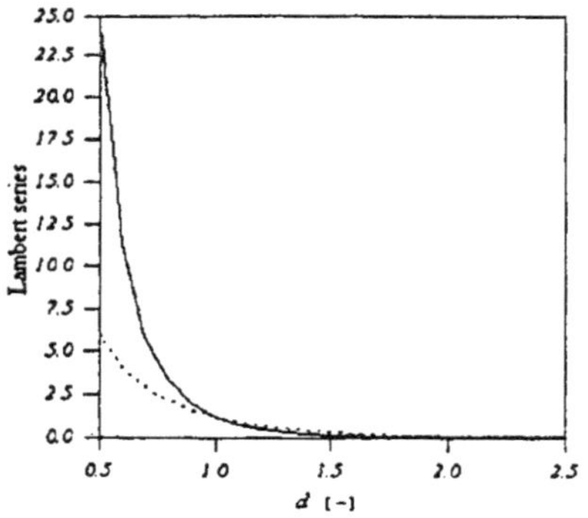

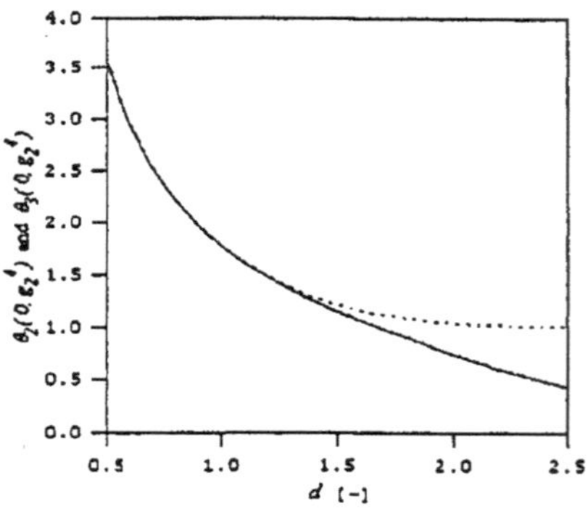

Fig.5, Lambert series.(bold line), that multiplied by d^2 (dashed line)

Fig.6, θ_3 and θ_2 functions with respect to interaction kernel.

The kinetic and interaction energies with differential on $\mathbf{q}$ are eventually ascribed to a Lambert series as shown in Fig.5, which is nicely convergent to zero as increasing distance d. The usual Gaussian operator can lead to elliptic θ_3 (related to dn) and θ_2 (given by cn) functions which correspond to direct and exchange terms respectively. Such characters are illustrated in Fig.6 which can obviously distinguish their exchange effects for the long distance, d. Here, the component in the elliptic functions is fixed to zero due to the orthogonality of the Cartesian meshes of the simple lattice.

5 Some results and discussions

The four kinds of nucleons with a definite spin-isospin state make a different sub-lattice, and a super-lattice including four sub-lattices can express a uniform aggregation of nucleonic sea and α-cluster matter simultaneously. To see the outline of the features of α-clustering in nucleonic sea, we refer to Fig.7 focusing upon the binding energy B_v and B_l per nucleon where suffices stand for the deep-sea part and the surface part in sea.[5] Here, the Skyrme force [6,7] and Brink-Boeker force [8] are employed as the effective internucleon force which reproduces the saturation properties of nuclear matter. The b-parameter on the left side can correspond to that leading to the minimum value of the total binding energy of aggregation. On the other hand, the right side shows the case of that reproducing the minimum binding energy of α particle. The parameter s on the lower line means the distance between neighboring α-clusters, and the parameter t on the upper line does that between neighboring nucleons in uniform density. The correspondence is done in the common density of nucleons. The arrows in the lower part show a watershed which goes to α-clustering phenomena in nucleonic sea. This value cannot reproduce binding energy for nuclear matter because only compact closed packing reaches to the minimum by removing zero-point oscillation energy. However, the critical value is not extremely changed from that pointed out with arrows in Fig.7.

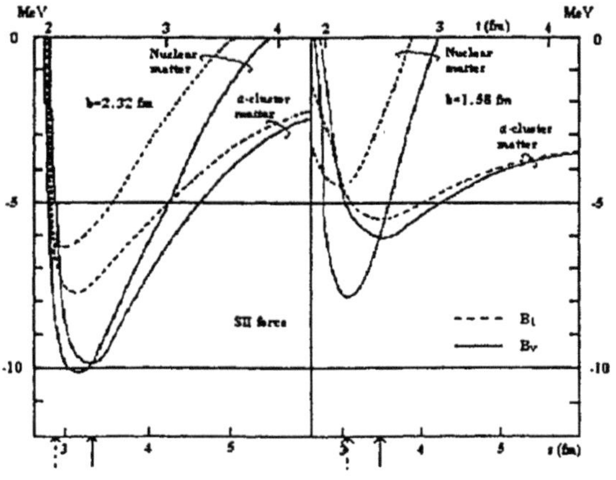

Fig.7, Energy curves per nucleon versus the neighboring distance s. Here, the parameter s has dimension of fm.

The bold arrows indicates the watershed point for deep-sea nucleon which corresponds to about $0.12 fm^{-3}$ (if this distance might be fixed for the ccp/fcc

configuration, the density is about $0.17fm^{-3}$). As for the surface, the critical value is pointed out by the dotted arrows, which are moved inward $0.4fm$. The energy minimum is in the α-clustering region for the surface nucleon, and on contrary, is in uniform density for the deep-sea nucleon. We can also point out that α-cluster becomes fuzzy in deep-sea as is seen from the broad b-parameter in the vicinity of stable energy.

6 Conclusions

A new idea will be expected to expand our knowledge on an aggregation with an infinite number of nucleons. In particular, it enables us to evaluate the effects of the Pauli exclusion principle as correctly as possible. In this talk, we predict that α-clusters float on the surface of nucleonic sea in the normal density as far as the Skyrme force is employed as the internucleon effective force. However, the property of the effective force strongly influences the position of the critical value between α-cluster favored sea and sea with uniform density. Anyhow, the density dependent terms in the effective force plays an essential role in accounting for the occurrence of such a phase transition.

We will suggest elsewhere that the spurious energy coming from the static model is about $-0.22MeV$ per nucleon and the Coulomb exchange energy round to the mass formula is estimated in $-0.09MeV$ by the use of the method mentioned in Sec. 2. Especially, the spurious energy in surface nucleon near the normal density gives an appreciable large value such as $-0.42MeV$ which is near to five times as large as that of deep-sea nucleon. After we add these values to the binding energy obtained from the compact packing (ccp/fcc) of nucleons, we will reproduce the volume and surface terms of Weizaecker's mass formula. Finally we emphasize the importance that various kinds of decays of heavy nuclei may be reinvestigated from a microscopic point of view.

References

1. A. Tohsaki, *Prog. Theor. Phys. Suppl. No.* **132**, 213 (1998).
2. R. Tamagaki, *Prog. Theor. Phys. Suppl. Extra No.* 242 (1968).
3. R.Tamagaki, S. Saito, H. Horiuchi, A. Tohsaki-Suzuki and H. Kamimura, *Prog. Theor. Phys. Suppl. No.* **62** (1977).
4. A. Tohsaki, *Prog. Theor. Phys.* **88**, 1119 (1992).
5. A. Tohsaki, *Prog. Theor. Phys.* **90**, 871 (1993).
6. T.H.R. Skyrme, *Nucl. Phys.* **9**, 615 (1959).
7. D. Vautherin and D.M. Brink, *Phys. Rev.* C **5**, 626 (1972).
8. D.M. Brink and E. Boeker, *Nucl. Phys.* A **91**, 1 (1967).

CLUSTERS, SYMMETRIES, AND PREDICTIONS

J. CSEH[1,2], G. LÉVAI[2], A. INTASORN[3], K. KATŌ[1], A. LÉPINE-SZILY[4], M. S. HUSSEIN[4] AND R. LICHTENTHÄLER[4]

[1] *Graduate School of Science, Hokkaido University, Sapporo 060, Japan*

[2] *Institute of Nuclear Research of the Hungarian Academy of Sciences, H-4001 Debrecen, Pf. 51, Hungary*

[3] *Srinakharinwirot University, Prasarnmit, Suhumvit 23, Bangkok 10110, Thailand*

[4] *IFUSP-Universidade de São Paulo, C.P. 66318, 14098 São Paulo, Brasil*

The basic features of the Semimicroscopic Algebraic Cluster Model are described, and the predictive power of the symmetry-based cluster studies is illustrated.

1 Introduction

Symmetry-based algebraic models proved to be very useful in various fields of nuclear physics. On the one hand side they are able to describe a great amount of experimental data in a consistent way, and on the other hand side they can illuminate the interrelation of different models.

In the present contribution we discuss these features from the viewpoint of cluster studies. The framework of our discussion is provided by the Semimicroscopic Algebraic Cluster Model (SACM) [1,2]. This approach combines a microscopically constructed model space with phenomenologic interactions. Both of them are treated by group theoretical means, and this fact gives rise to an extensive use of symmetries.

In particular, the internal structure of the clusters are described by the U(3) shell model [3], while their relative motion is treated in terms of the U(3) basis of the Vibron Model [4]. The Pauli-forbidden states are excluded from the cluster model space by making an intersection with the fully antisymmetric shell model space of the whole nucleus. In other words, the model space of a cluster configuration is obtained as an U(3) symmetry dictated truncation of the full shell model space. This procedure is based on the equivalence of the Hamiltonians of the shell model and cluster model in the harmonic oscillator approximation [5]. The relation of this approach to other cluster models, including the fully microscopic description [6], and the local potential model [7] have been established.

In the present contribution we concentrate on the predictive power based on the symmetry considerations in relation with cluster studies. The specific examples we discuss are as follows. i) Predictions of the U(3) selection rule on the appearance of exotic cluster configurations in the ground state of light nuclei, and a possible way for their experimental verification. ii) Predictions of a dynamically symmetric Hamiltonian for energy spectra. E.g. similar cluster configurations in different nuclei. iii) Predictions of a Hamiltonian, which is invariant under transformations from one cluster configuration to another one. E.g. different cluster configurations in the same nucleus. iv) The possible role of a Hamiltonian, being invariant with respect to the shell model to cluster model transformation.

2 The model

2.1 Selection rule

The matching or mismatching of the U(3) symmetry of a shell model basis state with those of the cluster model provides us with a selection rule for the presence or absence of this particular cluster configuration. Since many states, including the ground state, of light nuclei are dominated by a single U(3) symmetry, this is an easily applicable selection rule [8]. A two-cluster configuration is allowed if the $[n_1, n_2, n_3]$ Young pattern of the leading representation of the whole nucleus matches with one of the results from the direct product representations of the two clusters and the relative motion:

$$[n_1^{C_1}, n_2^{C_1}, n_3^{C_1}] \otimes [n_1^{C_2}, n_2^{C_2}, n_3^{C_2}] \otimes [n, 0, 0].$$

On the other hand the experimental verification of these predictions is highly nontrivial. E.g. rather exotic cluster configurations, like ^{12}C seem to be present in the ground state of light nuclei, from which the cluster emission is energetically forbidden. However, recent studies show that the interference between the elastic scattering and the elastic transfer processes may provide us with a detectable signature for the cluster structure defined by the entrance channels of these reactions [9]. Both the primary experimental data and the corresponding potentials of the ^{12}C scattering show rather different behaviour depending on the fact if the ground state of the target nucleus allow or not an interference between the elastic scattering and elastic transfer amplitude.

Experiments were carried out for the ^{24}Mg and ^{28}Si target nuclei, out of which the first one overlaps with the (^{12}C+*rest*) clusterization, while the other does not (in the leading order approximation). The oscillatory pattern of angular distribution below and around the Coulomb-barrier, as well as the absorption and notch-test sensitivity of the corresponding optical potentials are fairly different. For the ^{24}Mg case the potential obeys no dispersion relation, while for the ^{28}Si case it does.

Therefore, the interference between the elastic scattering and the elastic transfer seems to be a promising tool for the detection of exotic cluster configurations in the ground states of light nuclei. Further work along this line is highly desirable both from the experimental and from the theoretical sides.

(As for the extension of the U(3) selection rule for the investigation of exotic cluster radioactivity and spontaneous fission of heavy nuclei see Ref. [10], and references therein.)

2.2 *Dynamic symmetry*

The consequences of symmetry can be of great interest not only for specific states, but for a complete spectrum as well. In case the Hamiltonian has a symmetry, i.e. a dynamic symmetry is characterising the system, we may be able to predict the complete energy spectrum of a cluster configuration. An example of this is discussed in Ref. [11], for the core-plus-alpha-particle configuration, applying a Hamiltonian of U(3) dynamic symmetry. In this case the group structure of the problem is given by:

$$\begin{array}{l} U_C^{ST}(4) \otimes SU_C(3) \otimes U_R(4) \supset U_C^S(2) \otimes U_C^T(2) \otimes SU_C(3) \otimes SU_R(3) \\ \mid [f_1^C, f_2^C, f_3^C, f_4^C], (\lambda_C, \mu_C), [N,0,0,0], \quad S_C, \quad T_C, \quad (n_\pi, 0) \\ \supset U_C^S(2) \otimes U_C^T(2) \otimes SU(3) \supset SU_C^S(2) \otimes SU_C^T(2) \otimes O(3) \supset SU_C^T(2) \otimes SU(2) \\ \qquad (\lambda, \mu), \qquad K, L, \qquad J\rangle. \end{array} \quad (1)$$

Here we also indicated the quantum numbers belonging to the representations of the groups.

The following Hamiltonian (expressed in terms of invariant operators and quantum numbers of group-chain (1)) was applied for the description of the energy spectra:

$$H = \epsilon + \gamma_R \hat{n}_\pi + \delta_R C^{(2)}_{SU_R(3)} + \delta_C C^{(2)}_{SU_C(3)} + \delta C^{(2)}_{SU(3)} + \theta K^2 + \beta \hat{L}^2 + \kappa \hat{L} \cdot \hat{S}. \quad (2)$$

Here $C^{(2)}$ stands for the second order Casimir invariant of the indicated group. The mass-dependence of its parameters was found to be smooth for the ^{12}C, ^{14}C, ^{15}N and ^{16}O core nuclei. This fact allows one to make an interpolation of the parameters for another nucleus, which was not included in the fitting procedure. Therefore, we can obtain a Hamiltonian which has no free parameter at all. Similarly, the microscopic model space of the cluster configuration is fully determined by the procedure mentioned above. In such a way a complete energy spectrum can be predicted, and for the ^{13}C+α system the agreement of this kind of prediction with the experimental data proved to be fairly satisfactory.

2.3 Multichannel dynamic symmetry

A more complex symmetry arises, when the U(3) dynamically symmetric Hamiltonian of a cluster system is invariant with respect to the transformation, which takes from one cluster cofiguration to another one. This kind of symmetry is called multichannel dynamic symmetry [12]. If this very strict symmetry holds, then one can predict the energy spectrum of a culster configuration from that of another one, or predict e.g. level densities for a specific reaction channel from the other part of the energy spectrum.

Recently a calculation of this kind was carried out for the ^{32}S nucleus [13]. ^{28}Si+α and ^{16}O+^{16}O cluster configurations were taken into account. Three energy regions were described within the same framework: the ground-sate region, including both positive and negative parity states, the molecular resonance region of the ^{16}O+^{16}O reaction, and the α-particle resonant scattering energy region in between. (Previous theoretical studies were dealing only with one energy-range, and no unified treatment was done.) In our calculation the number of resonances with definite spin-parity were predicted by the multichannel symmetry, based on the (unified) description of the two other regions. The densities are determined for the energy regions, where the resonant scattering experiments were made, and the agreement turned out to be remarkably good.

2.4 Shell model invariance

The invariance of the Hamiltonian with respect to transformation between different configurations raises a very interesting question. Namely, that of the shell-model invariant cluster-cluster interactions. In other words: whether or not one can find a cluster model Hamiltonian, which is, on the one hand side, completely invariant with respect to the transformation to the shell model, and on the other hand side, realistic enough to reasonably describe the cluster spectrum.

Similar studies regarding the connection between the spherical shell model and quadrupole collective model have been carried out extensively. They resulted in a well-established derivation of the collective model from the shell model, based on realistic interactions. However, the connection between the shell model and the cluster model was discussed only for harmonic oscillator forces in the early days [5]. This is completely satisfactory for relating the U(3) basis states of the two approaches, however, it is very far from being realistic for energy spectrum calculation. To find a realistic and yet shell-model invariant cluster Hamiltonian is a very challenging task. This would establish the connection of the cluster model not only to the shell model, but via the shell model also to the quadrupole collective model for realistic interactions. The symmetry-based algebraic cluster model seems to be a very suitable framework for this purpuse as well. Work is in progress along this line [14].

3 Conclusions

The Semimicroscopic Algebraic Cluster Model provides us with a suitable framework for carrying out symmetry-based cluster studies. It gives a systematic description of a large amount of experimental data. The model seems to perform fairly well in dealing with the gross features of the cluster spectrum in a large energy interval. Typical examples are the unified description of excitation spectra starting with the ground state and including the region of the quasimolecular resonances [15,16]. It is less suited for the detailed studies of specific states. The predictive power of the model is especially remarkable.

On the other hand side it can also contribute considerably to our understanding of the relations between the basic nuclear structure models, i.e. between the cluster model, shell model and quadrupole collective model.

Acknowledgments

This work was supported in part by the OTKA (Grant No. T22187).

References

1. J. Cseh, *Phys. Lett.* B **281**, 173 (1992).
2. J. Cseh and G. Lévai, *Ann. Phys.* (N.Y.) **230**, 165 (1994).
3. J. P. Elliott, *Proc. R. Soc. London* A **245**, 128 (1958); **245**, 562 (1958).
4. F. Iachello and R. D. Levine, *J. Chem. Phys.* **77**, 3046 (1982).
5. K. Wildermuth and Th. Kanellopoulos, *Nucl. Phys.* **7**, 150 (1958).
6. K. Varga and J. Cseh, *Phys. Rev.* C **48**, 602 (1993).
7. P. O. Hess, G. Lévai and J.Cseh, *Phys. Rev.* C **54**, 2345 (1996).
8. J. Cseh, W. Scheid, *J. Phys.* G **18**, 1419 (1992), *ibid.* **19**, L97 (1993).
9. A. Lépine-Szily *et al*, *Phys. Rev. Lett.* **82**, 3972 (1999).
10. P. O. Hess *et al*, contribution to this conference.
11. G. Lévai and J. Cseh, *Phys. Lett.* B **381**, 1 (1996).
12. J. Cseh, Phys. Rev. C **50**, 2240 (1994).
13. J. Cseh, G. Lévai, A. Ventura and L. Zuffi, *Phys. Rev.* C **58**, 2144 (1998).
14. J. Cseh, A. Intasorn, K. Katō, work in progress.
15. J. Cseh, G. Lévai, W. Scheid, *Phys. Rev.* C **48**, 1724 (1993).
16. J. Cseh *et al*, to be published in *Few-Body Systems*.

INTERACTING BOSON PLUS BROKEN PAIR AND INTERACTING BOSON FERMION FERMION MODEL APPROACH TO HIGH-SPIN STATES

S. BRANT, D. VRETENAR, V. PAAR, AND V. KRSTIĆ
Department of Physics, Faculty of Science, University of Zagreb, 10000 Zagreb, Croatia

Nuclear structure of high-spin states is described in the Interacting boson plus broken pair and Interacting boson fermion fermion model. Main features of the models and recent results are presented.

1 Introduction

The interacting boson approximation (IBA) model[1] in its various versions has been remarkably successful in the description of a variety of nuclear structure phenomena. In its basic versions (IBM-1[1], IBM-2[2], IBFM[3] and IBFFM[4]), this approach has been applied mostly in the description of low-spin states in medium heavy and heavy nuclei. Various extensions that include selective noncollective two-fermion states in addition to s and d bosons can be used to describe the structure of high-spin states.

Here we present some recent results of our calculations in the Interacting boson plus broken pair model (IBBPM) for even-even[5] and Interacting boson fermion plus broken pair model (IBFBPM) for odd-even nuclei[6]. Both models are based on the IBM-1[1], i.e. the boson space consists of s and d bosons, with no distinction between protons and neutrons. Extensive calculations in IBBPM and IBFBPM have been performed in the Hg, neutron defficient Sr-Zr, Nd-Sm, neutron rich Y-Nb, Ga and Cd regions, i.e. for a variety of spherical, transitional and deformed nuclei. In our approach the ground state band and bands built on broken pairs are obtained simultaneously from a single set of parameters and a well defined Hamiltonian, in a consistent calculation that does not include any geometrical picture assumption. Polarization effects directly result from the model fermion-boson interactions. All calculations are performed in the laboratory frame, and therefore the results can be directly compared with experimental data. This is relevant for transitional and especially for spherical nuclei, where the cranking approach to high-spin physics is not adequate.

To generate high-spin states in even-even nuclei, the model allows one or two bosons to be destroyed and to form non-collective fermion pairs represented by two- and four-quasiparticle states which recouple to the boson core. High-spin states are described in terms of broken pairs. The model space for an even-even nucleus, in addition to boson states, contains two-proton, two-neutron, four-proton, four-neutron and two-proton-two-neutron configurations. In description of high-spin states, polarization effects can be expected, for which the standard form of the boson fermion interaction derived in the lowest seniority approximation [3] is inadequate. Therefore we have modified the boson-fermion dynamical interaction extending the standard boson quadrupole operator Q^B by the higher order term [7].

2 IBBPM, IBFBPM and IBFFM calculations

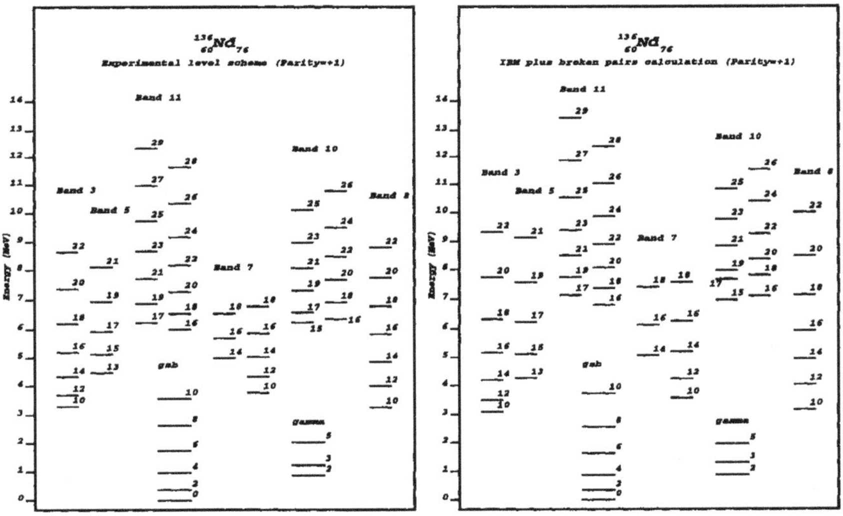

Figure 1: Experimental (left) and calculated IBBPM (right) positive-parity bands in ^{136}Nd.

In the beginning, IBBPM was applied to deformed and γ-soft nuclei. Later on, the interest was shifted to transitional and spherical nuclei (like Cd isotopes [8]). Recently we have used IBBPM to describe the excitation spectrum of ^{136}Nd. Details of calculations can be found in Ref. [9]. The IBBPM calculation (Figure 1) reproduces the experimental positions of states of the ground state band, the quasi γ-band, the $(\pi \tilde{h}_{11/2})^2$ bands, the $(\nu \tilde{h}_{11/2})^2$ band, and the two dipole bands with the configuration $(\pi \tilde{h}_{11/2})^2$ $(\nu \tilde{h}_{11/2})^2$. In the calculation of negative parity states, two $(\nu \tilde{d}_{3/2}, \nu \tilde{h}_{11/2})$ bands and one $\nu \tilde{d}_{3/2}(\nu \tilde{h}_{11/2})^3$ band have been identified. In IBBPM we interpret regular dipole bands ($\Delta J = 1$) as two-proton-two-neutron structures. We have found that the interplay of the proton-neutron surface δ-interaction and the additional term in the boson quadrupole operator play a crucial role in the excitation spectrum of these bands.

For odd-A nuclei, two-broken pair configurations are not included in the model space. They would generate five-quasiparticle configurations resulting in exhaustive numerical calculations. In IBFBPM [6] the two fermions in the broken pair can be of the same type as the unpaired fermion, resulting in a space with three identical fermions. If the fermions in the broken pair are different from the unpaired fermion, the fermion basis contains two protons and one neutron or vice versa.

Extensive calculations of excitation energies and electromagnetic transitions have been performed for low- and high-spin states (simultaneously, with the same set of parameters) in ^{139}Sm [10] and ^{137}Nd [7] where $(\nu\tilde{h}_{11/2})^3$ and $\nu\tilde{h}_{11/2}(\pi\tilde{h}_{11/2})^2$ bands were successfuly described.

The sequence of odd-even Ga nuclei is a challenge for the IBFBPM because broken-pair states can be yrast at a relatively low energy. Details of calculation for ^{65}Ga and ^{67}Ga can be found in Ref. [11]. IBFBPM calculations have been performed in complete 28 - 50 shells. The quasiparticle energies, occupation probabilities and interaction strengths have been consistently taken from a series of previous calculations for neighbouring nuclei (Figure 2).

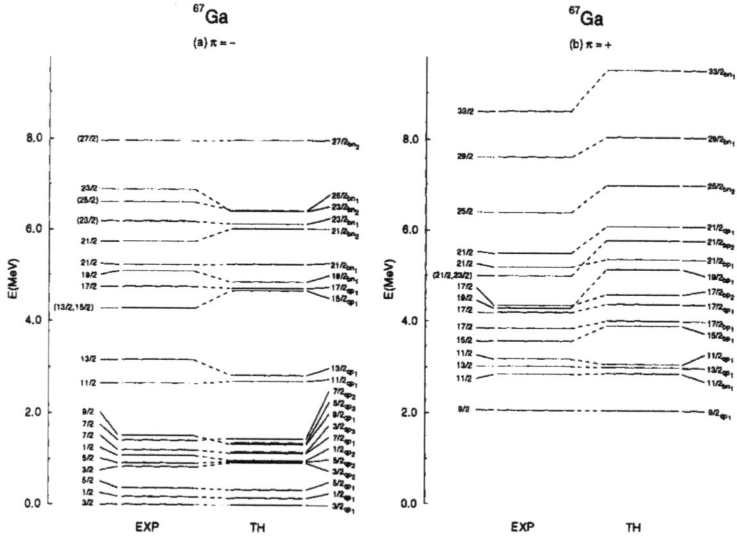

Figure 2: Calculated IBFBPM negative-parity (left) and positive-parity (right) states in ^{67}Ga, in comparison to data.

Recently the IBFBPM was applied to ^{97}Y [12] in order to find the structure and decay properties of the $27/2^-$ isomer at 3523 keV. As a result, the speculated $[\pi\tilde{g}_{9/2},(\nu\tilde{g}_{7/2},\nu\tilde{h}_{11/2})9]27/2^-$ configuration was confirmed. An isomer with the same structure is predicted in ^{99}Nb [12].

Up to now, IBFFM for odd-odd nuclei was not extended by including broken pairs. The model bases for realistic nuclei would be far too large. Nevertheless, IBFFM in its original form has been successfully applied for description of one-proton-one-neutron high-spin states [4].

As an example of IBFFM calculations of high-spin states, we present the results for ^{72}As [13] (Figure 3). The energy splittings of the observed multiplets are in accordance with the prediction of the IBFFM calculations using a transitional core lying between the vibrational SU(5) and γ-soft O(6) limits, except that of the $\pi\tilde{g}_{9/2}\nu\tilde{g}_{9/2}$ multiplet, which was described using the deformed prolate SU(3) core.

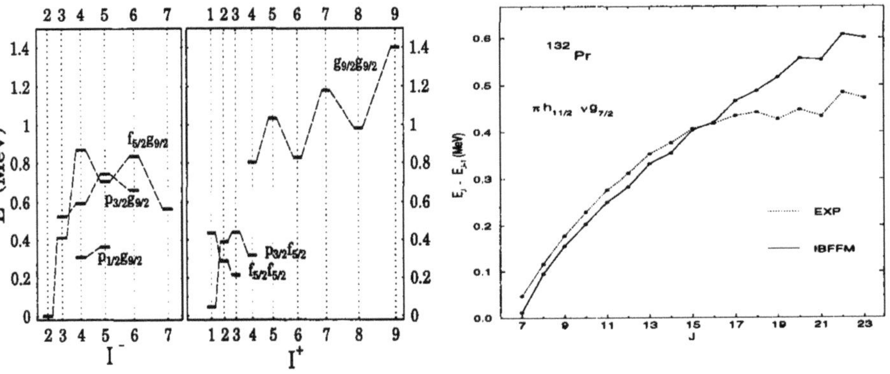

Figure 3: Proton-neutron multiplet states in ^{72}As (left). $E_J - E_{J-1}$ *vs* J plot for the $\pi\tilde{h}_{11/2}\nu\tilde{g}_{7/2}$ band in ^{132}Pr (right).

The formation of regular $\Delta J = 1$ high-spin band patterns in odd-odd nuclei with O(6) boson core, and even more the signature inversion in such bands, present a challenging problem for IBFFM. Namely, IBFFM predicts normal or decoupled band patterns with close-lying unfavored and favored bands, resulting in characteristic "doublet structures" and $\Delta J = 2$ transitions. This is because the standard IBFFM interactions can not induce a sufficiently strong effective interaction which would establish a regular $\Delta J = 1$ band pattern, without sizeably altering the low-energy part of the spectrum. In Ref. [14] detailed IBFFM calculations have been performed for ^{130}Pr and ^{132}Pr. The change from inverted to normal signature at high spin and the staggering pattern of all observed bands were well reproduced (Figure 3), being explained by the interplay between a new term introduced in the boson quadrupole operator and the residual quadrupole-quadrupole interaction. In our approach we have obtained the signature inversion in a consistent calculation that simultaneously gave all known states (both low- and high-spin) with the same parameters.

3 Conclusion

The extension of IBM-1 based models with broken pairs provide a framework for description of high-spin states in even-even and odd-even nuclei. In odd-odd nuclei the standard IBFFM provides a rather good simultaneous description of both low- and high-spin states.

References

1. F. Iachello and A. Arima, *The Interacting Boson Model* (Cambridge University Press, Cambridge, 1987); A. Arima and F. Iachello, *Phys. Rev. Lett.* **35**, 1069 (1975).
2. A. Arima *et al*, *Phys. Lett.* B **66**, 205 (1977).
3. F. Iachello and O. Scholten, *Phys. Rev. Lett.* **43**, 679 (1979); F. Iachello and P. Van Isacker, *The Interacting Boson-Fermion Model* (Cambridge University Press, Cambridge, 1991).
4. V. Paar, in *In-beam nuclear spectroscopy*, Vol.2, eds. Zs. Dombrádi and T. Fényes (Akadémiai Kiadó, Budapest, 1984) p.675; S. Brant *et al*, *Z. Phys.* A **319**, 355 (1984); V. Paar *et al*, A **327**, 291 (1987); S. Brant and V. Paar, **329**, 151 (1988).
5. F. Iachello and D. Vretenar, *Phys. Rev.* C **43**, R945 (1991); D. Vretenar, G. Bonsignori, and M. Savoia, *Phys. Rev.* C **47**, 2019 (1993).
6. D. Vretenar, G. Bonsignori, and M. Savoia, *Z. Phys.* A **351**, 289 (1995).
7. C.M. Petrache *et al.*, *Nucl. Phys.* **A617**, 228 (1997).
8. G. de Angelis *et al.*, *Nucl. Phys.* **A**, in press.
9. D. Vretenar *et al.*, *Phys. Rev.* C **57**, 675 (1998).
10. C. Rossi Alvarez *et al.*, *Phys. Rev.* C **54**, 57 (1996).
11. I. Dankó *et al.*, *Phys. Rev.* C **59**, 1956 (1999).
12. G. Lhersonneau *et al*, *Phys. Rev.* C **57**, 681 (1998); **57**, 2974 (1998).
13. D. Sohler *et al*, *Phys. Rev.* C **59**, 1328 (1999).
14. C.M. Petrache *et al*, *Nucl. Phys.* A **635**, 361 (1998).

SPONTANEOUS NUCLEAR CLUSTERING AND ATOMIC-ELECTRON EFFECTS IN DEEP INELASTIC HEAVY-ION COLLISIONS

SERGEY YU. KUN[1,2]

[1] *Department of Theoretical Physics, RSPhysSE, IAS, The Australian National University, Canberra ACT 0200, Australia*

[2] *MPI für Kernphysik, Heidelberg D-69029, Germany*

The statistical reactions with memory approach [1,2] suggests that, for high excitations corresponding to strongly overlapping resonances of the intermediate system, spontaneous correlations between the relevant matrix elements with different total spin and parity values result in the formation of strongly nonuniform clusters in the available phase space. The approach [1,2] provides an interpretation [3,4] of the universal phenomenon of the nonself-averaging of excitation function oscillations (EFO) in strongly dissipative heavy-ion collisions (DHIC) in terms of the spontaneous clustering, the spontaneous micro-channel correlations (MC), the slow decoherence and space-time delocalization of a highly excited dinucleus. This effect of nonself-averaging has been observed [5-7] in the ^{28}Si+^{64}Ni, ^{19}F+^{89}Y, ^{28}Si+^{48}Ti, ^{19}F+^{63}Cu, ^{19}F+^{45}Sc, ^{19}F+^{51}V, ^{28}Si+^{28}Si, ^{27}Al+^{27}Al, ^{19}F+^{27}Al DHIC.

In this contribution we argue that the spontaneous MC and slow spin decoherence results in the cross sections for DHIC being sensitive to an arbitrarily small perturbation. In particular, we predict that atomic electrons should influence energetic ($\simeq$ 100 MeV) heavy-ion reactions. These atomic-electron effects in DHIC are expected to be $\simeq$ 100% [8] of the magnitude of the nonself-averaged oscillating component of the nuclear cross sections and therefore should be easily measured.

We treat the DHIC in terms of the formation and decay of a highly excited intermediate system. The normalized S-matrix elements are taken to be of the form $\bar{S}^J_{\bar{a}\bar{b}}(E) = (\Gamma D/2\pi)^{1/2} \sum_\mu \bar{\gamma}^{J\bar{a}}_\mu \bar{\gamma}^{J\bar{b}}_\mu/(E - E^J_\mu + i\Gamma/2)$ [2]. Here, E is the total energy, J is the total spin, Γ and D are the total decay width and the average level spacing, and E^J_μ are the resonance energies of the highly excited ($D/\Gamma \ll 1$) intermediate system. The $\bar{a}(\bar{b})$ indices specify intrinsic microstates $\bar{a}(\bar{b})$ of the reaction partners in the entrance (exit) channels, and $\bar{\gamma}^{J\bar{a}(\bar{b})}_\mu = \sum_j B^J_{\mu_j} \bar{\xi}^{J\bar{a}(\bar{b})}_j$, where both $\bar{\gamma}^{J\bar{a}(\bar{b})}_\mu$ and $\bar{\xi}^{J\bar{a}(\bar{b})}_j$ are normalized random Gaussian variables with a mean value of zero. The quantities $B^J_{\mu_j}$ with a fixed J-value are the orthogonal matrices arising from the expansion $\phi^J_\mu = \sum_j B^J_{\mu_j} X^J_j$ of the eigenstates ϕ^J_μ of the intermediate nuclear system over the

Slater determinants X_j^J [2]. The $\bar{\gamma}_\mu^{J\bar{a}(\bar{b})}$ are the partial width amplitudes for the transitions between the entrance (exit) channel and the resonance states, ϕ_μ^J, due to the interaction V, and the $\bar{\xi}_j^{J\bar{a}(\bar{b})}$ are the corresponding amplitudes for the transitions between the entrance (exit) channel and the states X_j^J. The $\bar{\gamma}$'s, $\bar{\xi}$'s and $\bar{S}_{\bar{a}\bar{b}}^J(E)$ do not depend on the orbital momenta and channel spins due to the continuum-continuum correlation-coupling [1,2].

Let us consider the regime of relatively slow spin decoherence, $g\beta \leq \Gamma$ [3,4,7], where β is the spin decoherence width and g is the number of partial waves contributing to the reaction. Following [3] we use the decomposition

$$\bar{S}_{\bar{a}\bar{b}}^J(E) = \Delta S_{\bar{a}\bar{b}}^{J,I}(E) + \delta S_{\bar{a}\bar{b}}^{J,I}(E), \tag{1}$$

where I is the average spin and $\delta S_{\bar{a}\bar{b}}^{J,I}(E)$ and $\Delta S_{\bar{a}\bar{b}}^{J,I}(E)$ are given by Eqs. (12,13) in Ref. [3]. Due to the absence of correlations between $\Delta S_{\bar{a}\bar{b}}^{J,I}(E)$ and $\Delta S_{\bar{a}'\bar{b}'}^{J,I}(E)$ and between $\Delta S_{\bar{a}\bar{b}}^{J,I}(E)$ and $\delta S_{\bar{a}'\bar{b}'}^{J,I}(E)$, the correlation between $\bar{S}_{\bar{a}\bar{b}}^J(E)$ and $\bar{S}_{\bar{a}'\bar{b}'}^J(E)$ with $\bar{a} \neq \bar{a}'$ and/or $\bar{b} \neq \bar{b}'$ originates from the correlation between $\delta S_{\bar{a}\bar{b}}^{J,I}(E)$ and $\delta S_{\bar{a}'\bar{b}'}^{J,I}(E)$ [3]. From Eqs. (9,14,15,16) in Ref. [3] we obtain

$$\overline{\delta S_{\bar{a}\bar{b}}^{J,I}(E)\delta S_{\bar{a}'\bar{b}'}^{J,I}(E)^*} = \overline{|\delta S_{\bar{a}\bar{b}}^{J,I}(E)|^2} = \overline{|\delta S_{\bar{a}'\bar{b}'}^{J,I}(E)|^2} = 2|J-I|\beta/(\Gamma + 2|J-I|\beta), \tag{2}$$

where the overlines stand for ensemble averaging which is equivalent to energy averaging. Therefore, since the $\delta S_{\bar{a}\bar{b}}^{J,I}(E)$ are Gaussian stationary stochastic processes [1-3] we have $\delta S_{\bar{a}\bar{b}}^{J,I}(E) = \delta S_{\bar{a}'\bar{b}'}^{J,I}(E) \equiv \delta S^{J,I}(E)$ for all $\bar{a} \neq \bar{a}'$ and $\bar{b} \neq \bar{b}'$. This is obtained without any additional assumptions compared with those leading to the nonvanishing of the MC between $\bar{S}_{\bar{a}\bar{b}}^J(E)$ and $\bar{S}_{\bar{a}'\bar{b}'}^{J'}(E)$ [3]. This MC originates from the infinitesimally small ($|\lambda| \to 0$) entrance-exit correlation $\overline{\bar{\xi}_i^{J\bar{a}}\bar{\xi}_j^{J'\bar{b}}} = \lambda K_{ij}^{JJ'}(\bar{a},\bar{b})/N^{1/2}$ [1,2], where $K_{ij}^{JJ'}(\bar{a},\bar{b})$ are symmetric matrices whose elements are of the order of ± 1, and N is the dimension of the Hilbert space. It should be noted that the $K_{ij}^{JJ'}(\bar{a},\bar{b})$-matrices with different $(\bar{a},\bar{b})$ are different and independent. Yet the $\delta S_{\bar{a}\bar{b}}^{J,I}(E)$ with different $(\bar{a},\bar{b})$ are identical. Clearly this is possible only in the limit $|\lambda| \to 0$. Indeed then, and only then, the $K_{ij}^{JJ'}(\bar{a},\bar{b})$-matrices with different $(\bar{a},\bar{b})$ become indistinguishable so that we can consider these matrices to be $(\bar{a},\bar{b})$-independent: $K_{ij}^{JJ'}(\bar{a},\bar{b}) = K_{ij}^{JJ'}(\bar{a}',\bar{b}') = K_{ij}^{JJ'}$. The loss of information on the differences between $K_{ij}^{JJ'}(\bar{a},\bar{b})$ with different $(\bar{a},\bar{b})$ is a precondition for obtaining the $(\bar{a},\bar{b})$-independence of the correlation coefficient $\rho_{\mu\nu}^{JJ'} = \overline{\bar{\gamma}_\mu^{J\bar{a}}\bar{\gamma}_\nu^{J'\bar{b}}}$. On the other hand, switching off the correlation between $\bar{\xi}_i^{J\bar{a}}$ and $\bar{\xi}_j^{J'\bar{b}}$ does not result in the vanishing of $\rho_{\mu\nu}^{JJ'}$ if the limit $|\lambda| \to 0$ is taken simultaneously with the limit $N \to \infty$ [1,2].

Both the $(\bar{a}, \bar{b})$-independence and the nonvanishing of $\rho_{\mu\nu}^{JJ'}$ are necessary conditions to obtain the correlation between $\bar{S}_{\bar{a}\bar{b}}^{J}(E)$ and $\bar{S}_{\bar{a}'\bar{b}'}^{J'}(E)$ [3] which manifests itself in the universal phenomenon of the nonself-averaging of EFO for DHIC [5-7].

We observe that the correlation between $\delta S_{\bar{a}\bar{b}}^{J,I}(E)$ and $\delta S_{\bar{a}'\bar{b}'}^{J,I}(E)$ occurs spontaneously. This correlation reflects the instability of the noncorrelation condition between $\bar{\gamma}_{\mu}^{J\bar{a}}$ and $\bar{\gamma}_{\nu}^{J'\bar{b}}$ with respect to an arbitrarily small correlation between $\bar{\xi}_{i}^{J\bar{a}}$ and $\bar{\xi}_{j}^{J'\bar{b}}$ in the limit when this correlation vanishes, i.e. $\lim_{|\lambda|\to 0, N\to\infty}(\lambda/N^{1/2})K_{ij}^{JJ'}(\bar{a},\bar{b}) = 0$. This resembles the case of spontaneous magnetization below the Curie point (see, e.g., Ref. [9]). The spontaneous magnetization does not vanish if the limit of a vanishing external magnetic field is properly supplemented by the thermodynamic limit (infinite number of degrees of freedom and infinite volume of the system).

Let us ask: what is the preferential "direction" $\delta S^{J,I}(E)$ along which all the $\delta S_{\bar{a}\bar{b}}^{J,I}(E)$ with different $(\bar{a}, \bar{b})$ are spontaneously "oriented" and which determines the detailed energy dependence of the nucleus-nucleus cross section summed over a very large number of exit microchannels? It may seem that the $\delta S^{J,I}(E)$ are determined by a particular choice of the $K_{ij}^{JJ'}(\bar{a},\bar{b})$-matrices. However this is not so. Indeed, the genuine indistinguishability between the $K_{ij}^{JJ'}(\bar{a},\bar{b})$ with different $(\bar{a},\bar{b})$, which is necessary to obtain the MC relation (2) and thereby the nonself-averaging for EFO in DHIC [5-7], can only be achieved by taking the limits $N \to \infty$ and $|\lambda| \to 0$. But these very same limits inevitably result in the loss of information about the individual $K_{ij}^{JJ'}(\bar{a},\bar{b})$-matrices for any $(\bar{a},\bar{b})$, due to the vanishing of the correlations $\overline{\bar{\xi}_{i}^{J\bar{a}}\bar{\xi}_{j}^{J'\bar{b}}} = \lambda K_{ij}^{JJ'}(\bar{a},\bar{b})/N^{1/2} = 0$ in the limits $N \to \infty$ and $|\lambda| \to 0$. Consequently, after these limits have been taken, none of the observable physical phenomena such as the detailed energy dependence of the cross sections is determined by a particular choice of the $K_{ij}^{JJ'}(\bar{a},\bar{b})$-matrices. Therefore the preferential "direction" $\delta S^{J,I}(E)$ is set up at random reflecting the extreme sensitivity of the detailed energy dependence of the $\delta S^{J,I}(E)$ to an arbitrarily small ($|\lambda| \to 0$) perturbation. This leads to a sensitivity of the cross sections. However the decaying cross section energy autocorrelation functions are insensitive since $\overline{\bar{S}_{\bar{a}\bar{b}}^{J}(E+\varepsilon)\bar{S}_{\bar{a}'\bar{b}'}^{J'}(E)^{*}}$ with $\bar{a} \neq \bar{a}'$ and/or $\bar{b} \neq \bar{b}'$ are $(\bar{a} \neq \bar{a}', \bar{b} \neq \bar{b}')$-independent and stable [3]. Both the sensitivity of the cross sections and the decaying energy autocorrelation functions are manifestations of quantum chaos [10,11] in strongly dissipative nucleus-nucleus collisions.

The sensitivity of the cross sections for DHIC resembles the sensitivity of the direction of the spontaneous magnetization vector to the direction of an arbitrarily small external magnetic field in the limit when this field vanishes. However the absolute value of the spontaneous magnetization vector is stable and insensitive. Yet there is a difference between the sensitivity of the direction of the spontaneous magnetization vector and the sensitivity of the cross sections for DHIC. This is because spontaneous magnetization and the sensitivity of the direction of its vector can only be obtained in the thermodynamic limit. In contrast, the sensitivity of the cross sections for DHIC is obtained in the limit $N \to \infty$ but for a finite intermediate nuclear system.

The precondition for a sensitivity of the cross sections of DHIC is a finite β, i.e. the presence of quantum chaos and spin decoherence [1-3]. Indeed, in the limit $\beta/\Gamma \to 0$, $\overline{|\delta S^{J,I}(E)|^2} \propto \beta/\Gamma \to 0$ and $\bar{S}^{J}_{\bar{a}\bar{b}}(E) \to \Delta S^{J,I}_{\bar{a}\bar{b}}(E)$ resulting in both insensitivity and self-averaging of the cross sections. The cross sections are also insensitive and self-averaging in the compound nucleus limit $\Gamma/\beta \to 0$, since in this limit the MC vanishes [3] leading to the randomness conditions in the statistical model of compound nucleus scattering [12,13], theory of quantum chaotic scattering and random-matrix theory [13-18].

The sensitivity of the cross sections for DHIC implies that, e.g., atomic electrons should influence the nucleus-nucleus collision. In particular, we predict that the cross sections should be different for the same kinetic energy but for different charges of the ions in the incident beam. For example, Refs. [6,7] report measurements of the cross sections for $^{19}F^{+8}+^{89}Y$ strongly dissipative collisions. Let us consider the same collision, for the same range of kinetic energy of the incident ^{19}F-beam but with the charge of the ^{19}F ions being +7. The target consists of neutral atoms of Yttrium. Therefore the binding energy of an outer electron of the intermediate $^{19}F+^{89}Y$ system $\simeq$ 100 eV while the excitation energy of this intermediate nuclear system $\simeq$ 100 MeV. Consequently one does not expect a detectable difference between the cross sections for $^{19}F^{+7}+^{89}Y$ and $^{19}F^{+8}+^{89}Y$ strongly dissipative collisions. Yet, due to the extreme sensitivity of the nucleus-nucleus cross sections, this difference is predicted [8] to be $\simeq$ 100% of the magnitude of the nonself-averaged oscillating component of the cross section. For $^{19}F+^{89}Y$ DHIC the relative magnitude of this oscillating component is about $\pm$15% [6,7]. Therefore this atomic-electron effect should be detectable as reliably as the effect of the nonself-averaging of EFO for DHIC [5-7].

In conclusion, we have argued that spontaneous microchannel S-matrix correlations and slow spin decoherence lead to a sensitivity of the cross sections for DHIC to an arbitrarily small perturbation in the limit when this perturbation vanishes. Such a sensitivity implies that atomic electrons should influence the cross sections for DHIC. These atomic-electron effects are predicted to be $\simeq$ 100% of the magnitude of the nonself-averaged oscillating component of the nucleus-nucleus cross section. New experiments are being proposed to test the predicted possibility of controlling energetic ($\simeq$ 100 MeV) nucleus-nucleus reactions by changing the atomic electron environment.

Acknowledgments

This work has been supported in part by the Humboldt Fellowship and MPI Fellowship, Heidelberg. A Visiting Fellowship, Department of Theoretical Physics, RSPhysSE, IAS, ANU is acknowledged.

References

1. S.Yu. Kun, *Z. Phys.* A **357**, 255 (1997).
2. S.Yu. Kun, *Z. Phys.* A **357**, 271 (1997).
3. S.Yu. Kun and A.V. Vagov, *Z. Phys.* A **359**, 137 (1997).
4. S.Yu. Kun, V.Yu. Denisov, and A.V. Vagov, *Z. Phys.* A **359**, 257 (1997).
5. A. De Rosa *et al*, *Phys. Lett.* B **160**, 239 (1985); F. Rizzo *et al*, *Z. Phys.* A **349**, 169 (1994); Wang Qi *et al*, *Phys. Lett.* B **388**, 462 (1996); I. Berceanu *et al*, *Phys. Rev.* C **57**, 2359 (1998).
6. T. Suomijärvi *et al*, *Phys. Rev.* C **36**, 181 (1987).
7. S.Yu. Kun *et al*, *Z. Phys.* A **359**, 263 (1997).
8. S.Yu. Kun, submitted for publication.
9. N.N. Bogoliubov, *Lectures on Quantum Statistics*, Vol. 2, (Gordon and Breach, New York, 1970), p. 8.
10. R. Blümel, *Phys. Rev. Lett.* **73**, 428 (1994).
11. R. Blümel, *Nucl. Phys.* A **572**, 9 (1994).
12. T. Ericson, *Ann. Phys.* (N.Y.) **23**, 390 (1963).
13. J.J.M. Verbaarschot, H.A. Weidenmüller, and M.R. Zirnbauer, *Phys. Rep.* **129**, 367 (1985), and references therein.
14. R. Blümel and U. Smilansky, *Phys. Rev. Lett.* **60**, 477 (1988); *Phys. Rev. Lett.* **64**, 241 (1990).
15. V.V. Flambaum, A.A. Gribakina, and G.F. Gribakin, *Phys. Rev.* A **54**, 2066 (1996).
16. Y.V. Fyodorov and H.-J. Sommers, J. Math. Phys. **38**, 1918 (1997), and references therein.
17. Y.V. Fyodorov and Y. Alhassid, *Phys. Rev.* A **58**, R3375 (1998).
18. T. Guhr, A. Müller-Groeling, and H.A. Weidenmüller, *Phys. Rep.* **299**, 189 (1998), and references therein.

CLUSTERIZATION OF HEAVY NUCLEI: AN ALGEBRAIC ATTEMPT

P. O. HESS, L. HERNÁNDEZ DE LA PEÑA, W. SCHEID
Institut für Theoretische Physik, Justus-Liebig-Universität, Heinrich-Buff-Ring 16, D-35392 Giessen, Germany

A. ALGORA, J. CSEH, G. LÉVAI
Institute of Nuclear Research of the Hungarian Academy of Sciences, H-4001 Debrecen, Pf. 51, Hungary

A possible proceedure is presented for an extension of the Semimicroscopic Algebraic Cluster Model (SACM) to heavy nuclei without any restriction to an analytically solvable limit of dynamic symmetry. Each cluster is characterized by a linear combination of many $SU(3)$ irreducible representations (irreps), obtained via a diagonalization of the individual clusters. The relative motion still belongs to a definite $SU(3)$ irrep. The Pauli-exclusion principle is taken into account by matching the total $SU(3)$ irreps, contained in the product of the $SU(3)$ irreps of the two clusters and the relative motion, with the shell model space.

1 Introduction

The Semimicroscopic Algebraic Cluster Model (SACM) [1,2] was applied with success to light cluster systems. The structure of each cluster is taken into account through the $SU(3)$ shell model, while the relative motion is described by using π- and σ-bosons, the latter only in order to introduce a cutoff of the model space.

The Pauli-exclusion principle is taken into account by selecting those coupled SU(3) cluster model states, which are present also in the fully antisymmetric shell model basis of the whole nucleus. This prescription involves both the Wildermuth condition [3], and the selection of the Pauli-allowed $SU(3)$ irreps in a partially filled major shell.

The application of the SACM is limited to light nuclei mainly because the spin-orbit interaction is neglected. For heavy nuclei the spin-orbit interaction destroys the $SU(3)$ symmetry and the SACM cannot be used in its present form. Some attempts to extend the model to heavy nuclei have been made [4] using the pseudo-$SU(3)$ symmetry observed in nuclei [5]. However, the division of the orbitals into intruders and "normal" orbitals are different for the separated clusters and the united nucleus, creating basic problems in how to proceed.

Another attempt is presented in Ref. [6], based on the observation that the spectrum of deformed nuclei can be mapped to an $SU(3)$ [7,8], i.e. defining an effective $SU(3)$ irrep.

We present another attempt how to extend the model to heavy nuclei taking into account real $SU(3)$ irreps.

2 The model space and the Hamiltonian

We start with the states of the individual clusters. They are obtained via a diagonalization within the $SU(3) \times SU_S(2)$ basis, where S refers to spin. Note that the spin-orbit interaction mixes spin 0 with spin 1, so the extension of the SACM for light nuclei to heavy nuclei has this extra complication.

The resulting states for each cluster have the form

$$|J_{Ck} M_{Ck}> = \sum_{\Gamma_{Ck},\kappa_{Ck},L_{Ck},S_{Ck}} C^{J_{Ck}}_{\Gamma_{Ck},\kappa_{Ck},L_{Ck},S_{Ck}} \\ |\Gamma_{Ck}, \kappa_{Ck}, L_{Ck}, S_{Ck}, J_{Ck}, M_{Ck}> \tag{1}$$

where the $C^{J_{Ck}}_{\Gamma_{Ck},\kappa_{Ck},L_{Ck},S_{Ck}}$ are the expansion coefficients in the $SU(3)$ and spin basis obtained after diagonalizing the Hamiltonian of the individual clusters in the $SU(3)$ basis. We used partially the notation of J. Escher et al. [9]. There, Γ refers to the SU(3) irrep (λ, μ) and its multiplicity ρ if different from one. In addition to the ground state of each cluster we can take into account excited states as well, if required, which are obtained in the same way. Eq. (1) gives the expansion of the angular momentum basis states (sometimes called also as weak-coupling basis) in terms of the $SU(3)$ coupled basis (strong coupling basis). This latter one is used in the $SU(3)$ dynamic symmetry limit of the SACM, applied extensively to light nuclei.

Now we have to construct the cluster model state, including the relative motion. It is given by

$$|[J_{C1}, J_{C2}]J_C, (n_\pi, 0)l, JM >= \\ [\ [|\ J_{C_1} > \times \mid J_{C_2} >]^{J_C} \times \mid (n_\pi, 0)l >]^J_M \tag{2}$$

where the q_{Ck}'s can be set to 1, for the case taking only the lowest J_{Ck} state. This state is a linear combination of the states

$$|[\Gamma_{C1}, \kappa_{C1}, L_{C1}, S_{C1}, J_{C1}; \Gamma_{C2}, \kappa_{C2}, L_{C2}, S_{C2}, J_{C2}]J_C; (n_\pi, 0)l, JM > \tag{3}$$

where a $J - J$ coupling has been used. This has to be recoupled to a $L - S$ coupling form and afterwards to states with a definite $SU(3)$ irrep, first with respect to the coupling of the two cluster states to a cluster irrep Γ_C and then with the relative motion $((n_\pi, 0))$ to a total irrep Γ. Thus, it will be a linear combination in

$$|[\Gamma_{C_1} S_{C_1}, \Gamma_{C_2} S_{C_2}]\Gamma_C, (n_\pi, 0); \Gamma\kappa LS; JM> \quad . \tag{4}$$

The relative motion can still be described via the 3-dimensional harmonic oscillator basis, thus it has a definite $SU(3)$ irrep for the relative motion.

The result of the recoupling and coupling to the basis states of Eq. (4), applied to the states in Eq. (3) is given by

$$\begin{aligned} &|[\Gamma_{C1}, \kappa_{C1}, L_{C1}, S_{C1}, J_{C1}; \Gamma_{C2}, \kappa_{C2}, L_{C2}, S_{C2}, J_{C2}]J_C; (n_\pi, 0)l, JM >= \\ &\qquad \sum_{L_c, S, L} U \left\{ \begin{matrix} L_{C1} & S_{C1} & J_{C1} \\ L_{C2} & S_{C2} & J_{C2} \\ L_C & S & J_C \end{matrix} \right\} U \left\{ \begin{matrix} L_C & S & J_C \\ l & 0 & l \\ L & S & J \end{matrix} \right\} \\ &\qquad \times \sum_{\Gamma_C \kappa_C} \langle \Gamma_{C1}\kappa_{C1}L_{C1}, \Gamma_{C2}\kappa_{C2}L_{C2} || \Gamma_C \kappa_C L_C \rangle_{\rho_C} \\ &\qquad\qquad \times \sum_{\Gamma\kappa} \langle \Gamma_C \kappa_C L_C, (n_\pi, 0)1l || \Gamma\kappa L \rangle_\rho \\ &\qquad\qquad \times |[\Gamma_{C_1} S_{C_1}, \Gamma_{C_2} S_{C_2}]\Gamma_C, (n_\pi, 0); \Gamma LS; JM > \end{aligned} \tag{5}$$

The U-coefficients are the unitary 9-j symbols and are related to the normal 9-j symbols via a factor of square roots. The second U-coefficient recouples the $J_c, l \to J$ into L_c, S and l, S and the first U-coefficient does it for the cluster part. The next two isoscalar factors of $SU(3)$ couple first the cluster irreps to the cluster $SU(3)$ and then with the relative motion to the total $SU(3)$ irrep. Note that the multiplicity labels ρ and ρ_C are by definition contained in Γ and Γ_C.

Up to now we have only expressed the cluster states into an expansion of the cluster basis in $SU(3)$. The Pauli-exclusion principle has still to be implemented. We achieve it as follows: In order that our states of Eq. (5) obey the Pauli-exclusion principle, we have to apply to this state an operator which excludes all states not having a match to the shell model space. This includes also the way how the cluster irrep Γ_C is selected. This operation is simply obtained by changing the sum over Γ_C and Γ by a sum with a prime ($'$), indicating that all states with no match are excluded in the sum.

The resulting state has to be normalized and, when several states with the same J_{Ck}, J are used, eventually orthogonalized. Though the steps proposed seem to be complicated, they are, in fact, not because programs exist [10] which effectively calculate all coupling and recoupling coefficients.

For the Hamiltonian we propose

$$H = H_{U(3)} + H_{int} \tag{6}$$

where $H_{U(3)}$ is a $U(3)$ symmetry-preserving part (which is frequently applied in the studies of light nuclei), and H_{int} contains all $SU(3)$ violating interactions (like the spin-orbit interaction, pairing, etc.). The matrix elements are easy to calculate because $H_{U(3)}$ is diagonal in the $SU(3)$ coupled SACM basis of Eq. (4). To restrict the $H_{U(3)}$ to a dynamic symmetry is no restriction because the extension to the most general form is straightforward.

More complicted is the calculation of the matrix elements of H_{int}. For that we use the routines, developed by the Louisiana group [11], which can calculate the matrix elements of any one- and two-particle interaction with respect to a $SU(3)$ basis. A further simplification arises by using the results of Ref. [12]. There it was shown that for deformed nuclei the calculation can be simplified further by restricting to the "stretched" $SU(3)$ irreps in the product of proton and neutron $SU(3)$ irreps, i.e. $((\lambda, \mu) = (\lambda_\pi + \lambda_\nu, \mu_\pi + \mu_\nu))$ with spin $S = 0$ and 1. Using realistic parameters for the spin-orbit interaction, these basis states exhaust more than 80 percent of the low lying collective states. One condition is the dominance of the quadrupole-quadrupole interaction, or in other words, the nuclei are strongly deformed. Similar studies have been performed in Ref. [13].

3 Conclusions

We have presented a rough proposal how to make a general extension of the SACM to heavy nuclei, i.e. a generalisation without any restriction to an analytically solvable limit of dynamical symmetry. For the moment, only the schematic, technical method was presented. The main ideas of the SACM for light nuclei are maintained: a microscopically constructed model space is combined with phenomenological interactions. For the exclusion of the Pauli-forbidden states one can apply the matching of $SU(3)$ quatum numbers with those of the shell model space, or some other procedure used in microscopic cluster studies. The calculation of the matrix elements is carried out by applying grouptheoretical methods.

The size of calculation will be larger than for light nuclei. However, all necessary methods are available and easy to apply. Work is still in progess.

Acknowledgments

This work is supported by the DAAD and DGAPA and by the OTKA (No T22187). P.O.H. is on sabatical leave from the Instituto de Ciencias Nucleares, UNAM, Mexico; E-Mail: hess@nuclecu.unam.mx

References

1. J. Cseh and G. Lévai, *Ann. Phys., NY* **230**, 165 (1994).
2. J. Cseh, contribution to the present conference.
3. K. Wildermuth and Th. Kanellopoulos, *Nucl. Phys.* **7**, 150 (1958); K. Wildermuth, and Y.C. Tang, *A Unified Theory of the Nucleus* (New York:academic Press, 1977).
4. J. Cseh, R.K. Gupta and W. Scheid, *Phys. Lett.* B **299**, 205 (1993); A. Algora and J. Cseh, *J. Phys.* G **22**, L39 (1996).
5. K.T. Hecht and A. Adler, *Nucl. Phys.* A **137**, 129 (1969); A. Arima, M. Harvey and K. Shimizu, *Phys. Lett.* B **30**, 517 (1969).
6. A. Algora, J. Cseh and P.O. Hess, *J. Phys.* G **24**, 2111 (1998).
7. D.J. Rowe, *Rep. Prog. Phys.* **48**, 1419 (1985); O. Castaños, J.P. Draayer and Y. Leschber, *Z. Phys.* A **329**, 33 (1988).
8. M. Jarrio, J.L. Wood and D.J. Rowe, *Nucl. Phys.* A **528**, 409 (1991).
9. J. Escher and J. P. Draayer, *J. Math. Phys.* **39** (1998), 5123.
10. J. P. Draayer and Yoshimi Akiyama, *J. Math. Phys.* **14** (1973), 1904; Y. Akiyama and J. P. Draayer, *Comp. Phys. Comm.* **5** (1973), 405.
11. C. Bahri and J.P. Draayer, *Comp. Phys. Commun.* **83**, 59 (1994).
12. C. Vargas, J.G. Hirsch, P.O. Hess and J.P. Draayer, *Phys. Rev.* C **58**, 1488 (1998).
13. C. Bahri, J. Escher and J. P. Draayer, *Nucl. Phys.* A **592** (1995), 171.

FORTY YEARS OF NUCLEAR MOLECULES

R. R. BETTS[1,2]
[1]*Physics Division, Argonne National Laboratory, Argonne, IL 61439 USA*
[2]*Physics Department, University of Illinois at Chicago Chicago, IL 60607 USA*

1 Introduction

It is now almost four decades since the first indications of the existence of nuclear molecular states. Despite this considerable span of time, the idea has shown remarkable resilience and there is little doubt in the fundamental soundness of the concept of extended nuclear states with shapes reminiscent of diatomic and polyatomic molecules.

The phenomenon has shown itself in a wide range of nuclear systems and may well yet prove to be ubiquitous. The study of this intriguing topic has also always posed questions relating to or ahead of forefront issues in nuclear physics such as fission isomers, sub-barrier reactions and coupled-channel phenomena, super- and hyper-deformation and other manifestations of extended nuclear systems such as cluster and halo states.

Despite the solid basis underpinning the concept of the nuclear molecule, and the large body of circumstantial evidence in support of this view, it is nevertheless fair to say that there, as yet, exists no unequivocal proof of the nature of the observed nuclear states. This remains an outstanding challenge for the future.

2 The First Decade

The seminal experiments were carried out at Chalk River around 1960 [1]. Unexpected structure was observed in the energy dependence of cross-sections for elastic scattering and reactions of $^{12}C+^{12}C$ at energies close to the Coulomb barrier for this system. Notably, these experiments were carried out using the first examples of silicon diode charged particle detectors, the application of advances in which has continued to play a decisive role in the experimental study of nuclear molecular states. It is also remarkable that, already in the first papers, the notion that the observations might correspond to molecule-like nuclear excitations was proposed. It was even suggested that these might result from the existence of a secondary minimum in the $^{12}C+^{12}C$ scattering potential, although there was no known mechanism for producing such a minimum.

The existence of energy dependent structure in $^{12}C+^{12}C$ cross-sections with width intermediate between that expected for compound nuclear states and the expected width of potential scattering resonances posed a severe problem. Indeed, there was a long lasting debate as to whether or not the observed structures were indeed isolated resonances or simply fluctuations induced by many, random, overlapping resonances. These issues were resolved in favor of the former interpretation [2].

Another issue was the initial lack of observed structure in similar excitation functions for $^{16}O+^{16}O$. Later measurements for this system [3] at somewhat higher energies, however, showed similarity between $^{12}C+^{12}C$ and $^{16}O+^{16}O$ and suggested a possible wider occurrence of the phenomenon.

The observation [4] of narrow states at high excitation energy (16 MeV) in ^{24}Mg, excited in the $^{16}O(^{12}C, \alpha)$ reaction, provoked the suggestion that they might correspond to bound counterparts of the previously observed resonances.It was hypothesized that these states might be populated via direct transfer of massive clusters in a manner analogous to the, then extensively studied, direct transfer of nucleons to simple shell model configurations. Eventually, this proved not to be the case when the states were found to have high spin, their selective population then arising from the favoring of high spins by angular momentum matching conditions in compound nucleus evaporation.

On the theoretical side, many of the essential ideas relating to the structure of the resonances were already in place by the end of the first decade. As mentioned, the notion of a molecule-like nuclear state was proposed in the very first papers. This was expanded on by Michaud and Vogt [5] who proposed a more complex structure of alpha-particle clusters, an idea, which has provided the basis for much of our current thinking. Alternatively, from the viewpoint of resonance formation by dynamical reaction processes, the crucial importance of inelastic coupling to excited states of the colliding nuclei was first proposed by Nogami and Imanishi [6] which was subsequently extended to the double resonance mechanism of Scheid and Greiner [7]. Attempts to link reaction dynamics of the scattering process with the nuclear structure of the combined system were made in the calculations of potentials based on the two-center shell model of Preuss and Greiner [8].

3 The Second Decade

During this period, heavy-ion physics had moved into the mainstream of nuclear physics research and a number of new accelerators became available as well as existing tandems being converted to heavy-ion operation. The initial interest was on the, then poorly understood, gross features of the reaction cross sections such as fusion, fission and deep-inelastic scattering which were believed to be governed solely by the liquid drop aspects of the ion-ion interaction potential.

It was therefore a great surprise when, in the first precision measurements of excitation functions for fusion of light heavy ions, it was discovered that nuclear structure effects played a strong role. The discovery of oscillations in the energy dependence of the total fusion cross section [9] was shortly followed by the observation of similar features in the cross sections for inelastic scattering of target and projectile [10]. These energy dependent structures were of a magnitude, which implied the participation of entire partial waves with each structure corresponding to the successive angular momenta of a rotating molecule-like structure of the two colliding nuclei. This implied that perturbation approaches to calculations of the reaction dynamics were bound to fail and led to the successful development of full coupled channels models of the scattering process such as the Band-Crossing Model of Abe and collaborators [11]. It is interesting to note that these observations represent the first observation of sub-barrier fusion enhancements.

One interesting outcome of the coupled channels approach was due to Mosel [12] who used the scattering wavefunction from a full coupled channels calculation to determine the one dimensional potential of which it was an eigenstate. This "Trivially Equivalent Local Potential" showed a deep minimum at large separation similar to that in phenomenological molecular potentials such as the Morse Potential. The position of this minimum was close to the radius at which the single particle levels in a two-center shell model calculation start to cross, raising the issue of the influence of the structure of the di-nuclear system on the resonance formation process. Calculations of reaction processes using a basis of molecular orbitals were carried out by von Oertzen and Imanishi [13]. Enhanced promotion of nucleons was found to occur at level crossings - the Landau-Zener effect - giving rise to strong population of specific particle-hole states [14]. These ideas were extended by Harvey [15] who pointed out that multi-particle multi-hole cluster excitations should play an important role in the coalescence of the two colliding nuclei and that selection rules for the allowed partitions of the system could thus be derived. These multi-particle multi-hole configurations were investigated in the framework of the deformed shell model by Leander and Larsson [16] who calculated potential energy surfaces for a number of light nuclei which showed stable secondary minima akin to those responsible for fission isomers in Actinide nuclei.

The predictions of a much richer spectrum of resonances than expected on the basis of simple potential scattering led to a renewed experimental interest in the $^{12}C+^{12}C$ barrier resonances. A series of very detailed experiments led to the almost complete mapping of the resonance spectrum and spins [17] which were later quite well accounted for in models which assumed an underlying structure corresponding to a rotating, vibrating molecule [18,19].

4 The Third Decade

At this time, it became possible to make precision studies of heavy ion reactions with beams heavier than C and O. This was a consequence of accelerator upgrades and improvements in both magnetic spectrometers and silicon detectors. Once more experiment provided surprises.

It had generally been assumed that the vastly increased number of open reaction channels for heavy systems would lead to much stronger absorption in the scattering potentials which would suppress resonance formation. This belief was countered by the observation of large enhancements of the large angle elastic scattering cross sections for O and C scattering from s-d shell target nuclei [20,21] and, soon thereafter, for heavier symmetric systems involving Si and Mg [22,23]. These large angle yields showed striking energy dependence with structures as narrow as the precision of the measurements would allow [24]. Further, the angular distributions in the large angle region, in addition to the enhanced cross-sections, also showed highly oscillatory shapes, characteristic of single angular momenta which were, at the time, the highest spin nuclear excitations known.

It also became clear that resonances in these heavier systems appeared, in a correlated fashion, in many other inelastic and reaction channels than was the case for C+C and the other light systems. In fact, the dominant fraction of the resonance cross-sections appeared in quite strongly damped reaction channels which could not be included or accounted for in the coupled-channels approach at that time. This led to the suggestion [26] that it might be useful to think of the resonances as fissioning shape isomers - states in a secondary minimum in the potential-energy surface of the equilibrated composite system, produced by the stabilizing influence of shell closures at large deformations. Deformed Shell-Model calculations [27] did indeed show the existence of such secondary minima for all the nuclear systems for which resonance behavior had been observed, as had been noted for the lighter systems by Leander and Larsson. This could be though of as the first observation of high-spin superdeformed states, later observed in heavier nuclei in gamma-ray experiments.

A remarkable synthesis of many of these ideas came as result of cluster model calculations by Marsh and Rae [28]. These calculations, in the framework of the Cranked Cluster Model, identified the quasi-stable configurations of six alpha particles. These, in some cases quite complex configurations, could all be identified with the previously found, secondary minima in the Deformed Shell Model calculations and with closed-shell super- and hyper-deformed, multi-particle multi-hole cluster states. This result gave hope for a real spectroscopy of the molecular resonances, based on the selection rules which arise from the allowed partitions of each configuration, as had been originally suggested by Harvey.

In parallel with the new theoretical work and motivated by the results, new experiments studied the breakup of ^{24}Mg into its constituent clusters. These experiments [29] exploited new developments in silicon detector technology, which provided extremely precise measurement of the position of incident particles, and thus allowed reconstruction of the excitation energy spectrum of ^{24}Mg decaying into two ^{12}C nuclei in their ground states. This spectrum showed a number of narrow states which eventually were shown to correspond quite precisely with the long known ^{12}C+^{12}C resonances and to display the dependence of excitation energy with spin expected for a molecular configuration [30].

5 The Fourth Decade

One of the most striking expectations based on the Cluster and Deformed Shell Model calculations was the existence of quasi-stable states corresponding to linear chains of alpha particles. For ^{24}Mg, the corresponding six alpha-particle chain was thought likely to decay symmetrically to two three-alpha chains which, it was believed, is the configuration of the well-known 7.6 MeV state in ^{12}C. This state, responsible for the rate of He burning in stars, in turn decays sequentially through the ground state of 8Be into three alpha particles, thus resulting in six alpha particles with a unique set of kinematic correlations. The detection and characterization of this complex decay process presented a formidable experimental challenge but, using newly-developed double-sided Si strip detectors, it was possible to measure and reconstruct this complicated final state. Indeed, a striking resonance was found [31] in the yield of this process. Suggestively, this resonance occurred at precisely the energy predicted by the model calculations for the six-alpha chain state. This result is still the subject of study and the identification of the chain therefore not by any means firm.

In recent years there has been tremendous progress and new ideas in both experiment and theory. The experimental equipment has now evolved into highly sophisticated, dedicated arrays of detectors such as the charged particle arrays CHARISSA and MEGHA and gamma-ray arrays like EUROGAM and GAMMASPHERE. Used alone and in combination, they are proving to be powerful new tools in the study of nuclear molecular phenomena.

The charged particle arrays have allowed data of a quality and scope far in excess of what might only have been dreamed of a few years ago. It has been possible to study low cross section reactions such as cluster breakup with high resolution and good statistics. As an example, it has been possible to show that the states observed in the breakup of ^{24}Mg into ^{12}C+^{12}C are also observed in the transfer of an alpha particle to ^{20}Ne [32]. This result has implications for the proposed structure of the states and provides the first glimmerings of a real spectroscopy of the nuclear molecular states.

The notion of a nuclear molecule has been extended from the more familiar ones composed of alpha particles to those with an excess of neutrons, where the binding interaction now arises from the exchange of the excess neutrons between two core nuclei. It has been proposed [33] that such states have already been observed in the low-lying spectra of light nuclei such as the neutron rich Be and B isotopes, although definitive proof is still lacking.

Related to this idea, the study of cluster breakup has been extended to such nuclei by use of the newly available beams of unstable nuclei such as ^{12}Be. Indeed, in the breakup of ^{12}Be to ^{6}He+^{6}He and ^{8}He+^{4}He, resonance structures have been observed [34], which presumably correspond to these covalently-bound molecular states.

The theoretical basis for the existence of such states is quite strong, having been found in antisymmetrized molecular dynamics calculations by Kanada-Enyo and Horuichi [35]. More recently, ab initio Monte Carlo Shell Model calculations for Be, clearly show [36] a di-nuclear structure with the correct moment of inertia. Doubtless, this mass region will become an area of major activity, especially as it relates to the intensively recently studied phenomenon of halo nuclei.

The use of charged particle detectors in conjunction with state-of-the-art gamma ray arrays has added a new dimension to studies of nuclear molecular phenomena. The first experiments of this type were carried out [37,38] using low-resolution NaI arrays such as the Crystal Ball and the Spin Spectrometer. Nevertheless, important studies could be carried out such as searches for radiative transitions between molecular states and measurements of resonance spin assignments. More recently, the high resolution Ge arrays have been used [39] to determine the degree to which the intrinsic spin of the clusters in the molecule are aligned with the plane of molecular rotation. These experiments have provided interesting information relating to the predictions of models of the molecule, which include deformation of the clusters and complex modes of rotation and vibration [40].

Also of interest in this regard are new experiments, using GAMMASPHERE in conjunction with a Fragment Mass Analyzer [41], to search for and study radiative decays of highly unbound states in ^{24}Mg. Of course any highly unbound state with a significant radiative decay width is intrinsically interesting. However, it is hoped that these studies may somehow connect to the molecular resonances of which some might lie low enough to decay to the low-lying states of ^{24}Mg.

6 Conclusions and Outlook

It is clear that, after forty years of study, one cannot refer to the field of nuclear molecules as new and exciting. It must rather be said to be rather mature but still one which holds promise and challenge for the future. There is a well-documented and stable phenomenology associated with the molecule which seems to be well-backed by both empirical and fundamental models. Nevertheless, there remain large areas still to be explored such as heavier systems and the region of large neutron excess. It is also true that the definitive experiment still remains to be done, i.e. an experiment which conclusively demonstrates the shape of the system. This is a formidable challenge, but one from which we should not shy away.

Finally, it is important to note that areas of physics, which can be grasped intuitively by the intelligent lay person, are of special importance. It is these areas which can connect the general public and transmit our sense of adventure in science. I believe that the Nuclear Molecule falls into this category and one should not therefore underwrite its impact.

References

1. D.A. Bromley *et al*, *Phys. Rev. Lett.* **4**, 365 (1960).
2. J.P. Bondorf and R.B. Leachman, *Mat. Phys. Skr. Dan. Vid. Sel.* **34**, 1 (1965).
3. W. Reilly *et al*, *Nuovo Cimento*, **13**, 913 (1973); R.M. Wieland *et al*, *Phys. Rev. Lett.* **37**, 1458 (1976).
4. R. Middleton *et al*, *Phys. Rev. Lett.* **24**, 1436 (1970).
5. M. Michaud and E.W. Vogt, *Phys. Lett.* B **30**, 85 (1969).
6. B. Imanishi, *Phys. Lett.* B **27**, 267 (1968).
7. W. Scheid *et al*, *Phys. Rev. Lett.* **25**, 176 (1970).
8. K. Preuss and W. Greiner, *Phys. Lett.* B **33**, 197 (1970).
9. P. Sperr *et al*, *Phys. Rev. Lett.* **36**, 405 (1976).

10. T.M. Cormier *et al*, *Phys. Rev. Lett.* **38**, 940 (1977).
11. Y. Abe *et al*, *Phys. Rev.* C **19**, 1365 (1979).
12. O. Tanimura and U. Mosel, *Phys. Rev.* C **24**, 321 (1981).
13. B. Imanishi and W. von Oertzen, *Phys. Rep.* **155**, 29 (1987).
14. A. Theil *et al*, *J. Phys.* G **14**, L45 (1988).
15. M. Harvey, Proceedings 2nd International Conference On Clustering Phenomena in Nuclei (College Park, Maryland) 1975, USERDA Report ORO-4856-26 (1975) 549.
16. G. Leander and S.E. Larsson, *Nucl. Phys.* A **239**, 93 (1975).
17. K.A. Erb *et al*, *Phys. Rev.* C **22**, 507 (1980).
18. N. Cindro, *J. Phys.* G **4**, L23 (1978).
19. K.A. Erb and D.A. Bromley, *Phys. Rev.* C **23**, 2781 (1981).
20. P. Braun-Munzinger *et al*, *Phys. Rev. Lett.* **38**, 944 (1977).
21. M.R. Clover *et al*, *Phys. Rev. Lett.* **40**, 1008 (1978).
22. R.R. Betts *et al*, *Phys. Rev. Lett.* **43**, 253 (1979).
23. R.W. Zurmuhle *et al*, *Phys. Lett.* B **129**, 384 (1983).
24. R.R. Betts *et al*, *Phys. Rev. Lett.* **47**, 23 (1981).
25. R.R. Betts *et al*, *Phys. Lett.* B **100**, 117 (1981).
26. R.R. Betts, Proceedings Conference On Nuclear Physics With Heavy Ions (Stony Brook, New York) 1983, (Harwood, New York 1984) 347.
27. S. Aberg and M. Ploszajczak, (Private Communication).
28. S. Marsh and W.D.M. Rae, *Phys. Lett.* B **180**, 185 (1985).
29. B.R. Fulton *et al*, *Phys. Lett.* B **181**, 233 (1986).
30. B.R. Fulton *et al*, *Phys. Lett.* B **267**, 325 (1991).
31. A.W. Wuosmaa *et al*, *Phys. Rev. Lett.* **68**, 1295 (1992).
32. M.J. Leddy *et al*, *Nucl. Phys.* A **589**, 363 (1995).
33. W. von Oertzen, *Z. Phys.* A **354**, 37 (1996).
34. M. Freer *et al*, *Phys. Rev. Lett.* **82**, 1383 (1999).
35. Y. Kanada-Enyo *et al*, *Phys. Rev.* C **52**, 628 (1995).
36. S. Pieper and R. Wiringa, (to be published).
37. V. Metag *et al*, *Phys. Rev.* C **25**, 1486 (1982).
38. A.H. Wuosmaa *et al*, *Phys. Rev. Lett.* **58**, 1312 (1987).
39. R. Nouicer *et al*, *Phys. Rev.* C **60**, R041303 (1999).
40. E. Uegaki and Y. Abe, *Phys. Lett.* B **340**, 143 (1994).
41. C.J. Lister, (Private Communication).

THE TWENTIETH ANNIVERSARY OF THE ORBITING-CLUSTER MODEL OF RESONANCES[a]

UGO ABBONDANNO
Istituto Nazionale di Fisica Nucleare, Sezione di Trieste, via A. Valerio 2, 34127 Trieste, Italy

NIKOLA CINDRO
Institute Ruder Bošković, HR-10 001 Zagreb, Croatia

The historical evolution of the Orbiting-Cluster Model of resonances is presented. Particular care is devoted to describe the capability of the model of predicting the presence (or absence) of resonances, and to the comparison between nuclear parameters extracted from the experimental data and calculated by means of the model.

1 History

One of the most exciting and unexpected results of the first measurements performed when heavy-ion beams became available in Tandem accelerators was the discovery of resonance-like states in the $^{12}C+^{12}C$ system in the early 1960s[1]. Correlated structures were observed in p, n, α and γ outgoing channels: the interest in this behaviour arose from the fact that these structures appear in a region of excitation energy of the compound nucleus in which the statistical theory is expected to be valid. This theory predicts a monothonic decrease of the excitation functions as a function of the incident energy. Thus, this structured behaviour captured the interest of nuclear physicists; for the not-so young, I wish to recall that the same problem (presence of isolated structures in excitation energy regions in which a monothonic behaviour was expected) occurred in the same years for Isobaric Analogue Resonances.

Thus, the existence of these resonance-like objects stimulated interest in the interpretation of their physical nature; in turn, considerable effort was devoted to the search for similar phenomena in other nuclear systems. In this way, resonances in heavy-ion reactions (RHIR), considered in the 1960s as isolated phenomena present only in a few nuclear systems (such as $^{12}C+^{12}C$ and $^{12}C+^{16}O$), were shown in the 1970s to be present in many more systems, but still in reactions induced by ^{12}C and ^{16}O projectiles.

As for their characteristics, RHIR appear to have structural character, to be then individual states in the composite system: this assumption is based on (i) the fact that their spectroscopic characteristics correspond fairly well in

[a]Dedicated to Professor Giulio Cervani on the occasion of his 80th birthday

the same composite system created in two different ways and (ii) they have a well-defined L-value. The latter emerges clearly from an analysis of the experimental data. The "classical" way to study RHIR consists (i) in the measurements of the excitation functions in order to evidentiate the presence of nonstatistical structures and (ii) in the analysis of the angular distributions (measured at energies corresponding to the "bumps" in the excitation functions) in terms of Legendre polynomial analysis to extract the corresponding L-values. To observe the improvement of the quality of the fit and the sharp drop of the value of χ^2 obtained by adding the "right" value of L to the fit is namely very interesting; on the other hand, adding further terms did not substantially improve the quality of the fit.

2 Model

At the beginning of the 80's, resonant behaviour had been observed in six nuclear systems: ^{24}Mg from ^{12}C+^{12}C, ^{28}Si from ^{12}C+^{16}O, ^{30}Si from ^{12}C+^{18}O, ^{36}Ar from ^{12}C+^{24}Mg, ^{40}Ca from ^{12}C+^{28}Si and ^{16}O+^{24}Mg, and ^{44}Ti from ^{16}O+^{28}Si. On the other hand, the search for resonant behaviour in some other systems had given negative results. So, two main questions were under the attention of the physicists: (i) How can these resonances be described? In other words, what is the relation between the E_x (or E_{cm}) value of a resonance and its L-value? (ii) Why are resonances present in some nuclear systems and absent in others?

The basic idea underlying the Orbiting-Cluster Model (OCM) [2], derives from observation of data on the linear dependence of the slope of the measured excitation energy E_x of resonances on their spin L. The main features of the data can be summarised by the following statements:

(i) All resonance data lie in a narrow zone spread around a straight median line in the E_x vs. L(L+1) plane;

(ii) For each resonant system, the energy of the observed or projected L=0 resonance is close to E_B+E_C, where E_B is the binding energy of the projectile plus target in the composite system and E_C is the respective Coulomb barrier.

These two features, well-known from molecular physics, suggest a picture describing resonances as due to the orbiting of the two interacting nuclei forming a molecular-type configuration. In particular, the observed slope of E_x vs. L(L+1) yields effective moments of inertia $\mathcal{J}$ in good agreement with those calculated from the assumption of the two interacting nuclei as rigid spheres rotating around their centre of mass. The additional support to the orbiting-cluster interpretation provided by the energies of the projected resonance bandheads is directly related to the conjecture that the lowest resonance

(L=0) occurs when the two nuclei barely touch each other with no relative rotational energy. Hence, the corresponding excitation energy of the composite system is expected to be about $E_B + E_C$, as borne out by observation.

A serious problem arising from the adoption of such picture is that the above rotor mechanism yields resonances for any colliding pair of heavy-ions; yet, experimental evidence shows that resonant behaviour is, as a rule, observed in the so-called α-particle nuclei. Following experimental evidence, the presence or absence of resonances in the OCM is systematically related to the level density of composite systems. This is the so-called molecular-window concept; introduced by Greiner and Scheid [3], this concept provides a surprisingly consistent criterion for the observability of resonances. Hence, if we assume the simple orbiting configurations related to resonances to be doorway-like states preceding the dissolution of the composite system into more complex nuclear states, the presence or absence of resonances is simply related to the damping-width

$$\Gamma \downarrow = i\frac{\pi}{\hbar} \,|< \mathrm{cn} \mid V \mid \mathrm{el} >|^2 \, \rho(E, L)$$

where $\mid \mathrm{el} >$ and $\mid \mathrm{cn} >$ represent, respectively, the entrance (elastic) and compound nuclear states, and ρ(E,L) is the density of levels of spin L in the composite system at the excitation energy E. Neglecting the variations of the matrix element with mass and energy, the value of the damping-width is determined solely by the level density of the composite system: resonances will occur when this latter quantity has a minimum (no absorbtion, hence a long rotation period and the creation of a long-lived resonant state); otherwise, the intermediate structure is washed out.

3 Success

What was the situation in 1980? Firstly, the model describes the E_x vs. L(L+1) dependence rather well for resonances in all the colliding pairs in which these phenomena were observed at that time. Secondly, the values of the moment of inertia $\mathcal{J}$ calculated by the model and those extracted from the experimental slope of the straight line fitting the data in the E_x vs. L(L+1) representation are in fairly good agreement.

Another feature of the model and a very important one was that it gave the prediction of the nuclear composite systems in which resonant behaviour was likely to be observed. It must be underlined that at that time the possibility of experimentally testing the predictions of the OCM was favoured by the fact that a new generation of electrostatic accelerators became available to furnish more energetical and heavy beams, such as, for example, ^{24}Mg and ^{28}Si.

A quotation by Greiner from his 1991 Brioni Conference talk presents the evolution of our knowledge between 1980 and 1990 [4]: "Cindro and collaborators", said Greiner, "performed OCM calculations for a number of colliding systems. Their predictions of 10 years ago were tested against time in a recent article. They brillantly withstood the test of time! All of the newly found resonant systems were predicted as such by the OCM; in none of those forbidden by the OCM were resonances found, so far".

4 Extension

The success of the predictions of the OCM in describing the resonant behaviour of colliding nuclei leading to light- and medium-weight composite systems encouraged its application to medium- and high-mass systems. So, a systematic investigation was undertaken in this direction [5]. First of all, it was verified that the damping-width takes low values for the interaction of ^{12}C with targets such as ^{12}C, ^{16}O, ^{18}O, ^{20}Ne, ^{24}Mg, ^{28}Si and ^{32}S, where resonant behaviour was observed beyond reasonable doubt. Starting from this fact, calculations were extended to the interaction of ^{12}C with heavier targets; those systems for which a low value of $\Gamma\downarrow$ was obtained were classified as likely to show resonant behaviour. Repeating these calculations for target nuclei from A=12 to A=58 and for incident beams of ^{12}C, ^{16}O, ^{14}C, ^{18}O, ^{24}Mg, ^{28}Si, ^{32}S, ^{40}Ca and ^{46}Ti, a series of projectile-target pairs was identified in which the presence of resonances was found to be likely. Again, at that time the possibility of experimentally testing the predictions of the OCM was favoured by the fact that new electrostatic accelerators became available to furnish more energetical and heavy beams, such as, for example, ^{58}Ni.

In this case, unfortunately, there are very few experimental data to be used for a possible test of the predictions of the OCM. Data appeared in fact only on three heavy-ion systems candidate to exhibit resonant behaviour and on one system candidate not to exhibit them. In detail, in the ^{58}Ni+^{46}Ti reaction [6], two non-statistical structures were found in the excitation function at E_{cm}=99.0 and 100.6 MeV, to which a statistical analysis allowed the attribution of L-values of 36 and 47, respectively. In the frame of the OCM, that corresponds to an extracted moment of inertia $\mathcal{J} = 2.05{\cdot}10^{-41}$Mev s^2, to be compared to a calculated one of $\mathcal{J} = 3.54{\cdot}10^{-41}$Mev s^2. In the ^{58}Ni+^{58}Ni reaction [7], five non-statistical structures were found in the excitation function at E_{cm}=110.25, 111.25, 112.25, 113.25 and 114.50 MeV, to which a statistical analysis allowed the attribution of L-values of 60, 58, 59, 56 and 58, respectively. In the frame of the OCM, that corresponds to an extracted moment of inertia $\mathcal{J} = 3.28{\cdot}10^{-41}$Mev s^2, to be compared to a calculated one of $\mathcal{J}$ =

$4.27 \cdot 10^{-41}$Mev s^2. In the ^{58}Ni+^{62}Ni reaction [7], two non-statistical structures were found in the excitation function at E_{cm}=115.7 and 116.5 MeV, to which a statistical analysis allowed the attribution of L-values of 60 and 53, respectively. In the frame of the OCM, that corresponds to an extracted moment of inertia $\mathcal{J} = 2.55 \cdot 10^{-41}$Mev s^2, to be compared to a calculated one of $\mathcal{J} = 4.52 \cdot 10^{-41}$Mev s^2. Finally, no evidence was found of a resonant behaviour in the OCM unfavoured ^{32}S+^{62}Ni reaction, even though the quality of the data doesn't allow a cathegorical statement [8].

5 Conclusions

By way of conclusion, even though no definitive proof of the existence of RHIR in medium and heavy composite systems has been achieved, reaching such proof presents serious experimental problems. Among these problems, the greatest one is related to the required angular resolution. In fact, considering the "classical" data, they may be observed to be measured at beam energies of about twice the Coulomb barrier. So, in a simple comparison, for example, between the ^{28}Si+^{28}Si and the ^{58}Ni+^{58}Ni reactions, the respective center of mass energies corresponding to twice the Coulomb barrier are E_{cm}= 60 and 180 MeV, and the L-value predicted by simple OCM calculations are 40 and 134, respectively. That corresponds to a periodicity in the angular distributions of $\Delta\Theta = 4.5^0$ and 1.4^0: if six points are needed to have reasonably good information on the angular oscillations, the required angular resolution passes from 0.75^0 to 0.23^0: hence the heavier systems must be measured at lower beam energy, and that causes the presence in the angular distributions of a large Coulomb component, preventing the performance of a Legendre-polynomial analysis and obliging the use of statistical methods, that are less direct and less effective in an accurate L-value determination.

References

1. E. Almqvist *et al*, *Phys. Rev. Lett.* **4**, 515 (1960); D.A. Bromley *et al*, *Phys. Rev. Lett.* **4**, 365 (1960).
2. N. Cindro and D. Počanić, *J. Phys.* G **6**, 359 (1980).
3. W. Greiner and W. Scheid, *J. Physique* C **6**, 91 (1971).
4. W. Greiner in *Heavy-Ion Physics, Today and Tomorrow*, ed. Roman Čaplar and Walter Greiner (World Scientific, Singapore, 1991), page 405.
5. N. Cindro and M. Božin, *Ann. Phys.* (N.Y.) **192**, 307 (1989).
6. U. Abbondanno *et al*, *Int. J. Mod. Phys.* E **3**, 919 (1994).
7. L. Vannucci *et al*, *Z. Phys.* A **355**, 41 (1996).
8. U. Abbondanno *et al*, *Nuovo Cimento* A **106**, 541 (1993).

THE SIX ALPHA CHAIN STATE STORY - FROM STRASBOURG TO RAB

DR B R FULTON
School of Physics and Astronomy, The University of Birmingham, Edgbaston, Birmingham. B15 2TT. United Kingdom.,

E-mail: b.r.fulton@bham.ac.uk

1 Introduction

At the last Cluster Conference at Strasbourg in 1994, we learned of evidence for the existence of a six alpha chain state in ^{24}Mg - a linear arrangement of alpha particles. This report generated great interest at the conference and gave rise to a considerable amount of experimental and theoretical activity. My task in this talk is to review what we have learnt about this unusual configuration over the intervening years.

2 Evidence for Chain States

To introduce the idea of chain states there is probably no better way to start than with a figure which has appeared many times in this series of Cluster conferences, the Ikeda diagram (figure 1). This encapsulates the philosophy [1] which has remained central to our field for over 30 years, that different types of clustering can be expected to appear in a nucleus in the region of excitation energy where it becomes energetically favourable to break up into those cluster fragments. One of the intriguing speculations contained in this picture is that highly elongated structures formed of linear arrangements of alpha particles might exist in all these nuclei.

Around the time that Ikeda presented this picture the Bloch-Brink alpha cluster model was being developed [2] and this provided theoretical backing for the occurrence of these so-called chain states. Calculations for ^{12}C revealed two stable arrangements for the alpha clusters, the lowest energy corresponding to a triangular arrangement and a higher energy configuration corresponding to the three alphas in a line. Similarly, in ^{16}O the linear arrangement again appeared, along with a tetrahedral and a kite arrangement. Of course, since the alpha cluster model presupposes the existence of alpha clusters in the nucleus, one could argue that the appearance of the chain configuration was somehow built into the model. However, the results of Nilsson-Strutinsky calculations, a microscopic single particle model which does not presuppose clustering, supported these ideas. For example, the potential energy surface for ^{16}O from such a calculation revealed 3 minina with different deformations and the density distributions of these correspond to the tetrahedral, kite and chain configurations observed in the alpha cluster model calculations.

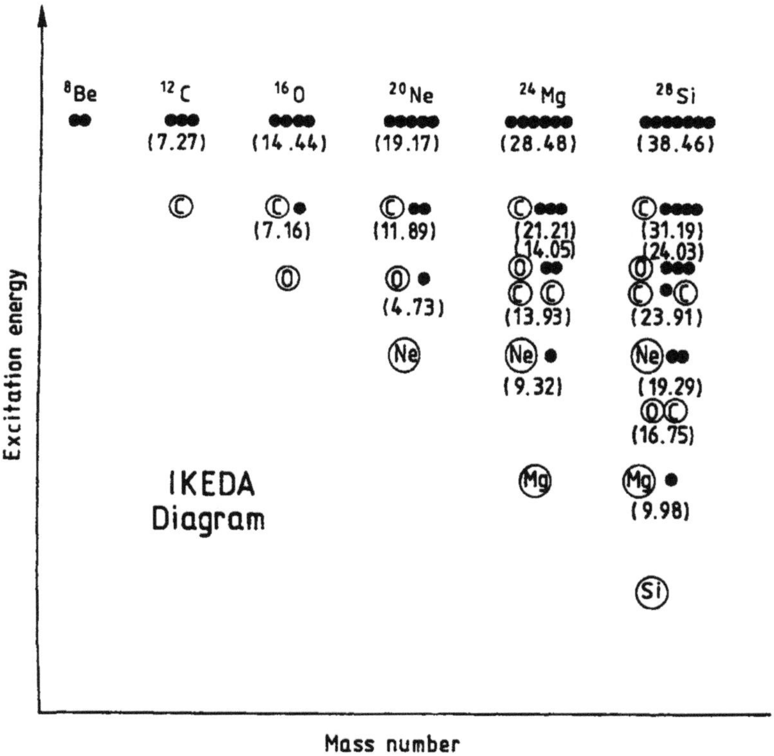

Figure 1. The Ikeda diagram, suggesting that chain states may be a common feature in alpha-conjugate nuclei.

In the 1980's, Marsh and Rae revisited the alpha cluster model. Their calculations, which now included full anti symetrisation and cranking of the wave functions, could also be applied to larger nuclei. One of their first calculations was for ^{24}Mg, and they identified 8 configurations which were stable against the disruptive Coriollis effects of rotation [3]. Once again, a beautiful overlap is observed between the results of the alpha cluster model and Nilsson-Strutinsky calculations (figure 2). Most importantly for our present discussion, both models support the existence of a six alpha chain configuration.

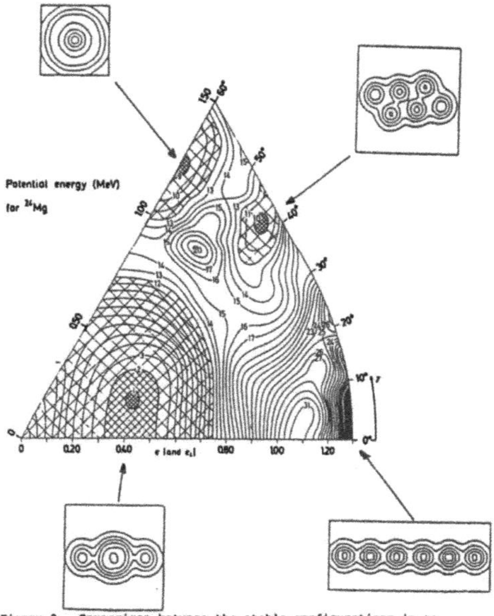

Figure 2. Comparison between the stable configurations in an alpha cluster model calculation for ^{24}Mg with the secondary minima in the potential energy surface as calculated in the Nilsson-Strutinsky model.

3 The Six Alpha Chain - New Results at Cluster '94

Evidence for the shorter chain states has been recognised for some time. For example, the ground state of ^{8}Be has an alpha decay width close to the Wigner limit for this configuration and 2^+ and 4^+ states are observed at the expected energy for a rotational band with a moment of inertia appropriate to such a configuration. In ^{12}C the 0_2^+ state at 7.6MeV has been suggested to be a chain state, based on its large alpha decay width and high deformation as revealed in electron scattering. However, there has been an ongoing debate in the literature as to whether this is truly a chain, or a more compact arrangement more like a bent chain, or acute triangle. In ^{16}O candidate states have been identified around 20MeV excitation in ^{12}C(α, ^{8}Be) ^{8}Be measurements by Chevallier et al. [4] and in ^{12}C(^{12}C,^{8}Be^{8}Be) ^{8}Be breakup measurements by Freer et al.[5].

At the Strasbourg meeting, Wuosmaa [6] showed the first experimental evidence for a six alpha chain state in ^{24}Mg. The presumption in this experiment was that if the six alpha chain could be formed, a likely decay mode would be for the chain to snap into shorter chains, for instance to two carbon nuclei each in the 7.6 MeV state. Each of these would subsequently decay to three alpha particles. Hence the experiment was designed to identify two jets, each of three alpha particles, emerging from the reaction in which excited ^{24}Mg was formed in ^{12}C + ^{12}C collisions. By varying the beam energy, and hence the excitation energy in ^{24}Mg, evidence for the formation of the chain state would show as an enhancement of the six alpha yield. As figure 3a shows, the experiment revealed a beautiful resonance within about one MeV of where the alpha cluster model suggested the six alpha chain state should exist. Further confirmation of the chain state association came from the on-resonance cross section in the mutual 0_2^+ channel relative to the single 0_2^+ excitation. This was inconsistent with the decay penetrabilities for these two channels, an observation which could be reconciled if the decay was occurring from a highly deformed structure. One rather unusual feature of the measurement was that the enhanced yield appeared mostly around 90° in the angular distribution.

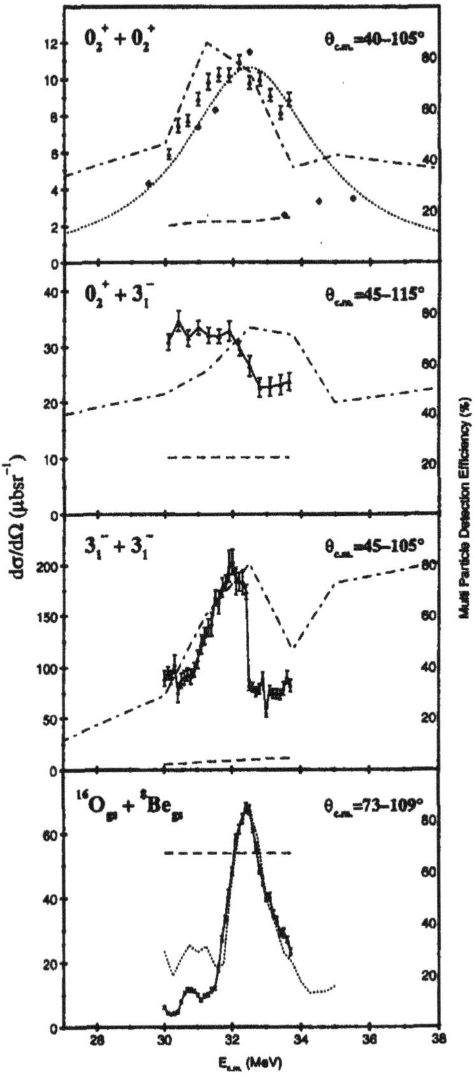

Figure 3. A high resolution measurement of the 32.5MeV resonance in the inelastic and alpha transfer channels.

We also had two theoretical talks relating to this experiment at Strasbourg. Rae reported on the "shape eigenstate" interpretation of the data [7] which is capable of reproducing the 90° enhancement. Since the six alpha configuration is so elongated, with a corresponding large moment of inertia, the states based on a rotation of this structure would be very close in energy. Since they are high above the decay threshold they would also be broad. In this interpretation the "resonance" is not the yield from a single state, but the combined yield from a series of almost degenerate, overlapping states of different spin. However, as the states are from a rotational sequence, there is a unique phase relationship between the partial waves exciting each state, and this coherence of the phases gives rise to a highly aligned shape at 90° in the scattering system. This can be understood in semi-classical terms through the uncertainty principle ($\Delta l.\Delta\theta \sim \hbar$) and can explain the 90° peaking of the breakup yield.

Abe [8] also showed an interpretation of the data in terms of his Band Crossing Model. In this coupled channels calculation, including single and mutual excitations of the $0_1^+ 2_1^+$ and 0_2^+ channels, the $0_2^+ 0_2^+$ excitation function enhancement could be reproduced.

Summarising the position at the end of the Strasbourg meeting, the experimental evidence looked good - a resonance was observed close to the energy expected for the chain state, it was observed to decay into fragments which were themselves believed to be chains and the cross section ratios suggested decay from a highly deformed state. However, there were two features which cast some doubt on the chain hypothesis, the large width of the resonance and the fact that it could be reproduced in the band-crossing model, since this description does not require the assumption of a deformed structure in ^{24}Mg.

4 New Results Since Cluster '94

A number of groups turned their attention to the six alpha chain and a great deal of experimental work has been reported since the Strasbourg meeting. One of the first were the results from the Charissa group (Chappell et al. [9]) from a very similar experiment to that of Wuosmaa et al. These new data confirmed the resonance in the mutual $0_2^+0_2^+$ channel, but showed that there were even stronger resonances at the same energy in the $0_2^+3_1^-$ and $3_1^-3_1^-$ channels. If this were the same resonance it would cast doubt on the chain interpretation, since the 3_1^- state has a compact structure which should have no overlap with a chain arrangement. More recent data, again by Chappell et al. [10], shows that not only is the resonance much larger in the $3_1^-3_1^-$ channel (by a factor of 10), but it is followed by two even more striking resonances at higher energies. Using a rather powerful angular correlation technique which can be used to extract spin assignments in non-zero spin channels (Chappell et al. [11]), these $3_1^-3_1^-$ channel resonances are shown to fall on a continuation of the long known resonance sequence observed in the single and mutual 2_1^+ channel. Taken together, these results may indicate that the $0_2^+0_2^+$ resonance is the result of a weak coupling to the dominant resonance configuration based on the collective 2_1^+ and 3_1^- states in ^{12}C as identified in the alpha cluster calculations. Indeed, in a contribution to this conference Bremner et al. show preliminary data indicating that the resonances in the $3_1^-3_1^-$ channel disappear at higher energies, consistent with the band termination for this structure in the alpha cluster model.

A further blow to the chain state hypothesis came from $^{12}C(^{12}C, ^{8}Be)^{16}O$ results by the Strasbourg group, (Freeman et al. [12]) which reveal a resonance in the ^{16}O + ^{8}Be channel at exactly the same energy as the $0_2^+0_2^+$ resonance, and with the same spin [13]. Again, the spherical structure of the ^{16}O nucleus is inconsistent with decay from a linear chain arrangement. These $^{16}O+^{8}Be$ results were confirmed by additional measurements by groups in Catania (Alliotta et al. [14]) and Bombay (Mirgule et al. [15]). In a contribution to this meeting Di Pietro et al. [16] show that the ^{11}B + ^{13}N entrance channel can also give rise to yield in the 6α decay channel in this energy region.

The resonance observed in the ^{16}O + ^{8}Be channel is much narrower than that in the $0_2^+0_2^+$ channel (1.5 MeV compared to 4.0 MeV). It is possible to argue that this may reflect a chance coincidence of different resonances. To explore this the Charissa group performed a fine step excitation function (100keV) with a thin target (15keV), picking up the inelastic channels, along with the ^{16}O + ^{8}Be channel (Le Marechal et al. [17]). There results (figure 3) show that the resonance in the $3_1^-3_1^-$ channel is indeed as narrow as the ^{16}O + ^{8}Be channel. The $0_2^+0_2^+$ resonance does still appear as a much wider structure, although as the authors' report the position and width of the resonance change depending on the angular interval over which the data are summed (presumably reflecting the effect of interference with the large underlying background).

A further line of investigation was to look for evidence of chain states in the neighbouring nuclei using a similar approach. The Charissa group (Simmonds et al. [18]) reported no evidence for the 7-alpha chain in ^{28}Si in the region around the energy predicted by the alpha cluster model. The Bombay group did report preliminary evidence for the 5-alpha chain (Eswaran et al. [19]), but this has not yet been confirmed.

5 What Have We Learned and Where Do We Go Next?

The consensus view is that the results gathered over the last few years indicate that the resonance in the $0_2^+0_2^+$ channel is not related to the 6-alpha chain configuration. Whether it may exist in a different excitation energy region, is, of course, still an open question. However, in the course of this work we have learned (or re-learned) a few things. The first is that the extraction of resonance widths is a difficult job in heavy ion reactions, where the resonance generally sits on a sizeable background from competing reaction processes. We must always remember that the apparent position and width of a resonance may depend on the angular range over which the data is summed, and can appear different in different channels depending on the magnitude and phase of the background yields. The second lesson is, however, more worrying. Our approach to identifying the longer chains is to assume that they will decay into channels involving shorter chain nuclei. But do we really understand these shorter chains? For example, let me quote the title of a paper "Why is there no rotational band based on the 7.65MeV 0_2^+ state in ^{12}C?" (Friedrich et al. [20]). Note that although this appeared back in 1971, we are no further forward to answering the question in the title some 28 years later! Not only is there an absence of hard experimental data to confirm the chain state in ^{12}C, but there exist many theoretical studies over the last 30 years (too numerous to quote) which variously describe the 7.65MeV state as the chain, or argue that it has some other property, often more like a bent chain. Suggestions have been made that perhaps the states 10.3MeV (0^+) and 11.2MeV (2^+) might be candidates for the chain state and first member of a rotational band, but this is open to question. Much of our trouble with addressing these questions is the incomplete nature of our knowledge of the states in the relevant excitation region in ^{12}C. There are many broad, overlapping states and often spin assignments and partial decay widths are missing. Much of this spectroscopic information was determined over 20 years ago and it is undoubtedly the case that a big improvement could be made by exploiting modern high segmentation, high resolution detectors. Perhaps we will see results on this in the next conference in this series, and perhaps a definitive answer to whether there really is a chain configuration in ^{12}C.

Finally, and to finish on a more up-beat note, it is quite possible that we have been looking for chain states in the wrong nuclei! The recent exciting work by von Oertzen [21] has suggested that, based on the two centre well produced by the two α cluster nature of ^{8}Be, additional neutrons occupy molecular type orbits in the two centre potential. These orbits are analogous to electron orbits in an atomic molecule (e.g. H_2^+). They provide a concentration of the neutron wavefunction in the region between the alphas and give additional stability to the nucleus. This picture is confirmed by recent antisymmetrised molecular dynamics calculations. Von

Oertzen has identified candidate bands of states based on these configurations in ^{9}Be, ^{10}Be and ^{11}Be and a recent experiment on the breakup of ^{12}Be to ^{6}He + ^{6}He (M. Freer et al. [22]) may have revealed similar structures in this isotope.

Based on these ideas, it is possible to speculate that similar configurations might exist in longer chains, with the additional neutrons providing additional binding and stability to the structures. Hence it may well be easier to identify the chain configuration in more neutron rich isotopes, for example ^{16}C, ^{22}O or may be even ^{34}Mg. Such studies will become possible in the next few years as the next generation of radioactive beam facilities come on line and perhaps we will hear of some exciting new results at the next Cluster conference.

May I acknowledge the contributions of many colleagues in the Charissa collaboration, both past and present, with whom I have enjoyed exploring this topic.

References

1. K. Ikeda, N. Takigawa and H. Horiuchi, Suppl. Prog. Phys. Jpn. (1969) 464.
2. D.M. Brink, Nucl. Phys. A91 (1967) 1.
3. S. Marsh and W.D.M. Rae, Phys. Lett. B180 (1986) 185.
4. P. Chevallier et al., Phys. Rev. 160 (1967) 827.
5. M. Freer et al., Phys. Rev. C51 (1995) 1682.
6. A. Wuosmaa et al., Phys. Rev. Lett. 68 (1992) 1295 and A. Wuosmaa et al., Phys. Rev. C50 (1994) 2909.
7. W.D.M. Rae, A.C. Merchant and B. Buck, Phys. Rev. Lett 69 (1992) 3709.
8. Y. Hirabayashi, Y. Sakuragi and Y. Abe, Phys. Rev. Lett. 74 (1995) 4141.
9. S.P.G. Chappell et al., Phys. Rev. C51 (1995) 695.
10. S.P.G. Chappell et al., Phys. Lett. B444 (1998) 260.
11. S.P.G. Chappell et al., Phys. Rev. C53 (1996) 2879.
12. R.M. Freeman et al., Phys. Rev. C51 (1995) 3504.
13. S. Szilner et al., Phys. Rev. C55 (1997) 1312.
14. M. Aliotta et al., Z. Phys. A353 (1995) 43 and Nucl. Phys. A583 (1995) 281.
15. E.T. Mirgule et al., Nucl. Phys. A583 (1995) 287 and Phys. Rev. C56 (1997) 1943.
16. Di Pietro, Contributed Talk at this Conference.
17. R.A. Le Marechal et al., Phys. Rev. C55 (1997) 1881.
18. P. Simmonds et al., Phys. Rev. C51 (1995) 3500.
19. M.A. Eswaran et al., Contribution to INPS-95, India 1995.
20. H. Friedrich, L. Satpathy and A. Weiguny, Phys. Lett. B36 (1971) 189.
21. W. von Oertzen, Z. Phys. A354 (1996) 355 and A357 (1997) 37.
22. M. Freer et al., Phys. Rev. Lett. 82 (1999) 1383-1386.

A DEEP POTENTIAL DESCRIPTION OF $^{12}C+^{12}C$, $^{16}O+^{12}C$, AND $^{16}O+^{16}O$ SCATTERING: GROSS RESONANT STRUCTURES AND NUCLEAR RAINBOWS

YOSIO KONDŌ

Department of Natural Sciences, Kyoto Women's University,
35 Kitahiyoshi-cho, Imakumano, Higashiyama-ku, Kyoto 605-8501, Japan
E-mail: kondo@yukawa.kyoto-u.ac.jp

Since the observation of *nuclear molecules* nearly forty years ago, low-energy heavy-ion scattering has been described mainly by optical potentials with a very-*shallow* real depth. The purpose of this review is to demonstrate that deep optical potentials can provide a satisfactory account of not only scattering at intermediate energies but also the low energy observations for $^{12}C+^{12}C$, $^{16}O+^{12}C$ and $^{16}O+^{16}O$ scattering, thus providing a unified mean field description of the interaction between these nuclei over the whole energy range.

1 Introduction

Nearly forty years ago, measurements of $^{12}C+^{12}C$ scattering at energies close to the top of their mutual Coulomb barrier revealed the existence of sharp resonant structures in the excitation functions. [1] These structures were interpreted immediately as *nuclear molecules* in analogy with atomic-molecules, and an explanation using a shallow interaction potential was proposed. [2] Further measurements of the excitation functions at energies up to 40 MeV for light heavy-ion systems, like $^{12}C+^{12}C$, $^{16}O+^{12}C$, $^{16}O+^{16}O$ and $^{16}O+^{20}Ne$, showed the existence of a hierarchy of structures that could be loosely classified as *gross structures*, having characteristic widths of a few MeV, and superimposed on these *intermediate structures* with smaller widths. For the $^{16}O+^{16}O$ system, these two structures can be distinguished with little ambiguity. On the other hand, the $^{12}C+^{12}C$ system shows considerably more prominent intermediate structure which makes it less easy to see the gross structure. Finally, there may be an overall weak *fine structure*, perhaps of statistical origin. Gross and intermediate structures were also observed in total fusion and total reaction cross sections for these systems. [3,4,5]

It was long ago suggested [6] that the gross structure could be understood as a mean field phenomenon, represented by an overall optical potential of interaction between the two nuclei when in their ground states. The intermediate structure could then arise from couplings between this ground-state configuration and states in which one or both of the nuclei is excited. [7] The ground state configuration may be expected to dominate in the case of doubly closed-shell nuclei like ^{16}O, while the intermediate structure would be expected to be much more prominent in the case of strongly deformed and easily excited nuclei like ^{12}C. Indeed, this is what is observed when comparing the scattering of ^{16}O+^{16}O and ^{12}C+^{12}C. [8] In this report, our discussion will be concentrated on the gross structure which could be described by a mean field or an optical potential.

Traditionally, low-energy heavy-ion scattering has been described by optical potentials with a very *shallow* real depth. For example, work at Yale on ^{16}O+^{16}O scattering at energies up to about E_{cm}=36 MeV used a Woods-Saxon potential with a real part of depth 17 MeV. [6,8] Analyses of similar data for the ^{12}C+^{12}C system led to a Woods-Saxon potential with a depth of only 14 MeV. [9] For the ^{16}O+^{12}C system, [10] the energy-dependent depth of $7.5MeV + 0.4E_{cm}$ was used. These potentials are so shallow that the resonating partial waves have no nodes ($n = 0$) in their associated radial wavefunctions, which violates the requirement of the Pauli principle.

During the last decade, significant progress has been achieved in our knowledge and understanding of the optical potential between two light heavy-ions. Refractive effects, particularly nuclear rainbows, are especially evident in the angular distribution of ^{16}O+^{16}O elastic scattering at 350 MeV. [11] Nuclear rainbows are also observed at other energies and in other light heavy-ion systems such as ^{12}C+^{12}C and ^{16}O+^{12}C. Their angular locations and cross sections have led to the determination of the gross features of the local optical potentials and in many cases have removed ambiguities in the depths of the real parts of the optical potentials. [12,13,14,15] These potentials have real parts that are strongly attractive (typically several hundred MeV *deep*) with relatively-weak absorption. A good understanding of the theoretical basis of these features has been attained. [16]

The striking contrast between the shallow and the deep potentials has been the main subject of our research for the past ten years. The purpose of this report is to demonstrate that deep optical potentials [17,18,19,20] can also provide a satisfactory account of the low-energy observations for ^{12}C+^{12}C, ^{16}O+^{12}C, ^{16}O+^{16}O and ^{16}O+^{20}Ne scattering, thus providing a unified mean field description of the interaction between these nuclei over the whole energy range. In contrast to the use of shallow potentials, such a deep potential generates partial-wave radial functions with more realistic numbers of nodes, allowing us to consider that the Pauli principle has been satisfied in a microscopic way when the system is antisymmetrized.

2 Symmetric Systems

The symmetric $^{12}C+^{12}C$ and $^{16}O+^{16}O$ systems have been the subject of extensive analyses for many years. Since the systems consist of two identical nuclei, angular distributions are symmetric with respect to 90° and only the even partial waves contribute to the scattering amplitudes. Because of this symmetry, an important difference between deep and shallow potentials is likely to be overlooked in those systems. Since the space is rather limited for this report, I will refer to a review article [16] which has a good coverage of symmetric systems up to 1996.

2.1 The $^{12}C+^{12}C$ System

A pioneering study of refractive (rainbow) phenomena and its extension to lower energies have been performed by Brandan. [21,22] Recently, we have extended her potential to the low-energy region, and have achieved a moderate success in describing the gross structures for this system. [19]

2.2 The $^{16}O+^{16}O$ System

In ref. [17], we have demonstrated that the gross resonant structures in this system can be described successfully in terms of three ion-ion potentials, all of which are sufficiently deep to support wave functions with the proper number of radial nodes, consistent with the Pauli principle. The systematics observed in the analyses [11,12,13,14,23] of nuclear rainbow phenomena at E_{lab}=350, 250, 145 and 124 MeV favour the optical potential set [17] with the real volume integral $j_v \approx 340 MeV fm^3$.

3 Unsymmetric Systems

3.1 The $^{16}O+^{12}C$ System

Recently significant progress has been achieved for this system both in the rainbow [15] and in the resonance [18] regions. In ref. [15], we found a clear example of supernumerary rainbows for this system at E_{lab}=132 MeV. On the other hand, the previous analysis [21] was based upon inferior data.

In ref. [18], we found a deep optical potential which can describe gross structures in the low-energy region for this system. In figure 1, the calculated total absorption cross sections σ_{abs} are shown for both the shallow Siemssen potential (thin solid curve) and the deep optical potential (dashed curve). The predicted absorption cross section is rather smooth for the shallow potential

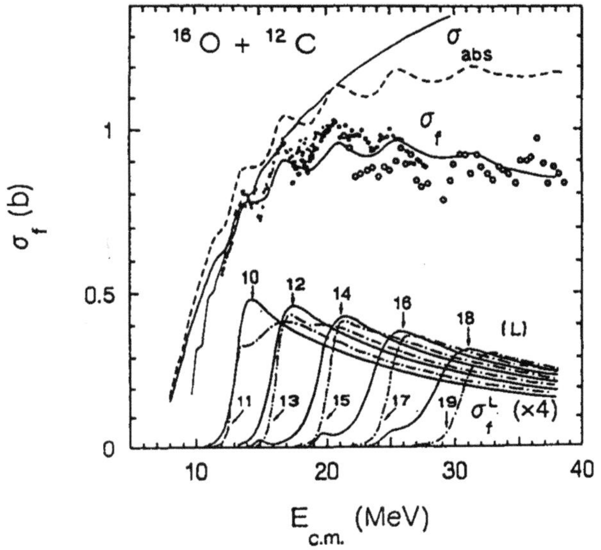

Figure 1. Fusion excitation function for the $^{16}O+^{12}C$ system. The thin solid and the dashed curves represent the total absorption cross sections calculated with the shallow optical potential [10] and the deep optical potential, [18] respectively. The thick solid curve represents the total fusion cross section calculated with the deep potential. In the lower part of the figure, partial fusion cross sections are shown by the solid (even L) and dash-dotted (odd L) curves (from ref. [18]).

except for the low-energy region, where it shows a narrowly-spaced step-like increase reflecting the existence of the $n = 0$ potential resonance band. On the other hand, the result for the deep optical potential shows regular oscillations over the whole energy region. The fusion cross section σ_f (the thick solid curve) predicted by the deep optical potential [18] shows regular oscillations with peak positions and magnitudes comparable to the data. In the lower part of the figure, the fusion cross sections from each grazing angular momentum are shown by solid (for even L) and dash-dotted (for odd L) curves. It can be seen that the gross fusion peaks appear as a manifestation of at least two overlapping potential resonance bands. For example, the peak at about $E_{cm} \sim 21MeV$ originates not only from the $L = 14$ member of the $N(= 2n+L) = 24$ band but also the $L = 15$ member of the $N = 23$ band. It also contains a small contribution from the $L = 16$ member of the $N = 22$ band. The peak at about $E_{cm} \sim 26MeV$ is mainly composed of $(L = 16, N = 24)$ and $(L = 17, N = 23)$ resonances. Thus, the observed spacing, $\Delta L \sim 2$, of the fusion oscillations is

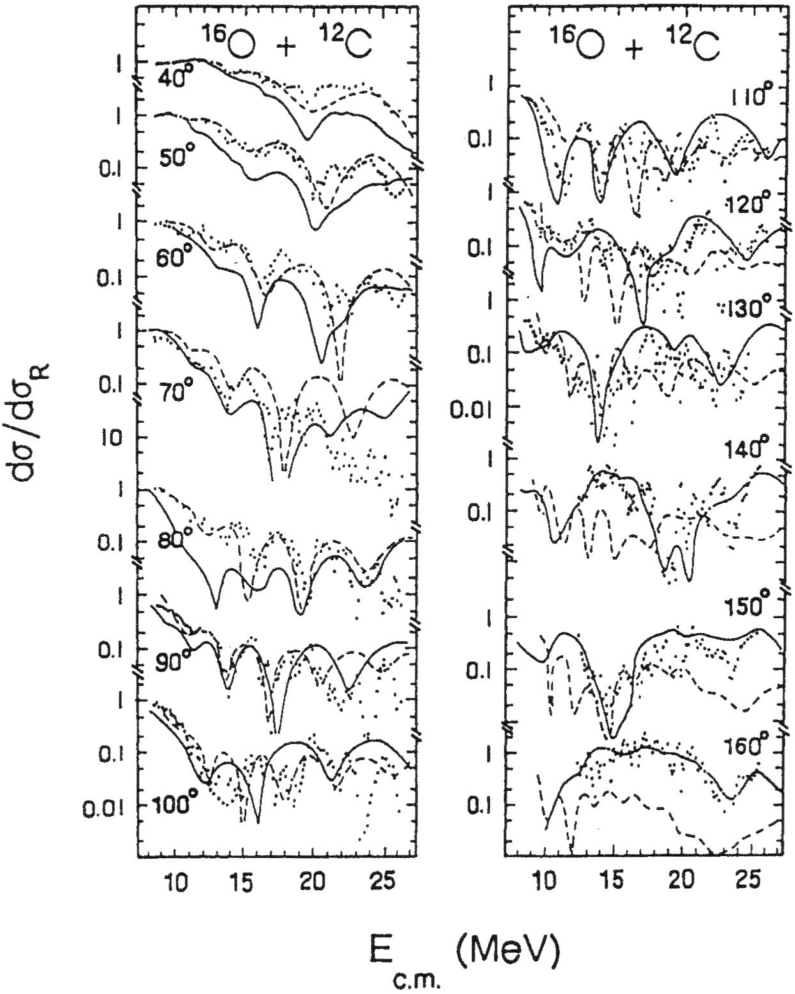

Figure 2. The elastic-scattering excitation functions for the $^{16}O+^{12}C$ system. The dashed curves represent the results for the shallow optical potential, [10] whereas the solid curves represent the results for the deep optical potential [18] (from ref. [18]).

reproduced naturally by using the deep optical potential. On the other hand, this spacing cannot be reproduced by the shallow potential, since the $n = 0$ band results in contributions from both even- and odd-L.

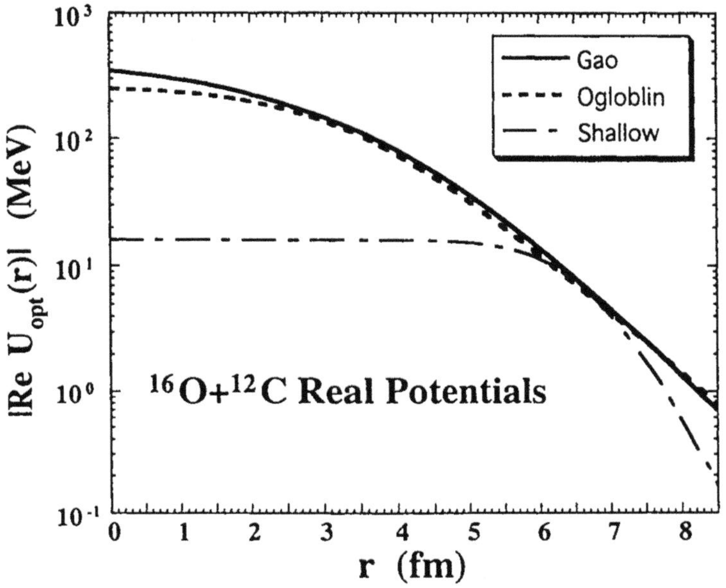

Figure 3. Real parts of optical potentials that fit $^{16}O+^{12}C$ scattering. The thick solid and the dashed curves represent the deep optical potential [18] and WSII of Ogloblin, [15] respectively. The dash dotted curve represents the shallow potential of Siemssen [10] at E_{cm}=21 MeV.

Figure 2 shows the elastic-scattering excitation functions for the $^{16}O+^{12}C$ system. The results using the deep potential (solid curves) and the shallow potential (dashed curves) are compared with the corresponding experimental data. As can be seen from the figure, the gross features of the energy dependence of the elastic-scattering excitation functions at backward angles are well reproduced by using the deep optical potential, while the shallow potential tends to underestimate the cross sections at backward angles. On the other hand, the deep potential tends to underestimate the cross sections at forward angles.

Figure 3 compares three real potentials for the $^{16}O+^{12}C$ system. The WSII potential (dashed curve) was obtained in the analysis of the rainbow data. [15] On the other hand, the shallow potential of Siemssen [10] (dash-dotted curve) and the deep optical potential [18] (solid curve) were obtained in the analysis of low-energy data. All of them have a similar value at $r \approx 6.5 fm$. It should be noted that the two deep potentials obtained from the rainbow data and the low-energy resonances are rather similar.

3.2 The ^{16}O+^{20}Ne System

Gross resonant structures observed in excitation functions for elastic ^{16}O+^{20}Ne scattering have been described successfully using a deep optical potential. [20] In this calculation, effects of alpha-particle transfer between two ^{16}O nuclei are taken into account through the LCNO formalism. [24] Since nuclear rainbow phenomena have not been observed for this system, the discrete ambiguity in the real potential depth has not yet been removed.

4 Summary and Discussion

Observation of refractive (rainbow) phenomena at intermediate energies allowed the unique determination of deep optical potentials for light heavy-ion systems. An extension of these deep potentials to low energies, as is demonstrated in this review, provides a natural and consistent description of scattering phenomena over the whole energy range.

For the ^{16}O+^{16}O system, a supersymmetric transformation [25] has demonstrated that the deep optical potential of ref. [17] can be simulated by a shallow (but singular at the origin) optical potential. This method is providing a powerful tool for the investigation of phase equivalent potentials. It should be noted, however, that for the unsymmetric ^{16}O+^{12}C system the deep potential [18] and the shallow potential [10] cannot be connected by the supersymmetric transformation, since they predict rather different cross sections at backward angles.

The WKB-RGM method [26] of Wada and Horiuchi provided a successful description of equivalent local potentials for α+^{16}O and α+^{40}Ca scattering. However, its application to the ^{16}O+^{16}O system was not very successful in the range of the real potential [17] and in reproducing the energy dependence of the interaction. [27] These defects are likely to originate from the poor choice of the effective interaction rather than the method itself. It is desirable to see improvements in this respect.

Acknowledgments

The author thanks C. Gao, Brian Robson, Ray Satchler and Y. Sugiyama for fruitful discussions and collaborations on this subject during the past ten years. I am grateful to Fred Barker for carefully reading the manuscript.

References

1. D.A. Bromley, J.A. Kuehner and E. Almquist, *Phys. Rev. Lett.* **4**, 365 (1960); E. Almquist, D.A. Bromley and J.A. Kuehner, *ibid.* **4**, 515 (1960).
2. E. Vogt and H. McManus, *Phys. Rev. Lett.* **4**, 518 (1960).
3. P. Sperr *et al.*, *Phys. Rev. Lett.* **36**, 405 (1976).
4. B. Fernandez *et al.*, *Nucl. Phys.* A **306**, 259 (1978).
5. J.J. Kolata *et al.*, *Phys. Rev.* C **21**, 579 (1980).
6. J.V. Maher *et al.*, *Phys. Rev.* **188**, 1665 (1969).
7. Y. Abe, Y. Kondō and T. Matsuse, *Prog. Theor. Phys.* Suppl. **68**, 303 (1980).
8. A. Gobbi *et al.*, *Phys. Rev.* C **7**, 30 (1973).
9. W. Reilly *et al.*, *Nuovo Cimento* A **13**, 897 (1973).
10. R.H. Siemssen, in *Proceedings of the Symposium on Heavy-Ion Scattering*, Argonne National Laboratory, 1971, edited by R.H. Siemssen (Argonne National Laboratory, Argonne, Ill., 1971) P. 145.
11. E. Stiliaris *et al.*, *Phys. Lett.* B **223**, 291 (1989); H.G. Bohlen *et al.*, *Z. Phys.* A **346**, 189 (1993)
12. Y. Kondō, F. Michel and G. Reidemeister, *Phys. Lett.* B **242**, 340 (1990).
13. Y. Sugiyama *et al.*, *Phys. Lett.* B **312**, 35 (1993).
14. Y. Kondō *et al.*, *Phys. Lett.* B **365**, 17 (1996).
15. A.A. Ogloblin *et al.*, *Phys. Rev.* C **57**, 1797 (1998).
16. M.E. Brandan and G.R. Satchler, *Phys. Rep.* **285**, 143 (1997).
17. Y. Kondō, B.A. Robson and R. Smith, *Phys. Lett.* B **227**, 310 (1989); *J. Phys. Soc. Jpn.* Suppl. **58**, 597 (1989).
18. C. Gao and Y. Kondō, *Phys. Lett.* B **408**, 7 (1997).
19. Y. Kondō, M.E. Brandan and G.R. Satchler, *Nucl. Phys.* A **637**, 175 (1998).
20. C. Gao, Y. Kondō and B.A. Robson, *Nucl. Phys.* A **529**, 234 (1991).
21. M.E. Brandan, *Phys. Rev. Lett.* **60**, 784 (1988).
22. M.E. Brandan, M. Rodríguez-Villafuerte and A. Ayala, *Phys. Rev.* C **41**, 1520 (1990).
23. M.E. Brandan, K.W. McVoy and G.R. Satchler, *Phys. Lett.* B **281**, 185 (1992).
24. C. Gao and Y. Kondō, *Phys. Lett.* B **219**, 40 (1989).
25. J.-M. Sparenberg and D. Baye, *Phys. Rev.* C **54**, 1309 (1996).
26. T. Wada and H. Horiuchi, *Prog. Theor. Phys.* **80**, 488 (1988).
27. S. Ait-Tahar *et al.*, *Nucl. Phys.* A **562**, 101 (1993).

MICROSCOPIC COUPLED-CHANNEL STUDY OF HEAVY-ION RESONANCES IN MULTI-CLUSTER EXIT CHANNELS

Y. SAKURAGI, M. ITO, M. KATSUMA, M. TAKASHINA, Y. KUDO, Y. HIRABAYASHI [†], S. OKABE [†] AND Y. ABE [‡]

Department of Physics, Osaka City University, Osaka 558-8585, Japan

[†] *Center for Information and Multimedia Studies, Hokkaido University, Sapporo 060-0810, Japan*

[‡] *Yukawa Institute for Theoretical Physics, Kyoto University, Kyoto 606-8502, Japan*

Heavy-ion resonances recently discovered in nuclear collisions of $^{12}C+^{12}C$ and $^{16}O+^{16}O$ systems leading to multi-cluster exit channels, $^{12}C+^{12}C \to 3\alpha+^{12}C$, $3\alpha+3\alpha$, $2\alpha+^{16}O$ and $^{16}O+^{16}O \to (\alpha+^{12}C)+^{16}O$, $(\alpha+^{12}C)+(\alpha+^{12}C)$, are studied theoretically by the coupled-channel (CC) and coupled-channel Born approximation (CCBA) calculations using microscopic cluster-model wave functions of the colliding nuclei, ^{12}C and ^{16}O, and nucleus-nucleus interactions based on the realistic G-matrix interaction. These resonance states correspond to highly excited states of the compound systems, ^{24}Mg and ^{32}S nuclei, at excitation energies more than 40$\sim$55 MeV. The the microscopic CC and CCBA calculations well reproduce the characteristic features of these new resonance states.

§1. Introduction

The study of heavy-ion resonances has entered into a new phase since the discovery [1] of a broad resonance around E_{cm}=32.5 MeV in the $^{12}C+^{12}C$ inelastic scattering leading to the $3\alpha+3\alpha$ exit channel via the mutual excitation of the ^{12}C nuclei to the 0_2^+ state at 7.65 MeV. This resonance energy corresponds to the excitation energy of about 46 MeV with respect to the ground state of ^{24}Mg. Since the 0_2^+ state is a famous state having a typical 3α-cluster structure, the observed resonance is discussed [1,2,3] as an evidence for a quasi-stable molecular state consisting of six α particles in this highly-excited region of the compound system, ^{24}Mg.

This resonance was initially discussed in connection to the so-called *6α linear-chain state*.[1,2] Later, another interpretation was proposed [3] by three of the present authors based on a microscopic coupled-channels (CC) calculation combined with an extended band-crossing model (BCM), in which they employed microscopic 3α-cluster-model wave functions and the density-dependent double-folding-model (DFM) interactions. The CC calculation was successful in reproducing the angular distributions and the excitation function of the 0_2^+–0_2^+ channel. The calculation also reproduced the famous resonance in the single-0_2^+ channel[4] and predicted more resonances in other inelastic channels, such as 3_1^-–0_2^+ and 0_3^+–0_2^+ ones, which were discovered in subsequent experiments by British group[5] at the right position. Recently, similar resonances have also been observed in the same enerygy range in other multi-cluster reaction channels, such as $^{12}C+^{12}C \to {}^8Be+^{16}O_{g.s.}$ and $^{16}O+^{16}O \to {}^{16}O(0_2^+)+^{16}O_{g.s.}$, $^{16}O(0_2^+)+^{16}O(0_2^+)$. Here, the 6.05 MeV 0_2^+ state of ^{16}O is known to have a well developed $\alpha+^{12}C$ cluster structure and thus the latter two exit channels have $(\alpha+^{12}C)+^{16}O_{g.s.}$ and $(\alpha+^{12}C)+(\alpha+^{12}C)$ cluster structures, respectively.

In the present paper, we present the latest results of our microscopic CC and CCBA study of heavy-ion resonances observed in those multi-cluster exit channels induced by the $^{12}C+^{12}C$ and $^{16}O+^{16}O$ collisions. In the study, we take into account the excitation of the ^{12}C, ^{16}O and ^8Be nuclei to their excited states having 3α, $\alpha+^{12}C$ and 2α cluster configurations, respectively, in addition to the normal collective states. We also analyze the "classical" resonances having the di-nuclear molecular structures.

§2. Formalism of the Microscopic Coupled-Channels Method

When we discuss the resonances observed in the inelastic channels, we solve coupled-channels equations by taking into account the excitations of the colliding nuclei. The real part of the diagonal (α=β) or coupling ($\alpha \neq \beta$) potential between channels α and β is given by the double-folding model. More specifically, the potential between the channel α in which the projectile and the target are in the ith and jth states, respectively, and the channel β in which the projectile and the target are in the kth and ℓth state, is calculated as

$$V_{\alpha(i,j),\beta(k,\ell)}(\boldsymbol{R}) = \int \rho_{ik}^{(P)}(\boldsymbol{r}_1)\,\rho_{j\ell}^{(T)}(\boldsymbol{r}_2)\,v_{\rm NN}(\boldsymbol{r}_1,\boldsymbol{r}_2,\boldsymbol{R})\,d\boldsymbol{r}_1\,d\boldsymbol{r}_2 \quad ,$$

where $\rho_{ik}^{(P)}(\boldsymbol{r}_1)$ ($\rho_{j\ell}^{(T)}(\boldsymbol{r}_2)$) is the transition density of the projectile (target) nucleus between the states i and k (j and ℓ), while $v_{\rm NN}$ is the density-dependent effective nucleon-nucleon interaction called the DDM3Y interaction.[6] In the case of identical nuclei, the channel wave functions are symmetrized with respect to the exchange of two identical nuclei in the standard way.[7]

For ^{12}C, we consider the coupling among the ground state (0_1^+) and the 2_1^+, 3_1^-, 0_2^+, 2_2^+ and 0_3^+ excited states in the CC calculation and we adopt the transition densities calculated by Kamimura with 3α RGM.[9] The latter three states are known to have a well-developed 3α-cluster structure. For ^{16}O, we consider the coupling among the ground state (0_1^+) and the 3_1^-, 0_2^+, 2_1^+,

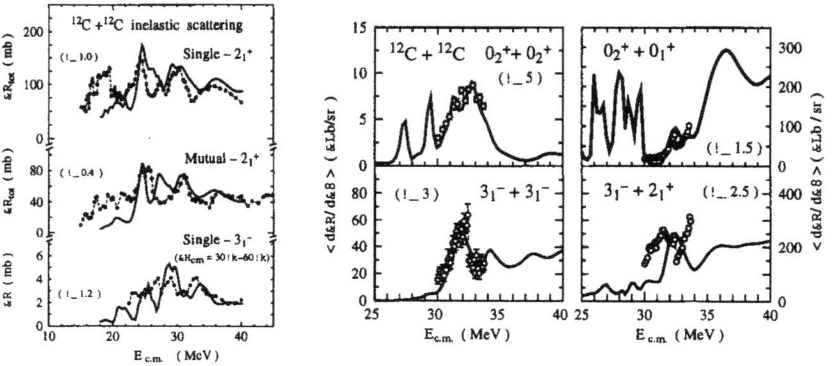

Fig. 1: Observed excitation functions of the ^{12}C+^{12}C inelastic scattering.[11–13] The left pannel shows the integrated cross sections of the single- and mutual-2_1^+ and single-3_1^- channels. The right one shows the differential cross sections averaged over $\theta_{\rm c.m.} = 70^\circ \sim 105^\circ$. The solid curves show the results of the CC calculation.

2_2^+ and 2_3^+ excited states and the transition densities are obtained by a CC-OCM calculation based on the α+^{12}C cluster model.[10] The latter four state are known to have a developed α+^{12}C cluster structures. For ^{8}Be, we include the 0^+ ground state and the 2^+ and 4^+ excited states and the transition densities are calculated in the 2α-cluster model with the bound-state approximation.

In addition to the real potentials calculated by DFM above, an Woods-Saxon-form imaginary part is added to the diagonal potential. The parameters are chosen so as to optimize the CC calculation to the experiments. However, the energy dependence of the parameters is taken to be smooth enough over a wide energy range and, hence, the resonance structure of the calculated excitation function is not due to any artificial energy dependence of the adopte imaginary potential. In the case of ^{8}Be+^{16}O systems, we switch-off the imaginary potential to avoid the ambiguity due to the choice of the parameters.

§3. Results and discussion

3.1. Resonances in the ^{12}C + ^{12}C and ^{16}O + ^{16}O inelastic scattering

Figure 1 shows the result of CC calculation for the ^{12}C+^{12}C inelastic scattering. Here we have adopted the J-dependent imaginary potential[8] supplemented by a surface-absorption potential, which simulates the absorption due to fusion and other direct-reaction channels. In order to avoid the ambiguity of the result due to the choice of the imaginary potential, we fixed the imaginary potential so as to reproduce the total cross sections of the low-lying collective-state channels (left pannel). No additional adjustment was

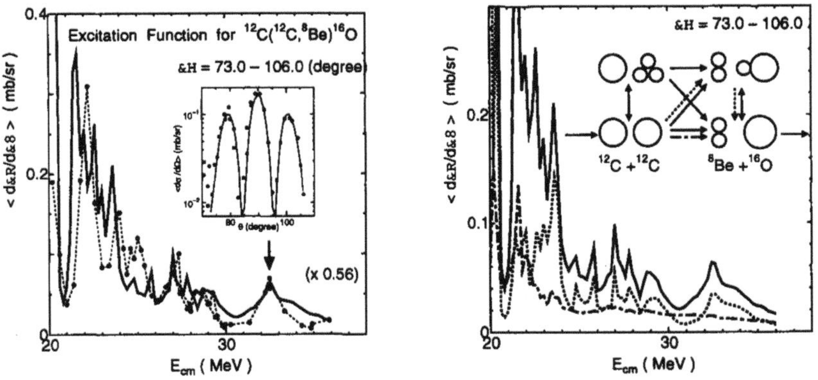

Fig. 2: Calculated and observed excitation functions of the differential cross section averaged over $\theta_{c.m.}$=73° ~106° for the ^{12}C+^{12}C → ^{8}Be$_{g.s.}$+^{16}O$_{g.s}$ α-transfer reaction (left) and contributions of the multi-step transitions (right). The dot-dashed curve shows the contribution of the direct transition and the dotted one shows that of the multi-step transition via the 2α+(α+^{12}C)-type four-cluster channels. All the calculation are shifted to low-energy side by 4 MeV and divided by 0.56 to be compared with the experimental data.[11,15,16]

made to optimize the fit to the cross sections in other channels shown in the right pannel. Nevertheless, the calculation reproduces the observed resonances not only in the shell-model-like collective-state channels (left) but also in the multi-cluster channels (right) reasonably well.

A similar but more impressive fit was obtained in the case of ^{16}O+^{16}O inelastic scattering to the single- and mutual-3_1^- channels and to the single-0_2^+ and mutual-0_2^+ ones, although no figure is presented here due to the lack of space. The result is given in the contribution paper to this conference.[14]

3.2. Resonances observed in ^{12}C + ^{12}C → ^{8}Be + ^{16}O reactions

In order to understand the resonances observed in the ^{12}C+^{12}C → ^{8}Be$_{g.s.}$+ ^{16}O$_{g.s}$ α-transfer reaction,[11,15,16] we perform a finite-range (FR) CCBA calculation, in which we first perform CC calculations for the ^{12}C+^{12}C and ^{8}Be+^{16}O systems separately to generate the distorted waves and calculate the transition amplitude for the α-transfer reaction. We adopt the post-form formalism and the interaction between α and ^{8}Be in ^{12}C and that between α and ^{12}C in ^{16}O are taken to be the same as those used in a previous FR-DWBA study.[17] The spectroscopic factors for the ^{12}C and ^{16}O states are taken from the microscopic cluster-model calculations.

The distorted waves for the ^{12}C+^{12}C channels are obtained by the same CC calculation as that gave the results of Fig. 1. For the ^{8}Be+^{16}O system, we perform a CC calculation by taking into account the excitations of ^{8}Be and ^{16}O to generate the distorted waves for the exit channels. In the ^{8}Be+^{16}O system, the ^{8}Be excitation to the 2^+ and 4^+ states is found to play an essential role in generating the resonances in this system. The coupling between the 0_2^+ and 2_1^+ states in ^{16}O also plays an important role in generating a resonance structure in the ^{8}Be+^{16}O(0_2^+) channel.

In the left pannel of Fig. 2, we compare the FR-CCBA calculation with the experimental data.[11,15,16] The calculation well reproduces the essential feature of the observed resonance structure in the entire energy range. The calculation also reproduces the angular distribution, which is only available at $E_{\rm c.m.}$=32.5 MeV (indicated by arrow in Fig. 2). The dominant partial wave at this energy is found to be J=18, although other lower partial waves also have non-negligible contributions to produce a characteristic enhancement at $\theta_{\rm c.m.}$=90°.

The right pannel of Fig. 2 shows the contributions of the direct and multi-step transitions together with the full calculation. It is seen that most resonance peaks in this energy range are generated by the multi-step transition through the ^{8}Be$_{\rm g.s.}$+^{16}O(0_2^+, 2_1^+) channels. Since the 0_2^+ and 2_1^+ states in ^{16}O have an (α+^{12}C)-type cluster structure, the present result gives an evidence for the existence of (α+α)+(α+^{12}C) type four-cluster resonance states in the highly-excited region of the compound system, ^{24}Mg.

References

1. A. Wuosmaa *et al.*, *Phys. Rev. Lett.* **68**, 1295 (1992).
2. W.D.M. Rae, A.C. Merchant and B. Buck, *Phys. Rev. Lett.* **68**, 3709 (1992).
3. Y. Hirabayashi, Y. Sakuragi and Y. Abe, *Phys. Rev. Lett.* **74**, 4141 (1995).
4. B. Fulton *et al.*, *Phys. Lett.* B **136**, 322 (1984).
5. S. Chappell *et al.*, *Phys. Rev.* C **51**, 695 (1995).
6. M. El-Azab Farid and G. R. Satchler, *Nucl. Phys.* A **438**, 525 (1985).
7. Y. Kondo, D.A. Bromley and Y. Abe, *Phys. Rev.* C **22**, 1068 (1980).
8. Y. Abe, Y. Kondo and T. Matsuse, *Prog. Theor. Phys. Suppl.* **68**, 303 (1980).
9. M. Kamimura, *Nucl. Phys.* A **351**, 456 (1981), and private communication.
10. S. Okabe, unpublished.
11. R.A.Le Marechal *et al.*, *Phys. Rev.* C **55**, 1881 (1997).
12. T.M. Cormier *et al.*, *Phys. Rev. Lett.* **40**, 924 (1978).
13. B. Fulton, T.M. Cormier and B.J. Herman, *Phys. Rev.* C **21**, 198 (1980).
14. M. Katsuma *et al.*, contribution paper to this conference, p35.
15. R.M. Freeman, *et al.*, *Phys. Rev.* C **51**, 3504 (1995).
16. M. Aliotta, *et al.*, *Z. Phys.* A **353**, 43 (1995); M. Aliotta, *et al.*, *Z. Phys.* A **354**, 119 (1996). S. Cherubini, *et al.*, *Z. Phys.* A **357**, 291 (1997).
17. E. Mathiak *et al.*, *Nucl. Phys.* A **259**, 129 (1976).

'A NEW COUPLING POTENTIAL' FOR THE SCATTERING OF DEFORMED LIGHT HEAVY-ION NUCLEI

I. BOZTOSUN and W. D. M. RAE

Department of Physics, University of Oxford, Keble Road, Oxford OX1 3RH, United Kingdom

E-mail: i.boztosun1@physics.ox.ac.uk

A new approach has been used to fit the elastic, inelastic and mutual inelastic scattering data as well as the observed resonances in the light heavy-ion scattering reactions. This Coupled-Channels-based approach involves replacing the first order coupling potential by a new empirical coupling potential. It has been applied to the study of four different systems $^{12}C+^{12}C$, $^{16}O+^{28}Si$, $^{12}C+^{24}Mg$ and $^{16}O+^{24}Mg$ where there remain many unsolved problems in the conventional CC and Optical Model Calculations. Excellent fits to the experimental data have been obtained.

1 Introduction

The surprising observations in the elastic and inelastic scattering of light heavy-ions have stimulated a great deal of interest over the last few decades. A variety of theoretical interpretations, based on dynamical models or purely phenomenological assumptions, have been proposed to explain or merely reproduce these observations [1]. However, although most of those descriptions give reasonably good fits to the angular distributions, and a few fit the elastic excitation functions, there is as yet no satisfactory physical explanation of all the observed phenomena. In particular for the highly deformed systems, it has been impossible to fit the measured elastic and inelastic scattering data as well as the resonances and the excitation functions simultaneously over a wide energy range.

So far although the elastic scattering of the heavy-ions has been studied in detail, the inelastic scattering of heavy-ions has received little attention and no detailed systematic studies have been carried out. Whereas it has been shown that the inclusion of the excited states has a significant effect on the scattering and when one looks at the calculations performed for these systems [15,8,16,13,5], many problems are immediately seen such as magnitude problems for the predicted 2^+ cross-sections, the lack of the correct oscillatory structure, the backward rise in the cross sections and out of phase problems at large angles and in the excitation functions *etc.*

Therefore, here we are concerned with the measured experimental data for $^{12}C+^{12}C$, $^{16}O+^{28}Si$, $^{12}C+^{24}Mg$ and $^{16}O+^{24}Mg$ in an attempt to find a global model which fits the experimental data for the elastic and the excited state data as well as the resonances and the single angle excitation functions (at 90° or 180° in the centre of mass) simultaneously using a slightly energy dependent or a fixed real potential over a significant wide energy range for all these different systems.

The first system we considered for this purpose was that of $^{16}O+^{28}Si$. The elastic and inelastic scattering as well as 180° excitation functions for the ground and first excited states of ^{28}Si have been studied from E = 29 MeV to 81 MeV in the laboratory system [20]. The experimental data show a strong backward rise depending on bombarding energy and manifest "*anomalous large angle scattering*" (ALAS). There is also a discrepancy between experimental data and theoretical predictions for the 180° excitation function of the ground and the first excited states, which exhibit pronounced and regular gross structures over the whole energy range [7].

2 The Standard Coupled-Channels Model

Although we have considered and studied four different reactions in detail, we are only going to show some of the results for the $^{16}O+^{28}Si$ and $^{12}C+^{12}C$ reactions. The details of the models and a complete set of the results for all four reactions will be published in the forthcoming papers.

We describe the interaction between ^{16}O and ^{28}Si nuclei with a deformed optical potential. The real potential is assumed to have the square of a Woods-Saxon shape and the parameters fixed to reproduce the folding model potential of Ref. 2 i.e.

$$V_N(r) = \frac{-V_0}{(1+exp(r-R)/a)^2} \tag{1}$$

with V_0=706.5 MeV, R=$r_0(A_p{}^{1/3}+A_t{}^{1/3})$ with r_0=0.7490 fm and a=1.40 fm. The parameters of the real potential were fixed as a function of energy and were not changed in the present calculations although it was observed that small changes could improve the quality of the fits. The Coulomb potential was assumed to be that of a uniformly charged nucleus with radius 5.556 fm deformed in the usual way.

The imaginary part of the potential was taken as in Ref 2. as the sum of a Woods-Saxon volume and surface potential i.e.

$$W(r) = -W_V f_{(r,R_V,a_V)} + 4W_S a_S df_{(r,R_S,a_S)}/dr \tag{2}$$

$$f_{(r,R,a)} = \frac{1}{(1+exp((r-R)/a))} \tag{3}$$

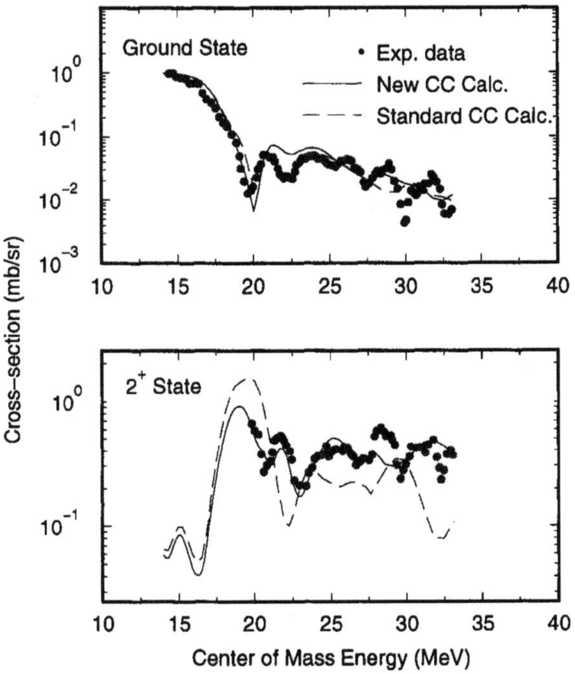

Figure 1. The *dashed* lines are the results of *standard CC calculations* and *solid* lines are the results of *new coupling potential* for $^{16}O+^{28}Si$ for the 180° excitation functions.

with W_V=59.9 MeV, a_V=0.127 fm and W_S=25.0 MeV, a_S=0.257 fm. These parameters were also fixed in the calculations and only their radii increased linearly with energy according to the following formulae.

$$R_V = 0.06084 E_{cm} fm/MeV - 0.442 fm \tag{4}$$

$$R_S = 0.2406 E_{cm} fm/MeV - 2.191 fm \tag{5}$$

Since the target nucleus ^{28}Si is well known to be strongly deformed, it is essential to treat its collective excitation explicitly in the framework of the coupled-channels formalism. It has been assumed that the target nucleus has a static quadrupole deformation, and that its rotation can be described in the framework of the collective rotational model. It is therefore taken into

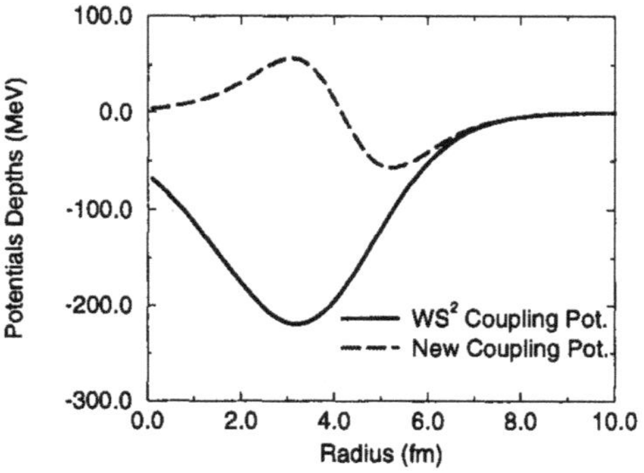

Figure 2. For $^{16}O+^{28}Si$, the comparison of the ***standard coupling potential*** which is the first derivative of the central potential and our ***new coupling potential*** which is parameterised as the 2nd derivative of Woods-Saxon shape and which has V=155.0 MeV, R=4.1561 fm and a=0.81 fm.

account by deforming the real optical potential in the following way

$$R(\theta, \phi) = r_0 A_p^{1/3} + r_0 A_t^{1/3}[1 + \beta_2 Y_{20}(\theta, \phi)] \tag{6}$$

where β_2=-0.64 is the deformation parameter of ^{28}Si. This value is actually larger than the value calculated from the known BE(2). However this larger β_2 was needed to be able to fit the magnitude for the 2^+ data.

In the present calculations the first two excited states of the target nucleus ^{28}Si: 2^+ (1.78 MeV) and 4^+ (4.62 MeV) were included and the 0^+-2^+-4^+ coupling scheme was employed. The reorientation effects for 2^+ and 4^+ excited states were also included. The calculations were performed with an extended version of the code CHUCK [19].

Some of the results for the ground and the first excited states of the $^{16}O+^{28}Si$ reaction for the 180° excitation functions are shown in figure 1. The magnitude of the cross-sections and the phase of the oscillations are given correctly at most angles. However when one looks at the 2^+ state results in detail, it is apparent that towards large angles at higher energies the experimental data and our predictions are out of phase. This problem is clearly seen in the 180° excitation functions which is shown in the figure. In the past a number of models have been proposed, ranging from isolated resonances [12,14] to cluster exchange between the projectile and target nucleus [13,17] (see Ref. 1 for a detailed discussion). We have attempted to overcome these problems by

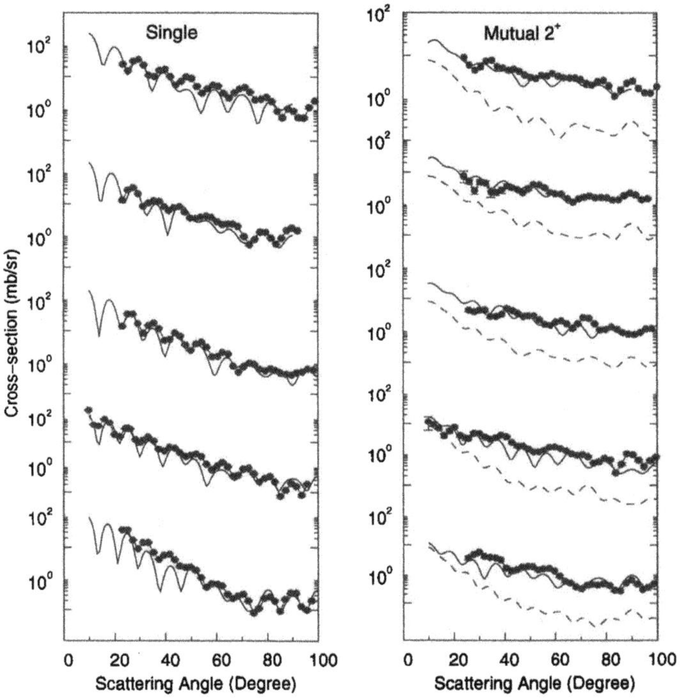

Figure 3. The single and mutual 2^+ inelastic scattering data for the $^{12}C+^{12}C$ reaction. The ***solid*** lines are the results of ***the new coupling potential***, while the ***dashed*** lines are the results of ***standard CC model.***

considering: (i) Changes in the real and imaginary potentials, (ii) inclusion of 6^+ excited state, (iii) changes in the β_2 value and the inclusion of β_4 and some other ideas. These attempts failed to solve the problems at all [13]. We were unable to get an agreement with the elastic and the 2^+ inelastic data as well as the 180° excitation functions simultaneously.

3 New Coupling Potential

The solution to these problems comes from a new empirical double-peaked coupling potential as shown in figure 2. Here we employed the same method with the same potential parameters except the depth of real potential and β_2

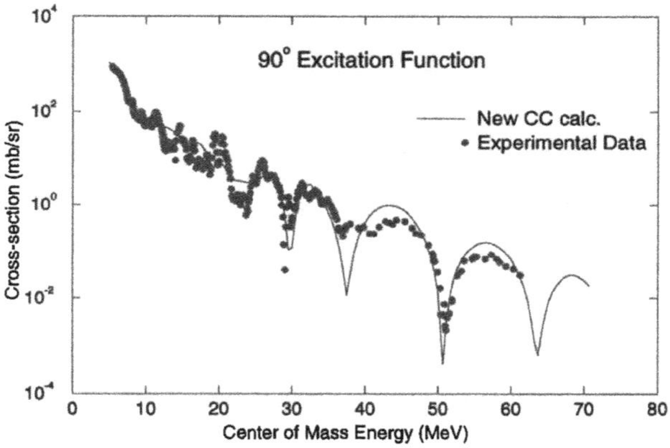

Figure 4. The $^{12}C+^{12}C$ 90° excitation function for the elastic scattering using smoothed averaged parameters with little dependence on energy except for the depth of the imaginary potential.

value. We needed to readjust them as V_0=750.5 MeV and β_2=-0.338, which is the exact value as derived from life time of the 2^+ state [8,6]. (However it should be noted that since the depth of the new coupling potential is arbitrary the meaning of β_2 is not well defined.) As shown in figure 1, this new coupling potential has solved the out of phase problem and fits the ground state data and the 180° excitation functions simultaneously.

Other systems we have considered are $^{12}C+^{24}Mg$ and $^{16}O+^{24}Mg$. Fifteen complete angular distributions of the elastic scattering of $^{12}C+^{24}Mg$ system were measured at energies around the Coulomb barrier and were published recently [18]. We have studied these fifteen complete elastic scattering angular distributions as well as some inelastic scattering data measured by Carter et al. some twenty years ago [4,10]. Excellent agreement to the data was obtained by using this new coupling potential for these data. Our model has also solved some problems in $^{16}O+^{24}Mg$ scattering [17].

The final system we have considered is that of $^{12}C+^{12}C$, which has been studied extensively over the last 40 years. There has been so far no model, which fits the elastic, inelastic scattering data and mutual excited state data as well as the resonances and excitation functions simultaneously. Another problem is the predicted magnitude of excited state cross sections, in particular for the mutual 2^+ channel. The conventional CC model underestimates its magnitude by a factor of at least two and often much more [15,21,22]. We have also observed this in our conventional CC calculations as shown in figure 2 with dashed lines. There are also resonances [3,9] observed at low energies, which have never been fitted by a potential, which also fits either the angular distributions or the excitation functions. Therefore, the experimental data at many energies between 10.0 and 160.0 MeV in the laboratory system have been studied simultaneously to attempt to find a global potential. We have been able to fit the energy average of all the available ground, 2^+, mutual 2^+ and the backgrounds in the integrated cross-sections of Cormier [9] as well as the main gross features of the 90° excitation function [11] as shown in the figure 4. simultaneously by using this new coupling potential. Initial calculations of resonances using no imaginary potential are hopeful but there are problems with the widths of the resonances.

To summarise, as we have indicated, while these four systems show quite different properties and problems, a unique solution has come from a new *empirical* coupling potential. Although the origin of this new coupling potential is still speculative and needs to be understood from a microscopic viewpoint, the approach outlined here is, we believe, universal and applicable to all the systems. We believe that this work represents an important step in understanding the elastic and inelastic scattering of light deformed heavy-ion systems.

Acknowledgments

I. Boztosun would like to thank the Turkish Council of Higher Education (YOK) and Erciyes University, Turkey, for their financial support.

References

1. P. Braun-Munzinger and J. Barrette, *Phys. Rev.* **87**, 209 (1982).
2. A. M. Kobos and G. R. Satchler, *Nucl. Phys.* A **427**, 589 (1984).
3. E. Almqvist *et al*, *Phys. Rev. Lett.* **4**, 515 (1960).
4. J. Carter *et al*, *Nucl. Phys.* A **273**, 523 (1976).
5. P. Braun-Munzinger *et al.*, *Phys. Rev. Lett.* **38**, 944 (1977).
6. D. Schwalm *et al*, *Nucl. Phys.* A **293**, 467 (1977).
7. J. Barrette *et al.*, *Phys. Rev. Lett.* **40**, 445 (1978).
8. A. Dudek-Ellis *et al*, *Phys. Rev.* C **18**, 1039 (1978).
9. T. M. Cormier *et al*, *Phys. Rev. Lett.* **40**, 924 (1978).
10. J. Carter *et al*, *Nucl. Phys.* A **297**, 520 (1978).
11. R. G. Stokstad *et al*, *Phys. Rev.* C **20**, 655 (1979).
12. J. Barrette *et al.*, *Phys. Rev.* C **20**, 1759 (1979).

13. M. A. Franey *et al.*, *Phys. Lett.* B **81**, 132 (1979).
14. S. J. Sanders *et al.*, *Phys. Rev.* C **21**, 1810 (1980).
15. R. Wolf *et al*, *Z. Phys.* A **305**, 179 (1982).
16. S. Kubono *et al.*, *Phys. Rev.* C **31**, 2082 (1985).
17. R. L. Filho *et al*, *Phys. Rev.* C **39**, 884 (1989).
18. W. Sciani *et al*, *Nucl. Phys.* A **620**, 91 (1997).
19. P. D. Kunz, CHUCK, An optical model search code, unpublished.
20. I. Boztosun, *IOP Nuclear Physics Conference*, Liverpool 15-17 April 1998, abstract D3; to be published.
21. Y. Sakuragi, *invited paper, ICC '99*, Croatia and private communication.
22. P. E. Fry, *PhD Thesis*, University of Oxford (1997), unpublished.

REFRACTIVE EFFECTS IN COLLISIONS BETWEEN CARBON AND OXYGEN NUCLEI: FROM RESONANCES TO AIRY STRUCUTRES

M.P. NICOLI[a,f], S. SZILNER[a,b], F. HAAS[a], R.M. FREEMAN[a], N. AISSAOUI[a], C. BECK[a], A. ELANIQUE[a], R. NOUICER[a], A. MORSAD[c], Z. BASRAK[b], M.E. BRANDAN[d], G.R. SATCHLER[e], and CHARISSA COLLABORATION[f,g,h]

[a]*Institut de Recherches Subatomiques, UMR7500, CNRS-IN2P3 and Université Louis Pasteur, BP 28, F-67037 Strasbourg Cedex 2, France*

[b]*Ruđer Bošković Institute, Zagreb, Croatia*

[c]*Faculté des Sciences Ben M'Sik, Université Hassan II, Casablanca, Morroco*

[d]*Instituto de Física, Universidad Nacional Autónoma de México A.P. 20-364, México 01000 DF, Mexico*

[e]*Department of Physics & Astronomy, University of Tennessee Knoxville, TN 37996, USA*

[e]*Physics Division, Oak Ridge National Laboratory Oak Ridge, TN 37831-6373, USA*

[f]*School of Physics & Astronomy, University of Birmingham, Edgbaston, Birmingham, B15 2TT, UK*

[g]*Department of Nuclear Physics, University of Oxford, Keble Road, Oxford OX1 3RH, UK*

[h]*Department of Physics, University of York, Heslington, York YO1 5DD, UK*

Complete angular distributions of the $^{16}O+^{16}O$ elastic channel have been measured at bombarding energies between 5 and 10 MeV per nucleon at forward angles ($\theta_{lab} = 5°$ to $20°$) using the Q3D spectrometer and at backward angles ($\theta_{lab} = 20°$ to $50°$) using two Position Sensitive Silicon detectors in kinematical coincidence. Refractive effects giving rise to nuclear rainbows and the associated Airy structures have been observed for this system. An optical model analysis has been done where the real part of the potential has been determined with no ambiguity over the whole energy range whereas the absorption (imaginary part) changed drastically around $E_{lab} = 90$ MeV. We will then give a brief review of the rainbow supernumeraries observed in the elastic scattering of $^{12}C+^{12}C$ system and evidence for the observation of these effects in the inelastic scattering will be also presented.

1 Introduction

The number of open channels[1] in collisions involving the light heavy ions ^{12}C and ^{16}O is small and the absorption is thus weaker than for other heavy ion collisions. The signatures of this interaction "transparency" are the presence of resonant phenomena in the low energy domain (E ≤ 5 MeV/u) and at higher energies, the occurence of refractive effects like nuclear rainbows and the associated Airy structures. In the higher energy regime, the measurement of accurate and extensive $^{16}O+^{16}O$ elastic scattering differential cross sections, obtained by the Jaeri[2] and Berlin[3] groups, have allowed the observation of strong refractive effects (the first A_1 and second A_2 Airy minima have been identified), which in turn have permitted a better understanding of the interior part of the interaction potential and to extract "unique" optical potentials: strongly attractive (deep) with relatively weak energy dependent absorption. The aim of our work was to find a link between the high energy regime (≥ 10 MeV/u) where deep potentials have been obtained and the low energy regime (≤ 5 MeV/u) where the data are usually described with unrealistic shallow potentials[4] and also to see how far down does this high energy description work by using an optical model analysis.

2 Experimental methods

The $^{16}O+^{16}O$ reaction was carried out at the Strasbourg Tandem accelerator Vivitron which is well suited for our studies in the intermediate energy domain between 5 and 10 MeV/u. The 9 following bombarding energies have been chosen: $E_{lab} = 75.0$, 80.6, 87.2, 92.4, 94.8, 98.6, 103.1, 115.9 and 124 MeV which correspond to maxima and minima of the $\theta_{cm} = 90°$ elastic scattering excitation function published by the Oak Ridge group[5] and which is the only detailed data available for this reaction in that energy range. The oxygen targets used were self supporting BeO films of $\sim 20\mu g/cm^2$ thickness[6]. Detailed and complete angular distributions have been measured at all energies. The most forward angles ($5° \leq \theta_{lab} \leq 20°$) of a given angular distribution were covered by a Q3D magnetic spectrometer having a proportional counter at its focal plane. Spectrometer measurements were taken every $\Delta\theta_{lab} = 0.5°$. The cross sections at larger angles were measured by a fixed kinematical-coincidence setup consisting of two position-sensitive silicon detectors (area = 5x1 cm^2) placed 7.8 cm from the target, on both sides of the beam, and covering angles between 15° and 50°, and -35° and -70°, respectively. All details concerning the data reduction, in particular the identification of the elastic channel in both detection systems and the extraction of cross sections can be found in Ref. 7.

3 Optical model analysis

Figure 1a shows the elastic angular distributions and fits obtained with the code Ptolemy for the higher energies of our work. They correspond to the combined results obtained with the Q3D (forward) and the Si detectors (backward). At larger angles, angular distributions are characterized by a modulation which can be explained by peripheral partial waves interfering with waves refracted within the interior of the nucleus-nucleus interaction. The optical potential ($U = V + iW$) used has a real part (V) which describes the nuclear interaction and an imaginary part (W) which describes the absorption (the flux carried away by all reaction channels). This analysis has been done in two steps: in the first, a squared Woods-Saxon ($WS2$) potential has been used to describe the real and imaginary part, where the absorption is composed of a volume plus a surface term. In the second step, the real part has been described by the folded potential characterizing the partial overlap of the two nuclei (all details concerning this analysis can be found in Ref. 8). For this analysis, a monotonic variation of the volume integrals J_v and J_w [9] as a function of energy has been imposed and we have also adopted the convention that the experimental errors should be the same (~ 10 %) over the entire angular domain in order to increase the data weight of the backward angles which are essential for the observation of refractive effects. It can be seen in Figure 1a that the fits (obtained from the two descriptions) seem to be satisfactory both for the amplitude and for the position of the structures. The shape of the real and imaginary parts as a function of interaction radius are represented in Figures 1b and 1c. To obtain reasonable fits, we had to change the absorption while keeping the real part as constant as possible, this is shown on Figure 1c. The real part is practically constant throughout the energy domain and is strongly attractive, but the shape of the absorption changes with energy. In Figure 1b, the imaginary potential at 124 MeV is mostly of the volume type. At 75 MeV, quite a different situation has set in. The absorption is separated into two parts: a volume plus a surface term. The potential shown in Figure 1b for 95 MeV, a medium energy in this study, is intermediate between the two extreme situations. As can also be observed, the imaginary potential is weaker than the real one at all radii. A change in the shape of the absorption in this energy region could be anticipated, as being the signature of reaction mechanisms evolving from fusion + direct reactions at the lowest energies (represented by the "box + surface-peak" imaginary potential) to a more spatially-uniform absorption at energies near $E/A = 10$ MeV. Binary reactions were measured simultaneously with these elastic scattering data, and once their analysis is completed [10], they might help to explain the results encountered in this work. At the present time we do not have any suggestion why the change in the character of the imaginary potential near 90 MeV should occur so abruptly, particularly since fusion cross sections seem to have reached saturation at about 60 MeV [11].

4 $^{12}C+^{12}C$ System

The $^{12}C+^{12}C$ reaction is certainly the one which has been the most studied. In the 90° excitation function for the $^{12}C+^{12}C$ elastic scattering at bombarding energies between the Coulomb barrier and 10 MeV per nucleon, complex structures of narrow and broad peaks and valleys have been observed [12,13]. The optical potentials used have a deep real part and a shallow imaginary part, and the broad structures observed in the 90° excitation function of the $^{12}C+^{12}C$ elastic scattering [12] have been interpretated in terms of refractive effects [14]. The remarkable deep minimum at $E_{lab}=102.1$ MeV has been identified as the second Airy minimum A_2 (Figure 2a). No experimental results exist for A_1 which is predicted to occur at ~ 130 MeV [14]. The presence of the precise energy of this minimum is essential to confirm the validity of the interaction potential. It has been noted by Marty [15] that the deep minimum (A_2) at 90° in the elastic scattering (0_1^+,0_1^+) at $E_{lab}=102.1$ MeV was correlated with minima in the inelastic channels: at an incident energy of E_{lab} - 2×4.4 MeV for the (2^+,0_1^+) channel and at an energy of E_{lab} - 4×4.4 MeV for the (2^+,2^+) channel where 4.44 MeV is the excitation energy of the ^{12}C first excited 2^+ state. For $E_{lab}=102.1$ MeV, this would correspond to minima at energies of 93.3 and 84.5 MeV for the inelastic (2^+,0_1^+) and (2^+,2^+) channels respectively (Figure 2a). This correlation is not understood, and its explanation remains an experimental and theoretical challenge. The effects in the 90° excitation function should be particularly strong for reactions with channel spin 0 in the entrance and exit channels. For other exit channels like the $^{12}C+^{12}C$ (2^+,0_1^+) and (2^+,2^+) channels, the effects at 90° should be partially washed out due to the possible contributions of different intrinsic spin projections. This would not be the case for the $^{12}C+^{12}C$ (0_1^+,0_2^+) inelastic channel in which the 0_2^+ state at $E_x=7.65$ MeV is excited. If for this channel the correlation noted by Marty [15] holds, its 90° excitation function should present an Airy minimum A_2 at $E_{lab}=86.8$ MeV corresponding to A_2 at 102.1 MeV in the (0_1^+,0_1^+) channel and an Airy minimum A_1 at $E_{lab}=114.7$ MeV corresponding to the predicted A_1 at 130 MeV in the elastic channel [14]. Figure 2b shows preliminary results from an experiment carried out in Australia (ANU) in April '99, which involved measuring the excitation function of the inelastic channel $^{12}C(0_1^+)+^{12}C(0_2^+)$ at 20 energies between $E_{lab}=80$ to 120 MeV. Preliminary results show two deep

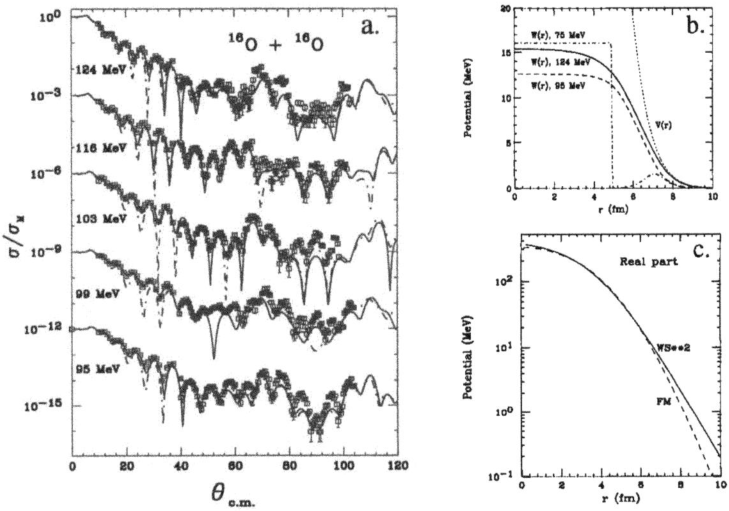

Figure 1: (a) Elastic scattering data and optical model calculations with phenomenological (dash) and microscopic (solid) potentials. (b) The imaginary potential at the three bombarding energies. The dotted curve shows the real potential at $E_{\mathrm{lab}} = 124$ MeV. (c) Comparison between the folding model (dash curve) and the WS2 (full curve) real parts at $E_{\mathrm{lab}} = 124$ MeV.

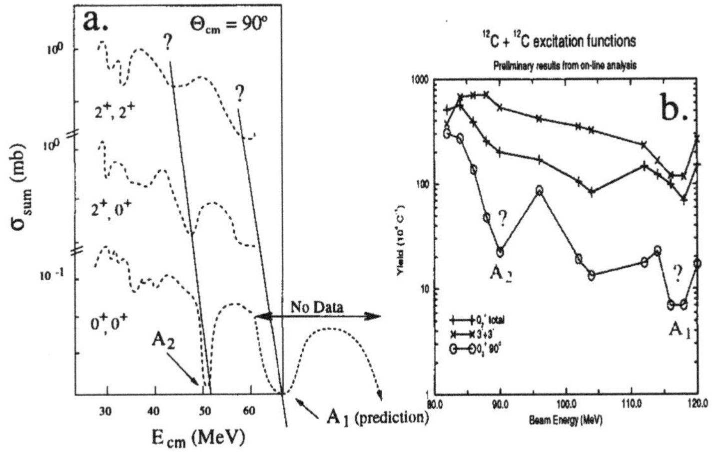

Figure 2: (a) Excitation functions at $\theta_{\mathrm{cm}} = 90°$ for the elastic $(0_1^+, 0_1^+)$ and inelastic $(0_1^+, 2^+)$ and $(2^+, 2^+)$ channels of the $^{12}C + ^{12}C$ system for bombarding energies between $E_{\mathrm{lab}} = 60$ to 120 MeV [12]. (b) Excitation functions at $\theta_{\mathrm{cm}} = 90°$ for the inelastic $(3^-, 3^-)$ and $(0_1^+, 0_2^+)$ channels of the $^{12}C + ^{12}C$ system for energies between $E_{\mathrm{lab}} = 80$ to 120 MeV (April '99 Data).

mimima in agreement with the Marty's correlation (see Figure 2b).

5 Conclusions

For the $^{16}O+^{16}O$ system, refractive effects dominant above 10 MeV/u, are still important at 5 MeV/u. The real part of the interaction is well described by present folding model calculations, while the reported sudden change in the absorption requires, and deserves, further investigation. Qualitatively similar results have been reached in studies of $^{12}C+^{12}C$ [16] and $^{16}O+^{12}C$ [17] at approximately the same energies. From the $\theta_{cm} = 90°$ excitation function of the inelastic channel $^{12}C(0_1^+)+^{12}C(0_2^+)$, there is the possibility of observing for the first time the first Airy minimum (A_1). The precise energy of A_1 will be a crucial test of the accuracy of the interaction potential. The preliminary results (Figure 2b) show two deep minima perhaps due to refractive effects.

References

1. F. Haas and Y. Abe, *Phys. Rev. Lett.* **46**, 1667 (1981).
2. Y. Kondō *et al,Phys. Lett.* B **365**, 17 (1996).
3. G. Barnitzky *et al*, *Phys. Lett.* B **365**, 23 (1996).
4. J.V. Maher *et al*, *Phys. Rev.* **188**, 1665 (1969).
5. M.L. Halbert *et al*, *Phys. Lett.* B **51**, 341 (1974).
6. A. Meens, M.P. Nicoli and D. Raiser, *Nucl. Instrum. Methods* A **397**, 64 (1997).
7. M.P. Nicoli, Thesis, Univ. Louis Pasteur, Strasbourg, France (1998); Institut de Recherches Subatomiques, Report IReS 98-16.
8. M. P. Nicoli *et al*, submitted to *Phys. Rev.* C.
9. M.E. Brandan and G.R. Satchler, *Phys. Rep.* **285**, 143 (1997).
10. S. Szilner *et al*, *J. Phys.* G, in press.
11. F. Saint-Laurent *et al*, *Nucl. Phys.* A **327**, 517 (1979).
12. A. Morsad *et al*, *Z. Phys.* A **338**, 61 (1991).
13. R.G. Stokstad *et al*, *Phys. Rev.* C **20**, 655 (1979).
14. K.W. McVoy at M.E. Brandan, *Nucl. Phys.* A **542**, 295 (1992).
15. C. Marty *et al*, Proceedings of the Symposium on Heavy-Ion Elastic Scattering, University of Rochester 507 (1977).
16. M. E. Brandan, M. Rodríguez-Villafuerte and A. Ayala, *Phys. Rev.* C **41**, 1520 (1990).
17. M.P. Nicoli *et al*, to be published.

HOT HYPERDEFORMED COHERENTLY ROTATING CLUSTERS IN ^{24}MG+^{24}MG AND ^{28}SI+^{28}SI ELASTIC AND INELASTIC SCATTERINGS

S.YU. KUN, B.A. ROBSON

Department of Theoretical Physics, RSPhysSE, IAS, The Australian National University, Canberra ACT 0200, Australia

The current understanding of heavy-ion scattering implies that the presence of intermediate ($\simeq 0.5 - 1$ MeV) regular correlated structures in the excitation functions is an unambiguous indication of the excitation and decay of isolated molecular resonances [1]. In this contribution we demonstrate [2] that the non-statistical oscillating-correlated intermediate structures can also be generated by fluctuating S-matrix elements even in the absence of isolated resonances, provided the heavy-ion scattering can be treated in terms of the formation of a hot strongly deformed dinucleus. Yet, in the region of strongly overlapping resonances of the intermediate system, strong deformation alone is obviously not sufficient to reproduce the non-statistical oscillating-correlated structures observed in heavy-ion scattering. We show that the additional necessary conditions for these non-statistical effects to occur in the region of strongly overlapping resonances are (i) correlations between fluctuating S-matrix elements with different spin (J) values and (ii) slow but finite spin decoherence. These off-diagonal spin correlations are generated spontaneously [3], allowing one to interpret [4] the nonself-averaging of the excitation function oscillations in strongly dissipative heavy-ion collisions [5]. We demonstrate that our approach reproduces both the channel-channel correlations and the non-statistical oscillating intermediate ($\simeq 0.8$ MeV) structures observed in the excitation functions for ^{24}Mg+^{24}Mg [6] and ^{28}Si+^{28}Si [7] elastic and inelastic scattering.

In calculating the cross section energy autocorrelation function (EAF) $C(\varepsilon)$ we take into account the S-matrix spin autocorrelation [3] $< S^J(E+\varepsilon)S^{J'}(E)^* >= \Gamma/(\Gamma+\beta|J-J'|+i\hbar\omega(J-J')-i\varepsilon)$, where ω is the angular velocity of the coherent rotation, β is the spin decoherence width and $\hbar/\Gamma$ is the average life-time of the dinucleus. We obtain [2]

$$C(\varepsilon) \propto \mathrm{Re}\{\exp[i\pi\varepsilon/(\hbar\omega - i\mathrm{sgn}(\varepsilon)\beta)]/[1-\exp[i\pi(\varepsilon+i\Gamma)/(\hbar\omega - i\mathrm{sgn}(\varepsilon)\beta)]]\}.$$

Putting $\beta \to 0$ yields $[\omega - i\mathrm{sgn}(\varepsilon)\beta/\hbar] \to \omega$. This indicates that $\beta/\hbar$ has the physical meaning of the imaginary part of the angular velocity signifying the space-time delocalization of the dinucleus and the damping of the coherent nuclear rotation [3]. For $\beta = 0$, $C(\varepsilon)$ is an oscillating periodic function with

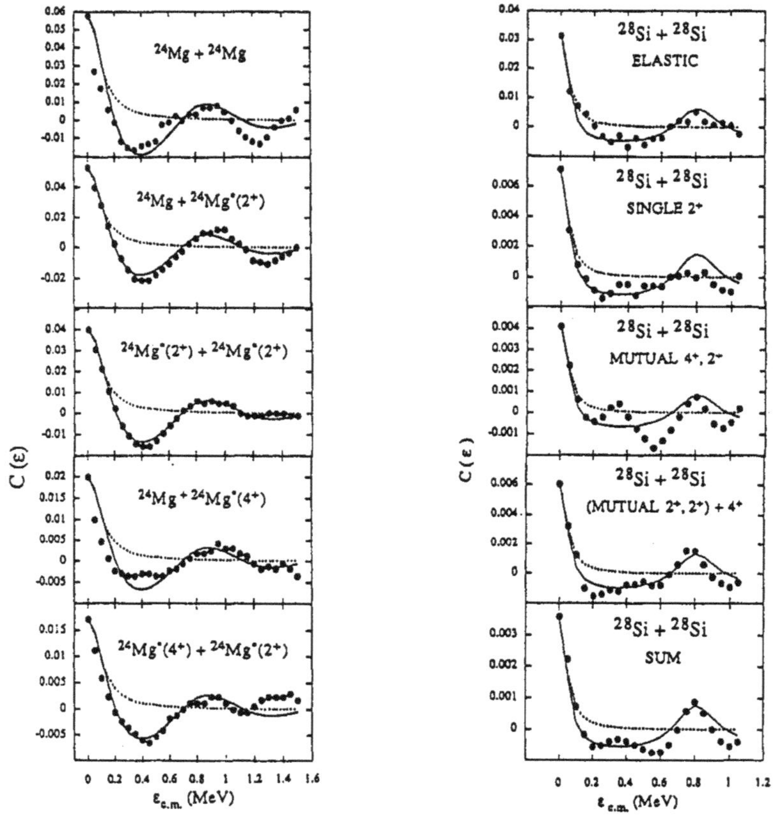

Figure 1. Experimental and calculated $C(\varepsilon)$'s for ^{24}Mg+^{24}Mg and ^{28}Si+^{28}Si elastic and inelastic scattering. Dotted lines are Lorentzians.

period $2\hbar\omega$. For finite β, the amplitude of the oscillations in $C(\varepsilon)$ decreases with increasing $|\varepsilon|$: the larger β the stronger the damping of the oscillations.

In Fig. 1 we present EAFs for ^{24}Mg+^{24}Mg elastic and inelastic scattering on the $E_{c.m.} = 42 - 56$ MeV interval [6]. The experimental $C(\varepsilon)$'s are not Lorentzian but oscillate with a period $\simeq 0.85$ MeV. The fit of the experimental $C(\varepsilon)$'s is obtained with $\Gamma = 0.2$ MeV, $\beta = 0.1$ MeV, $\hbar\omega = 0.42$ MeV. The extracted value of $\hbar\omega$ suggests an anomalous deformation of the intermediate dinucleus. Indeed, for $J \simeq 34 - 38$ [6], using the moment of inertia of a $\simeq (2:1)$ superdeformed dinucleus yields $\hbar\omega \simeq 1.7 - 1.9$ MeV. This suggests the excitation of $\simeq (4:1)$ hyperdeformed coherent rotational states of ^{48}Cr.

We calculate the channel-channel correlation coefficient, k, following the same approach of Refs. [3,4] which has been used for the derivation of $C(\varepsilon)$. Using the above values of Γ, β, $\hbar\omega$ we obtain $k = 0.11$ while $k_{exp} = 0.19 \pm 0.05$. On the other hand, $k_{exp} \simeq 1$ would support the interpretation of the non-statistical correlated structures in terms of partially overlapping ($\Gamma \simeq D$) or isolated ($\Gamma < D$) resonances.

In Fig. 1 we also present EAFs for ^{28}Si+^{28}Si elastic and inelastic scattering on the $E_{c.m.} = 52.5 - 60.5$ MeV interval [7]. The fit of the experimental $C(\varepsilon)$'s is obtained with $\Gamma = 0.05$ MeV, $\beta = 0.05$ MeV, $\hbar\omega = 0.4$ MeV. One can see that the agreement between the data and the calculated EAFs is not as consistent as for the ^{24}Mg+^{24}Mg system. The data are rather scattered around the calculated curves. In order to further check our interpretation we construct the EAF for the cross section summed over the elastic and three inelastic channels. Such a summation is expected to enhance the correlated non-statistical structure if it is present in all the channels. One can see in Fig. 1 that $C(\varepsilon)$ for the summed cross section does show a noticeable maximum at $\varepsilon = 0.8$ MeV supporting our interpretation. For $J \simeq 36 - 40$ [8] calculation of the moment of inertia of a superdeformed $\simeq (2:1)$ ^{56}Ni dinucleus yields $\hbar\omega \simeq 1.6 - 1.8$ MeV. Since our analysis gives $\hbar\omega = 0.4$ MeV this suggests the excitation of $\simeq (4:1)$ hyperdeformed coherent rotational states of ^{56}Ni in ^{28}Si+^{28}Si scattering. We also calculate the channel-channel correlation coefficient with the above values of Γ, β, $\hbar\omega$ and obtain $k = 0.14$ while $k_{exp} = 0.18 \pm 0.06$. The nonvanishing of the channel-channel correlation is due to the finite spin decoherence width. This correlation is predicted [4] to vanish in both (i) the limit of the regular coherent rotation $\beta/\Gamma \to 0$, and (ii) the compound nucleus limit $\Gamma/\beta \to 0$.

References

1. R.R. Betts and A.H. Wuosmaa, *Rep. Prog. Phys.* **60**, 819 (1997).
2. S.Yu. Kun, B.A. Robson and A.V. Vagov, *Phys. Rev. Lett.*, in press.
3. S.Yu. Kun, *Z. Phys.* A **357**, 271 (1997).
4. S.Yu. Kun and A.V. Vagov, *Z. Phys.* A **359**, 137 (1997).
5. G. Pappalardo, *Nucl. Phys.* A **488**, 395c (1988); T. Suomijärvi *et al*, *Phys. Rev.* C **36**, 181 (1987); F. Rizzo *et al*, *Z. Phys.* A **349**, 169 (1994); Wang Qi *et al*, *Phys. Lett.* B **388**, 462 (1996); S.Yu. Kun *et al*, *Z. Phys.* A **359**, 263 (1997); I. Berceanu *et al*, *Phys. Rev.* C **57**, 2359 (1998).
6. R.W. Zurmühle *et al*, *Phys. Lett.* B **129**, 384 (1983).
7. S. Saini and R.R. Betts, *Phys. Rev.* C **29**, 1769 (1984).
8. R.R. Betts *et al*, *Phys. Lett.* B **100**, 117 (1981).

SPIN ALIGNMENTS IN HEAVY-ION RESONANCES

E. UEGAKI
Physics Division, Department of Mechanical Engineering, Akita University, Akita 010-8502, Japan
E-mail: ue@uws13.phys.akita-u.ac.jp

Y. ABE
Yukawa Institute for Theoretical Physics, Kyoto University, Kyoto 606-8502, Japan
E-mail: abey@yukawa.kyoto-u.ac.jp

Recent particle-particle-γ coincident measurements on a ^{28}Si+^{28}Si resonance have suggested "vanishing spin alignments". New analyses for spin alignments by using a molecular model are reported. Different aspects between di-nuclear systems with prolate deformed nuclei and those with oblate deformed nuclei are clarified.

1 Introduction

Resonances observed in heavy-ion scattering have offered intriguing subjects in nuclear physics. ^{24}Mg+^{24}Mg and ^{28}Si+^{28}Si resonances exhibit very narrow decay widths, with prominent peaks correlated among the elastic and inelastic channels, which suggest rather long lived compound nuclear systems.[1] ¿From the viewpoint of di-nuclear molecules, the present authors have studied normal modes around the equilibrium configurations, which are expected to be responsible to the observed resonances.[2,3]

Very recently, ^{28}Si + ^{28}Si scattering experiment has been done on the resonance at $E_{\rm CM}$ = 55.8MeV at IReS Strasbourg.[4] Figure 1 shows angular distributions for the elastic and inelastic channels 2^+, $(2^+, 2^+)$, respectively. The oscillating patterns are found to be in good agreement with $L = 38$, which suggests $L = J = 38$ dominance in the resonance, namely, *misalignments* of the fragment spins. Angular distributions of γ-rays from ^{28}Si first excited state to the ground state have been also measured in coincidence with two ^{28}Si nuclei detected at $\theta_{\rm CM} = 90°$. Figure 2 displays γ-ray intensities in the three panels, where quantizations are made for z-axis to be parallel to the beam direction in (a), z-axis parallel to normal to the scattering plane in (b) and z-axis parallel to the fragment direction in (c). The angular distribution in (b) shows characteristic "m=0" pattern, which suggests fragment spins $\mathbf{I_1}$ and $\mathbf{I_2}$ are on the scattering plane and is consistent with the misalignments observed in ^{28}Si angular distributions. Those features are much different from ^{12}C + ^{12}C and ^{24}Mg+ ^{24}Mg systems, which exhibit spin alignments.[5] The aim

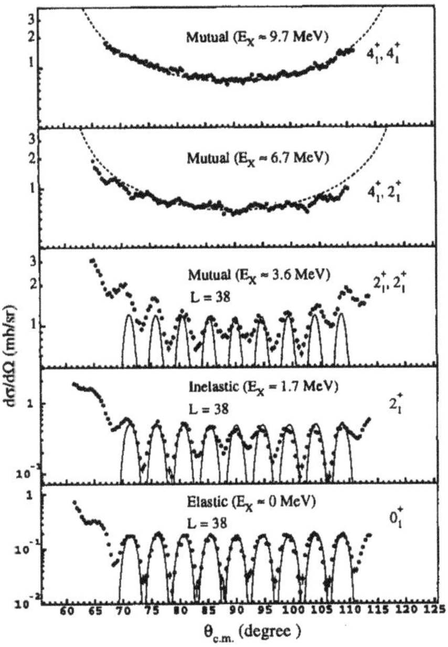

Figure 1. Experimental angular distributions for the elastic and inelastic scattering for ^{28}Si + ^{28}Si at $E_{\rm CM}$ = 55.8MeV. Solid curves show L=38 Legendre fits for comparison.

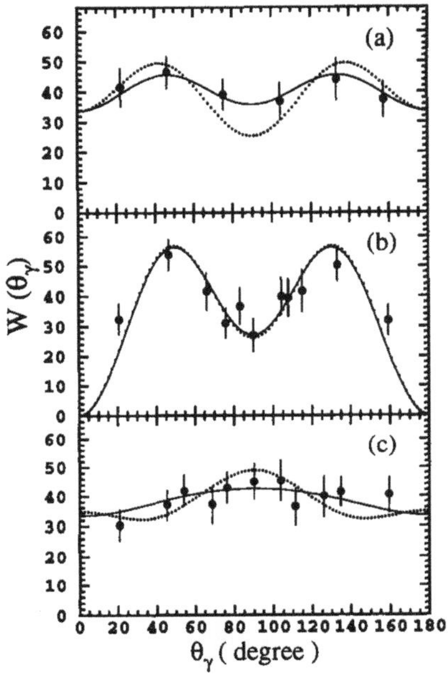

Figure 2. γ-rays distributions obtained from particle-particle-γ coincident measurements. Solid curves show χ^2-fits. Dotted lines show the theoretical prediction obtained by the molecular model with tilting mode (K-mixings).

of the present paper is to clarify the mechanism of the appearance of spin *misalignments* in the resonance of ^{28}Si + ^{28}Si system.

2 Di-nuclear Structure of ^{28}Si + ^{28}Si System

First, structures of the resonance states of ^{28}Si + ^{28}Si system and their normal modes around a stable configuration are briefly revisited. For simplicity, assuming a constant deformation and axial symmetry of the constituent nuclei, we have seven degrees of freedom $(q_i) = (\theta_1, \theta_2, \theta_3, R, \alpha, \beta_1, \beta_2)$, as illustrated in Fig. 3(a). Consistently with the coordinate system, we introduce a rotation-vibration type wave function as basis one,

$$\Psi_\lambda \sim D^J_{MK}(\theta_i)\chi_K(R, \alpha, \beta_1, \beta_2), \tag{1}$$

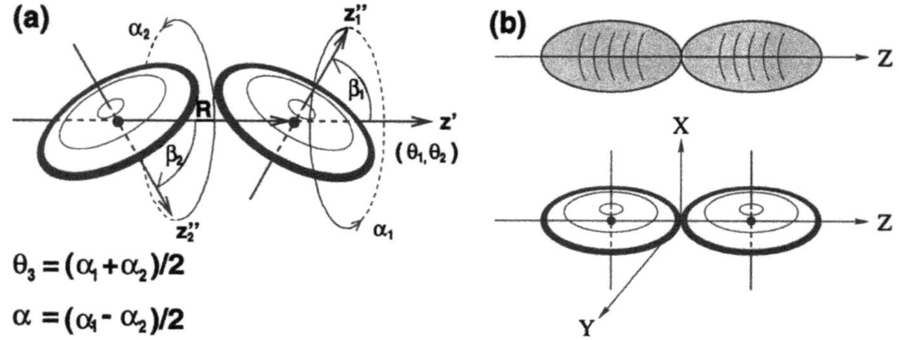

Figure 3. (a) Coordinate system of $^{28}Si + ^{28}Si$. (b) Equilibrium configurations of di-nuclear systems. Upper panel is for $^{24}Mg + ^{24}Mg$ and lower panel for $^{28}Si + ^{28}Si$.

where χ_K describes internal motions. Dynamics of the internal motions have been solved around the equilibrium and various normal modes such as butterfly vibrations have been obtained,[3] as is shown in Fig. 4(a).

Each mode has a characteristic feature with respect to the spin alignments. For example, butterfly motion shows the total intrinsic spin $I = 0$ dominance, namely, anti-alignment, which is very consistent with the fragments angular distributions.[6] However the indication by the γ-ray measurements that the spin vectors $\mathbf{I}_1$ and $\mathbf{I}_2$ are both on the scattering plane provides more detailed information. In the following, we introduce a new mode in order to explain γ-rays data suggesting the fragments spin components "$m = 0$".

With extremely high spins such as $30 \sim 40\hbar$, stable di-nuclear configurations tend to be "elongated systems by the strong centrifugal force". In the prolate-prolate systems, a stable configuration is pole-to-pole one, while in the oblate-oblate systems, it is an equator-to-equator one. The former has axial symmetry as a whole, but the latter has axial asymmetry, as is displayed in Fig. 3(b). *What would be expected from the difference of the symmetries?* Approximately a triaxial system rotates around the axis with the maximum moments of inertia. In the oblate-oblate systems, thus two pancakes-like objects touching side-by-side as in the lower panel of Fig. 3(b) rotate around x-axis which is parallel to the normal to the reaction plane. The axial asymmetry, however, gives rise to a mixing of K-quantum numbers,[7] which is different from the axial symmetric prolate-prolate cases. In the high spin limit $(K/J \sim 0)$, the diagonalization in the K-space is found to be equivalent to solving a differential equation of the harmonic oscillator with parameters

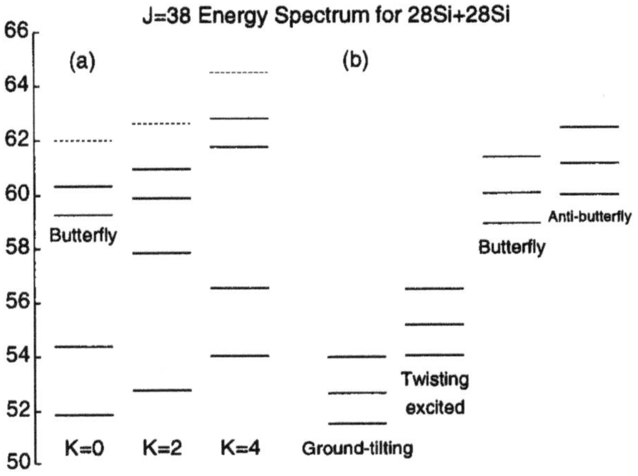

Figure 4. Energy spectra for $^{28}\mathrm{Si} + ^{28}\mathrm{Si}$. (a) Without K-mixing and (b) with K-mixing.

given by the moments of inertia. Thereby, the solution is a gaussian, or a gaussian multiplied by an Hermite polynomial,

$$f_n(K) = H_n(\frac{K}{b}) \exp\left[-\frac{1}{2}\left(\frac{K}{b}\right)^2\right], \tag{2}$$

with the width of $b = (2J^2 I_K/\Delta)^{1/4}$, where $I_K^{-1} = I_z^{-1} - I_{\mathrm{av}}^{-1}$ and $\Delta^{-1} = I_y^{-1} - I_{\mathrm{av}}^{-1}$ with $I_{\mathrm{av}}^{-1} = \frac{1}{2}(I_x^{-1} + I_y^{-1})$. The resultant energy spectrum is displayed in Fig.4(b) compared with the spectrum without K-mixings in Fig.4(a). Consequences of the K-mixings to fragment spin orientations and then to the γ-ray distributions are discussed in the next section.

3 Spin Alignments

We define scattering waves and the collision matrix $U_{c'c}$ such as

$$\psi \sim (G_c - iF_c) - \sum_{c'} U_{c'c}(G_{c'} + iF_{c'}). \tag{3}$$

By using the R-matrix formula with one level approximation, we obtain $U_{c'c} = e^{-i\phi'_{l'}}(-2\sqrt{2P_{c'l'}}\gamma_{c'l'}\sqrt{2P_{cl}}\gamma_{cl})e^{-i\phi_l}/\Gamma_{\mathrm{total}}$ for the inelastic scattering, where the reduced widths γ_{cl} are calculated from the model wave functions.

The scattering amplitudes with specified magnetic substates are given by

$$A_{m_1 m_2}(\mathbf{k}', \mathbf{k}) = \frac{2\pi}{ik} \sum_{L', S', M} (22 m_1 m_2 | S' M'_S)(L' S' m' M'_S | JM)$$

$$\times i^{J-L'} e^{i(\sigma_J+\sigma'_{L'})} U^J_{L'S'} Y^*_{JM}(\hat{k}) Y_{L'm'}(\hat{k}'). \tag{4}$$

The transition amplitudes for the γ-ray emissions from the polarized nuclei are discussed by several authors (see for examlpe Ref. 8). Note that in the experiment, only one of two emitted photons is detected in most cases, even with EUROGAM. Note also that the intensities of the detectors are averaged over the azimuthal angle ϕ_γ in the data shown in Fig. 2 and then are expressed as $W(\theta_\gamma) = \sum_m P_m W_m(\theta_\gamma)$, where P_m denotes the probability in the m magnetic substate.

Now we calculate P_m's and $W(\theta_\gamma)$'s with the molecular model. Combining with the lowest state of the tilting mode in Eq. (2), i.e., $f_0(K) \sim \exp(-K^2/2b^2)$, we introduce a refined wave function,

$$\Psi^{JM}_\lambda \sim \sum_K \exp(-K^2/2b^2) D^J_{MK}(\theta_i) \chi_K(R, \alpha, \beta_1, \beta_2), \tag{5}$$

where in general, χ_K can be any excitation mode. A simple choice is that all the internal motions are zero-point ones. In Fig. 2, theoretical results for $b = 1.3$ are shown by dotted lines, which are seen to be in good agreement with the data, in all the three axes (a), (b) and (c). This value of b is consistent with the di-nuclear configuration of $^{28}Si + ^{28}Si$ system.

4 Concluding Remarks

Differences between the oblate-oblate system(^{28}Si + ^{28}Si) and the prolate-prolate system (^{24}Mg+^{24}Mg) has been manifested. Corresponding to the axial asymmetry by the equilibrium shape of ^{28}Si+^{28}Si system, we have introduced K-mixings and have succeeded to explain the characteristic features for the spin misalignments. Those K-mixings, namely, the *tilting mode* would be a new facet in nuclear resonance phenomena. Further experimental study is strongly desired.

Acknowledgments

The authors are grateful to Dr.'s R. Nouicer and C. Beck for the informations on the experimental data and stimulating discussion.

References

1. R.R. Betts, in *Proc. 4th Intern. Conf. on Clustering Aspects of Nuclear Structure and Nuclear Reactions (Chester, July 1984)*, p. 133.

2. E. Uegaki and Y. Abe, in *Proc. Fifth Intern. Conf. on Clustering Aspects in Nuclear and Subnuclear Systems (Kyoto, July 1988)*, p. 330; *Phys. Lett.* B **231**, 28 (1989).
3. E. Uegaki and Y. Abe, *Phys. Lett.* B **340**, 143 (1994).
4. R. Nouicer, C. Beck, *et al.*, *Phys. Rev.* C, to be published.
5. R.W. Zurmühle, in *Proc. Fifth Intern. Conf. on Clustering Aspects in Nuclear and Subnuclear Systems (Kyoto, July 1988)*, p. 37.
   D. Konnerth et al., ibid., p. 325.
6. E. Uegaki, *Prog. Theor. Phys.* Suppl. No. **132**, 135 (1998).
7. A. Bohr and B.R. Mottelson, *Nuclear Structure vol. II* , p. 175.
8. F. Rybicki, T. Tamura and G.R. Satchler, *Nucl. Phys.* A **146**, 659 (1970).

NEW OBSERVATION METHODS FOR ALPHA-PARTICLE ELASTIC SCATTERING. MICROSCOPIC PROPERTIES OF ALPHA-CLUSTER STATES

T. LÖNNROTH[1], V.Z. GOLDBERG[2], M. GUTTORMSEN[3], E. INDOLA[1], K-M. KÄLLMAN[1], R. LIOTTA[4], E. MELBY[3], G.V. ROGATCHEV[2], M.V. ROZHKOV[2], A. SCHILLER[3], T. SIISKONEN[5], S. TORILOV[2], W.H. TRZASKA[5], R. WOLSKI[6]

[1] *Department of Physics, Åbo Akademi, Turku, Finland*
[2] *Russian Research Center "Kurchatov Institute", Moscow, Russia*
[3] *Department of Physics and Cyclotron Laboratory, University of Oslo, Norway*
[4] *Manne Siegbahn Laboratory, Raoyal Institute of Technology, Stockholm, Sweden*
[5] *Department of Physics, University of Jyväskylä, Finland*
[6] *Henryk Niewodniczański Institute of Nuclear Physics, Kraków, Poland*

We report on a new, fast method to perform α-scattering on solid and gaseous targets. Narrow states are observed at high excitation, and quantum characteristics can be extracted. First experiments to observe γ-ray emission from these states in $\alpha+^{28}$Si are done. We discuss a new approach to describe these narrow resonances.

1 Introduction

A large number of narrow states with Γ ~30-100 keV are found at high excitation in the mass region with A~16-44. Experimentally the best studied case is $\alpha+^{28}$Si [1,2,3,4]. We are developing a fast method, the reverse-geometry method, using the α-particles as the target [5]. The question of γ-ray emission from α-cluster states [6] is addressed. Finally, we make a new approach to address the *interpretation* of the experimentally observed structures.

2 Experimental Methods

Elastic Scattering. In order to speed up the recording of excitation functions and angular distributions for (elastic) scattering reactions, we have developed a resolution-maintaining method for measurements in *large* energy steps, the thick-target technique [4]. However, in the mentioned region the majority of the target nuclides are either gases or highly volatile substances. Therefore it was realized that using the bombarding particle as target [5,7], i.e. helium gas, and the heavy specimen as the beam particle, could be a feasible way to proceed [5]. This comprizes the so-called reverse-geometry method, exemplified in Fig. 1. See also ref. [2] for recent details. Note the good resolution maintained in

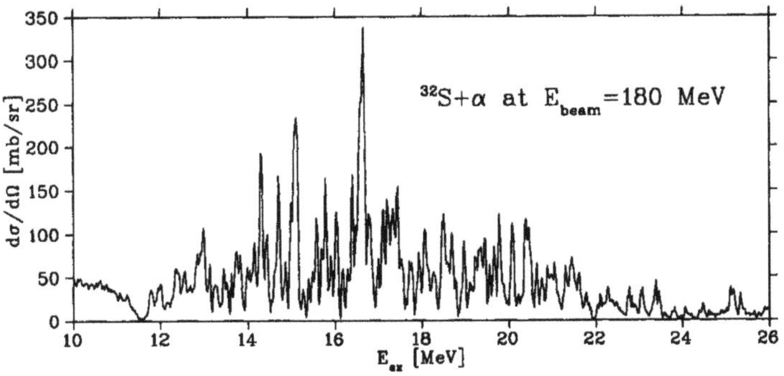

Figure 1. Excitation function at 0°/180° for ^{32}S+α in reverse geometry. Note that the data are recorded as one single spectrum during about 2 hours! A background has been subtracted.

fig. 1. We further develop the method at the K-130 cyclotron in Jyväskylä.

Gamma-Ray Emission. First attempts to observe γ-rays from cluster states in α+^{28}Si, cf. Fig. 2a, were made at the Oslo University K-36 cyclotron at five beam energies 14.8 – 24.0 MeV, with preselected entry states of $\sim$700 keV widths by use of thin target. We use the multi-module array CACTUS [8], which has 30 detector positions in six rings – twenty-seven 5" × 5" NaI and three Ge counters with 50-70% efficiency were used. Energy ranges were 9.5 MeV and 4.2 MeV, respectively.

3 Experimental Results

Scattering on Helium Gas. Results from earlier measurements of α-scattering on ^{28}Si have been reported elsewhere [1,2,3,9]. Excitation functions with the reverse-geometry method have been presented [10] for states populated in the reactions ^{12}C,^{16,18}O,^{20}Ne+α. New results from excitation function- and angular-distribution measurements of the ^{32}S+α reaction are obtained. The method is able to separate protons (and random particles) from alpha particles by time-of-flight, and in favourable cases the elastic from the inelastic channels. The alpha-particle spectra from the ^{32}S+α reaction indicate narrow highly excited states in the composite system ^{36}Ar up to $E_{exc} > 26$ MeV, see Fig. 1. Their properties ($E_{exc}, \ell, \pi, \Gamma_\alpha$) are similar to those of α+^{28}Si (fig. 2). Illustrating the efficiency of the method: an α-particle energy spectrum from the reaction ^{18}O+α, up to $E_{exc} > 22$ MeV in ^{22}Ne is recorded in a few hours, to be compared to a measurement of α+^{16}O [12], with about 800 energy

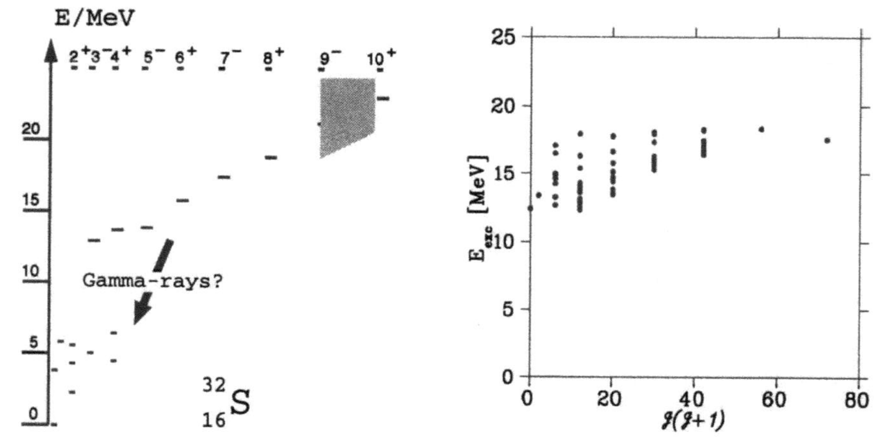

Figure 2. a) Search for γ-rays connecting α-cluster states (from $\alpha + ^{28}Si \rightarrow ^{32}S$ elastic scattering) to "spectroscopic" states. b) States in $^{32}S + \alpha \rightarrow ^{36}Ar$.

settings!

Gamma-Ray Emission. The γ-ray coincidence data were sorted separately for all 5 energies into matrices Ge-NaI in 2048×512 channels and Ge(gates)-Ge (111×2048), which are accessed by normal gating procedure. The preliminary results from these measurements are the following. Figure 3a shows the intensity contour of a Ge-NaI matrix (Ge on the horizontal axis) recorded at 15 MeV (entry in ^{32}S at $\sim$ 19.6 MeV). Although there are interesting high-energy coincidences, e.g. 2839 keV(4^+) – 7.0 MeV in ^{28}Si, a ($4^+, 5^-, 6^+$) state at 11.62 MeV and 2234 keV(4^+) – 8.4 MeV in ^{32}S (giving $4^+, 5^-, 6^+$ at 12.4 MeV), final conclusions of a would-be cluster origin cannot be drawn yet.

4 Theoretical Interpretations

Cluster Models. Phase space diagrams $E_{exc} = \hbar^2/2\mathcal{J}[\ell(\ell+1)]$ for the ℓ-centroids, see fig. 2, seem to indicate rotational motion of "molecular type". The members within a specific ℓ-group are often interpreted as α-vibrations built on the rotation. Models along these lines are the Orbiting Cluster Model [13,14], the Anharmonic Vibration-Rotation Model [15], the Bonding Potential model [16] and the Semimicroscopic Algebraic Cluster Model [17]. The energy relation is

$$E_{nlN} = A[\ell(\ell+1)] + B(n+\frac{1}{2}) - C(n+\frac{1}{2})^2 + D(N+\frac{1}{2}) + E(n,\ell),$$

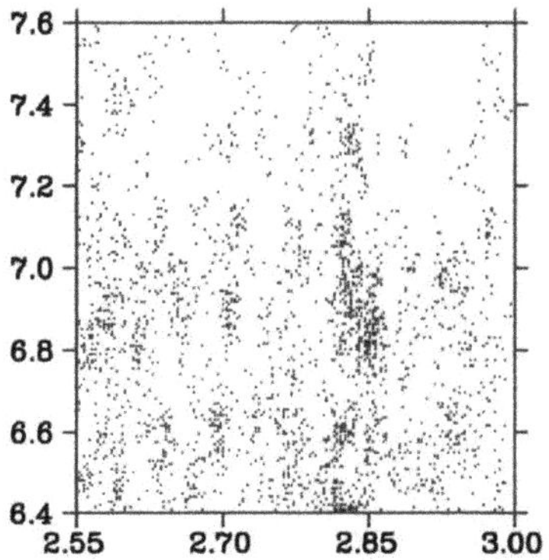

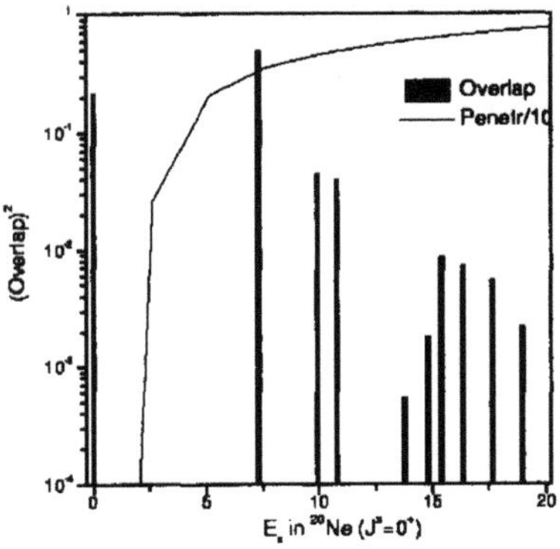

Figure 3. a) Illustration to $\gamma - \gamma$ coincidence data. b) Calculation of clusterization/overlap in the process ^{20}Ne$\rightarrow \alpha+^{16}$O. All 0^+ states below 20 MeV in ^{20}Ne from the $p_{1/2}$, $d_{5/2}$ and $s_{1/2}$ orbitals are shown.

with slight differences, in all models. The terms represent rotation, harmonic and (soft) anharmonic vibration, the D-term arises (a breather mode) in a non-linear approach, see [18] and references therein, and the last term, $E(n, \ell)$, describes the rotation-vibration coupling.

A Microscopic Attempt. A serious shortcoming in all these "macroscopic" models is that radial vibrational amplitudes deduced from experimental spectra are unphysical, largely in excess of 10 fm. Theoretical inconsistensies have also been pointed out [19]. A novel attempt to a microscopic approach to describe the observed narrow high-lying resonances is proposed. It origins from the similarity of elastic α-scattering to α-decay [20,21].

The Shell Model is microscopic (it has nucleon degrees of freedom) and contains all states. Results from a calculation with a standard program [22] (OXBASH) for "α-cluster-like" states in ^{20}Ne are shown in Fig. 3b. Two facts are conspicuous: the level density is *low*, and some of the states show fairly large clusterization, i.e. states in ^{20}Ne have overlaps of the form $\alpha+^{16}O_{g.s.}$.

Inclusion of a larger number of main shells does not notably increase the state density *at these excitation energies.* However, single-particle states at higher excitation may become states just at the binding limit, i.e. so-called Gamow states [23], which acquire a width. It has been shown [24] that these states can be handled as regular states, and they can thus be incorporated in a suitable basis, from which α-particle structures may be projected [25]. We have started work along these lines.

Acknowledgments

Experimental contribution from L. Bergholt, A. Bjerve, M. Brenner, V.I. Dukhanov, K. Ingeberg, P. Manngård, A.E. Pakhomov, V.V. Pankratov, S. Rezazadeh and I.N. Serikov is gratefully acknowledged. The work was supported in part by the Academy of Finland, the Jubileumsfonden 1968 of Åbo Akademi, the Research Institute of the Foundation of Åbo Akademi and the Russian Foundation of Basic Science under grant 97-02-17113.

References

1. P. Manngård, *PhD thesis*, Åbo Akademi, (1996); *Z. Phys.* A **349**, 335 (1994).
2. K.-M. Källman, *PhD thesis*, Åbo Akademi, (1998).
3. M. Brenner, *Z. Phys.* A **349**, 233 (1994).
4. K.-M. Källman *et al*, *Nucl. Instrum. Methods* A **338**, 413 (1994).
5. V.Z. Goldberg, *Clustering Phenomena in Atoms and Nuclei*, Turku, Finland, eds. M. Brenner *et al*, (Springer, 1992) p. 366.
6. F. Haas *et al*, *Nuovo Cimento* A **110**, 989 (1997).
7. K.P. Artemov *et al*, *Yad. Fiz.* **52**, 634 (1990).
8. M. Guttormsen *et al*, *Report 89-14*, University of Oslo, 1989.
9. K.-M. Källman *et al*, *Eur. Phys. J.* A, in preparation.
10. V.Z. Goldberg *et al*, *Yad. Fiz.* **60**, 1186 (1997).
11. V.Z. Goldberg *et al*, *Nucl. Instrum. Methods* A, in preparation.
12. R. Abegg and C.A. Davies, *Phys. Rev.* C **43**, 2523 (1991).
13. N. Cindro and D. Počanić, *J. Phys.* G **6**, 359 (1980).
14. U. Abbondanno *et al*, *Nuovo Cimento* A **110**, 955 (1997).
15. N. Cindro and W. Greiner, *J. Phys.* G **9**, L175 (1983).
16. L. Satpathy, P. Sarangi and A. Faessler, *J. Phys.* G **12**, 201 (1986).
17. J. Cseh and G. Levai, *Ann. Phys.* (N.Y.) **230**, 165 (1994).
18. A. Ludu *et al*, *J. Phys.* G **21**, L41 (1995).
19. K. Kato and Y. Abe, *Prog. Theor. Phys.* **80**, 119 (1988).
20. R.J. Liotta *et al*, *Phys. Lett.* B **367**, 1 (1996).
21. R.G. Lovas *et al*, *Phys. Rep.* **294**, 265 (1998).
22. B.A. Brown *et al*, *MSU-NSCL Report*, 524 (1988).
23. G. Gamow, *Z. Physik* **51**, 204 (1928).
24. T. Berggren, *Nucl. Phys.* A **109**, 265 (1968).
25. F. Janouch and R.J. Liotta, *Phys. Rev.* C **27**, 896 (1983).

INELASTIC NUCLEAR ORBITING IN DIFFERENT MASS REGIONS AND ITS CONNECTIONS WITH NUCLEAR STRUCTURE

A. RAY, A. MITRA, S. R. BANERJEE AND M. SENGUPTA

Variable Energy Cyclotron Centre, 1/AF, Bidhannagar, Calcutta-700064 , India

A. DE

Ranigung Girls' College, Dt. Burdwan, West Bengal, India

A. CHATTERJEE, S. KAILAS, H. S. PATEL AND M. G. BETIGIRI

Bhabha Atomic Research Centre, Mumbai-400 085, India

S. K. DUTTA

Nuclear Science Centre, New Delhi - 110076 , India

The studies of back-angle binary decay of projectile and target-like fragments from the fully damped nuclear reactions in the energy region E/A = 4 MeV to 10 MeV have shown that in the majority of the cases, a fully equilibrated statistical compound nucleus is formed. However in a few cases near A = 40 mass region, the experimental results point to the formation of a long-lived (10^{-20} s) orbiting complex where the nuclei are in excited states and maintain their identities. So the complex never reaches full equilibration and remembers how it is formed. We report here our recent observations regarding the formation of such orbiting complexes near A = 100 mass region. Our results show that an explanation of nuclear orbiting in terms of the small number of open channels which seems to be valid in the lower mass region cannot be extended to the A = 100 mass region. We think that for the formation of such an orbiting complex, the nuclei must have collective states and in the dinuclear configuration, the energy and momentum sharing should take place through the excitation of those collective states rather than through the nucleon exchange mechanism.

1 Introduction

The concept of nuclear orbiting, although an old idea is continuously being used under different names such as fast fission, dynamical fission [1,2], orbiting [3] etc. to give qualitative explanations of different observed anomalies in nuclear reactions over a wide energy region and projectile and target combinations. However it is still not understood under which conditions nuclear orbiting phenomenon takes place and why it has been seen in certain cases only. The idea of nuclear orbiting originated [4,5] from the studies of back-angle binary decay of target-like and projectile-like fragments from the fully damped nuclear reactions in the energy region E/A = 4 MeV to 10 MeV. The nuclear orbiting means that the two nuclei instead of fusing into a fully equilibrated compound nucleus form a long-lived rotating dinuclear complex maintaining the identities of each nuclei. According to the classical ideas of orbiting [6], for certain grazing angular momenta and below a certain bombarding energy, the effective potential in the radial motion shows a maximum at the turning point, both the time and the azimuthal angle diverge, meaning that the projectile performs a spiral motion towards the distance of closest approach. This phenomenon is called orbiting and corresponds to the situation where the deflection function asymptotically tends towards $-\infty$. However this description does not consider friction or any other phenomena that may develop when the two nuclei come in contact for relatively long time to undergo nuclear orbiting. In reality, when the two nuclei come in contact with each other, they usually experience frictional force. A neck may develop between the nuclei and nucleon exchange can take place between them. These possibilities complicate very much the above-mentioned simple picture of classical orbiting and lead to compound nucleus formation in most cases. So when the two nuclei come in contact with each other for a relatively long time period (of the order of 10^{-20}s), then instead of undergoing classical orbiting maintaining the identities of each nuclei, in most cases statistical equilibrium sets in and a compound nuclear description becomes very useful and relevant. As a result, nuclear orbiting has been seen for certain systems only.

The experimental signatures of nuclear orbiting are the observation of $1/\sin\theta$ angular distribution of the fragments implying the formation of a long-lived intermediate complex and very large entrance channel yields leading to strong entrance channel dependence when the same system is formed in various different ways. Earlier works on the binary decay of fragments from fully damped nuclear reactions showed that for the majority of the systems such as $^{12}C + ^{40}Ca$ [7], $^{9}Be + ^{40}Ca$ [7], $^{6}Li + ^{40}Ca$ [7], $^{35}Cl + ^{12}C$ [3], $^{31}P + ^{16}O$ [3] etc. reactions, compound nuclear picture is adequate to quantitatively understand the data. However back-angle studies of target-like fragments[4,5,8] from certain α-conjugate nuclei such as $^{28}Si + ^{12}C$, $^{24}Mg + ^{16}O$, $^{24}Mg + ^{12}C$ etc. reactions have shown strong entrance channel dependence and entrance channel yields in addition to having $1/\sin\theta$ angular distributions of the fragments in the center of mass frame. The studies of γ ray -angular correlation for $^{28}Si + ^{12}C$ reaction also showed [9] certain non-statistical features which cannot be explained from compound nucleus theory, but can be understood in the framework of orbiting models.

C. Beck and Y. Abe proposed [3] that the nuclear orbiting should be seen when the number of open channels for the systems is small and compound nucleus reaction should dominate when the number of open channels for the system is large. Their qualitative ideas seem to be valid in the lower mass region. However the back-angle studies [10] of alpha and proton particles from $^{4}He + ^{40}Ca$ and $^{4}He + ^{27}Al$ reactions show that the data can be understood from a compound nucleus picture, although the numbers of open channels for the systems are small. But the very small amount of available phase space for these reactions may be the reason for not seeing orbiting there. Since the nuclear orbiting has so far been seen for α-conjugate nuclei only, alpha cluster transfer may also be responsible for nuclear orbiting. It is interesting to look for inelastic nuclear orbiting in heavier non-α

conjugate systems. Such studies can help us to get new insight about nuclear orbiting and establish the underlying physics of orbiting.

2 Experimental Procedure and Results

We chose to study the back-angle emission of fragments and search for nuclear orbiting near A = 100 mass region. Earlier studies [1,2] fission fragment mass distributions near A = 100 mass region showed that the variances of total kinetic energy and mass distributions of the fission fragments are larger than the predictions of liquid drop model. However no significant entrance channel dependence [1] of the fission fragment mass distribution was observed. These results were qualitatively understood [2] broadly within the framework of compound nucleus theory with the additional consideration that the stiffness of the potential energy associated with the mass-asymmetry degree of freedom at a saddle point decreases with orbital angular momentum. However there has been no study regarding the very small back-angle cross-section of projectile-like fragments.

We form the same compound nucleus (^{105}Ag) at the same excitation energy (E_X = 76 MeV) and with very similar spin distributions using two different reaction channels — ^{16}O + ^{89}Y reaction at E(^{16}O) = 95.9 MeV and ^{12}C + ^{93}Nb reaction at E(^{12}C) = 85.6 MeV. A 10 pnA beam of ^{16}O particles from TIFR-BARC pelletron machine at Mumbai was used to bombard a 1mg/cm^2 thick ^{89}Y foil. Four ΔE-E particle telescopes were placed from 140° to 170° to record different fragments from alpha to oxygen particles. The oxygen and carbon particles were found to follow a 1/sinθ angular distribution in the center of mass frame. Similarly a 10 pnA ^{12}C beam from the pelletron machine at Nuclear Science Center, New Delhi was used to bombard a 1mg/cm^2 thick ^{93}Nb foil and 3 telescopes were placed at back-angles from 130° to 160°. The angular distributions of oxygen and carbon particles were found to follow a 1/sinθ function. The entrance channel dependence was studied from the ratio of back-angle oxygen to carbon yields as a function of exit channel energies for both ^{16}O + ^{89}Y and ^{12}C + ^{93}Nb reactions.

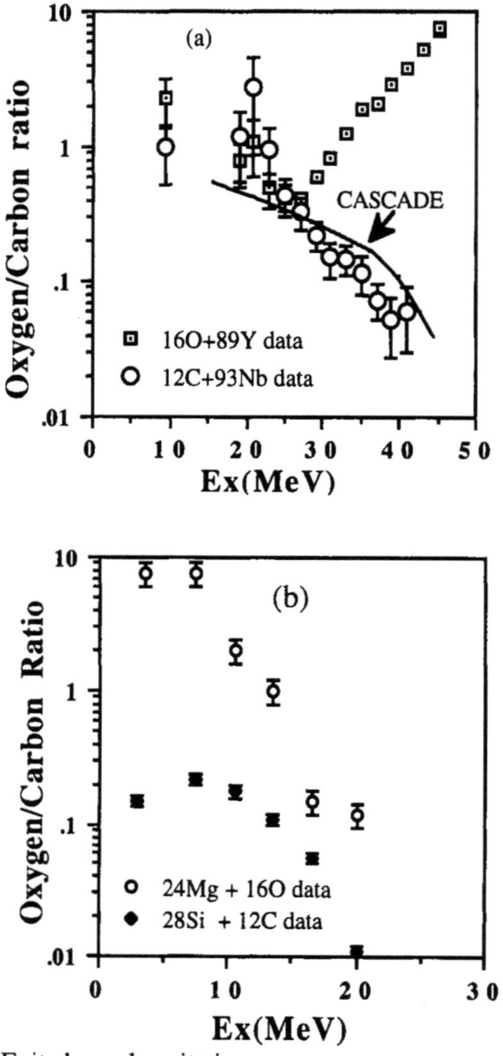

Fig.1 Exit channel excitation energy versus oxygen to carbon ratio

In Fig. 1(a), we have plotted exit channel excitation energy versus oxygen to carbon ratio for ^{16}O + ^{89}Y and ^{12}C + ^{93}Nb reactions and find a strong entrance channel dependence in the strongly damped region. The absolute back-angle integrated oxygen yield(1 mb) for ^{16}O + ^{89}Y reaction is about two orders of magnitude larger than the statistical compound nucleus model calculations. So we are certainly seeing the orbiting effect in $^{16}O + ^{89}Y$ reaction in the highly inelastic region. However in the lower exit channel excitation energy region ($E_X <$ 20 MeV), $^{16}O + ^{89}Y$ and $^{12}C + ^{93}Nb$ reaction data sets overlap and we see no entrance channel dependence or orbiting effect. Another interesting effect is that both the absolute yields of carbon and oxygen particles and their ratio for $^{12}C + ^{93}Nb$ reaction are in reasonable agreement with statistical compound nucleus predictions (CASCADE code) implying no orbiting reaction for $^{12}C + ^{93}Nb$ system. In Fig. 1.(b), we have plotted the exit channel excitation energy versus the oxygen to carbon ratio for $^{28}Si + ^{12}C$ and $^{24}Mg + ^{16}O$ reactions forming the same compound nucleus (^{40}Ca) at the same excitation energy and with very similar spin distributions. A comparison of Fig. 1(a) and Fig. 1(b) reveals that for A=40 systems, the entrance channel dependence and orbiting effect are dominant in the lower exit channel excitation energy region, whereas for A = 100 mass region, those effects are dominant in the higher exit channel excitation energy region. We have also searched for similar effects in the transfer channels, but have not seen any significant entrance channel dependence there.

3 Discussion

We have seen for the first time the evidence of nuclear orbiting in heavier non -α-conjugate system. As discussed in ref. 11, if we consider a long-lived dinuclear complex, then usually after a few nucleon exchanges between the two nuclei, the sticking limit configuration is reached and then a purely statistical compound nucleus reaction mechanism becomes dominant. In such a situation, there is no reason for entrance channel to be the dominant exit channel and phase space considerations should determine which exit channels should be dominant. So there should be no strong entrance channel dependence as is expected from the equilibrated dinuclear model developed by Shivakumar et.al [12].

We think that in the case of inelastic nuclear orbiting, the incident kinetic energy and orbital angular momentum are transferred to the two nuclei through the excitation of collective states (rotational, vibrational, compressional etc.) of the nuclei. Then the nuclei can maintain their identities in the dinuclear complex and the complex should have a large probability to break into its entrance channel. No significant nucleon exchange should take place in the case of nuclear orbiting, because complete statistical equilibrium is reached only after a few such exchanges. We can qualitatively understand our data using these ideas and are undertaking more experiments for further clarifications of these ideas.

4 References

1. H. Oeschler, P. Wagner, J. P. Coffin, P. Engelstein and B. Heusch, *Phys. Lett.* **B87** (1979) 193.
2. Y. Nagame, H. Ikezoe and T. Ohtsuki, *Phys. Rev.* **C47** (1993) 1586.
3. S. J. Sanders, A. Szanto de Toledo and C. Beck, *Phys. Rep.* **311** (1999) 487.
4.. D. Shapira, R. Novotny, Y. D. Chan, K. A. Erb, J. L. C. Ford Jr, J. C. Peng and J. D. Moses, *Phys. Lett.* **B114** (1982) 111.
5. D. Shapira, J. L. C Ford Jr and J. Gomez del Campo, *Phys. Rev.* **C26** (1982) 2470.
6. R. A. Brogila and A. Winther, *Heay Ion Reactions Lecture Notes*, ed. by D. Pines, Addison-Wesley Publishing Co., 1990.
7. K. Grotowski et.al., *Phys. Rev.* **C30**, (1984), 1214.
8. A. Ray, S. Gil, M. Khandaker, D. D. Leach, D. K. Lock and R. Vandenbosch, *Phys. Rev.* **C31**(1985) 1573.
9. A. Ray, D. D. Leach, R. Vandenbosch, K. T. Lesko and D. Shapira, *Phys. Rev. Lett.* **57** (1986) 815.
10. A. Ray, S. R. Banerjee, P. Das, A. Mitra, S. K. Basu, P. Bhattacharyya, *Phys. Rev.* **C51** (1995) R1604.
11. A. Ray and D. Shapira, *AIP conference Proceedings* **250** (1991) 55.
12. B. Shivakumar, S. Ayik, B. A. Harmon and D. Shapira, *Phys. Rev.* **C35** (1987) 1730.

6-α EMISSION IN THE ^{13}N+^{11}B REACTION

A. DI PIETRO, W. BRADFIELD-SMITH, T. DAVINSON, A.C. SHOTTER AND P.J. WOODS
Department of Physics and Astronomy, University of Edinburgh UK
E-mail: dipietro@ph.ed.ac.uk

A. AMORINI, G. CARDELLA, P. FIGUERA, G. PAPPALARDO, F. RIZZO AND S. TUDISCO
Dipartimento di Fisica, Universitá di Catania 95129 Catania, Italy
INFN-Laboratorio Nazionale del Sud and INFN-Sezione di Catania, Catania, Italy

W. GALSTER, P. LELEUX, A. MUSMARRA AND A. NINANE
Institut de Physique Nucléaire, Centre de Recherches du Cyclotron, Université Catholique de Louvain B-1348 Louvain la Neuve, Belgium

C. SÜKÖSD
Department of Physics, University of Budapest, Budapest, Hungary.

The ^{24}Mg decay into six α particles has been studied in Louvain la Neuve via the reaction ^{13}N+^{11}B at two beam energies, 29.5 MeV and 45 MeV. Large detector arrays allowed this reaction to be studied with low intensity radioactive ^{13}N beam. The excitation energy in the ^{24}Mg* was in the region of the 46 MeV resonance in ^{12}C+^{12}C scattering. The analysis revealed that most of these events originate from $^{12}C_{3_1^-}+^{12}C_{3_1^-}$.

The ^{12}C+^{12}C scattering has been widely investigated in the last thirty years, showing a large number of resonances even at high excitation energy [1]. A broad resonance has recently been discovered in the $^{12}C_{0_2^+}+^{12}C_{0_2^+}$ exit channel at E_{cm}=32.5MeV [2]. It was postulated to be associated to the six-α linear chain in ^{24}Mg [3]. Similar structures in the cross section have also been observed in the same excitation energy region but for different exit channels. In particular much narrower resonances have been found for the $^{16}O_{gs}+^{8}Be_{gs}$ [4] and $^{12}C_{3_1^-}+^{12}C_{3_1^-}$ exit channels [5]. In this contribution, the six-α channel, resulting from the reaction ^{13}N+^{11}B which populates the ^{24}Mg compound nucleus (CN), will be discussed. The experiment was performed at the radioactive beam facility in Louvain la Neuve. It was undertaken at ^{13}N laboratory energies of 45 and 29.5 MeV. The radioactive ^{13}N beam had an intensity of $\sim 10^8$ pps. Self supporting ^{11}B targets having thicknesses 1300 and 800 $\mu g/cm^2$ were used for the two runs respectively. At the higher bombarding energy the excitation energy of the CN was around the maximum for the previously reported broad state at 46.4 MeV[2].

The experimental set-up consisted of two large area annular silicon strip detectors 300μm thick. The forward detector array (LEDA Louvain-Edinburgh-Detector-Array) consisted of 8 sector detectors each containing sixteen strips and covering the angular range $6° \leq \theta \leq 14°$. The detector array placed at larger angles (LAMP Lampshade) consisted of six sectors angled forward at 45°, to give larger solid angle acceptance. In this case the angular range covered was $21° \leq \theta \leq 69°$. The total solid angle (LEDA+LAMP) in the laboratory system was $\Omega \sim$3.6 sr.

In the analysis, six-α particle events were selected from the total reaction Q value. The total Q value spectrum shows a peak at around -500 keV which corresponds to the reaction ^{13}N+^{11}B$\rightarrow 6\alpha$. In order to select α particles coming from the break-up of ^{24}Mg into two ^{12}C* nuclei and to eliminate the sequential evaporation of α particles from the compound nucleus, a further selection was performed. Knowing that the break-up of the 0_2^+ and 3_1^- states in ^{12}C proceeds largely via the ^{8}Be$_{g.s}$+α channel[6], two such events were selected from the 6 α data set. The two ^{8}Be were selected by looking for alpha particles detected in adjacent strips. The peak at $E_{rel} \sim$90 keV was used to select ^{8}Be$_{gs}$ events. By measuring energies and angles of these ^{8}Be and α-particles the intermediate states leading to the six α decay of ^{24}Mg were reconstructed. Fig.1 shows the double excitation energy spectrum for the two ^{12}C for the 45 MeV run. The arrows mark different possible combinations of ^{12}C excited states from the ^{12}C*-^{12}C* pair. The energy resolution of this spectrum is dominated by the angular resolution and by the choice of the binning due to low statistics. Because of the ambiguity arising from a wrong association between the two ^{8}Be and the two αs, an incorrect ^{12}C excitation assignment for some of the events is possible. Monte Carlo simulations have then been performed to understand the contribution of miscorrelated events. They show that these latter contribute only to the high excitation energies of fig.1 (hatched region).

In fig.1 one can observe that experimental yield seems to correlate with states in ^{12}C. They are mainly the double excitation of the 3_1^- at 9.64 MeV with a contribution of the 3_1^--1_1^- and 0_2^+-3_1^-. The Monte Carlo program UNIMONTE was also used to deduce the efficiency of our set-up which is shown as a dashed curve in fig.1.

The total cross section for the selected six-α events has also been extracted for the two beam energies. We obtained σ=28.3$\pm$7.1 μb for E_{lab}=45 MeV and σ=6.6$\pm$1.6 μb for E_{lab}=29.5 MeV. In conclusion, the break-up into six α particles from ^{24}Mg populated with the reaction ^{13}N+^{11}B at an excitation energy around the 46 MeV resonance found in ^{12}C+^{12}C scattering has been discussed. The analysis showed that many of the six-α events originate from

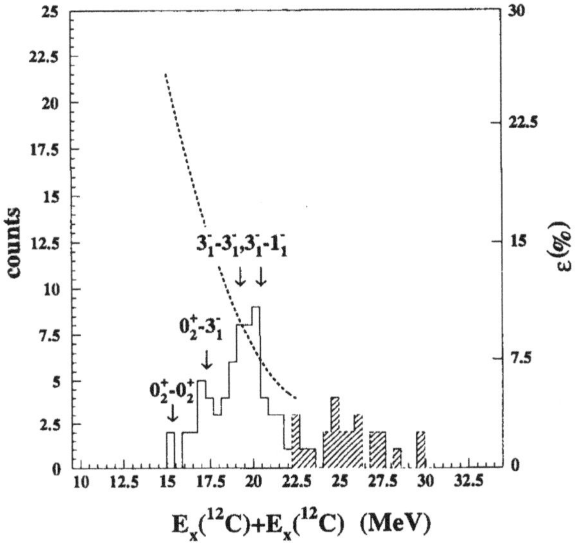

Figure 1: Total excitation energy spectrum for six-α coincidence data for E_{lab}=45 MeV uncorrected for the efficiency of the detector. The dashed curve indicates the efficiency (ε) of the apparatus for each combination of excited states. The hatched region shows where miscorrelated events contribute to the spectrum. The arrows mark the position of different possible combinations of ^{12}C states.

the break-up of two ^{12}C excited in the 9.64 MeV 3_1^- state, but also decays from the 0_2^+-3_1^- are observed. Within the limited statistics there is not strong evidence for the 0_2^+-0_2^+ decay taking place. There is reasonable consistency between the present $^{13}N+^{11}B$ results and the $^{12}C+^{12}C$ inelastic scattering [5]. Bearing in mind the single particle nature of the $^{13}N+^{11}B$ channel compared to the α cluster nature of $^{12}C+^{12}C$ this finding is surprising. Experimentally this experiment is an important development because it demonstrates that complicated reaction scenarios can be measured with low intensity radioactive beam provided highly efficient and segmented detector arrays are used.

References

1. R.R. Betts and A.H. Wuosmaa, *Rep. Prog. Phys.* **60**, 819 (1997) and references therein.
2. A.H. Wuosmaa *et al*, *Phys. Rev. Lett.* **68**, 1295 (1992).
3. A.H. Wuosmaa, *Z. Phys.* A **349**, 249 (1994).
4. M. Aliotta *et al*, *Z. Phys.* A **354**, 119 (1996).
5. R.A. Le Marechal *et al*, *Phys. Rev.* C **55**, 1881 (1997).
6. M. Freer *et al*, *Phys. Rev.* C **49**, R1751 (1994).

REFRACTIVE EFFECTS AND DIFFERENCE IN ABSORPTION BETWEEN $^{12}C+^{16}O$ AND $^{12}C+^{18}O$

S. SZILNER[1,2], M. P. NICOLI[2], Z. BASRAK[1], R. M. FREEMAN[2], F. HAAS[2], AND A. MORSAD[3]

[1] *Ruđer Bošković Institute, P.O.B. 1016, HR-10 001 Zagreb, Croatia*
[2] *Institut de Recherches Subatomiques, CNRS-IN2P3/ULP, Strasbourg, France*
[3] *Faculté des Sciences Ben M'sik, Université Hassan II, Casablanca, Morocco*

Binary channels of the $^{12}C+^{16}O$ and $^{12}C+^{18}O$ reactions have been measured at oxygen bombarding energies between 62 and 124 MeV. Detailed elastic scattering angular distributions have been obtained between 10° and 140° in the c.m. The data at 100 MeV are presented and described in terms of optical-model potentials. The difference in absorption in the two systems is evidenced in the elastic large angle yields and can be explained by an increased surface imaginary part of the potential in $^{12}C+^{18}O$ compared with $^{12}C+^{16}O$.

1 Introduction

Heavy-ion interactions are usually dominated by strong absorption. This is not the case for certain light heavy-ion collisions where the number of open channels is small [1] and the absorption weak. Consequently, resonance phenomena are observed in these systems at bombarding energies lower than ~ 5 MeV per nucleon. Indeed, strong resonances have been observed in $^{12}C+^{16}O$, they are also present in $^{12}C+^{18}O$ but are less pronounced [2]. At higher energies, the weak absorption should result in the observation of refractive effects, which are very important to determine the nucleus-nucleus potential. In the present study, the refractive effects are searched for in $^{12}C+^{16}O$ and $^{12}C+^{18}O$ reactions. In these reactions, the exit binary channels are measured and compared to get information on the absorption and its difference due to the addition of two extra neutrons.

2 Experimental details

The experiments were performed at the Strasbourg Vivitron accelerator at beam energies $E_L(^{16}O)$ = 62, 75, 80, 94.8, 100, 115.9, 124 MeV and $E_L(^{18}O)$ = 66.12, 85, 100, 120 MeV. These energies correspond to an incident energy range where the experimental results reported in the literature on binary channels are rather meager and where the fusion cross sections saturate. For the elastic scattering, angular distributions are measured at small lab. angles between 5° and 20° using a Q3D spectrometer and its associated detection system. At the same time, data at the larger angles between 20° and 50° were recorded for all binary exit channels using two Si position sensitive detectors in the kinematical coincidence mode which allowed complete mass and Q-value identifications of the emitted fragments. Detailed angular distributions were obtained for the elastic scattering at all energies.

3 Optical-model potential

In recent years, significant progress have been reported in our understanding of the main features of the optical potential describing light heavy-ion collisions [3]. Discernible refractive effects in the reported angular distributions at various energies and in different systems give crucial information about the character of the optical potential. To analyze our data, we adopted the optical-model prescriptions given in Refs. [3,4]. We used the squared Woods-Saxon shape, WS2, for the real part of the potential and the sum of a WS2 "volume" term, plus a derivate Woods-Saxon WSD "surface" term, for the imaginary part of the potential. The phenomenological nuclear potential is given by $U_{ph}(r) = -V[f_V(r)]^2 - iW[f_W(r)]^2 - iW_D f_D(r)$, where the V,W, and W_D are the potential depths, $R_{V,W,D}$ and $a_{V,W,D}$ their radii and diffusenesses. The form factors are $f_{V,W}(r) = [1 + e^{(\frac{r-R_{V,W}}{a_{V,W}})}]^{-1}$ and $f_D(r) = -4a_D\frac{d}{dr}[1 + e^{(\frac{r-R_D}{a_D})}]^{-1}$. For an interaction potential $U_{\rm E}(r)$ between nuclei with nucleon numbers A_1 and A_2, the volume integral $J_{\rm U}(E) = -\frac{4\pi}{A_1 A_2}\int U_{\rm E}(r)r^2dr$ is a sensitive measure of the potential strength. This definition also applies to the real and imaginary parts of $U_{\rm E}(r)$. The fits were obtained through the automatic search option in the program PTOLEMY. The uncertainties were set to be 10 % of the experimental cross sections.

4 Results of the analysis

In the context of our contribution to the Conference, we have chosen to present the elastic scattering results at a bombarding energy of 100 MeV. Figure 1 shows the data and the optical-model fits (solid lines) for the two systems. The fit parameters of the optical model are listed in Table 1. Most of the previous studies of refractive effects were based on identical boson systems (^{12}C+^{12}C and ^{16}O+^{16}O) where angular distributions were symmetrical around 90° (c.m.). Nonidentical systems allow one to directly explore the scattering at larger angles beyond 90°. It is quite easy to reproduce the forward diffractionlike behavior up to about 70°, while the fits at larger angles require the introduction of an imaginary surface term. After having obtained the elastical scattering fits for the $^{16}O+^{12}C$ [5], the same optical-model parameters were used to calculate the angular dependence for the $^{18}O+^{12}C$ system. The result of the calculation is presented as a dashed line superimposed on the $^{18}O+^{12}C$ data in Figure 1. The agreement between the measured and calculated distributions is reasonable at forward angles, but the calculated distribution cannot describe the two orders of magnitude less yield in $^{18}O+^{12}C$ at the larger angles.

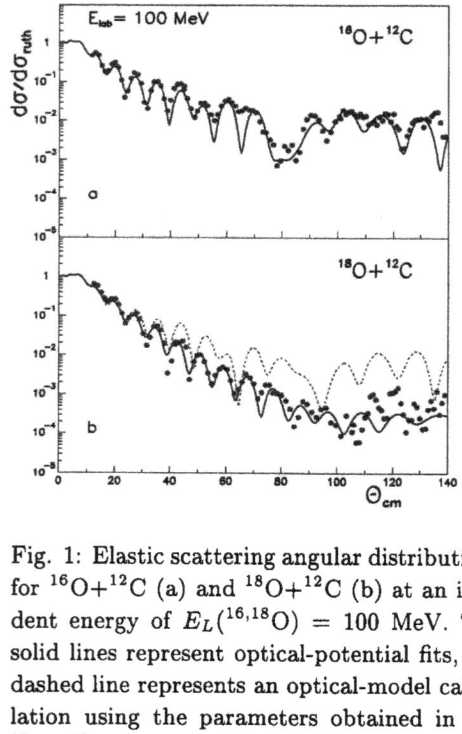

Fig. 1: Elastic scattering angular distributions for $^{16}O+^{12}C$ (a) and $^{18}O+^{12}C$ (b) at an incident energy of $E_L(^{16,18}O) = 100$ MeV. The solid lines represent optical-potential fits, the dashed line represents an optical-model calculation using the parameters obtained in the $^{16}O+^{12}C$ fit.

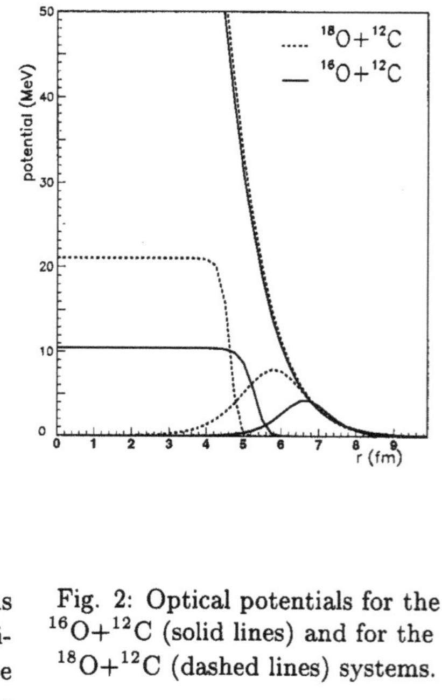

Fig. 2: Optical potentials for the $^{16}O+^{12}C$ (solid lines) and for the $^{18}O+^{12}C$ (dashed lines) systems.

It has to be noted that even without the fitting procedure the number of oscillations agrees with the measurements over the whole angular range which spans eight orders of magnitude in absolute cross sections. Figure 2 shows the real and imaginary ("volume" + "surface") parts of the potential for the two systems studied in this work.

	V	R_V	a_V	W	R_W	a_W	W_D	R_D	a_D
$^{16}O+^{12}C$	288.	4.00	1.40	10.45	5.466	0.19	3.8	6.64	0.44
$^{18}O+^{12}C$	288.	4.08	1.38	20.97	4.744	0.13	9.3	5.80	0.59

Table 1. Woods-Saxon squared potentials. V, W, and W_D are in [MeV], and R_V, R_W, R_D, a_V, a_W and a_D are in [fm].

The resulting potentials are found to have real parts that are strongly attractive ("deep") and imaginary parts that are relatively weakly absorbing. The real part remains almost unchanged for the two systems, and the real volume integral depends very weakly on the system ($\sim$ 300 MeV fm^3). The imaginary term has an almost squared shape with a surface part peaked at larger radii. The imaginary volume integral does not change from system to system, while the imaginary surface part increases rapidly from $^{16}O+^{12}C$ to $^{18}O+^{12}C$. The surface term contributes as much as 50% of the total imaginary volume integral.

In conclusion, the difference of the optical potential imaginary part reflects a much weaker absorption in $^{16}O+^{12}C$ system. While the fusion cross sections in both systems are almost the same, the larger imaginary surface integral in $^{12}C+^{18}O$ is a signature of a larger number of open direct (surface type) reaction channels owing to the contribution of two extra neutrons to the incident flux absorption.

References

1. F. Haas and Y. Abe, *Phys. Rev. Lett.* **46**, 1667 (1981).
2. K.A. Erb and D.A. Bromley in *Treatise on Heavy-Ion Science*, vol. 3, ed. D.A. Bromley (Plenum Press, New York, 1985) and references therein.
3. M.E. Brandan and G.R. Satchler, *Phys. Rep.* **285**, 143 (1997) and references therein.
4. M.E. Brandan and K.W. McVoy, *Phys. Rev.* C **55**, 1362 (1997).
5. M.P. Nicoli, F. Haas, R.M. Freeman, A. Morsad, S. Szilner, Z. Basrak, G.R. Satchler and M.E. Brandan, to be published.

COHERENT EFFECTS IN BINARY DISSIPATIVE REACTIONS AND DINUCLEAR MOLECULES

M. PAPA[1], F. AMORINI[2], G. CARDELLA[1], M. CAVALLARO[2], P. FIGUERA[2], A. MUSUMARRA[3], G. PAPPALARDO[2], F. RIZZO[2], S. ROMANO[2], S. TUDISCO[2], B. HEUSCH[4], WANG QI[5], LI SONGLIN[5], LU JUN[5], TIAN WENDONG[5] AND HU PENGYU[5]

[1] *Istituto Nazionale Fisica Nucleare-Sezione di Catania, Corso Italia 57, 95129 Italy*

[2] *Istituto Nazionale Fisica Nucleare-Laboratorio Nazionale del Sud, Via Santa Sofia 44, 95123 Catania, Italy*

[3] *Institut de Physique Nucleaire, Universite Catholique de Louvain, B-1348 Louvain-la-Neuve, Belgium*

[4] *Centre de Recherches Nucleares, F 67037 Strasbourg, Cedex 2, France*

[5] *Institute of Modern Physics, Academia Sinica, Lanzhou, PR China*

The energy fluctuations of the excitation functions in Dissipative Binary Heavy Ion Collisions (DBHIC), in which an huge of final channels are populated, have been established since 1985. In the following we recall the main steps in the description of the phenomenon including some new and interesting informations extracted from a preliminary analysis of the ^{27}Al+^{27}Al system at $E_{lab} \simeq 118 MeV$ recently studied.

1 Coherent effects in DBHIC

The theoretical approaches developed for the phenomenon description up to now have tried to include modification with respect to the T. Ericson theory[1] in such a way to take into account some particular characteristics of the IDS. In some of the experimental results collected [2,3] the energy autocorrelation functions (EAF) of the deep inelastic processes show a strong angular dependence which can be interpreted as an effect of the time space localization of the dinuclear system arising from the coherent effect of many partial waves. In 1987 D.M. Brink and K. Dietrich[4], proposed the following pole expansion of the S-matrix connecting the entrance channel α to the final channel β.

$$S^J_{\alpha,\beta}(E) = \sum_\mu \frac{b^\beta_{\alpha,\mu} W(J,\overline{J})^{1/2} e^{i\phi(J)} P_J(\theta)}{(E - E_\mu - E_{rot} + i\Gamma/2)} \tag{1}$$

This expression already contains the concept of coherent rotation being smooth the dependence on the total angular momentum J. The appearance of the collective rotational energy $E_{rot}(J)$ models also the degree of coherent rotation. In that work however no really new results respect to the Ericson theory were obtained because the authors worked out the final expression in the framework of a semiclassical approximation. Practically the same expansion was used by others authors[5,6]. By expressing the scattering amplitude for one final channel using Eq. 1., performing an integration in the complex energy plane and by using the Poisson's sum rule for the sum over the partial waves, the EAF, the angular correlation functions, and the associated time power spectra showed an interesting angular dependence able to fit the main behaviours of the experimental data. Also the so called "light house effect" can rise from this description when the rotation time is smaller than the mean life time of the IDS.

The same kind of experimental results triggered also an improvement of the above one final channel approach by taking into account the effect of the total angular momentum dissipation [7]. The dissipation effect can be the source of a spin dechoerence producing a delocalization of the dinuclear system due to the spread of the angular velocity. We want to note that the delocalization effect is also a rather general quantistic behaviour of waves packets propagating in an uniform field (this is our case because in this description the deformed dinuclear system can be seen like a free rotating wave packet in the c.m angle coordinate). This effect is not present in the above mentioned approaches because of some approximation performed to work out analytical expressions ($\overline{J} \simeq \frac{J+J'}{2}$). This spontaneous spread becomes prominent in time interval $\delta t(fm) > \frac{\hbar c}{\pi \hbar \omega} \overline{J} d^2$ where $\overline{J}$,ω,d are the average total angular momentum ,the average angular velocity and the initial angular dispersion of the wave packet, respectively. Therefore this effect should be seriously considered for the description of long lived dinuclear systems. To conclude this section we stress that in all the analytical expressions using the polar expansion (1) the concept of molecular states is present through the rotational energy in the real part of the poles, through the coherence of the partial waves but nothing is describing the structure of intrinsic excitations whose energy levels E_μ are supposed to have a very low spacing D.

2 Partially overlapped molecular level model and the ^{27}Al + ^{27}Al reaction

Until 1990 the main problem remained not solved: the fluctuations in the excitation functions are not washed out in spite of the high intrinsic excitation of IDS and of the large number of the final states contributing to the observable cross section in DBHIC.

Since 1991 the hypothesis that the fluctuating process proceeds through a selective excitation of low density "quasi-molecular " states gave rise to the so called Partially Overlapped Molecular Level Model POMLM[8,3] The main ingredients of this approach are:

- i) the polar expansion of equation (1);

-ii) the absence of an average amplitude correlation between the final channels;

-iii) the hypothesis that the levels are partially overlapped ($\frac{\Gamma}{D} = 10 - 30$).

The condition ii) is related to:

a) $\langle b^{\beta}_{\alpha,\mu} \rangle = 0; \overline{b^{\beta}_{\alpha,\mu}} = 0$ (the bar means ensemble average)

b) $\overline{b^{\beta}_{\alpha,\mu} b^{\beta'}_{\alpha,\mu'}} = \delta_{\mu,\mu'} \delta_{\beta,\beta'} \overline{|b^{\beta}_{\alpha,\mu}|^2}$

With this ingredients it is possible to work out a closed expression for the total cross section after the summation over large number of final channels:

$$\sigma_{tot} \simeq \sum_{J,J',\mu} (2J+1)(2J'+1) \frac{[W(J)W(J')]^{1/2} \overline{|b^{\beta}_{\alpha,\mu}|^2} e^{i(\phi(J) - \phi(J'))} P_J(\theta) P_{J'}(\theta)}{(E - E^{pole}_{\mu}(J))(E - E^{pole,*}_{\mu}(J'))} \tag{2}$$

Eq. 2. shows a fluctuating behaviour comparable with the experimental one if the level spacing satisfy the condition iii). The numbers $\overline{|b^{\beta}_{\alpha,\mu}|^2}$ have the following structure: $\overline{|b^{\beta}_{\alpha,\mu}|^2} = |\gamma_{\alpha,\mu}|^2 \overline{|\gamma_{\mu,\beta}|^2}$. The expressions (2) is not equivalent to the analytical expressions derived from (1) for different reasons:

-1) due to the conditions a) and b) and for a very large number of final channels, only the diagonal terms in the μ indexes give a non vanishing contribution and likewise to the expression (1) no coupling effects between different rotational bands are considered.

-2) it has been evaluated numerically due to the small value of the the ratio $\frac{\Gamma}{D}$. This allows also to take into account in a proper way the errors connected to the finite range of data (FRD) and the spontaneous spread of the wave packet. The physical idea underling this approach is simple: when a dinucleus is formed most of the total available energy is absorbed by rotation and deformation. The intrinsic excitation energy can be of the order of 15 MeV at $E/A \simeq 4MeV$. At this energy simple particles-holes configurations can be excited or probably in an equivalent way "molecular states"[9,10] can be excited. At the involved excitation energy these kind of "doorway states" can have low density of the order of some tents of Kev^{-1} and spreading width to more complicated configuration of the order of 500 Kev[11].

With this approach we explained the visibility of the fluctuations and their main behaviours[3] in different systems.

In the picture presented in the above section the necessary correlations between the final channels comes from the common entrance channel and is obtained through the condition of partial overlap of the levels. No amplitude correlation is present i.e. conditions 1) are valid. In ref.[3,12] however it has been shown that the conditions 1) produce for $\frac{\Gamma}{\hbar\omega} \leq 1$ a very strong average angular correlation $C(\theta, \theta') \simeq 1$.

To test the conditions 1) the fluctuations in the excitation functions of the dissipative collision induced on the $^{27}Al + ^{27}Al$ system have been studied in the laboratory energy range 114-123 MeV and in the c.m. angular range $40^\circ - 84^\circ$ [13].

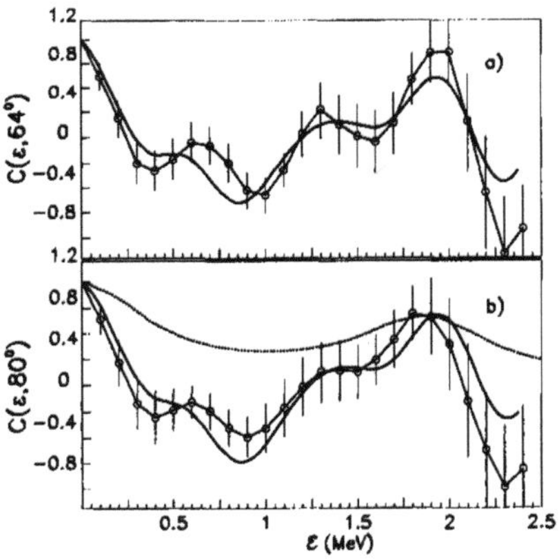

Fig.1 Experimental EAF (points) at $\theta = 64^\circ$ and $\theta = 80^\circ$ compared with the theoretical calculation (thick lines)

In Fig.1 we report as an example the experimental EAF (points) at two c.m. angles. The solid continuous lines represent the calculations which nicely reproduce the details of the experimental data (effects due to the FRD). The extracted parameter for the long lived dinuclear component are $\frac{\Gamma}{D} \simeq 13$, $\Gamma \simeq$ 450 Kev, $\hbar\omega \simeq 1$ MeV and $\Delta \simeq 2$.

For the same system, even if the global behaviour of the experimental angular correlation coefficients are well reproduced by the model, the calculations can not reproduce a slightly decreasing trend observed at backward angles[13]

This discrepancy suggests that in this case it is necessary to to go beyond the approximation scheme of the conditions 1) That is:

-I) coupling effect between different total angular momenta could be eventually considered. This can produce a further space time delocalization[13] of the system difficult to be experimentally distinguished from the others two sources briefly discussed in the previous section.

-II) the off-diagonal contributions $\gamma_{\alpha,\mu}\gamma^{\star}_{\alpha,\mu'}$ play a proper role. This means that an amplitude correlation has to be considered.

A natural and simple understanding of the nature of this correlation can then obtained, as already suggested by T.Ericson, by considering the following decomposition for the reduced amplitude: $\gamma_{\mu,\beta} = \overline{\gamma_{\mu,\beta}} + \delta\gamma_{\mu,\beta}$ The first term on the right hand side of this expression represents the common part between the resonant states and the outgoing channels. This term is already present in expression (2) through the squared modulus and it is connected to the degree of overlap (memory) between the IDS wave functions and the outgoing fragments wave functions.

The second term represents the true fluctuating part. The terms coming from the product $\delta\gamma_{\alpha,\mu}\overline{\gamma_{\mu,\beta}}$ in the S matrix expansion are able to give the necessary correlation that will vanish for $\frac{\Gamma}{D} \to \infty$.

Further work is in progress in this direction. Anyway it is already clear at this level that the relative weight p of the numbers $|\delta\gamma|$ and $|\gamma|$ together with the the ratio $\frac{\Gamma}{D}$ will characterize the coherence degree of the process giving also interesting insight on the microscopic and/or macroscopic description of highly deformed, excited, fast rotating intermediate dinuclear mulecules.

References

1. T. Ericson, *Ann. Phys* **23**, 390 (1963).
2. G. Pappalardo, *Nucl. Phys.* A **488**, 395c (1988) and reference therein.
3. M. Papa *et al*, *Z. Phys.* A **353**, 205 (1995) and references therein.
4. D.M. Brink and K.Z. Dietrich, *Z. Phys.* A **326**, 7 (1987).
5. S.Yu. Kun, M. Papa and D.K. Sunko, *Phys. Lett.* B **249**, 1 (1990).
6. S.Yu. Kun, *Phys. Lett.* B **257**, 247 (1991).
7. S.Yu. Kun, W. Nörenberg and M. Papa, *Phys. Lett.* B **298**, 273 (1993).
8. G. Pappalardo and M. Papa, in *Multistep Direct Reactions*, ed. R.H. Lemmer (World Scientific, Singapore, 1991), p. 89.
9. N. Cindro, *Ann. Phys.* **13**, 289 (1988).
10. E. Uegaki and Y. Abe, *Phys. Lett.* B **340**, 143 (1994).
11. M. Herman, G. Reffo and H.A. Weidenmuller, *Nucl. Phys.* A **356**, 124 (1992) and reference therein.
12. M. Papa *et al*, submitted for publications.
13. S.Yu. Kun, *Z. Phys.* A **357**, 255 (1997).

CLUSTER EMISSION OF ^{8}BE IN THE ^{28}SI+^{12}C FUSION REACTION AT LOW TEMPERATURE

M. ROUSSEAU[1], C. BECK[1], C. BHATTACHARYA[1,5], V. RAUCH[1], S. BELHABIB[2], A. DUMMER[3], R. M. FREEMAN[1], A. HACHEM[2], R. NOUICER[1], D. MAHBOUB[1], E. MARTIN[2], S. J. SANDERS[3], O. STEZOWSKI[1] AND A. SZANTO DE TOLEDO[4]

[1]*IReS, UMR7500, IN2P3-CNRS-ULP, F-67037 Strasbourg Cedex 02, France*
[2]*Universié de Nice-Sophia-Antipolis, Nice, France*
[3]*University of Kansas, Lawrence, Kansas 66045, USA*
[4]*Instituto de fisica da universidade de São Paulo, São Paulo, Brazil*
[5]*Variable Energy Cyclotron Centre, 1/AF Bidhan Nagar, Calcutta 700064, India*

Inclusive as well as exclusive energy spectra of the light charged particles emitted in the ^{28}Si (E_{lab}=112.6 MeV) + ^{12}C reaction has been measured using the **ICARE** multidetector array. The data have been analysed by statistical-model calculations using a spin-dependent level density parametrization. The results suggest significant deformation effects at high spin and cluster emission of ^{8}Be.

1 Introduction

In recent years, extensive efforts have been made to understand the decay of light di-nuclear systems (A<60) formed through low-energy heavy-ion reactions [1]. In most of the reactions studied, the properties of the observed fully energy damped yields have been successfully explained in terms of either a fusion-fission (FF) mechanism or a deep-inelastic (DI) orbiting mechanism behavior. The strong resonance-like structures observed in elastic and inelastic excitation functions of ^{24}Mg+^{24}Mg and ^{28}Si+^{28}Si have indicated the presence of shell stabilized, highly deformed configurations in the ^{48}Cr and ^{56}Ni compound systems respectively [1]. The present work aims to investigate the possible occurence of highly deformed configurations in the ^{40}Ca di-nucleus produced in the ^{28}Si+^{12}C reaction through the study of light charged particle (LCP) emission.

2 Experimental procedure

The experiment was performed at the IReS Strasbourg VIVITRON tandem facility using 112.6 MeV ^{28}Si beams on a ^{12}C (160 μg/cm^2) target. Both the heavy ions and their associated LCP's were detected using the **ICARE** charged particle multidetector array [2]. The heavy ions were detected in eight telescopes, each consisting of an ionisation chamber (IC) followed by a 500

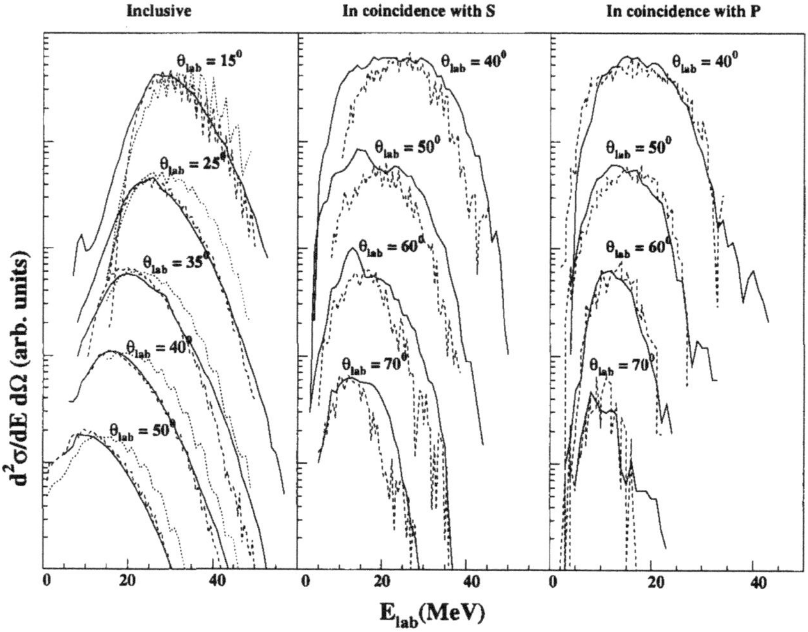

Figure 1. Inclusive (1.a) and exclusive experimental (solid line) α energy spectra in coincidence with S (1.b) and P (1.c). The dotted and dashed lines are CACARIZO calculations with and without deformation.

μm Si detector. The in-plane coincident LCP's were detected using four triple telescopes (Si 40 μm, Si 300 μm, 2 cm CsI(Tl)), 16 double telescopes (Si 40 μm, 2 cm CsI(Tl)) and two double telescopes (IC, Si 500 μm) located at the most backward angles. Typical inclusive α energy spectra are shown in Fig.1.a.

3 Experimental results and statistical-model calculations

The data analysis was performed using CACARIZO [3], the Monte Carlo version of the statistical-model code CASCADE. The angular momenta distribution, needed as the main input to constrain the calculation, was taken from $^{28}Si+^{12}C$ complete fusion data [4,5]. The other ingredients such as the nuclear level densities and the barrier transmission coefficients, are usually deduced from the study of the evaporated LCP spectra. Standard statistical-model calculations are not able to reproduce the shape of experimental α-particle energy spectra satisfactorily [3]. Several attempts have been made to explain this anomaly either by changing the emission barrier or by using a spin-dependent level density. In hot rotating nuclei forme density at higher angular momentum should be spin dependent. In CACARIZO, the level density, $\rho(E,J)$, for a given angular momentum J and energy E is given by the well known Fermi gas expression:

$$\rho(E,J) = \frac{(2J+1)}{12} a^{1/2} \left(\frac{\hbar^2}{2\Im_{eff}}\right)^{3/2} \frac{1}{(E-\Delta-E_J)^2} exp(2[a(E-\Delta-E_J)]^{1/2}),$$

Where a is the level density parameter, Δ is the pairing correction, $E_J = \frac{\hbar^2}{2\Im_{eff}} J(J+1)$ and $\Im_{eff} = \Im_0 \times (1 + \delta_1 J^2 + \delta_2 J^4)$ with $\Im_0$ the rigid body moment of inertia and δ_1, δ_2 the deformability parameters. By changing the deformability parameters one can simulate the deformation effects on the level densities. For the ^{28}Si + ^{28}Si reaction [6], the shape of the inclusive and exclusive α energy spectra are well reproduced by using large deformation effects [6]. Similarly the experimental inclusive α energy spectra for ^{28}Si + ^{12}C of Fig.1.a are better described by using deformation effects (dotted lines) than with the standard liquid-drop deformation (dashed lines).

The exclusive energy spectra of α-particle in coincidence with individual S and P ER's shown in Figs.1.b and 1.c are quite interesting. The dotted lines are the predictions of CACARIZO using non-zero values of the deformability parameters. The energy spectra associated with S are completely different from those associated with P. The latter are reasonably well reproduced by the CACARIZO curves whereas the model could not predict the shape of the spectra obtained in coincidence with S (Fig.1.b). This is due to the fact that an additional component might be significant in this case. One could suggest the hypothesis of a contribution arising from the decay of unbound ^{8}Be produced in a binary reaction ^{40}Ca $\rightarrow$ ^{32}S+^{8}Be. In order to determine the sources of both the α emission and ^{8}Be breakup the invariant cross sections in coincidence with P and S are plotted in Fig.2 in the ($V_\perp$,$V_\parallel$) plane. Fig.2.b shows the invariant cross sections in coincidence with P which maxima are centered on the compound nucleus velocity as expected for a fusion-evaporation mechanism. Fig.2.a presents two additional contributions (in circles) for angles close to 30° and 60° arising from the binary decay of unbound ^{8}Be. This conclusion is also consistent with the ER kinematical analysis of the S and P exclusive energy spectra (not shown here). The question of the real nature (FF or orbiting) of this decay process remains to be explored.

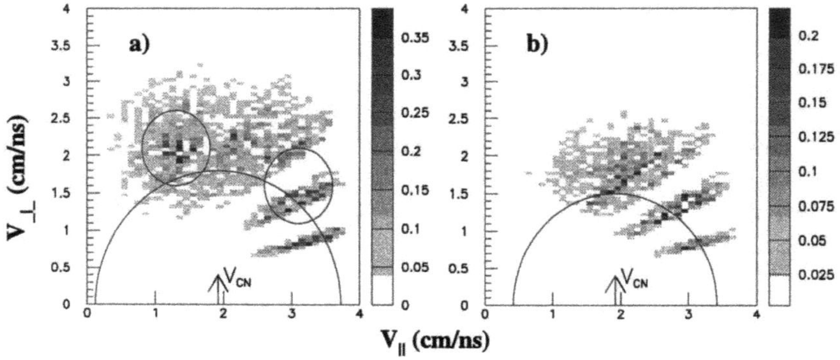

Figure 2. α invariant cross section in the $(V_{\perp},V_{\parallel})$ plane in coincidence with S (2.a) and P (2.b), the two circles in (2.a) show the additionnal contribution arising from the decay of ^{8}Be.

4 Summary

The α-particle energy spectra measured in coincidence with S have an additional component which may come from the decay of ^{8}Be, which is unbound and produced through the binary decay of $^{40}Ca \rightarrow ^{32}S + ^{8}Be$. Work is in progress to analyse the proton energy spectra as well as the in-plane angular correlations of both the α-particles and the protons.

References

1. S.J. Sanders *et al*, *Phys. Rep.* **311**, 487 (1999).
2. T. Bellot, PhD Thesis, Strasbourg University, Report **IReS 97-34**.
3. G. Viesti *et al*, *Phys. Rev.* C **38**, 2640 (1988).
4. B.A. Harmon *et al*, *Phys. Rev.* C **34**, 552 (1986).
5. M.F. Vineyard *et al*, *Phys. Rev.* C **47**, 2374 (1993).
6. C. Bhattacharya *et al*, *Nucl. Phys.* A, in press.

CLUSTERING CONFIGURATIONS IN INTERMEDIATE MEDIUM-MASS SYSTEMS

L. VANNUCCI, A. BOLOGNA, R.A.RICCI
INFN, Laboratori Nazionali di Legnaro, 35020 Legnaro, Italy

U. ABBONDANNO, P.M. MILAZZO
INFN, Sezione di Trieste, 34127 Trieste, Italy

M. BRUNO, M. D'AGOSTINO, G. VANNINI
Dipartimento di Fisica dell'Universita' di Bologna, 40126 Bologna, Italy
INFN, Sezione di Bologna, 40126 Bologna, Italy

W. SCHEID
Institut für Theoretische Physik der Justus-Liebig-Universität, Giessen, Germany

N. CINDRO
Ruđer Bošković Institute, HR-10 001 Zagreb, Croatia

A preliminary study of the $^{32}S+^{63}Cu$ collisions at 5.6 A·MeV has evidenced the presence of three-body processes in the reaction. Peripheral collisions produce ternary events characterized by non equilibrium effects and by preferential decay of the system in α-like fragments. These phenomenological aspects suggest the formation of macroscopic quasi-molecular intermediate states which decay in three-body final states.

1 Resonances in two body decay of medium-mass systems

The resonant-like behaviour is a general characteristic of the heavy-ion collisions [1–5] and has mostly been observed in the light mass region. The occurrence of resonant states is also predicted in the medium-mass region ($A\simeq100$ u) by the Orbiting Cluster Model (OCM) [6]. In particular are considered as favourable candidates some colliding systems in which the composite nucleus or the colliding nuclei have close $g_{\frac{7}{2}}$ shell. So, following the indications of the OCM, some systems [7–9] of the $A\simeq100$ mass region have been experimentally studied looking for resonant states.

Resonant structures were found in the excitation function of the $^{58}Ni+^{46}Ti$ elastic scattering [7] and in the elastic and inelastic excitation functions of the $^{58}Ni+^{58}Ni$ and $^{58}Ni+^{62}Ni$ scattering [9].

The corresponding experimental angular distributions are not oscillating and asymmetric. This behaviour is not compatible either with the formation of a compound nucleus or with the excitation of a prominent single-wave state. A similar phenomenology has been observed also in other systems of the medium-mass region (in particular in the two-body channel of $^{19}F+^{89}Y$ [10–12] and $^{28}Si+^{48}Ti$ [13,14]) and has been interpreted in the framework of the theory proposed by Kun [10]. In the Kun model, the formation of the intermediate state is due to the preferential excitation of a large number of rotational levels. The level overlapping and their common rotational character originate an angular-momentum coherence of pole form and focusing effects in the in-plane angular distributions like the ones observed in deep-inelastic collisions. The anisotropy of the angular distributions vanishes for a large rotation of the system and then for long-living intermediate states which have enough time to equilibrate.

Therefore, the observed structures in the excitation functions are not due to the excitation of isolated resonances with single spin and parity values but to the formation of *quasi – molecular* states in which the system behaves as a *rotating macroscopic* object characterized by collective angular velocity.

2 Perspectives for observation of three-body decay of resonant states formed in medium-mass systems

A phenomenology similar to the one described in the previous section was also observed in ternary final states produced in $^{100}Mo+^{100}Mo$ and $^{120}Sn+^{120}Sn$ dissipative collisions at 18.7 and 18.4 A·MeV respectively [15]. The authors had found an anisotropy of the in-plane angular distribution particularly evident for asymmetric mass divisions and had interpreted this non-equilibrium effect, in the framework of the Kun model, as due to formation of a "quasi-molecular" rotating system.

The result of [15] suggests that the decay of non-equilibrated quasi-molecular states could occur in three-body channels also at low bombarding energy. In the light-mass region this phenomenon was already observed, for example in [16–18]. For this reason we looked into three-body final states for a system ($^{32}S+^{63}Cu$) considered favourable candidate for resonant behaviour by the OCM because of its close $g_{\frac{7}{2}}$ shell.

In this first experiment we had look for three-body final states to check if they exist also at low bombarding energy and to evaluate the magnitude of their cross section. Thus we have not measured excitation functions which are the straightforward method to evidence non-statistical structures. The three-body data, however, show non-equilibrium effects that are probably due to the formation of an intermediate "macroscopic quasi-molecular" configuration in the sense described by Kun.

In the experiment we measured, event-by-event, the emission angle, the atomic number and the kinetic energy for pairs of fragments detected in coincidence. Two groups of events (B and T) are distinguishable in the scatter plot (Fig. 1a) of the total measured atomic number ($Z_1 + Z_2$) versus the Q-value.

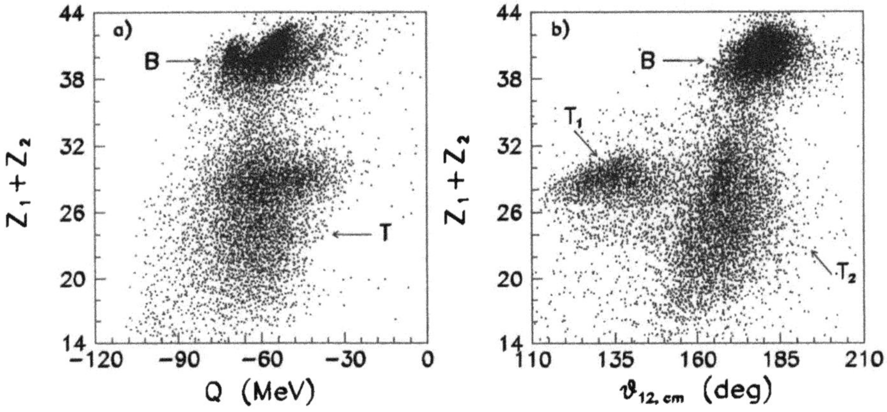

Figure 1: scatter plots for couple of fragments of the $^{32}S+^{63}Cu$ reaction detected in coincidence: **a** total measured atomic number *vs.* Q-value, **b** total measured atomic number *vs.* the relative angle of the detected fragments in the c.m..

Although of different magnitude, both the groups of events are characterized by charge deficit (ΔZ) with respect to the compound nucleus atomic number.

For the B group of events we observe $\Delta Z \simeq 5$ charge units. This signature suggests that the B events are due to binary processes coupled to light-particle emission. This hypothesis is also supported by the evidence that the two detected fragments (Z_1 and Z_2) have a back-to-back motion in the c.m. (Fig. 1b).

In comparison, for the T set of events the charge deficit varies from 9 until 31 charge units (roughly corresponding to 18÷62 missing nucleons). To our knowledge, no evidence is reported in literature of processes, induced by low-energy collisions, which produce so relevant an emission of light particles. So at least a third massive body should be present, although undetected, in the final state. We assumed (in the kinematical calculation) that this fragment corresponds to the center of mass of the undetected nuclear matter. This assumption seems realistic because the emission of some nucleons from the system does not significantly influence the kinematics. This can be seen for the binary events (group B) where the relative angle of the two detected fragments in the c.m. remains close to 180° in spite of the missing nucleons. So, from the pure three-body kinematics it is possible to deduce approximate values of the third body linear momentum and energy with an uncertainty that does not invalidate the qualitative conclusions drawn in this article.

All the observed three-body final states (T events) are due to dissipative processes (see Fig. 1a) but they appear to be differently characterized. In fact, in the scatter plot (Fig. 1b) of the total measured atomic number *vs.* the relative angle in the c.m. ($\theta_{12,cm}$) of the two detected fragments, two groups (T_1 and T_2) are clearly distinguishable besides the group of the binary events.

The first group of events (T_1), peaked at $\theta_{12,cm} \simeq 140°$, has previously been studied [19–22] and characterized as due to a sequential binary mechanism. As the two steps of the reaction are uncoupled, the outgoing direction of the two secondary fragments is randomly oriented with respect to the outgoing direction of the primary fragments.

In comparison, in the second group, events (T_2), the two detected fragments have a back-to-back relative motion ($\theta_{12,cm} \simeq 180°$ in Fig. 1b), This implies that also the linear momentum of the third fragment is aligned to the other two, and that the system breaks up in collinear configuration. The fragment alignment indicates that the three bodies produced in the final state are strongly interacting at the scission and that the decay of the intermediate system occurs in a *fast break-up process*.

Therefore the $^{32}S+^{63}Cu$ reaction produces intermediate states, decaying in the three-body channel, with characteristics (fast reaction time scale and collinear configuration at the scission) compatible with the "macroscopic quasi-molecular states" depicted by Kun in his model. However, the proof that such a resonant state decay also in the three-body channel could not be reached in this work because excitation functions were not measured. So, from the present data, can be obtained only useful indications for future studies.

The presence of non statistical intermediate states in the reaction can be indicated by correlations and non equilibrium effects in the measured observables (energy, atomic number, emission angle).

Correlation between fragments in three-body final states can be evidenced in the scatter plot of the kinetic energy of one fragment versus the kinetic energy of the other one. In fact, narrow horizontal or vertical bands in this plot indicate energy correlation because if the energy of one fragment is constant also the relative energy of the other two is constant. This topology is observed for some of the T_2 events (see the P_1 and P_2 groups in Fig. 2a).

Moreover the $^{32}S+^{63}Cu$ reaction produces ternary events in which one of the fragments is systematically a Calcium nucleus (P_1 final states) or a Sulphur nucleus (P_2 final states).

Since the atomic number of one of the fragments is fixed, the other two fragments are strongly Z-correlated. However, not all the possible Z-partitions between these two fragments are observed, only some of those which produce

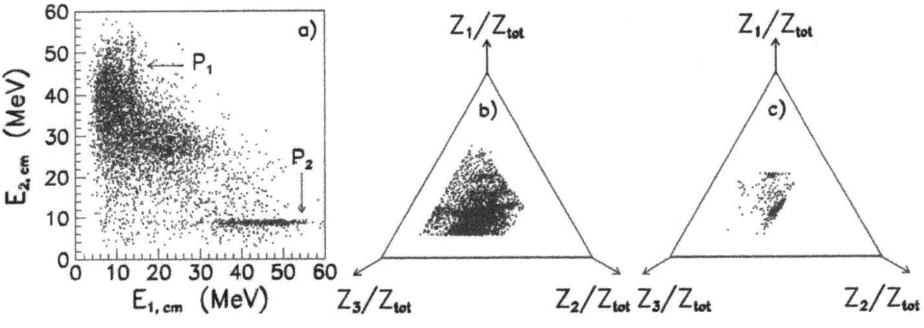

Figure 2: **a** scatter plot of the kinetic energy in the c.m. of one detected fragment versus the kinetic energy of another for the T_2 events. P_1 and P_2 indicate classes of events which show a strong fragment-fragment correlation. **b** Dalitz-plots of the fragment atomic number for the full set of T_2 events and **c** the subsets P_1 and P_2.

Table 1: atomic number of the three fragments produced in the collinear fragmentations.

ev.	Z_1	Z_2	Z_3	ev.	Z_1	Z_2	Z_3	ev.	Z_1	Z_2	Z_3
	20	11	14		11	16	18				
P_1	20	12	13	P_2	12	16	17	T_2	6÷26	6÷22	10÷30
					13	16	16				

one nucleus close to Magnesium (see Tab. 1).

The effect does not seem due to kinematics cuts. In fact, the Dalitz-plot of the fragment atomic number (Fig. 2b), relative to the T_2 events, present regions not covered by the apparatus acceptance only in the corners Z_1/Z_{tot} and Z_2/Z_{tot}. The other evident cuts at low atomic number are due to software event selection required to have three massive fragments ($Z_i > 6$) in the final state. These cuts, however, have no influence on the distribution of the P_1 and P_2 events that appear located in the middle of the Dalitz-plots and not near the borders (see Fig. 2c).

Therefore, while the majority of the events has fragment atomic number distributed in a wide range, the process producing the P_1 and P_2 events selectively populates final states in which at least one fragment is an α-like nucleus indicating the possible presence of clustering phenomena in the reaction.

Associated to this structure effect there are also non equilibrium effects.

This can be seen in the distribution of the flow-angle (Fig. 3). The flow-

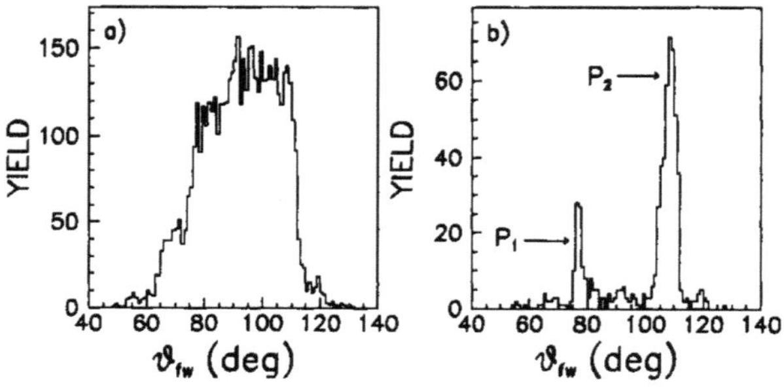

Figure 3: flow-angle distribution **a** for the full set (T_2) of collinear events produced in the $^{32}S+^{63}Cu$ reaction and **b** for the subset P_1 and P_2.

angle (θ_{fw}) is given by the orientation of the largest axis of the linear momentum ellipsoid in the c.m..

The events (T_2) due to collinear fragmentation of the system completely fill the apparatus angular acceptance with a broad distribution (see Fig. 3a). In this experiment, however, the angular acceptance was too restricted to establish if the distribution of the T_2 events is isotropic or not. In comparison, the subset of events P_1 and P_2, which shows structure effects in Z-partition and fragment-fragment correlation, shows narrow peaks in the θ_{fw} distribution (see Fig. 3b), indicating the formation of a short-living intermediate state. This focusing in the in-plane angular distribution indicates the persistence of some memory of the entrance channel and probably also coherence effect between the waves involved in the process.

At low bombarding energy anisotropy in the angular distributions is usually associated to non equilibrium processes induced by peripheral collisions which lead only to partial damping of the kinetic energy.

The centrality of the collision can be valued through the $Y_3 =< v_{rel} > -(v_{rel})_{min}$ parameter [23] which estimates the dispersion of the relative velocities of the fragments. Indeed low values of Y_3 are correlated with small impact parameters whereas high Y_3-values indicate peripheral collisions. The kinetic energy dissipated is taken into account by the reaction Q-value.

The three-body reactions which produce the T_2 events are strongly dissipative because the Q-value, which ranges between -40 and -70 MeV, is lower than for two-body dissipative processes (see Figs. 4a,b). Fig. 4b also shows that the collinear fragmentation of the system is due to non equilibrium processes because the dissipated energy clearly depends on the centrality of the collision. In comparison (see Fig. 4a) the two-body reactions show a flat behaviour for

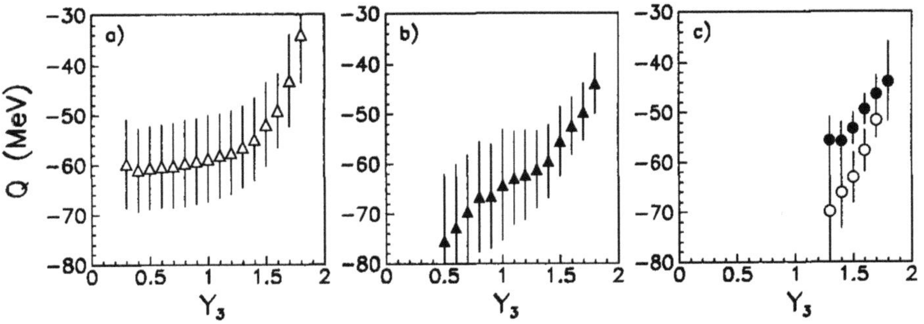

Figure 4: Q-value *vs.* Y_3, **a** for two-body events, **b** for the full set of ternary collinear events (T_2), **c** for the subsets P_1 (circles) and P_2 (dots). Error bars represent the FWHM of the distributions, the error on the mean is smaller than the symbol size.

central collisions (until $Y_3 \simeq 1.3$), as expected for fully equilibrated processes (like the fusion-fission), and dependence on the impact parameter for peripheral collisions which induce non equilibrium processes (like the deep-inelastic).

It also appears (Fig. 4c) that the process which produces the P_1 and P_2 events is peripheral and seems induced by collisions with impact parameter in the region which is typical of the most peripheral dissipative binary process (deep-inelastic). This is in agreement with the observed anisotropy of the angular distribution. All these evidences suggest that the preferential break-up of the system in fragments similar to α-like nuclei could be favoured by the preformation of a ternary clustered configuration which does not evolve in mono-nuclear regime.

Anyway to establish the nature (statistical or unstatistical) of the process which induces the collinear decay of the system (T_2 events) additional specific measurements need. In particular, it seems important to study the events (P_1 and P_2) which show fragment-fragment correlation and clustering effects that could be clues of new (ternary) reaction mechanisms.

3 Conclusions

The study of the two-body decay of medium-mass ($A \simeq 100$ u) systems evidenced the presence of non-statistical structures in the elastic and inelastic excitation functions. They seem due to the formation of dinuclear "macroscopic quasi-molecular" configurations.

Preliminary measurements of the $^{32}S+^{63}Cu$ reaction at 5.6 $A\cdot$MeV has evidenced the formation of intermediate states which decay in three bodies with aligned linear momenta. This process, when induced by peripheral collisions, produces ternary events characterized by non equilibrium effects in the energy dissipation and in the angular distribution. These effects are associated to strong energy and atomic number correlation and to preferential production of some particular α-like fragments in the final state.

These phenomenological aspects suggest the formation of macroscopic quasi-molecular intermediate states which decay in three-body final states.

The proof of this hypothesis, however, needs measurements of the excitation functions, evidence of both cross-channel correlation and periodicity in the energy auto-correlation functions.

References

1. R.R. Betts *et al*, *Phys. Lett.* B **100**, 117 (1981).
2. R.W. Zurmühle *et al*, *Phys. Lett.* B **129**, 384 (1983).
3. R.R. Betts, *Nucl. Phys.* A **447**, 257c (1985).
4. S. Saini *et al*, *Phys. Lett.* B **185**, 316 (1987).
5. U. Abbondanno, *Phys. Rev.* C **43**, 1484 (1991).
6. N. Cindro *et al*, *Ann. Phys.* (N.Y.) **192**, 307 (1989).
7. U. Abbondanno *et al*, *Int. J. Mod. Phys.* E **3**, 919 (1994).
8. S. Yu. Kun *et al*, *Z. Phys.* A **359**, 145 (1997).
9. L. Vannucci *et al*, *Z. Phys.* A **355**, 41 (1996).
10. S. Yu. Kun, *Phys. Lett.* B **257**, 247 (1991).
11. S. Yu. Kun *et al*, *Z. Phys.* A **357**, 367 (1997).
12. S. Yu. Kun, *J. Phys.* G **23**, 1359 (1997).
13. S. Yu. Kun, W. Nörenberg, *Z. Phys.* A **343**, 215 (1992).
14. F. Rizzo *et al*, *Z. Phys.* A **349**, 169 (1994).
15. G. Casini *et al*, *Phys. Rev. Lett.* **71**, 2567 (1993).
16. E. Costanzo *et al*, *Europhys. Lett.* **14**, 221 (1991).
17. E. Costanzo *et al*, *Phys. Rev.* C **44**, 111 (1991).
18. E. Costanzo *et al*, *Phys. Rev.* C **49**, 985 (1994).
19. P. Boccaccio *et al*, *Eur. Phys. J.* A **1**, 399 (1998).
20. G. Vannini *et al*, *Europhys. Lett.* **7**, 311 (1988).
21. P. Boccaccio *et al*, *Nuovo Cimento* A **106**, 379 (1993).
22. P. Boccaccio *et al*, *Z. Phys.* A **354**, 121 (1996).
23. R. Bougault *et al*, *Nucl. Phys.* A **488**, 255c (1988).

PRODUCTION OF NUCLEI CLOSE TO ^{100}Sn: THE ^{58}Ni+^{58}Ni FUSION REACTION FOLLOWED BY EMISSION OF NUCLEAR CLUSTERS

M. LA COMMARA, M. GLOGOWSKI, P. JARILLO-HERRERO,
R. BORCEA, M. GÓRSKA, H. GRAWE, R. KIRCHNER,
M. REJMUND, E. ROECKL, K. SCHMIDT
*Gesellschaft für Schwerionenforschung, Planckstrasse 1,
Darmstadt, D-64291, Germany*

A. D'ONOFRIO, V. ROCA, M. ROMANO, F. TERRASI
*Istituto Nazionale Fisica Nucleare, Sezione di Napoli,
Complesso Universitario M.S.A., Naples, I-80126, Italy*

N. BELCARI, A. GADEA, G. DE ANGELIS
*Istituto Nazionale Fisica Nucleare, Laboratori Nazionali di Legnaro,
via Romea 4, Legnaro (Pd), I-35020, Italy*

C. FAHLANDER, M. HELLSTRÖM
*Division of Cosmic and Subatomic Physics, Lund University,
Lund, S-22100, Sweden*

J. GÓMEZ DEL CAMPO, K. RYKACZEWSKI
*Physics Division, Oak Ridge National Laboratory,
Oak Ridge, TN 37831, USA*

The production of neutron-deficient isotopes near ^{100}Sn has been investigated by using on-line mass separation of evaporation residues produced by heavy-ion induced complete fusion reactions. We measured β-delayed protons and β-delayed γ-rays, deduced the mass-separated beam intensities for ^{99}Cd, ^{100}In, ^{101}Sn and ^{102}In, and determined the corresponding production cross-sections σ for ^{58}Ni+^{58}Ni reactions followed by nuclear cluster emission, and for ^{58}Ni+^{50}Cr fusion-evaporation reactions accompanied by emission of protons, neutrons or α particles. The preliminary results indicate that the σ values for the former reactions are lower than those for the latter ones. The comparison of the data with the results of Hauser-Feschbach calculations is discussed.

Studies of N=Z and neighbouring nuclei, especially in the region of a double shell closure, are important for testing and improving nuclear models. Especially isotopes near ^{100}Sn are of great interest since they exhibit particularly interesting nuclear-structure phenomena: (*i*) the isospin symmetry is connected to the question of proton-neutron pairing and isospin mixing; (*ii*) nuclei in this region lie near the proton drip-line; (*iii*) in addition to the known islands of α and proton radioactivity beyond ^{100}Sn, even the spontaneous emission of intermediate-mass fragments ("clusters") from nuclear ground states may occur in this region of the chart of nuclei. Moreover, the double shell closure at N=Z=50 involves features related to the astrophysical rp-process [1], which is expected to end in this region.

It has been recently observed [2] that, due to enhanced emission of clusters, ^{58}Ni+^{58}Ni reactions with projectile energies above 5·A MeV result in high production rates of nuclei below although not very close to ^{100}Sn. This was attributed to a complete fusion reaction followed by cluster emission. The latter process was described within the framework of the Hauser-Feschbach formula extended to multiple reaction channels [3]. In order to reach smaller cross sections and to thus investigate the production of nuclei closer to ^{100}Sn a more sensitive technique is necessary. We tried to fulfil this requirement by using the GSI mass separator [4] on-line to the heavy-ion accelerator UNILAC. We investigated reactions between a 30-40 particle-nA beam of 320-380 MeV ^{58}Ni and a 3.6 mg/cm^2 thick target of highly enriched ^{50}Cr, and between a ^{58}Ni-beam with the same intensity and with energies ranging from 380 to 430 MeV and a 3.8 mg/cm^2 thick target of highly enriched ^{58}Ni. The former projectile-target combination served to study proton, neutron and α emission, while the latter aimed at the investigation of cluster emission. The recoiling reaction products were stopped in the graphite catcher inside a FEBIAD-E ion source [5]. After ionization the radioactive isotopes were extracted, accelerated to 55 keV, and separated according to their mass in a sector magnet. The A=99, 100 and 101 beams were implanted into a thin carbon foil mounted in front of a ΔE-E silicon telescope for the measurement of β-delayed proton (βp) activities. The A=102 beam was collected on a movable tape and transported to a counting position which contained a ΔE-E silicon telescope and a HP-Ge detector for the measurement of βp and β-delayed γ ($\beta\gamma$) activities, respectively.

The production cross-sections σ for individual isotopes were deduced from mass-separated beam intensities, which were determined on the basis of the experimental βp and $\beta\gamma$ activities. The collection efficiency of the recoils in the ion source catcher was preliminarily estimated to be 89% for the ^{58}Ni+^{58}Ni reactions, whereas it was assumed to be 100% for the ^{58}Ni+^{50}Cr ones. The total separation efficiency, including release and ionization efficiency of the ion source as well as the transmission of the mass-separator, is 15% for ^{101}Sn [6], and was estimated to be the same (within a factor of 2 uncertainty) for ^{99}Cd, ^{100}In and ^{102}In. The βp and $\beta\gamma$ branching ratios were taken from previous experiments in the case of ^{99}Cd and ^{102}In; for ^{100}In and ^{101}Sn an experimental lower limit [7] and a theoretical prediction [6], respectively, were used.

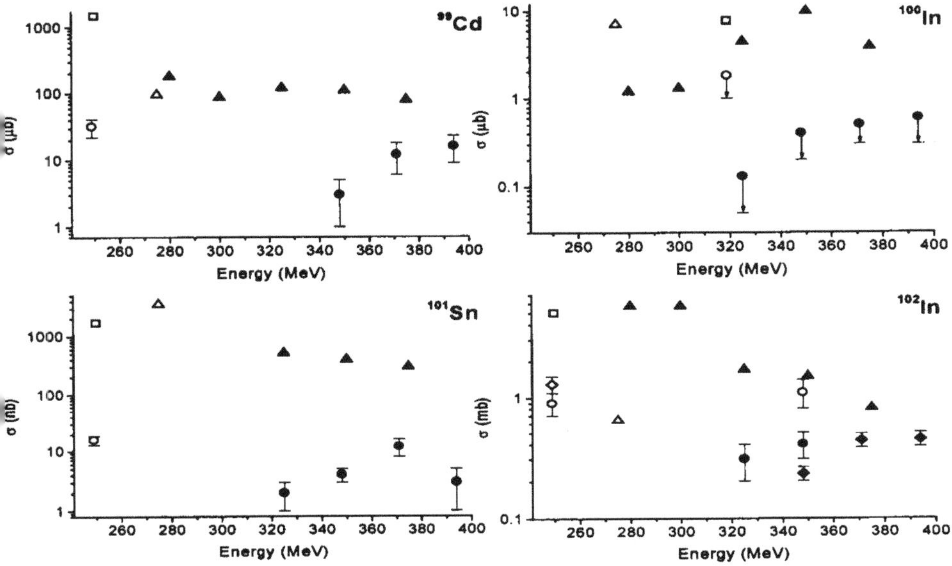

Figure 1: Comparison of measured and calculated cross-sections σ versus ^{58}Ni-beam energy in the middle of the targets. Experimental σ values for ^{99}Cd, ^{100}In, ^{101}Sn and ^{102}In produced in the reactions ^{58}Ni+^{50}Cr (open circles) and ^{58}Ni+^{58}Ni (solid circles) were determined from βp activities. Note that the experimental results for ^{100}In represent upper limits. Experimental σ values for ^{102}In produced in the reactions ^{58}Ni+^{50}Cr (open diamonds) and ^{58}Ni+^{58}Ni (solid diamonds) were determined from $\beta\gamma$ activity. BUSCO predictions of σ for ^{99}Cd, ^{100}In, ^{101}Sn and ^{102}In in the reactions ^{58}Ni+^{50}Cr (open triangles) and ^{58}Ni+^{58}Ni (solid triangles) and HIVAP predictions of σ for ^{99}Cd, ^{100}In, ^{101}Sn and ^{102}In in the reaction ^{58}Ni+^{50}Cr (open squares) were averaged over the corresponding target thickness.

The dependence of the preliminary σ values upon the projectile energy is displayed in Fig. 1. It can be clearly seen that the experimental σ values corresponding to reactions accompanied by emission of clusters are lower than those involving emission of protons, neutrons or α particles. Thus the fusion-evaporation reactions followed by cluster emission appear to be less suitable for producing exotic nuclei near ^{100}Sn if compared to the mechanism of light particle emission after complete fusion. The corresponding reduction factors range from 1.3±0.6 for ^{101}Sn to 14±5 for ^{100}In. For the experimental ^{58}Ni+^{58}Ni data, there seems to be a slight trend of increasing σ values with increasing beam energy. This trend is, however, not statistically significant except for the case of ^{100}In. The σ values measured for ^{99}Cd and ^{101}Sn from ^{58}Ni+^{50}Cr reactions are consistent with those obtained from previous experiments[6,8].

The experimental σ values are compared to theoretical predictions obtained by using the Hauser-Feschbach code BUSCO[3] which takes into account the emission of light particles ($Z \leq 2$) and clusters ($Z \geq 3$) up to calcium with as many as 400 decay channels. In the calculation, a channel, determined by Z, A and by the excitation energy of the emitted cluster, contains discrete as well as continuum states. BUSCO predicts that the emission of ^{12}C dominates over that of others cluster decay channels. As can be seen from Fig. 1, the predicted σ values for ^{58}Ni+^{58}Ni reactions are considerably larger than the measured ones. The calculation apparently yields better agreement with the experimental data for large σ values than for smaller ones. Moreover, the calculations overestimate σ also in the case of ^{58}Ni+^{50}Cr reactions. The prediction of the code HIVAP[9] for the ^{58}Ni+^{50}Cr reactions yields higher σ values than those obtained both from the experiment and from the BUSCO calculations (see Fig. 1). The common trend of the two codes to predict cross sections for the weak reaction channels to be higher than the measured ones has to be related to the model assumptions that enter in the calculations. Shell effects, for example, could play a non-negligible role at the excitation energies of the compound nuclei, that are of the order of 100 MeV. Further investigations however, which involve both more refined calculations (especially concerning decays through weak channels, that are very interesting for the applications in nuclear structure at the limits of the stability) and improved experiments, are clearly needed to confirm the preliminary results obtained in this work, and to thus enhance the knowledge about the cluster emission reaction mechanism.

References

1. H. Schatz *et al.*, *Phys. Rep.* **294**, 167 (1998).
2. J. Gómez del Campo *et al.*, *Phys. Rev.* C **57**, 2 (1998).
3. J. Gómez del Campo *et al.*, *Phys. Rev.* C **43**, 2689 (1991).
4. K. Burkard *et al.*, *Nucl. Instrum. Methods* B **126**, 12 (1997).
5. R. Kirchner *et al.*, *Nucl. Instrum. Methods* **186**, 295 (1981).
6. Z. Janas *et al.*, *Phys. Scripta* **56**, 262 (1995).
7. J. Szerypo *et al.*, *Nucl. Phys.* A **584**, 221 (1995).
8. M. Lipoglavšek *et al.*, *Phys. Rev. Lett.* **76**, 888 (1996).
9. W. Reisdorf *et al.*, *Z. Phys.* A **300**, 227 (1981).

THREE-CLUSTER NUCLEAR MOLECULES

D. N. POENARU AND B. DOBRESCU
Horia Hulubei National Institute of Physics and Nuclear Engineering,
P.O. Box MG-6, RO-76900 Bucharest, Romania
E-mail: poenaru@ifin.nipne.ro

W. GREINER
Institut für Theoretische Physik der Universität, Postfach 111932,
D-60054 Frankfurt am Main, Germany
E-mail: greiner@th.physik.uni-frankfurt.de

A three-center phenomenological model able to explain, at least from a qualitative point of view, the difference in the observed yield of a particle-accompanied fission and that of binary fission was developed. It is derived from the liquid drop model under the assumption that the aligned configuration, with the emitted particle between the light and heavy fragment is obtained by increasing continuously the separation distance, while the radii of the light fragment and of the light particle are kept constant. During the first stage of the deformation one has a two-center evolution until the neck radius becomes equal to the radius of the emitted particle. Then the three center starts developing by decreasing with the same amount the two tip distances. In such a way a second minimum, typical for a cluster molecule, appears in the deformation energy. Examples are presented for ^{240}Pu parent nucleus emitting α-particles and ^{14}C in a ternary process.

1 Introduction

Fission approach[1] to the cluster radioactivities[2] and α-decay has been systematically developed during the last two decades (see Ref. 1 and the references therein) as an alternative to the many-body theory.[3] One has to stress the quantum nature of these decay modes and of the fission process as well. The three groups of binary phenomena are taking place by tunneling through a potential barrier. Fission theory has also been extended toward extremely large mass asymmetry[4] to study the evaporation of light particles from a hot excited compound nucleus, going over the barrier. In a cold binary fission[5,6,7] the fragments and the parent are neither excited nor strongly deformed, hence no neutron is evaporated; the total kinetic energy of the fragments equals the released energy.

A more complex phenomenon, the particle-accompanied fission (or ternary fission) was observed both in neutron-induced and spontaneous fission. It was discovered[8,9] in 1946. Several such processes, in which the charged particle is a proton, deuteron, triton, $^{3-6,8}$He, $^{6-11}$Li, $^{7-14}$Be, $^{10-17}$B, $^{13-18}$C,

$^{15-20}$N, $^{15-22}$O, have been detected. Many other heavier isotopes of F, Ne, Na, Mg, Al, Si, P, S, Cl, Ar, and even Ca were mentioned.[10]

A very powerful technique, based on the fragment identification by using triple γ coincidences in the large arrays of Ge-detectors, like GAMMASPHERE, was employed to discover new characteristics of the fission process,[11,12] and new decay modes[13] (emission of an alpha particle and of ^{10}Be, accompanying the cold fission of ^{252}Cf, the double fine structure, and the triple fine structure in binary and ternary fission, respectively).

The possibility of a whole family of new decay modes, *the multicluster accompanied fission*, was recently envisaged.[14,15] Besides the fission into two or three fragments, a heavy or superheavy nucleus spontaneously breaks into four, five or six nuclei of which two are asymmetric or symmetric heavy fragments and the others are light clusters, e.g. α-particles, ^{10}Be, ^{14}C, ^{20}O, or combinations of them. Examples were presented for the two-, three- and four cluster accompanied cold fission of ^{252}Cf and ^{262}Rf, in which the emitted clusters are: 2α, $\alpha+^{6}$He, $\alpha+^{10}$Be, $\alpha+^{14}$C, 3α, $\alpha+^{6}$He + ^{10}Be, $2\alpha+^{6}$He, $2\alpha+^{8}$Be, $2\alpha+^{14}$C, and 4α.

The strong shell effect corresponding to the doubly magic heavy fragment ^{132}Sn was emphasized. ¿From the analysis of different configurations of fragments in touch, we concluded that the most favorable mechanism of such a decay mode should be the cluster emission from an elongated neck formed between the two heavy fragments. The fact that the potential barrier height is lower, suggests that in a competition between aligned and compact configurations, the former should prevail.

This idea is further exploited in the following for ternary fission, by suggesting a formation mechanism of the touching configuration, based on a three-center phenomenological model, able to explain the difference in the observed yield of a particle-accompanied fission and that of binary fission. It is derived from the liquid drop model under the assumption that the aligned configuration, with the emitted particle between the light and heavy fragment is obtained by increasing continuously the separation distance, while the radii of the heavy fragment and of the light particle are kept constant. During the first stage of the deformation one has a two-center evolution until the neck radius becomes equal to the radius of the emitted particle. Then the three center starts developping by decreasing with the same amount the two tip distances. We shall show that in such a way a second minimum, typical for a cluster molecule, appears in the deformation energy.

2 Shape Parametrization

The basic condition to be fulfilled in the ternary decay process, ${}^{A}Z \rightarrow \sum_1^3 {}^{A_i}Z_i$, concerns the released energy (Q-value)

$$Q = M - \sum_1^3 m_i \tag{1}$$

which should be positive and high enough in order to assure a relatively low potential barrier height. The hadron numbers are conserved. We took[17,18] the masses (in units of energy), entering in the above equation, from the compilation of measurements.[16] We make the convention $A_1 \geq A_2 \geq A_3$.

For the first stage of the process, we adopt the shape parametrization of two intersected spheres with radii R_1 and R_2. By placing the origin in the center of the large sphere, the surface equation can be written in a cylindrical system of coordinates as:

$$\rho_s^2 = \begin{cases} \rho_{sl}^2 = R_1^2 - z^2 & , \ -R_1 \leq z \leq z_{s1} \\ \rho_{sr}^2 = R_2^2 - (z-R)^2 & , \ z_{s1} \leq z \leq R + R_2 \end{cases} \tag{2}$$

in which z_{s1} is the position of the separation plane, and R is the distance between the two centers. This equation is valid as long as $R \leq R_{ov3}$ defined below. The fragment radius, R_1, is kept constant during the deformation, and for a given separation distance, R, the radius R_1 is derived from the volume conservation and matching conditions. The final fragments and the initial

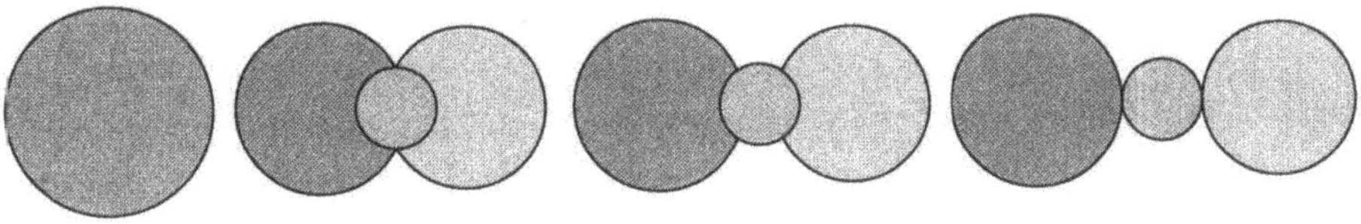

Figure 1. The assumed sequence of aligned shapes for the ternary fission of ^{240}Pu, leading to ^{14}C accompanied cold fission with ^{132}Sn and ^{98}Sr fragments.

parent nucleus are assumed to posses spherical shapes with radii R_1, R_2, R_3, and R_0, where $R_j = 1.2249 A_j^{1/3}$ fm ($j = 0, 1, 2, 3$). Within the range of R from $R_i = R_0 - R_1$ up to R_{ov3} one has a configuration of two overlapping spheres.

At $R = R_{ov3}$ (see the second position in Fig. 1) the neck radius $\rho_{neck1} = R_3$; R_3 is also kept constant. From that moment, the third fragment comes into play and one has two necks and two separating planes instead of one, hence:

$$\rho_s^2 = \begin{cases} \rho_{sl}^2 = R_1^2 - z^2 & , \ -R_1 \le z \le z_{s1} \\ \rho_{sc}^2 = R_3^2 - (z - z_3)^2 & , \ z_{s1} \le z \le z_{s2} \\ \rho_{sr}^2 = R_2^2 - (z - R)^2 & , \ z_{s2} \le z \le R + R_2 \end{cases} \tag{3}$$

In order to arrive safely at the final aligned configuration of fragments in touch with a corresponding decrease of the neck radii ρ_{neck1} and ρ_{neck2}, we assume a further elongation with a corresponding decrease of the neck radii ρ_{neck1} and ρ_{neck2} in a particular way, allowing to have the same (smaller and smaller) tip distance between the (overlapping) fragments 13 and 32 when R increases from R_{ov3} to $R_t = R_1 + R_{2f} + 2R_3$. In such a way the geometry is perfectly determined by giving one independent shape parameter, R, and the mass numbers of the parent and fragment nuclei.

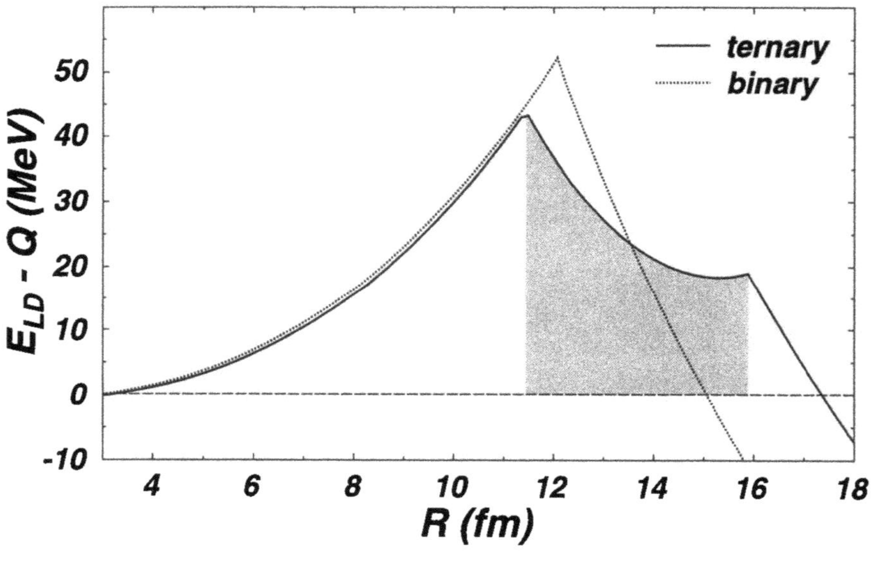

Figure 2. Deformation energy for the binary and ternary (accompanied by α emission) fission of ^{240}Pu. In both cases the heavy fragment is the double magic nucleus $^{132}_{50}Sn_{82}$. The light fragments for binary- and ternary processes are ^{108}Ru and ^{104}Mo, respectively. The region due to the devolopment of the light particle, from R_{ov3} to R_t is emphasized.

3 Deformation Energy

According to the liquid-drop model (LDM),[19] by requesting zero energy for a spherical shape, the deformation energy is defined as

$$E_{def} = (E_s - E_s^0) + (E_C - E_C^0) = E_s^0[B_s - 1 + 2X(B_C - 1)] \qquad (4)$$

where $E_s^0 = a_s(1-\kappa I^2)A^{2/3}$ and $E_C^0 = a_c Z^2 A^{-1/3}$ are energies corresponding to spherical shape. The relative surface and Coulomb energies $B_s = E_s/E_s^0$, $B_C = E_C/E_C^0$ are only functions of the nuclear shape. The dependence on the neutron and proton numbers is contained in E_s^0 and in the fissility parameter $X = E_C^0/(2E_s^0)$. The constants are $a_s = 17.9439$ MeV, $\kappa = 1.7826$, $a_c = 3e^2/(5r_0)$, $e^2 = 1.44$ MeV·fm, $r_0 = 1.2249$ fm.

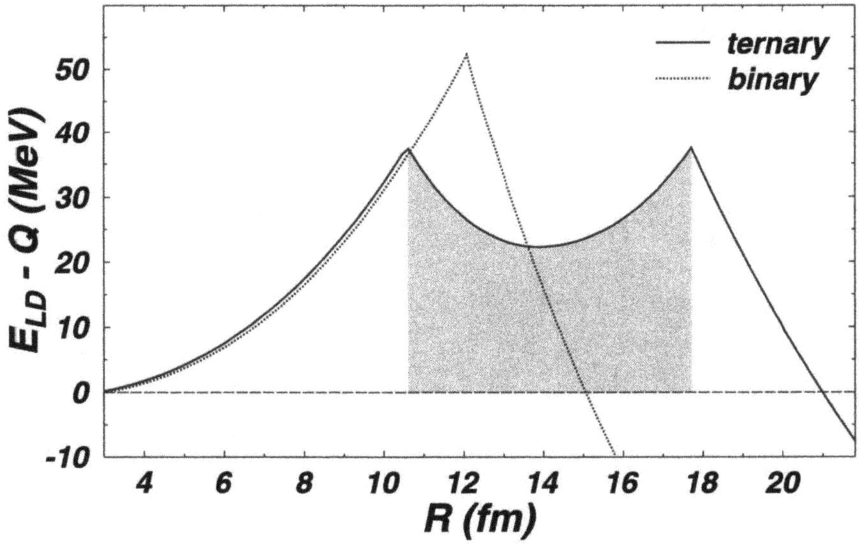

Figure 3. Deformation energy for the binary and ternary fission (accompanied by ^{14}C emission) of ^{240}Pu. In both cases the heavy fragment is the double magic nucleus $^{132}_{50}Sn_{82}$. The light fragments for binary- and ternary processes are ^{108}Ru and ^{94}Sr, respectively. The region due to the devolopment of the light particle, from R_{ov3} to R_t is emphasized.

To the deformation energy expressed in eq. (4), we add a small phenomenological shell correction, allowing to reproduce, at $R = R_i$, exactly the experimental Q-value in a system in which the origin of energy is taken as the sum of self energies of the fragments separated at infinity.

$$E_{LDM}(R) = E_{def}(R) + Q_{th} + (Q_{exp} - Q_{th})[1-(R-R_i)/(R_t-R_i)] - Q_{exp} \quad (5)$$

which is $E_{def}(R) + (Q_{th} - Q_{exp})(R - R_i)/(R_t - R_i)$, where $Q_{th} = E^0 - (E_1^0 + E_2^0 + E_3^0) = E_s^0 + E_C^0 - \sum_1^3 (E_{si}^0 + E_{Ci}^0)$. In this manner the barrier height increases if $Q_{exp} < Q_{th}$ and decreases if $Q_{exp} > Q_{th}$. The correction is increased gradually with R up to R_t and then remains constant for $R > R_t$. Apart this correction, after the touching point configuration, $R \geq R_t$, one is left with the Coulomb interaction energies. For spherical fragments this has the same expression as that would be obtained for points placed into the fragment centers and carying their whole charge.

Both the surface and Coulomb energies are calculated by performing numerical integration.[20,21] The relative surface energy is proportional to surface area. By expressing the nuclear surface equation in cylindrical coordinates $\rho = \rho(z, \varphi)$, one has

$$B_s = \frac{1}{4\pi R_0^2} \int_{z'}^{z''} dz \int_0^{2\pi} \rho \left[1 + \left(\frac{\partial \rho}{\partial z} \right)^2 + \left(\frac{1}{\rho} \frac{\partial \rho}{\partial \varphi} \right)^2 \right]^{1/2} d\varphi \qquad (6)$$

where z', z'' are the intersection points of the nuclear surface with Oz axis. Generally speaking, the Coulomb energy, E_C, for a system of three fragments with different charge densities, is defined by the following six fold integrals

$$E_C = \sum_1^3 \frac{\rho_{ie}^2}{2} \int_{V_i} d^3 r_1 \int_{V_i} \frac{d^3 r_2}{r_{12}} + \sum_{j \neq k} \rho_{je} \rho_{ke} \int_{Vj} d^3 r_1 \int_{Vk} \frac{d^3 r_2}{r_{12}} \qquad (7)$$

where the first three terms belong to individual fragments and the other three represent their interaction. Here $r_{12} = |\mathbf{r}_1 - \mathbf{r}_2|$. The charge densities of the compound nucleus and of the three fragments are denoted by ρ_{0e}, ρ_{1e}, ρ_{2e} and ρ_{3e} respectively. The six-fold integral is reduced to a four-fold one of the following kind.

$$E_C = \frac{\rho_e^2}{10} \int_{z'}^{z''} dz \int_{z'}^{z''} dz_1 \int_0^{2\pi} d\varphi \int_0^{2\pi} d\varphi_1 \left(\rho^2 - \frac{z}{2} \frac{\partial \rho^2}{\partial z} \right) \left[\rho_1^2 - \rho\rho_1 \cos(\varphi - \varphi_1) + \rho \frac{\partial \rho_1}{\partial \varphi_1} \sin(\varphi - \varphi_1) + \frac{(z - z_1)}{2} \frac{\partial \rho_1^2}{\partial z_1} \right] [\rho^2 + \rho_1^2 - 2\rho\rho_1 \cos(\varphi - \varphi_1) + (z - z_1)^2]^{-1/2} \qquad (8)$$

for a general shape without axial symmetry. One can get three-fold integrals for shapes possesing a symmetry axis, as for example:

$$B_{c1} = b_c \int_{-1}^{x_c} dx \int_{-1}^{x_c} dx' F(x, x') \tag{9}$$

where $b_c = 5d^5/8\pi$, $d = (z'' - z')/2R_0$, and x_c is the position of separation plane between fragments with -1, +1 intercepts on the symmetry axis (surface equation $y = y(x)$ or $y_1 = y(x')$). In the integrand

$$\begin{aligned} F(x, x') = \{ & yy_1[(K - 2D)/3] \cdot \\ & \left[2(y^2 + y_1^2) - (x - x')^2 + \frac{3}{2}(x - x')\left(\frac{dy_1^2}{dx'} - \frac{dy^2}{dx}\right)\right] + \\ & K\left\{y^2y_1^2/3 + \left[y^2 - \frac{x - x'}{2}\frac{dy^2}{dx}\right]\left[y_1^2 - \frac{x - x'}{2}\frac{dy_1^2}{dx'}\right]\right\}\}a_\rho^{-1} \end{aligned} \tag{10}$$

K and K' are the complete elliptic integrals of the first and second kind, respectively:

$$K(k) = \int_0^{\pi/2} (1 - k^2 \sin^2 t)^{-1/2} dt \tag{11}$$

$$K'(k) = \int_0^{\pi/2} (1 - k^2 \sin^2 t)^{1/2} dt \tag{12}$$

and $a_\rho^2 = (y + y_1)^2 + (x - x')^2$, $k^2 = 4yy_1/a_\rho^2$, $D = (K - K')/k^2$. In our computer program the elliptic integrals are calculated by using Chebyshev polynomial approximation. For $x = x'$ the function F is not determined. In this case, after removing the indetermination, we get $F(x, x') = 4y^3/3$.

4 Results

Two examples of deformation energies are presented in Figures 2 and 3. They were obtained for the α-particle-(Fig. 2) and ^{14}C (Fig. 3) accompanied fission of ^{240}Pu, by assuming a double-magic heavy fragment $^{132}_{50}Sn_{82}$. The corresponding deformation energy fvor the binary cold fission of the same nucleus is also shown.

We would like to stress two striking features of these plots. Besides the first (ground state) minimum there is a *second minimum*, proving the nuclear molecule character of the aligned configuration of three fragments in touch [(^{132}Sn, ^{4}He, ^{104}Mo) and (^{132}Sn, ^{14}C, ^{94}Sr, respectively].

On the second hand, by comparing the surface areas under the deformation energy curve of the binary and ternary pocesses, one can see the difference explaining at least qualitatively the increased yield of the binary relative to that of the ternary cold fission.

References

1. D. N. Poenaru and W. Greiner, in *Nuclear Decay Modes*, (IOP Publishing, Bristol, 1996), Chap. 6, pp. 275–336.
2. A. Sandulescu, D. N. Poenaru and W. Greiner, *Sov. J. Part. Nucl.* **11**, 528 (1980).
3. R. Blendowske, T. Fliessbach, and H. Walliser, in Ref. 1, Chap. 7 pp. 337–349.
4. L. G. Moretto, *Nucl. Phys.* A **247**, 211 (1975).
5. C. Signarbieux *et al.*, *J. Phys. Lett. Paris* **42**, L437 (1981).
6. D. N. Poenaru, M. Ivaşcu and W. Greiner in *Particle Emission from Nuclei* Vol III (CRC Press, Boca Raton, FL, 1989) 203.
7. F. Gönnenwein in *The Nuclear Fission Process* (CRC Press, Boca Raton, FL, 1991).
8. T. San-Tsiang, R. Chastel, H. Zah-Way and L. Vigneron *C. R. Acad. Sci. Paris* **223**, 986 (1946).
9. M. Mutterer and J. P. Theobald, in Ref. 1, Chap. 12, pp. 487–522.
10. F. Gönnenwein *et al.*, *Nuovo Cimento* A **110**, 1089 (1997).
11. J. H. Hamilton *et al.*, *J. Phys. G* **20**, L85 (1994).
12. G. M. Ter-Akopian *et al.*, *Phys. Rev. Lett.* **77**, 32 (1996).
13. A. V. Ramayya *et al.*, *Phys. Rev.* C **57**, 2370 (1998). *Phys. Rev. Lett.* **81**, 947 (1998).
14. D. N. Poenaru and W. Greiner, *J. Phys.* G. **25**, L7 (1999).
15. D. N. Poenaru, W. Greiner, J. H. Hamilton, A. V. Ramayya, E. Hourany, and R. A. Gherghescu, *Phys. Rev.* C **59**, 3457 (1999).
16. G. Audi and A. H. Wapstra, *Nucl. Phys.* A **595**, 409 (1995).
17. D. N. Poenaru, W. Greiner and R. A. Gherghescu, *Atomic Data Nucl. Data Tables* **68**, 91 (1998).
18. D. N. Poenaru and W. Greiner, in *Handbook of Nuclear Properties*, (Oxford University Press, 1996), Chap. 5, pp. 131–182.
19. W. D. Myers and W. J. Swiatecki, *Nucl. Phys.* **81**, 1 (1966).
20. K. T. R. Davies and A. J. Sierk, *J. Comput. Phys.* **18**, 311 (1975).
21. D. N. Poenaru and M. Ivaşcu, *Comput. Phys. Commun.* **16**, 85 (1978).

CLUSTER AND ALPHA RADIOACTIVITIES AND ASYMMETRIC FISSION

G. ROYER
SUBATECH, Université et Ecole des Mines de Nantes, IN2P3-CNRS, 44072 - Nantes Cedex 03, France
E-mail:royer@subatech.in2p3.fr

The emission of clusters like C, O, Ne, Mg, Si and alpha decay has been described within a very asymmetric fission leading to two spherical touching nuclei before crossing the potential barrier. The energy has been calculated with a generalized liquid drop model taking into account the proximity effects between the cluster or alpha and the daughter nucleus. The microscopic corrections have been introduced empirically to reproduce the experimental Q values. The theoretical partial half-lives obtained within the WKB barrier penetration probability are in good agreement with the experimental data. For alpha decay, data for 373 e-e, e-o, o-e and odd-odd nuclei can be fitted by $log_{10}[T_{1/2}(s)] = -26.06 - 1.114A^{1/6}\sqrt{Z} + \frac{1.5837\ Z}{\sqrt{Q_\alpha}}$ with a rms deviation of 0.42.

1 Introduction

In addition to the alpha decay, events of spontaneous decay of ^{223}Ra in a light ^{14}C fragment and a ^{209}Pb nucleus was observed in 1984 with a branching ratio to the alpha emission of around 6.10^{-10} and a half-life of $2.10^{15}s$. Since this experiment, the decay of 13 radioactive isotopes of Ra, Th, Pa, U and Pu via the spontaneous emission of ^{14}C, ^{20}O, ^{24}Ne, ^{28}Mg and ^{32}Si has been observed with branching ratios relative to the alpha decay from 10^{-9} down to 10^{-16} and partial half-lives from 10^{11} up to 10^{28}s [1].

The question is whether both alpha and cluster decays correspond to the emission of a preformed cluster while the spontaneous fission follows roughly the picture of the liquid drop model or, alternatively, the alpha and cluster emission and the spontaneous fission correspond to the break-up of a nuclear liquid drop leading to two fragments for three different values of the asymmetry. In the approach of the preformed cluster models [2] , the decay constant λ is the product of three terms : the barrier probability P, the assault frequency ν_0 and the cluster preformation probability P_0. The first contribution depends on the potential barrier while the two last ones vary with the size of the preformed cluster. In the approach of the unified fission models [3], the decay constant is simply the product of the barrier penetrability and of a constant assault frequency. Then, the half-lives are mainly governed by the position, width and height of the potential barriers.

An intermediate approach has been investigated [4 5] assuming that these decays are very asymmetric fission processes occuring in a deformation valley through compact and creviced shapes leading to the separation point before reaching the barrier top. In this fusion-like deformatiom valley, the data on the symmetric and asymmetric fission can be reproduced in the whole mass range [6 7 8].

2 Generalized liquid drop model and shape sequence

Within our generalized liquid drop model, the energy of a deformed nucleus is the sum of the volume, surface, Coulomb and proximity energies. The proximity energy term takes into account the attractive nuclear forces between the surfaces in regard in the neck or the gap between the alpha or cluster and the daughter nucleus. The proximity function is integrated in the neck and then depends on the selected shape sequence. This additional term to the surface energy is not a small correction like curvature energy. For a parent plutonium nucleus, the proximity energy reaches - 13.4 MeV. This correction appears only for necked shapes since the surface diffuseness is not taken into account within the liquid drop model.

The selected shape sequence simulates the transition from one spherical nucleus to two tangent spherical fragments in introducing the formation of a neck while keeping quasi-spherical ends. Later on, the separated nascent fragments remain spherical and turn away from each other. For a given decay asymmetry, the shape sequence is one parameter dependent.

3 Asymmetric and symmetric fission through quasi-molecular shapes

In this deformation valley the top of the potential barrier corresponds to a dinuclear molecule maintained in unstable equilibrium by the balance between the Coulomb forces and the surface and proximity forces. The potential barrier heights are similar to the fission barrier heights for the symmetric break-up and for the asymmetric one [6 7]. The Businaro-Gallone picture which tells us the fissility for which the fission mode becomes unstable with asymmetry is described [8]. The kinetic energies can be reproduced assuming that the proximity energy is less than 0.1-0.2 MeV at the scission point. If the shell effects are introduced double humped barriers appear for actinides. The first peak is due to the microscopic corrections but the outer peak is governed by the balance between the Coulomb and proximity forces. It disappears for the heaviest actinides as experimentally observed. As a conclusion, there is no experimental data allowing to exclude that the decay path through quasi-molecular shapes is really taken by most of the fissionning nuclei.

Analytic expressions allowing to calculate the main characteristics of the fission process in this decay path have been provided and a simple fortran code fisfus.ana is available by e-mail [9].

4 Theoretical half-lives

In such an unified fission model, the decay constant of the parent nucleus is simply defined as

$$\lambda = \nu_0 P \tag{1}$$

The assault frequency has been chosen as $\nu_0 = 10^{20} s^{-1}$.

The barrier penetrability P is calculated within the action integral,

$$P = \exp\{-\frac{2}{\hbar}\int_{R_a}^{R_b} [2B(r)(E(r) - E(R_a))]^{1/2}\, dr\} \tag{2}$$

with $E(R_a) = E(R_b) = Q_{exp}$. To take into account the microscopic corrections the difference between the experimental and the theoretical Q-value deduced from the generalized LDM has been added at the macroscopic potential energy of the initial spherical nucleus with a linear attenuation factor vanishing at the contact point of the nascent fragments. The expression proposed in Ref. [10] for the inertia B(r) in this new fission valley has been used.

5 Cluster radioactivity

The partial half-lives of the 13 radioactive isotopes : ^{222,223,224,226}Ra, ^{228,230}Th, ^{231}Pa, ^{232,233,234,235}U and ^{236,238}Pu decaying via the spontaneous emission of ^{14}C, ^{20}O, ^{24}Ne, ^{28}Mg and ^{32}Si have been calculated in this very asymmetric fission valley leading rapidly to two spherical touching nuclei at the early stage of the tunneling process. The agreement with the recent experimental data is quite correct [4].

Therefore, the emission of clusters by heavy nuclei may be viewed as the limiting case of very asymmetric fission via compact and creviced shapes. The main explanation of the good reproduction of the experimental data is the ability of the present model to reproduce the height and width of the potential barriers with the help of the experimental Q-value. The physical process which leads to two tangent spherical fragments may be an adiabatic fission process or the emission of a preformed cluster. Whatever the physical process is, the role of the proximity energy and the microscopic corrections is emphasized since their introduction is sufficient to reproduce the potential barrier characteristics which govern the half-lives of the compound nuclei.

6 Alpha radioactivity

The same approach (generalized liquid drop model and barrier tunneling) has been used to determine the alpha decay half-lives of 373 nuclei [5]. The only input data are the mass A and charge Z of the alpha emitter and the experimental alpha decay Q value. No parameter of the model has been adjusted.

The rms deviation of the theoretical values of $log_{10}[T_{1/2}(s)]$ from the experimental ones is 0.63.

A fitting procedure leads to the following formula for the 373 nuclei with a rms deviation of 0.42.

$$\log_{10}[T_{1/2}(s)] = -26.06 - 1.114A^{1/6}\sqrt{Z} + \frac{1.5837\ Z}{\sqrt{Q_\alpha}} \tag{3}$$

For the subset of the 131 even-even nuclei the rms deviation of the theoretical values from the experimental ones is 0.35.

Within the following formula the rms is 0.285.

$$\log_{10}[T_{1/2}(s)] = -25.31 - 1.1629A^{1/6}\sqrt{Z} + \frac{1.5864\ Z}{\sqrt{Q_\alpha}} \tag{4}$$

For the subset of the 106 even-odd nuclei the rms deviation of the theoretical values from the experimental ones is 0.71.

Within the following formula the rms is 0.39.

$$\log_{10}[T_{1/2}(s)] = -26.65 - 1.0859A^{1/6}\sqrt{Z} + \frac{1.5848\ Z}{\sqrt{Q_\alpha}} \tag{5}$$

For the subset of the 86 odd-even nuclei the rms deviation of the theoretical predictions from the experimental ones is 0.57.

The following formula leads to a rms of 0.36.

$$\log_{10}[T_{1/2}(s)] = -25.68 - 1.1423A^{1/6}\sqrt{Z} + \frac{1.592\ Z}{\sqrt{Q_\alpha}} \tag{6}$$

For the subset of the 50 odd-odd nuclei the rms deviation is 0.99.

The following formula leads to a rms of 0.35.

$$\log_{10}[T_{1/2}(s)] = -29.48 - 1.113A^{1/6}\sqrt{Z} + \frac{1.6971\ Z}{\sqrt{Q_\alpha}} \tag{7}$$

The experimental data are correctly reproduced by the model without introducing artificial variation of the assault frequency and artificial cluster preformation probability. This points out the accuracy of the proximity energy term, of the selected nuclear radius and the main role of the experimental alpha decay Q value. The proposed new formulas complete the formula set already given [11 12].

7 Conclusion

Within a generalized liquid drop model taking into account the effects of the proximity forces and the asymmetry degree of freedom and with the help of the experimental Q value, the alpha decay and the emission of clusters like C, O, Ne, Mg and Si can be described within a spontaneous tunneling process under the fusion-like fission valley. The introduction of a cluster preformation probability is not necessary. The same model reproduces also the main characteristics of the fission and fusion processes without adjustment of the parameters.

Acknowledgments

I thank G. Legoux, F. Roy, B. Viaud and K. Zbiri who determined several tables and analytic fits.

References

1. R.K. Gupta and W. Greiner, *Int. J. Mod. Phys.* E **3**, 335 (1994).
2. S. Kumar, R.K. Gupta and W. Scheid, *Int. J. Mod. Phys.* E **3**, 195 (1994).
3. D.N. Poenaru, M. Ivascu, A. Sandulescu and W. Greiner, *Phys. Rev.* C **32**, 572 (1985).
4. G. Royer, R.K. Gupta and V.Y. Denisov, *Nucl. Phys.* A **632**, 275 (1998).
5. G. Royer, to be submitted to *Nucl. Phys.* A.
6. G. Royer and B. Remaud, *J. Phys.* G **10**, 1057 (1984).
7. G. Royer and F. Haddad, *Phys. Rev.* C **51**, 2813 (1995).
8. F. Haddad and G. Royer, *J. Phys.* G **21**, 1357 (1995).
9. G. Royer, C. Normand and E. Druet, *Nucl. Phys.* A **634**, 267 (1998).
10. P. Moller, J.R. Nix and W.J. Swiatecki, *Nucl. Phys.* A **492**, 349 (1989).
11. B.A. Brown, *Phys. Rev.* C **46**, 811 (1992).
12. D.N. Poenaru and M. Ivascu, *J. Phys. (Paris)* **44**, 791 (1983).

DINUCLEAR CONCEPT – CLUSTER MODEL OF FUSION

G. G. ADAMIAN[1,2,3], N. V. ANTONENKO[1,2], E. A. CHEREPANOV[2], A. K. NASIROV[2,3], W. SCHEID[1] AND V. V. VOLKOV[2]

[1]*Institut für Theoretische Physik, D-35392 Giessen, Germany*
[2]*Joint Institute for Nuclear Research, 141980 Dubna, Russia*
[3]*Institute of Nuclear Physics, Tashkent 702132, Uzbekistan*

The synthesis of superheavy elements is analysed within the dinuclear system concept of compound nucleus formation. The perspectives for using radioactive beams in complete fusion reactions are discussed.

The existing fusion models are distinguished by the choice of the relevant collective degree of freedom which is mainly responsible for the complete fusion. For example, many models assume a melting of the nuclei along the relative distance. It was demonstrated that the adiabatic scenario of fusion in the relative distance leads to an overestimation of the fusion probability P_{CN} [1] and mostly gives an incorrect isotopic trend of P_{CN}. In the dinuclear system (DNS) concept [2] the compound nucleus is reached by a series of transfers of nucleons from the light nucleus to the heavy one. The dynamics of the DNS is considered as a combined diffusion in the degrees of freedom of the mass asymmetry $\eta = (A_1 - A_2)/(A_1 + A_2)$ (A_1 and A_2 are the mass numbers the DNS nuclei) and of the relative distance describing the formation of the compound nucleus and the quasifission process (decay of the DNS), respectively [3]. The competition between the complete fusion and quasifission processes is taken into consideration in the DNS model and leads to a strong reduction of the fusion cross section [3,4]. This cluster fusion model is justified by the structural forbiddenness effect [5] which hinders the nuclei to melt together along the relative distance. In the DNS concept [3] the evaporation residue cross section is calculated as $\sigma_{ER} = \sum_J \sigma_c P_{CN} W_{sur}$, where σ_c is the capture cross section for angular momentum J. The stabilizing shell effects of the formed superheavy compound nucleus against fission in the de-excitation process are thoroughly studied by the theory and surviving probabilities W_{sur} of compound nuclei are derived. The dependence of the probability of complete fusion P_{CN} on nuclear structure effects during the fusion process starting from the entrance channel and ending with the compound nucleus formation is the crucial factor for the correct calculation of σ_{ER}. In the reactions ^{90}Zr+^{90,92,96}Zr, ^{90,96}Zr+^{100}Mo, ^{86}Kr+^{99,102,104}Ru, ^{90,92,94,96}Zr+^{124}Sn and ^{86}Kr+^{130,136}Xe the fusion probabilities are decreased [6] when the neutron number of projectile or target deviates from the magic

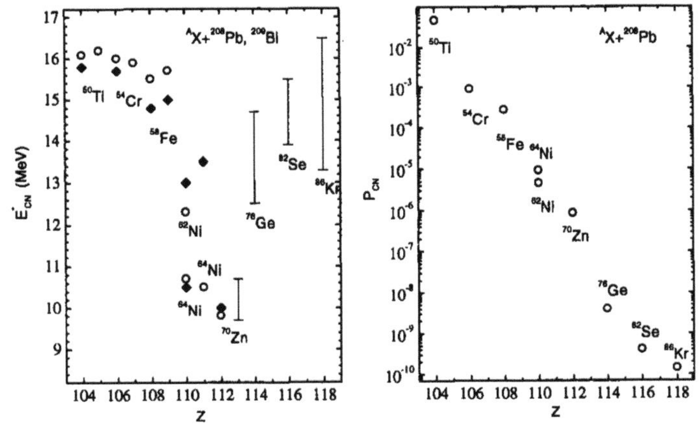

Figure 1. a) Optimal excitation energies of the compound nuclei. b) Calculated fusion probabilities P_{CN} for cold fusion (HI,1n) reactions. The experimental data [7] are shown by solid diamonds. The projectiles are indicated. For compound nuclei with Z=104-112, the calculations were performed with Q-values from Ref.[8]. For the elements with $Z \geq 113$ the lower (upper) limit of bars was calculated with Q-values from Ref.[8] (Ref.[9]).

number. In the DNS model this behaviour is simply explained taking the deformation of nuclei and shell effects in dependence of the DNS potential energy on η into account. For example, the value of the energy threshold for fusion, which determines P_{CN}, is larger in the ^{86}Kr+^{130}Xe reaction than in the ^{86}Kr+^{136}Xe reaction. So, the values of P_{CN} and W_{sur} are larger in the reaction with ^{136}Xe than with ^{130}Xe, which leads to a difference of the about 3 orders of magnitude in σ_{ER} in these reactions. In the fusion reactions leading to actinides, for example the ^{66,76}Zn+^{174}Yb reaction, the increase of W_{sur} with increasing neutron number of the system is stronger than the decrease of P_{CN}. This gives evident benefit to the neutron-rich projectiles for producing actinides.

In contrast to other models, the optimal excitation energy E^*_{CN} of the compound nucleus formed in cold fusion reactions is reproduced in the DNS concept. The value of E^*_{CN} increases after Z=113 (Fig. 1a). The difference between the Q-values of Refs. [8,9] for elements till Z=113 is small. The strong decrease (few orders of magnitude) of the cold fusion cross section with increasing charge number Z of the compound nucleus [7] is mainly caused by a decrease of the fusion probability P_{CN} due to a strong competition between complete fusion and quasifission in the DNS (Fig. 1b). Therefore, in reactions ^{76,74}Ge+^{208}Pb $\rightarrow^{283,281}$114+1n we expect a value of σ_{ER} which is smaller than 0.2 pb. The σ_{ER} for the Z=116 and 118 elements formed in the ^{80,82}Se,^{84,86}Kr+^{208}Pb reactions are smaller than the value for Z=114. In actinide-based reactions ^{48}Ca+^{232}Th, ^{238}U, ^{242,244}Pu, ^{248}Cm, ^{249}Cf, the P_{CN} also decrease with increasing Z, but they are larger than in Pb-based reactions. For ^{48}Ca+^{244}Pu→^{289}114+3n, the P_{CN} is 6×10^{-4} which is about 10^5 times larger than in ^{76}Ge+^{208}Pb→^{283}114+1n. The gain in fusion and capture probabilities for actinide-based reactions with respect to cold fusion reactions is not compensated by loss in the survival probability of the compound nucleus. So, our comparison of the formation cross section of element Z=114 in Pb- and actinide-based reactions shows that the latter one is larger (σ_{3n}=1.5 pb) [4]. The σ_{ER} of the ^{48}Ca+^{248}Cm,^{249}Cf reactions are smaller than experimental σ_{ER}=1 pb of the ^{48}Ca+^{244}Pu reaction.

In the Pb-based reactions with neutron-rich nuclei ^{70,74,78}Ni, ^{80}Zn,^{86}Ge and ^{92}Se the increase of W_{sur} with the number of neutrons could be compensated by decreasing P_{CN}. For example, in the ^{62}Ni+^{208}Pb reaction the yield of the Z=110 element is comparable with the yields in the ^{70,74}Ni+^{208}Pb reactions. The larger lifetime of the neutron-rich superheavies will allow a detailed study of their properties.

Acknowledgments

One of the authors (G.G.A.) is grateful to the Alexander von Humboldt-Stiftung for the support. This work was supported in part by DFG and RFBR.

References

1. G.G. Adamian *et al*, *Nucl. Phys.* A **646**, 29 (1999).
2. V.V. Volkov, *Izv. AN SSSR ser. fiz.* **50**, 1879 (1986).
3. N.V. Antonenko et al., *Phys. Lett.* B **319**, 425 (1993); *Phys. Rev.* C **51**, 2635 (1995); G.G. Adamian *et al*, *Nucl. Phys.* A **618**, 176 (1997); A 627, 332 (1997); A 633, 154 (1998); R.V. Jolos *et al*, *Europ. Phys. J.* A **4**, 245 (1999).
4. E.A. Cherepanov, *Pramana* **23**, 1c (1999); Yu.Ts. Oganessian *et al*, Preprint JINR, E7-99-53 (1998).
5. G.G. Adamian *et al*, *Phys. Lett.* B **451**, 289 (1999).
6. K.H. Schmidt, W. Morawek, *Rep. Prog. Phys.* **54**, 949 (1991).
7. S. Hofmann, *Rep. Prog. Phys.* **61**, 570 (1998); G. Münzenberg, *Phil. Trans. R. Soc. Lond.* A **356**, 2083 (1998).
8. P. Möller, J.R. Nix, *At. Data Nucl. Data Tables* **39**, 213 (1988).
9. P. Möller *et al*, *At. Data Nucl. Data Tables* **59**, 185 (1995).

PARTICLE-UNBOUND CLUSTERS FAR-OFF STABILITY

P. G. THIROLF, M. J. CHROMIK
Ludwig-Maximilians-Universität München, Am Coulombwall 1, D-85748 Garching, Germany
E-mail: peter.thirolf@physik.uni-muenchen.de

M. THOENNESSEN, M. FAUERBACH, T. GLASMACHER, R. IBBOTSON, R. A. KRYGER, H. SCHEIT
National Superconducting Cyclotron Laboratory, Michigan State University, East Lansing, MI 48824-1321, USA

P. J. WOODS
University of Edinburgh, Edinburgh EH9 3JZ, United Kingdom

The study of nuclear matter under extreme conditions, e.g. particle-unbound nucleon clusters at or beyond the proton dripline may reveal new insights into nuclear structure at the limits of particle-stability. The search for Two-Proton-Radioactivity plays an intriguing role in this context. The first excited state of ^{17}Ne has been populated via intermediate energy Coulomb excitation with a radioactive beam of ^{17}Ne on a ^{197}Au target to search for the simultaneous Two-Proton-Decay of this state. This decay branch is in competition with the γ-decay back to the ground state of ^{17}Ne. The reconstructed invariant mass spectrum of the outgoing decay products ^{15}O in coincidence with two protons shows evidence for a simultaneous two-proton transition from the first excited state in ^{17}Ne as well as for transitions from higher excited states decaying via sequential two-proton emission.

1 Introduction

With the availability of radioactive nuclear beams new opportunities have opened up to study nuclear matter far-off the valley of β-stability. Investigating particle-unbound nucleon clusters at or beyond the proton dripline represents a unique tool to probe nuclear structure under extreme conditions. Unlike in the case of α-decay, where the pre-formation of the α-particle in the nucleus has to be considered, the decay rate of a potential two-proton decay is mainly governed only by barrier penetration properties.
The search for diproton (^{2}He) radioactivity has still not been successful although its prediction by Goldanskii almost 40 years ago [1] triggered several experimental attempts [2]. So far all measured two-proton decays, for example β-delayed two-proton emission [3,4] or the breakup of ^{6}Be [5] and ^{12}O [2] turned out to be sequential. In these cases an intermediate state exists through which the decay can proceed.

The proton dripline nucleus ^{17}Ne represents a promising candidate for the identification of simultaneous two-proton emission, since its first excited state ($J^\pi = 3/2^-$, $E^* = 1.288$ MeV) is bound by 168 keV with respect to one-proton emission but unbound by 344 keV with respect to two-proton emission. Therefore this state can decay via a simultaneous two-proton decay to ^{15}O, since the widths of the low-lying states in ^{16}F are too small ($\simeq$40 keV) for a sequential decay via their tails (see Fig. 1). However, the two-proton decay is in competition with the γ-decay to the ground state in ^{17}Ne.

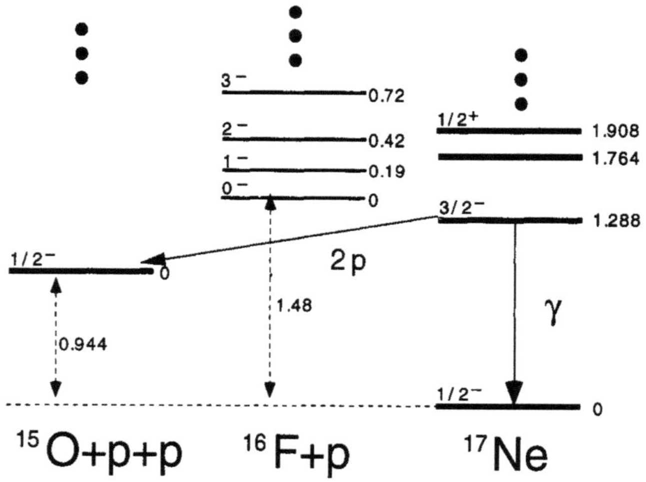

Figure 1. Energy level scheme of the lowest levels in ^{17}Ne, ^{16}F and ^{15}O. Also shown are the potential decay branches of the first excited state in ^{17}Ne, γ-decay to the ground state in ^{17}Ne and simultaneous two-proton-decay to the ground state of ^{15}O.

In a recent intermediate energy Coulomb excitation experiment the γ-decay from the first excited state to the ground state ($J^\pi = 1/2^-$) has been measured and the experimental yield has been compared to the theoretically expected cross section [7]. The measured γ-ray yield accounts for only 43% of the predicted yield from an excitation cross section of 18.7 mbarn, thus encouraging the investigation of a potential two-proton decay branch.

2 Experimental Procedure

The experiment to search for the two-proton decay of ^{17}Ne was performed at the National Superconducting Cyclotron Laboratory at Michigan State University. The radioactive ^{17}Ne beam with 60 MeV/u was produced from a ^{20}Ne primary beam (100 MeV/u) using the A1200 fragment separator and a Wien filter to further purify the secondary beam. We achieved a beam purity of 90% with a beam intensity of $\simeq$5000 ^{17}Ne particles per second. By scattering the secondary beam off a 184 mg/cm^2 ^{197}Au target excited states were populated via intermediate energy Coulomb excitation.

In order to identify the two-proton decay from the first excited state in ^{17}Ne a complete reconstruction of the decay kinematics in the rest frame is needed. Thus the interaction point on the target as well as the energies and directions of all outgoing decay particles had to be measured. Fig. 2 shows the experimental setup.

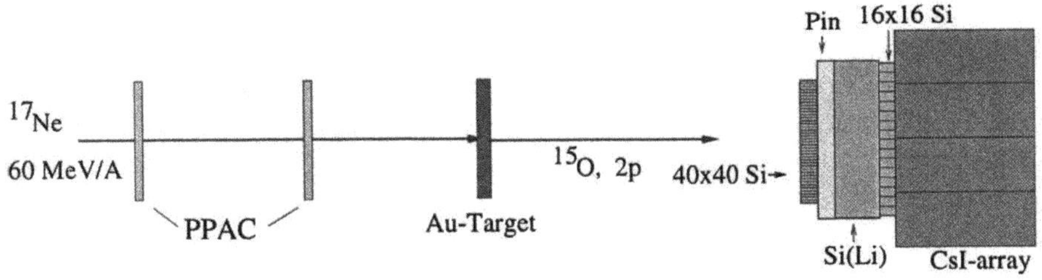

Figure 2. Experimental setup

Two position-sensitive parallel plate avalanche counters (PPAC) in front of the target served to determine the interaction point of the secondary beam with the target and a thin plastic scintillator 40 m upstream was used for starting time-of-flight measurements to identify the incoming particle.

The target was surrounded by the MSU-NaI-array [8] to simultaneously measure the γ-ray decay (not included in Fig. 2). The particle fragments were analyzed in a multiple stage particle telescope, which was positioned at 0° relative to the beam axis. The position of the heavy fragment was measured in a 300μm thick 40$\times$40 double-sided silicon strip detector and the ΔE in a Si-PIN diode (500μm), which also delivered a time signal. The heavy fragments were then stopped in a 5mm thick Si(Li)-detector. The light fragments penetrated these detectors and were then detected in a 300 μm thick 16$\times$16 double-sided silicon strip detector, which served as a ΔE-detector as well as for position measurements of the light reaction products.

Finally the protons were stopped in a 4×4 CsI array, consisting of 16 crystals (1.7 cm×1.7 cm ×5 cm), read out by photo diodes. These detectors were packed in a close geometry and placed 16.3 cm behind the target, thus covering an opening angle from 0° to 7°. A 6.5 cm thick lead collimator (with a central opening adapted to the size of the 40×40 strip detector) was placed in front of the detectors in order to shield the surrounding NaI-array from γ-rays originating from interactions with the detector material.

3 Data Analysis and Results

The incoming ^{17}Ne particles had to be identified and separated from the 10% of contamination which primarily consisted of ^{15}O. Fig. 3 (left side) shows the clear separation in the ToF spectrum between the upstream plastic scintillator and and the Si-PIN diode. The ^{17}Ne-projectiles were also spatially separated from the beam contaminant ^{15}O after a bending magnet at the end of the Wien filter, which is clearly visible in the 40×40 silicon strip detector (right panel of Fig. 3).

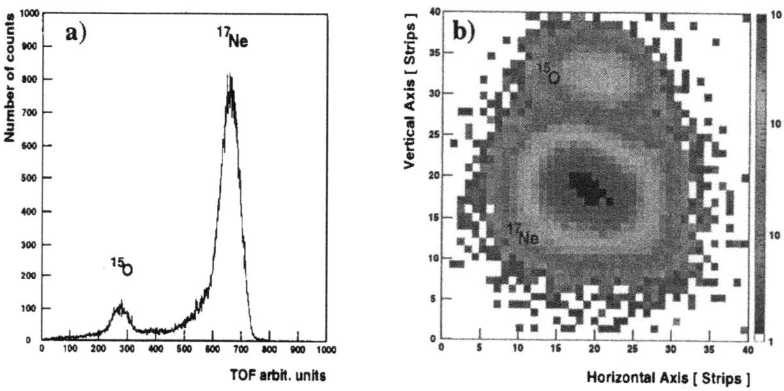

Figure 3. Particle identification of the incoming beam. The incoming ^{17}Ne was identified via a time-of-flight measurement (a). The position distribution in the 40×40 silicon strip detector displays the spatial separation between ^{17}Ne and the contaminant ^{15}O (b).

The identification of the ^{15}O fragments from the (^{17}Ne,^{15}O+p+p) reaction was done via the ΔE-E measurement using the PIN-diode and the 5 mm thick Silicon detector which stopped the heavy reaction products. Fig. 4 displays the resulting ΔE-E distribution a) without application of further conditions and b) under the requirement of identified incoming ^{17}Ne projectiles. While the first plot is dominated by non-reacting projectiles (^{17}Ne and the secondary beam contamination from ^{15}O), the component from ^{15}O reaction products is clearly visible after requiring ^{17}Ne projectiles in the entrance channel (right part of Fig. 4).

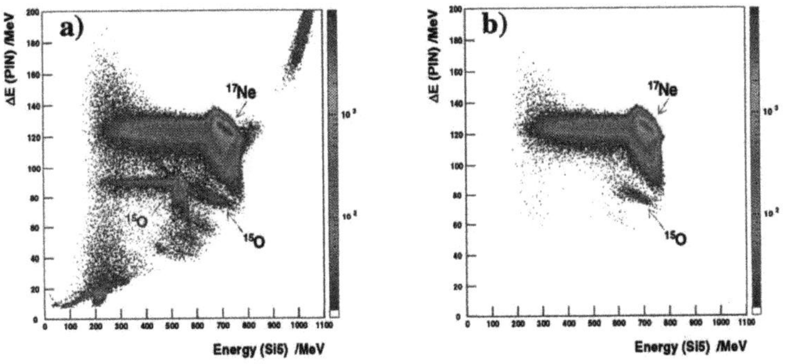

Figure 4. Particle identification of the heavy reaction products: The fragments were identified via a ΔE-E measurement using the PIN-diode and the thick (5 mm) Silicon detector. The distribution shown above was analyzed a) with no further conditions, b) under the condition of identified incoming ^{17}Ne projectiles.

After identifying the heavy reaction product ^{15}O events with two protons in the exit channel had to be selected. In Fig. 5 the sum of the energy loss ΔE_{tot} in all of the strips of the 16x16 strip detector is plotted against the sum of the deposited energy (E_{tot}) in all CsI detectors. In a spectrum gated on incoming ^{17}Ne the proton-, deuteron- and barely the triton-band are visible (left panel). An intensive band, consisting of events with twice the ΔE_{tot}- and E_{tot}-values of the proton band can be identified with two-proton events. At even larger energy losses one can also see the ^{3}He- and ^{4}He-bands (the band labeled with 'X' corresponds to events with 3 and more particle hits). An additional condition requiring only double-hit events in the strip detector verifies the identification of the two-proton band (right panel).

Using the geometric correlation between events in the 16×16 strip detector and the CsI-array one can then extract the energies and directions for each proton. Together with the information on the energies of the outgoing frag-

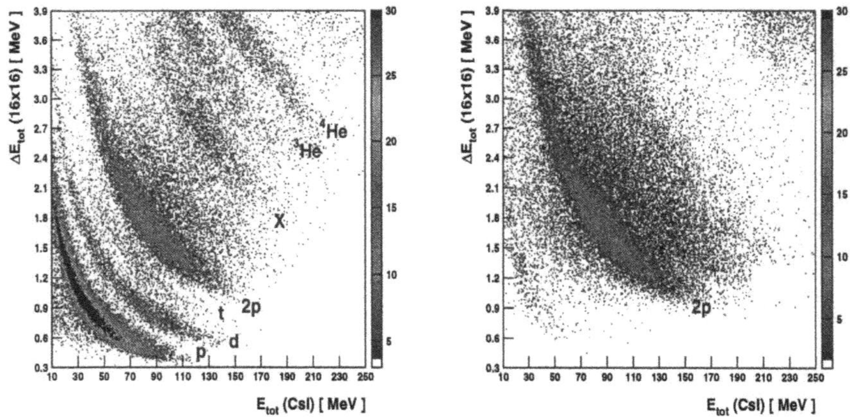

Figure 5. Particle identification of the light reaction products: The summed energy loss signal in the second silicon strip detector is plotted against the energy sum in the CsI detector array, gated on incoming ^{17}Ne. Panel a) shows the distribution without further conditions, while panel b) selects two-particle hits in the strip detector.

ments and their trajectories the transformation into the rest frame can be performed in order to obtain the invariant mass spectrum which is shown in Fig. 6. The measured decay energies agree well with known values [6], indicated by arrows in Fig. 6. Partly those transitions could not be resolved. The energy resolution is of the order of 250 keV, mainly influenced by the error in the determination of the interaction point on the target. The decays of the higher lying states proceed predominantly sequential via intermediate states in ^{16}F. Also shown in Fig. 6 is the result of a kinematical simulation for the decay of the first four states in ^{17}Ne.

The lowest peak at $340 \pm 40^{\text{stat.}} \pm 50^{\text{syst.}}$ keV corresponds to the simultaneous two-proton decay of the first excited state.

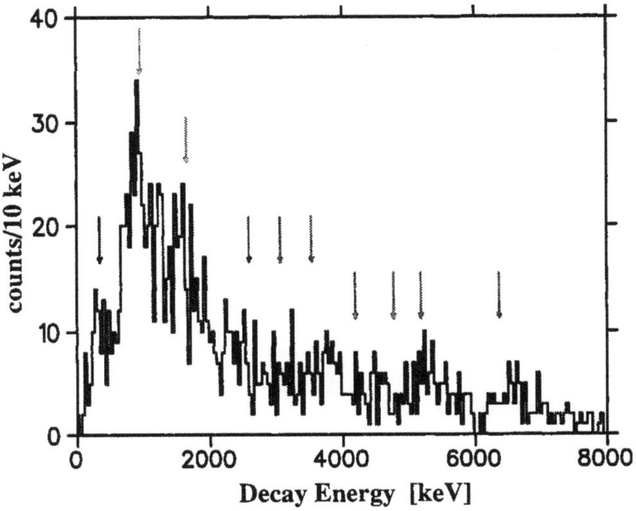

Figure 6. Reconstructed decay energy spectrum. The arrows indicate known transitions in ^{17}Ne. The lowest transition at ~340 keV gives experimental evidence for simultaneous two-proton decay from the first excited state.

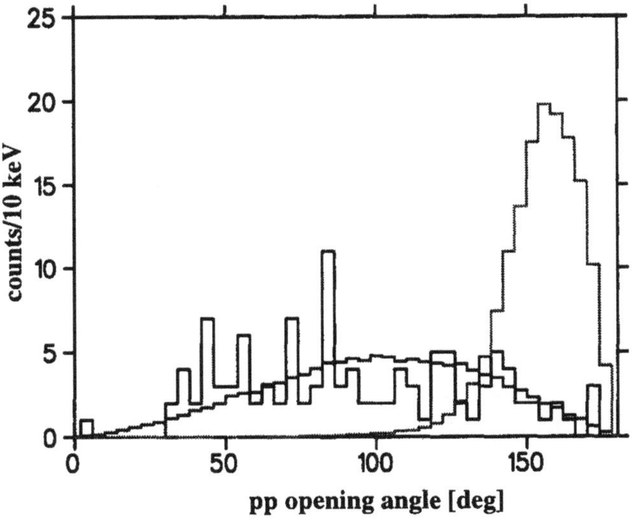

Figure 7. Two-proton opening angle distribution together with simulations for correlated (narrow distribution) and uncorrelated 2p-emission (broad distribution).

The analysis of the opening angle of the protons in the rest frame indicates an isotropic distribution, consistent with simultaneous uncorrelated emission of the two protons (Fig. 7). A preliminary analysis of the lifetime indicates a two-proton lifetime in the order of picoseconds, compared to a lifetime of $\simeq 0.1$ ps for the γ-decay from the first excited state in ^{17}Ne.

4 Conclusion

In conclusion, we observed several sequential two-proton transitions at their expected energies as well as for the first time evidence for a simultaneous two-proton emission from the first excited state in ^{17}Ne. The preliminary lifetime of picoseconds is too fast for the emission of two protons from the $1d_{5/2}$ shell, which is calculated to be a factor of $\simeq$300 slower [7]. The results of a recent experiment with optimized efficiency and energy resolution will help to clarify remaining uncertainties in the context of the reported first evidence for two-proton radioactivity.

Acknowledgments

One of us (MJC) acknowledges the support and hospitality of the NSCL and the support of the 'Studienstiftung des Deutschen Volkes'. We acknowledge the help of A. Azhari and S. Yokoyama during the experiment and thank J. Brown, D.J. Morrissey and M. Steiner for producing the radioactive ^{17}Ne beam. This work was supported by the National Science Foundation under grant PHY95-28844.

References

1. V.I. Goldanskii, *Sov. Phys. JETP* **39**, 497 (1960); *Nucl. Phys.* **19**, 482 (1960); **27**, 648 (1961).
2. R.A. Kryger *et al*, *Phys. Rev. Lett.* **74**, 860 (1995) and references therein.
3. D.M. Moltz *et al*, *Z. Phys. A* **342**, 273 (1992) and references therein.
4. C. Detraz, *Z. Phys.* A **340**, 227 (1991).
5. D.F. Geesaman *et al*, *Phys. Rev.* C **15**, 1835 (1977).
6. V. Guimaraes *et al*, *Nucl. Phys.* A **488**, 161c (1995).
7. M. Chromik *et al*, *Phys. Rev.* C **55**, 1676 (1997).
8. H. Scheit *et al*, *Nucl. Instr. Meth.* A **422**, 124 (1998).

NUCLEAR ASTROPHYSICS WITH INDIRECT METHODS

S. CHERUBINI[1], W. GALSTER[1], J. S. GRAULICH[1], P. LELEUX[1], J. VERVIER[1], Đ. MILJANIĆ[2], C. SPITALERI[3], M. ALIOTTA[3], M. LATTUADA[3], A. TUMINO[3], A. LAIRD[4], T. DAVINSON[4], A. DI PIETRO[4], A. N. OSTROWSKI[4], A. C. SHOTTER[4] AND P. J. WOODS[4]

[1]*Institut de physique nucléaire, Université catholique de Louvain, Louvain-la-Neuve, Belgium*

[2]*Ruđer Bošković Institute, Zagreb, Croatia*

[3]*Università di Catania and INFN-LNS, Catania, Italy*

[4]*Department of Physics and Astronomy, University of Edinburgh, Edinburgh, UK*

1 Introduction

The study of nuclear reactions at energies corresponding to stellar temperatures, important for the understanding of astrophysical processes such as the evolution of stars and the observed abundance of various elements in the Universe, is presently receiving a renewed interest from both astro- and nuclear physicists.

The main problem in performing accurate measurements of charged-particle cross-sections at energies relevant for nuclear astrophysics (typically ranging from tens of keV to hundreds of keV) is basically connected with the presence of the Coulomb barrier between the interacting nuclei. This lowers the values of the cross-sections down to the nano- and even picobarn domain. The experimental difficulties arising from this situation are enormous and require techniques which represent the state of the art in the field of low energy nuclear physics. As an example of this, one can consider that the effects of the atomic electrons in the target must be properly taken into account and that controversies on this point are still strong.

In order to overcome some of the experimental difficulties arising from such small values of the cross-sections, efforts have been done in developing methods which allow to extract the information of interest in an indirect way and to compare it with the available data. These are generally obtained by extrapolating the experimental value obtained at higher energies down to the relevant ones. The extrapolation often covers a range of various order of magnitude in the cross section values [1] and requires the assumption of theoretical models in order to get the astrophysical S-factor at stellar energies, which implies that some uncertainties are introduced. Moreover, the presence of unpredicted resonances in the Gamow window region [2] could change the situation giving rise to S-factors larger than the calculated ones.

The so called "Trojan Horse" (TH) method [3] -specialized to quasi-free reaction[4]- was applied to the indirect measurement of the $^6Li + d \rightarrow \alpha + \alpha$ reaction via the $^6Li + ^6Li \rightarrow 3\alpha$ one, where the ^6Li clusterization into $(\alpha + d)$ was exploited [5]. Results from that study and ongoing researches in nuclear astrophysics using indirect methods at the radioactive nuclear beam (RNB) facility in Louvain la Neuve are reviewed.

2 The Trojan Horse method.

Quasi-free reactions (QFR) are easily described in the framework of the Impulse Approximation (IA) . In this approach, the incident particle a is supposed to interact only with part b of the target nucleus A, while the other part of it, C, behaves as a spectator to the process [6]. The wave function of the target nucleus A is assumed to have a large amplitude for the C-b configuration: the nucleus A is said to have a high probability for C-b cluster structure. This is a necessary condition for the method to be applicable, as not all nuclei show such a strong clustering behaviour.
The three-body cross-section for the reaction shown in Fig.1, $A(a,\, cd)C$, can be expressed in Plane Wave Impulse Approximation (PWIA) as:

$$\frac{d^3\sigma}{dE_1 d\Omega_1 d\Omega_2} \propto (KF)\ |\Phi(\vec{p}_C)|^2 \left(\frac{d\sigma}{d\Omega}\right) \tag{1}$$

where KF is a kinematic factor containing the final state phase-space factor; $|\Phi(\vec{p}_C)|^2$ is the momentum distribution of the C-b relative motion in A, which is a function of the impulse of the spectator particle, p_C; $d\sigma/d\Omega$ is the off-shell cross-section for the $a + b \rightarrow c + d$ two-body (virtual) reaction. Assuming that $|\Phi(\vec{p}_C)|^2$ is known and calculating KF, it is then possible to derive $d\sigma/d\Omega$ from a measurement of $d^3\sigma/dE_1 d\Omega_1 d\Omega_2$.

In order to check these ideas, several measurements, involving reactions with either a charged particle or a neutron as spectator, have been performed. Excitation functions for the reactions $^2H(^7Li, \alpha\alpha)n$ and $^6Li(d,\ ^3He\ ^4He)n$ have been studied in the energy region $E_{^7Li} = 28 - 48\ MeV$ [7] and $E_{^6Li} = 21.6 - 33.6\ MeV$ [8] respectively. From these experiments, the behaviour of the cross-section of the $^7Li + p \rightarrow \alpha + \alpha$ and $^6Li + d \rightarrow \alpha + \alpha$ processes were extracted for energies above the Coulomb barrier. In both cases the energy dependance of the virtual reaction cross-sections agrees favourably with the known excitation functions of the corresponding two-body reactions and also the resonances are reproduced.

If the bombarding energy is high enough to overcome the Coulomb barrier in the incident channel of the reaction

$$a + A \rightarrow c + d + C \tag{2}$$

then particle b can be brought into the nuclear reaction zone and induce the reaction

$$b + a \rightarrow c + d. \tag{3}$$

If the Fermi motion of particle b inside A compensates for the initial projectile velocity $\vec{v}_a$ reaction (3) is induced at very low (even vanishing) relative energy between a and b, so as to match the relevant astrophysical energy [3]. It is then possible to extract the two-body cross-section as:

$$\frac{d\sigma_l^N}{d\Omega} \propto \frac{d^3\sigma}{dE_1 d\Omega_1 d\Omega_2}(KF\ |\Phi(\vec{p}_s)|^2)^{-1} \tag{4}$$

where l refers to the partial wave involved in the process. It has to be emphasized that in the present case the obtained cross-section, $d\sigma_l^N/d\Omega$, is almost a "pure" nuclear part, since the Coulomb barrier has already been overcome in the entrance channel and other effects are neglected in the PWIA.

In order to compare the cross-section given by (4) with that obtained in a direct measurement, the former has to be multiplied by a factor G_l:

$$\frac{d\sigma}{d\Omega} = \sum_l G_l \frac{d\sigma_l^N}{d\Omega} \tag{5}$$

where G_l represents the transmission coefficient for the $l-th$ partial wave and $d\sigma_l^N/d\Omega$ is the corresponding cross-section.

The TH method has been applied to the process $^6Li + d \rightarrow \alpha + \alpha$ as a test case. This was studied using the three-body reaction $^6Li(^6Li, \alpha\alpha)\alpha$ measured in a series of experiments carried out at the Demokritos Nuclear Research Center in Athens (experimental details are reported in [5] and references therein). A re-analysis of those data according to the method explained above brought to the indirect measurement of the S-factor for $^6Li + d \rightarrow \alpha + \alpha$ as also shown in [5].

3 Indirect methods and radioactive ion beams.

The difficulties in performing measurements of nuclear reactions at stellar energies already underlined above become even more serious if the process involves a short-lived (lifetime of the order of the seconds or less) radioactive nucleus. In this case the measurements can be performed -either directly or indirectly- only if a radioactive ion beam is available. In particular, the intensity of presently available RNB is often orders of magnitude lower than that of stable ones. In this conditions, high efficiency detection systems are necessary in order to maximize the acceptance of the experiments and indirect methods can play an interesting role once their reliability for the specific case has been checked.

A test experiment has recently been performed using the CYCLONE facility at Louvain-la-Neuve. The aim of the experiment was to show that the $^{15}O(\alpha,\gamma)^{19}Ne$ process, which is of interest in the so-called hot-CNO cycles [9,10,11], could be studied using the $^{18}Ne(d,p)^{19}Ne^{*}$ reaction. The basic idea is to populate the relevant ^{19}Ne states (4.033 MeV, which correspond to the first scattering resonance for the $^{15}O+\alpha$ system, and above), study their decay into $^{15}O+\alpha$ and derive from this the cross section for the inverse process, the radiative capture of α on ^{15}O. A detection system based on three LEDA array (for a total of 320 silicon detectors) was used to perform the measurements. The results of this preliminary experiment [12,13], whose data analysis is not yet completed, encouraged us to apply this or similar methods to the study of nuclear reactions of astrophysical interest involving radioactive beams.

References

1. W.A. Fowler, *Rev. Mod. Phys.* **56**, 149 (1984).
2. C. Rolfs and W. S. Rodney, in *Cauldrons in the cosmos* (Chicago University Press, 1988).
3. G. Baur, *Phys. Lett.* B **178**, 135 (1986).
4. C. Spitaleri in *Problems of fundamental Modern Physics* (World Scientific, Singapore, 1990).
5. S. Cherubini *et al*, *Ap. J.* **457**, 855 (1996).
6. G.F. Chew and G.C. Wick *Phys. Rev.* **85**, 632 (1952).
7. M. Zadro *et al*, *Phys. Rev.* C **40**, 181 (1989).
8. G. Calvi *et al*, *Phys. Rev.* C **41**, 1848 (1990).
9. P. Leleux, *Nucl. Phys.* A **621**, 183c (1997).
10. W. Bradfield-Smith *et al* in *Nuclear Astrophysics*, ed. M. Buballa, W. Nörenberg, J. Wambach and A. Wirzba, (Gesellschaft für Schwerionenforschung (GSI), Darmstadt, Germany, 1998).
11. J. Vervier in *Nuclear astrophysics*, (IOS Press, Amsterdam, 1998).
12. A.M. Laird *et al* in *5th Conference on Nuclei in the Cosmos* (Volos, Greece, 1998).
13. A.M. Laird *et al*, in preparation.

THE UTILITY OF A MICROSCOPIC CLUSTER DESCRIPTION IN LIGHT HALO AND HYPERNUCLEI

Y. SUZUKI
Department of Physics, Niigata University,
Niigata 950-2181, Japan
E-mail: suzuki@nt.sc.niigata-u.ac.jp

The cluster concept is employed to discuss the halo structure of ^{6}He and the overbinding problem in $^5_\Lambda$He and $^{\ 6}_{\Lambda\Lambda}$He. A realistic description of the α core is attempted in a $3N + N$ model and we find that it leads to the enhancement of the interaction of the halo neutrons with the core. This is vital to get enough binding in ^{6}He. The quark structure of baryons suggests a specific state which must be excluded from configurations available for hypernuclei. The binding energy reduction begins to be noticeable at a baryon size of 0.6 fm and becomes significantly large at 0.86 fm (the size of a free proton) for $^5_\Lambda$He and $^{\ 6}_{\Lambda\Lambda}$He, whereas it is much smaller for $^{\ 5}_{\Lambda\Lambda}$H and $^{\ 5}_{\Lambda\Lambda}$He.

1 Why a cluster model

There are a number of systems where the constituent particles are made up of more fundamental particles. Such constituents may be called clusters. A hypernucleus contains strange particles such as Λ and Σ besides nucleons. These baryons are composite particles of quarks. For s-shell hypernuclei both sizes of the baryon and the nucleus are different only by a factor of 2-4 so that it is interesting to examine the extent to which the compositeness of the constituent particles has an influence on the properties of the system.

The cluster concept is also useful when the constituents have no substructure but the system tends to form a subunit because of the dynamical consequence of correlations as in e.g. α-cluster model. The utility of the cluster concept is revived by the study of halo nuclei, where, for example, two halo neutrons in ^{6}He move in spatially extended region for most of the time. The ^{6}He nucleus is well described in a three-cluster model of $\alpha + n + n$.

In this talk I will focus on the utility of the cluster concept to discuss the halo structure of ^{6}He and the Λ separation energy of s-shell hypernuclei.

2 A Microscopic Cluster Model

Our basic goal is to solve the many-particle Schrödinger equation. Among various approaches to solve a many-body problem the variational method is the simplest. The crucial point of the variational approach is the choice of the trial function. We construct it as a combination of a finite number of correlated Gaussians (CG)

$$\Psi = \sum_k c_k \mathcal{A}\Big\{\phi_{\rm int}\exp\Big(-\frac{1}{2}\sum_{i,j} A_{kij}\mathbf{x}_i\cdot\mathbf{x}_j\Big)[\theta_{L_k}(\mathbf{x})\chi_{S_k}]_{JM}\eta_{TM_T}\Big\}, \quad (1)$$

where $\mathbf{x} = (\mathbf{x}_1, ..., \mathbf{x}_{N-1})$ is a set of independent coordinates between the particles but the center-of-mass coordinate, and $\phi_{\rm int}$ denotes a product of the internal wave functions of the clusters. When no clusters are present, $\phi_{\rm int}$ can be omitted. A_k is an $(N-1)\times(N-1)$ positive-definite, symmetric matrix of nonlinear parameters. The function $\theta_{LM}(\mathbf{x})$ describes the angular part and χ (η) the spin (isospin) part. The symbol $\mathcal{A}$ is an antisymmetrizer.

Table 1: Energies and rms radii of N-nucleon systems interacting via the Malfliet-Tjon potential V. See Ref. 2 for the detail. ($\hbar^2/m = 41.47\,\text{MeV fm}^2$).

N	$(L,S)J^\pi$	Method	E (MeV)	$\langle r^2\rangle^{1/2}$ (fm)
3	$(0,1/2)1/2^+$	Faddeev	−8.25273	
		ATMS	−8.26±0.01	1.682
		CHH	−8.240	
		GFMC	−8.26±0.01	1.682
		VMC	−8.2689 ±0.03	1.68
		SVM	−8.2527	1.682
4	$(0,0)0^+$	FY	−31.36	
		ATMS	−31.36	1.40
		CRCG	−31.357	
		GFMC	−31.3±0.2	1.36
		VMC	−31.3 ±0.05	1.39
		SVM	−31.360	1.4087
5	$(1,1/2)3/2^-$	VMC	−42.98 ±0.16	1.51
		SVM	−43.48	1.51
6 (^{6}He)	$(0,0)0^+$	VMC	−66.34±0.29	1.50
		SVM	−66.30	1.52
7 (^{7}Li)	$(1,1/2)3/2^-$	SVM	−83.4	1.68

The above microscopic multicluster model has the following advantages: (i) The Pauli principle is strictly obeyed. (ii) There is no need to introduce phenomenological cluster-cluster interactions, but basic interactions between the constituent particles are employed. (iii) If necessary, the distortion of the clusters can be taken into account. (iv) It is a calculation of "variation after projection" type. (v) No spurious center-of-mass motion is included.

The variational parameters are determined by the stochastic variational method (SVM).[1-3)] The basis element is increased one by one in the SVM until the solution reaches a practical convergence. There are many examples demonstrating that the SVM works very well.[2,3)] Table 1 is an example of calculations for few-nucleon systems.[2] One can see that the SVM energies agree with those of other precise calculations.

3 A Microscopic Description of Halo in ^{6}He

A good description of ^{6}He is obtained by the microscopic cluster model of $\alpha+n+n$, where the core α-particle is represented by a a single $(0s)^4$ harmonic-oscillator state.

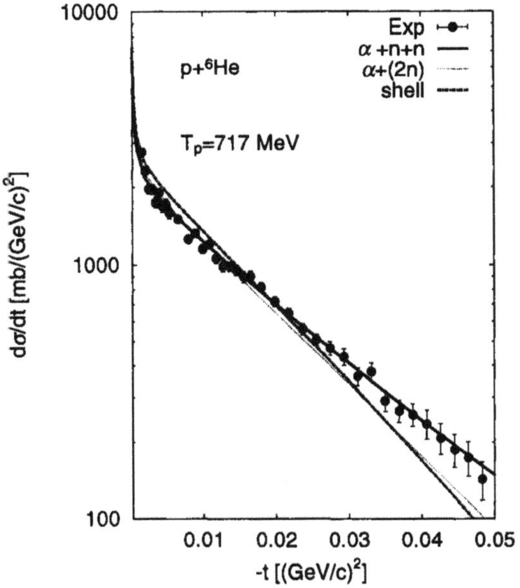

Figure 1: Comparison of p+^{6}He elastic differential cross sections calculated with different wave functions which give the same point nucleon radius of 2.51 fm for ^{6}He. The data are taken from Ref. 5.

This microscopic ^{6}He wave function reproduces the p+^{6}He elastic differential cross section[4] at high energy as shown in Fig. 1, where the angular distributions obtained in the Galuber model by using three different wave functions are compared with experiment. They include an α plus di-neutron cluster model and a simple shell model. These three wave functions all give the same root mean square radius of 2.51 fm. It is interesting to examine the sensitivity of the angular distribution on the size of ^{6}He. Figure 2 compares the angular distribution calculated by three $\alpha+n+n$ wave functions which give different sizes for ^{6}He. They are obtained in exactly the same model but by changing the strength of the two-nucleon potential. One can conclude from the two figures that the size of ^{6}He is around 2.5 fm and that the angular distribution can be very well reproduced by the $\alpha + n + n$ model.

Despite the success of the $\alpha + n + n$ model, the two-neutron separation energy of ^{6}He cannot be reproduced precisely. In fact several calculations using the rigid α core in the $\alpha + n + n$ model indicate the underbinding of the ^{6}He ground state[6-9] by 0.2-0.3 MeV. To resolve this underbinding we note that there are two defects in the $(0s)^4$ configuration when it is used to represent the α core in the $\alpha + n + n$ model. One is that its tail part falls off far too rapidly and it is expected to affect the binding of the halo nucleus ^{6}He because the halo neutrons has large probability even in the remote tail of the core.[10]

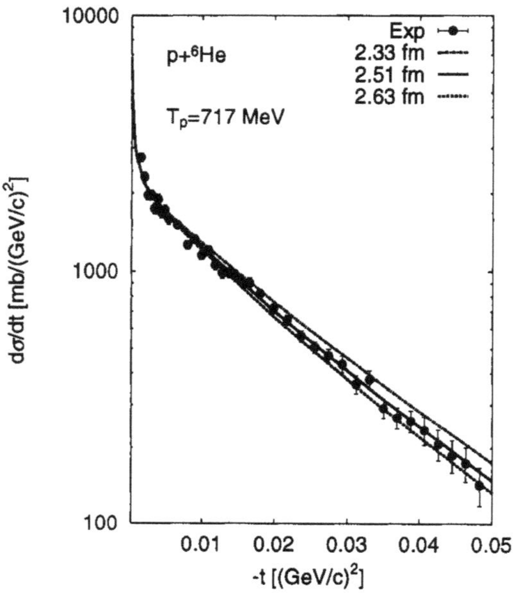

Figure 2: Comparison of p+^{6}He elastic differential cross sections calculated with the α+n+n-model wave functions which give the different radii for ^{6}He.

Another is that there is no possibility of including the first excited state of the α-particle which has the same spin-parity as the ground state, but a dominant $3N + N$ structure. Furthermore, the $3N + N$ configurations may be non-negligible even in the α ground state since these provide the first-order correction to the tail part.

To show this clearly, we compare four different models for the α-particle. In model (a) the α-particle ground state is a single $(0s)^4$ configuration with its parameter chosen to minimize the energy (ν=0.303 fm^{-2}). In model (b) the wave function is a superposition of three similar states of different oscillator parameters, which, again, minimize the ground-state energy. Model (c) is a $3N + N$ two-cluster model, the same as will be used below in the extended three-cluster model of ^{6}He. Model (d) is the result of a precise $N + N + N + N$

four-body calculation. The energies (in MeV) and radii (in fm) are obtained as follows: $(a) - 24.69$, 1.36, $(b) - 25.60$, 1.41, $(c) - 26.55$, 1.42, $(d) - 29.94$, 1.41. The experimental values are -28.30 MeV and 1.47 fm.

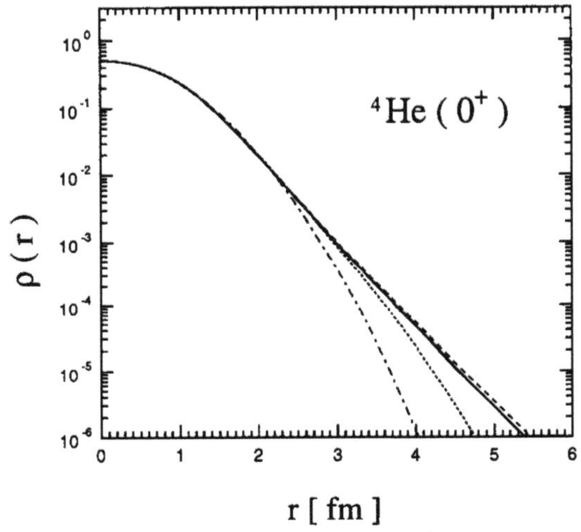

Figure 3: The density distribution of the α-particle ground state in four different models, (a) dot-dashed, (b) dotted, (c) dashed, and (d) solid. See text for the models. The central Minnesota potential[11] (with u=1.0) and the Coulomb force were used.

Figure 3 shows that the Gaussian density in model (a) falls off far too rapidly at large r and that even the superposition of the three functions with different ranges in b produces a too steep fall-off. The tail of the density in model (c) is, however, quite satisfactory.

There is one more reason why the internal wave functions of the α-particle and its excited states in the pure three-cluster model may have to be refined. It is due to the effect of the two extra neutrons of ^{6}He that the α excited states mix into the ground state of ^{6}He. This "cluster distortion" should allow for the breakup of the α-cluster as well, but model (b) is too restrictive in this respect: the first excited state is at excitation energy E_x=34.0 MeV, in contrast with the experimental value of E_x=20.2 MeV. In model (c), however, $E_x \approx 20.9$ MeV, which is indeed very near to the $t+p$ and $h+n$ thresholds (the model values of which are 20.53 and 21.27 MeV, respectively), and the tail behavior of the excited state is improved to a large extent.

An extended calculation including the $3N + N$ breakup of the α-particle in the $\alpha + n + n$ model is carried out and it is shown that this core distortion effect leads to a significant attraction. The detail of this calculation is reported by Arai[10] in this conference.

4 Quark Pauli Effects in s-Shell Hypernuclei

Λ hypernuclei consist of N's and Λ's. The N and Λ can be distinguished from each other at the baryon level but they are subjected to the Pauli principle at the quark level since the (p), the (n) and the Λ are three-quark systems of uud, udd and uds, respectively. The orbital motion of each quark is assumed to be described by a (0s) harmonic-oscillator function, $\exp(-\rho^2/2b^2)$, with a size parameter b. The orbital part of the baryon wave function is expressed as

$$\phi^{(o)} \sim \exp\left(-\frac{1}{2b^2}(\rho_1^2+\rho_2^2+\rho_3^2)\right) \sim \exp\left(-\frac{3}{2b^2}\mathbf{r}_1^2\right)\phi^{(\mathrm{int})}, \tag{2}$$

where $\mathbf{r}_1 = \frac{1}{3}(\rho_1+\rho_2+\rho_3)$ is the baryon's position vector. Therefore a 15-quark, five-baryon wave function for $ppnn\Lambda$, $ppn\Lambda\Lambda$, or $pnn\Lambda\Lambda$ system

$$\psi_{\mathrm{F}} = \left(\frac{\pi b^2}{3}\right)^{-3} \exp\left(-\frac{3}{2b^2}\sum_{i=1}^{5}(\mathbf{r}_i-\mathbf{R})^2\right) \tag{3}$$

is manifestly Pauli-forbidden, where $\mathbf{R}$ is the center-of-mass coordinate of the five baryons, because more than six identical (u or d) quarks turn out to occupy the same $0s$ orbit.

A sharp contrast occurs between the nuclear and baryon clusters. In the former case the equation of motion between the clusters can be solved by using the nucleon-nucleon interaction, while in the latter case we cannot yet determine the baryon dynamics from the quark-quark interaction, but must rely on the baryon-baryon interaction. The Pauli effect in the latter case is taken into account by restricting the configuration space available for the baryons. Namely the variational trial function must satisfy a subsidiary condition $\langle\psi_{\mathrm{F}}|\Psi\rangle = 0$.

Table 2: The Λ separation energies, in MeV, of $A = 5$ systems as a function of the baryon size b (in fm). The case of $b = 0$ indicates that no quark Pauli effect is imposed in the calculation.

	Theory			Expt.
	$b=0$	0.7	0.86	
$B_\Lambda({}^5_\Lambda\mathrm{He})$	4.98	4.70	3.63	3.12±0.02
$B_{\Lambda\Lambda}({}^5_{\Lambda\Lambda}\mathrm{H})$	5.63	5.52	5.12	
$B_{\Lambda\Lambda}({}^5_{\Lambda\Lambda}\mathrm{He})$	5.53	5.45	5.05	

Table 2 compares the Λ separation energies calculated for $A = 5$ systems (${}^5_\Lambda$He, ${}^5_{\Lambda\Lambda}$H, ${}^5_{\Lambda\Lambda}$He). Spin, parity and isospin (J^π, I) are ($\frac{1}{2}^+$, 0), ($\frac{1}{2}^+$, $\frac{1}{2}$), and ($\frac{1}{2}^+$, $\frac{1}{2}$), respectively. Experimental search for ${}^5_{\Lambda\Lambda}$H is under way.[12] The NN and ΛN potentials were the same as employed in Ref. 13. The $\Lambda\Lambda$ potential was taken from Ref. 14.

As shown in Table 2, the quark Pauli effect leads to a significant reduction of the binding energy for ${}^{5}_{\Lambda}$He at $b = 0.86$ fm. As discussed in Ref. 13, this effect plays an important clue in resolving the anomaly[15,16)] in the ${}^{5}_{\Lambda}$He binding energy provided this choice of baryon size is acceptable. The quark Pauli effect becomes much less in ${}_{\Lambda\Lambda}^{5}$H and ${}_{\Lambda\Lambda}^{5}$He, which reflects the fact that they are spatially more extended than ${}^{5}_{\Lambda}$He. In fact, the root mean square (rms) radius of ${}_{\Lambda\Lambda}^{5}$H (and also ${}_{\Lambda\Lambda}^{5}$He) turns out to be 2.0 fm, while the rms radius of ${}^{5}_{\Lambda}$He is 1.6 fm.

We have calculated the binding energy of ${}_{\Lambda\Lambda}^{6}$He in a complete six-body treatment. This system has three types of forbidden states (3), containing $ppnn\Lambda$, $ppn\Lambda\Lambda$, or $pnn\Lambda\Lambda$ five baryons. The result is given in Table 3. The energy obtained without the quark Pauli effect ($b = 0$) is apparently overbound. The elimination of the forbidden states leads to a surprisingly large reduction in the binding energy. In order to see whether or not this large effect is due to the choice of the $\Lambda\Lambda$ potential, we have repeated the calculation by using another $\Lambda\Lambda$ potential.[17] The result was very similar. The rms radius of ${}_{\Lambda\Lambda}^{6}$He is calculated to be 1.6 fm, considerably smaller than that (2.0 fm) of ${}_{\Lambda\Lambda}^{5}$H, which explains a larger quark Pauli effect in ${}_{\Lambda\Lambda}^{6}$He.

Table 3: The Λ separation energy $B_{\Lambda\Lambda}$, in MeV, of ${}_{\Lambda\Lambda}^{6}$He as a function of the baryon size b (in fm).

Theory				Expt.
$b = 0$	0.6	0.7	0.86	
14.3	13.9	12.9	9.4	10.9±0.6

5 Summary

We introduced an extended three-cluster model for ^{6}He in which the α-core internal function is improved by the $3N + N$ two-cluster description. The core breakup or distortion effects increase the intercluster binding energy compared to the pure $\alpha+n+n$ model, and give binding energies similar to the $\{\alpha+n+n;\ t+t\}$ model. A similar thing suggesting the importance of the core breakup effect must occur in other neutron-halo nuclei, such as ^{11}Li, ^{11}Be and ^{10}Li.

We studied the consequence of the quark structure of the baryons on the binding energies of Λ- and $\Lambda\Lambda$ hypernuclei. Without taking into account the quark Pauli effect, the ΛN and $\Lambda\Lambda$ interactions that are equivalent to the meson theoretical or the quark cluster model potentials led to a substantial overbinding in ${}^{5}_{\Lambda}$He and ${}_{\Lambda\Lambda}^{6}$He. By excluding the quark Pauli-forbidden states we found that the binding energies started to decrease at the baryon size of $b = 0.6$ fm and reduced significantly at the size of free baryons, $b = 0.86$ fm.

Acknowledgments

I would like to thank B. Abu-Ibrahim, K. Arai, K. Fujimura, H. Nemura and R. G. Lovas for the collaboration which led to the work presented here. This work was supported in part by Grant-in-Aid for Scientific Research (No. 10640255) of the Ministry of Education, Science, Sports and Culture (Japan).

References

1. V. I. Kukulin and V. M. Krasnopol'sky, *J. Phys.* G: Nucl. Part. Phys. **3**, 795 (1977).
2. K. Varga and Y. Suzuki, *Phys. Rev.* C**52**, 2885 (1995).
3. Y. Suzuki and K. Varga, *Stochastic Variational Approach to Quantum-Mechanical Few-Body Problems*, Lecture notes in physics, Vol. m54, (Springer-Verlag, Berlin Heidelberg, 1998).
4. B. Abu-Ibrahim, K. Fujmura and Y. Suzuki, submitted to Nucl. Phys.
5. G.D. Alkhazov *et al. Phys. Rev. Lett.* **78**, 2313 (1997).
6. A. Csótó, *Phys. Rev.* C**48**, 165 (1993).
7. S. Funada, H. Kameyama and Y. Sakuragi, *Nucl. Phys.* A**575**, 93 (1994).
8. S. Aoyama, S. Mukai, K. Katō and K. Ikeda, *Prog. Theor. Phys.* **93**, 99 (1995).
9. E. Hiyama and M. Kamimura, *Nucl. Phys.* A**588**, 35c (1995).
10. K. Arai, Y. Suzuki and R. G. Lovas, *Phys. Rev.* C**59**, 1432 (1999); see also the proceedings of this conference.
11. D. R. Thompson, M. LeMere, and Y. C. Tang, *Nucl. Phys.* A**286**, 53 (1977); I. Reichstein and Y. C. Tang, *ibid.* A**158**, 529 (1970).
12. T. Fukuda for the E906 collaboration, *Nucl. Phys.* A**639**, 355c (1998).
13. H. Nemura, Y. Suzuki, Y. Fujiwara and C. Nakamoto, *Prog. Theor. Phys.* **101**, 981 (1999); Y. Suzuki and H. Nemura, *ibid.* **102**, No. 1 (1999) .
14. E. Hiyama, M. Kamimura, T. Motoba, T. Yamada and Y. Yamamoto, *Prog. Theor. Phys.* **97**, 881 (1997).
15. R. H. Dalitz, R. C. Herndon and Y. C. Tang, *Nucl. Phys.* B**47**, 109 (1972).
16. B. F. Gibson and E. V. Hungerford III, *Phys. Rep.* **257**, 349 (1995).
17. C. Nakamoto, Y. Fujiwara and Y. Suzuki, *Proceedings of the KEK-Tanashi International Symposium on Physics of Hadrons and Nuclei*, Dec. 14-17, 1998, Tokyo, Japan, to be published in Nucl. Phys. A.

CLUSTERING IN BERYLLIUM NUCLEI

Đ. MILJANIĆ, S. BLAGUS, M. BOGOVAC, M. MILIN, D. RENDIĆ, N. SOIĆ, M. ZADRO
Ruđer Bošković Institute, Zagreb, Croatia

M. ALIOTTA, D. CALI, E. COSTANZO, P. FIGUERA, M. LATTUADA, S. ROMANO, C. SPITALERI, A. TUMINO
INFN – Laboratori Nazionali del Sud and Università di Catania, Catania, Italy

S. CHERUBINI, A. NINANE
Université Catholique de Louvain, Louvain-la-Neuve, Belgium

N. GODINOVIĆ
University of Split, Split, Croatia

T. DAVINSON, A. DI PIETRO, A.N. OSTROWSKI, A.C. SHOTTER
University of Edinburgh, Edinburgh, United Kingdom

Different aspects of clustering in beryllium nuclei, observed in experimental studies of the d+^{7}Li, ^{6}He+^{6}Li, ^{7}Li+^{7}Li and ^{7}Li+^{9}Be reactions, are discussed.

1 Introduction

Light nuclei very often exhibit strong clustering even in their ground states, the best examples being ^{6}Li, ^{7}Li, ^{7}Be and ^{8}Be with their pronounced α-d, α-t, α-^{3}He and α-α structure. Several recent experimental results are presented here which either shed new light on cluster structure of ^{8}Be, ^{9}Be and ^{10}Be at higher excitations or confirm their special features in an alternative way.

2 ^{8}Be

Ground state of ^{8}Be, with its dumbbell α-α structure, and the first two excited states, 2^+ at 3.0 MeV and 4^+ at 11.4 MeV, belong to the same rotational band. Next two states (both 2^+) are found just below the p+^{7}Li decay threshold, at 16.6 and 16.9 MeV in excitation. Marion [1] was the first to point out that no definite isospin can be assigned to either of these two levels and he argued that they should have pure single particle configurations: ^{7}Li$_{g.s.}$+p for the first and ^{7}Be$_{g.s.}$+n for the second state.

Ample evidence now exists supporting this assertion. One more is added here from our recent measurements of the ^{7}Li(d,$\alpha\alpha$)n reaction at deuteron energies between 3 and 6 MeV [2].

In the experiment two α-particles from the reaction were detected in coincidence. Fig. 1. shows the parts of the ^{10}Be excitation energy spectra corresponding to the energy region of the 2^+ isospin mixed states. The spectra were measured at E_0=4.35 MeV, and for two different kinematical conditions in the n-^{8}Be c.m. system, i.e. for ^{8}Be nuclei emitted backwards (a) and forwards (b) with respect to incident deuterons. Very intense 16.6 MeV peak for neutron forward angles,

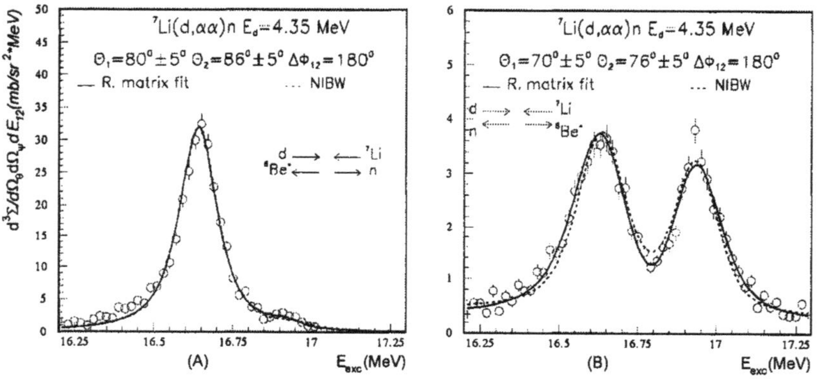

Figure 1. ^{8}Be excitation energy spectra from the ^{7}Li(d,$\alpha\alpha$)n reaction for (a) forward and (b) backward c.m. neutron emission angles. (Note the difference in cross section scales)

observed also at all other deuteron energies, reflects the direct nature of the process as well as the predominantly $^7\text{Li}_{g.s.}$+p(α+t+p) structure of the state. The 16.9 MeV state having the $^7\text{Be}_{g.s.}$+n structure could not be so easily reached in this process. Comparable and much weaker contributions of both states for neutron backward angles suggest that in this case the states are formed mainly by decay of highly excited compound nucleus, ^{9}Be. A discussion on interference of the states, as well as on possible relevance of the results to determination of the astrophysical $S_{17}(0)$ factor can be found in ref. [2].

3 ^{9}Be

^{9}Be is one of the most deformed and weakly bound stable nuclei, having no particle stable excited states. In recent antisymmetrized molecular dynamics [3] and microscopic multicluster model [4] calculations it was again shown that the ^{5}He+α configuration is a very good approximation for ground state wave function. Several members of its rotational band are now well established. Some of these features have been explored in a recent measurement of the $^{9}Be+^{7}Li$ reactions at E_0=52 MeV [5-7]. Three results from the study are mentioned here:

1. The $^{7}Li+^{9}Be$ scattering spectra, like one on Fig. 1a, exhibit four distinct groups corresponding to ^{9}Be excitation energies of 0.0, 2.4, 6.5 and 11.3 MeV. (Their weaker satellites correspond to simultaneous excitation of ^{7}Li into its first excited state). The spectra are quite similar to those from α-particle scattering

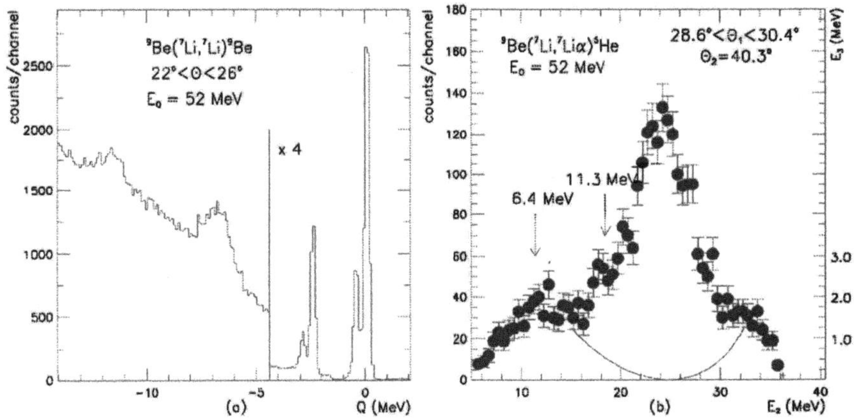

Figure 2. Spectra from the $^{7}Li+^{9}Be$ processes: (a) inelastic scattering and (b) ^{9}Be break-up.

   on the same nucleus [8] and represent an additional evidence for the first four members of the ground state rotational band.
2. The unbound states of the band, having structure similar to the ground state, are expected to decay into the $\alpha+^{5}He$ channel. This was confirmed by the measurement of the α-^{7}Li coincidence spectra from the reaction [5,6]. Two peaks were observed at excitation energies around 6.5 and 11.3 MeV, corresponding to the upper two levels of the band.
3. The α-^{5}He cluster structure of the ground state has often been revealed in the quasifree processes, especially in the (p,pα) and (α,2α) reactions at higher energies [9,10]. However, it was somewhat surprising to find out, that in the $^{7}Li+^{9}Be$ collisions at c.m. energies of only 4 MeV per nucleon the ^{7}Li nucleus quasi-free scatters off the α-particle in ^{9}Be [7]. Fig. 1b shows a ^{7}Li-α coincidence spectrum with a very strong $^{7}Li+\alpha$ quasi-free contribution corresponding to small energies ($0<E_s<1$ MeV) of the spectator nucleus.

The results on the $^{7}Li+^{7}Be$ reactions i.e. strong contributions of the inelastic scattering to unbound ^{9}Be states and of the $^{7}Li+\alpha$ quasi-free scattering, are in accordance with recent findings [11], that the cross section for fusion of light, weakly bound nuclei at low energies is significantly lower than the total reaction cross section, which could be interpreted to have origin in the "fragility" of the nuclei.

4 ^{10}Be

^{10}Be has the largest binding energy (6.8 MeV) of all beryllium nuclei. This does not make its ground state as good a candidate for a well developed cluster structure, as the states close to the n+^{9}Be and α+^{6}He break-up thresholds are. Von Oertzen [12] argued that all the states of the quartet arround 6 MeV in excitation have the two-center molecular structure. He presented some evidence for the rotational bands based on the 6.18 MeV (0^+) and 5.96 (1^-) states. Recently there have been several theoretical investigations concerning ^{10}Be states and all of them point out to the existence of molecular structure at these excitations [13-16].

In several measurements of the ^{7}Li+^{7}Li and ^{7}Li+^{9}Be reactions [5-7] the states of ^{10}Be decaying into the ^{6}He+α channel were also looked for. A ^{10}Be excitation energy spectrum from the ^{7}Li+^{7}Li experiment at E_0=30 MeV is shown on Fig. 3a. Contributions of only two states are clearly visible in this and other spectra from both reactions. The state at 10.2 MeV was previously seen only in the ^{7}Li(α,p) reaction study [17] and in our measurement of the ^{7}Li(^{7}Li,α^{6}He)^{4}He reaction at a lower energy [18]. It was then speculated that it could be the third, 4^+ member of the rotational band based on the second 0^+ state at 6.18 MeV. The small energy separation between them would imply their large deformations. However, in the spectra there is no indication of possible higher members of the band.

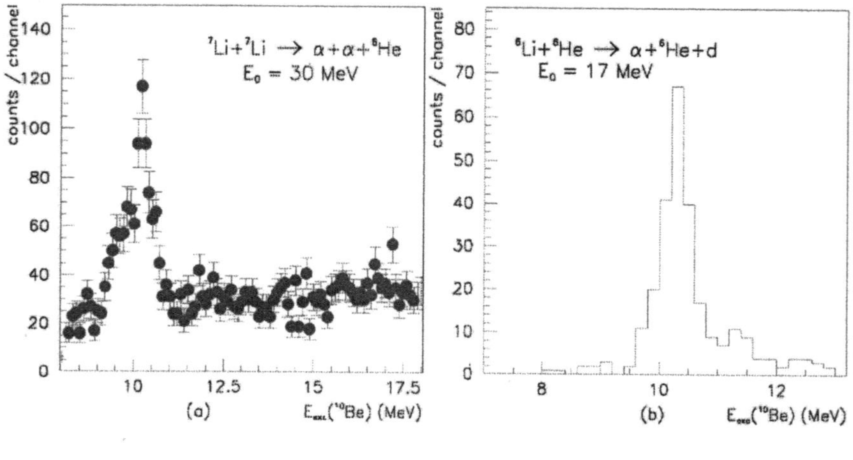

Figure 3. ^{10}Be excitation energy spectra from (a) ^{7}Li+^{7}Li and (b) ^{6}He+^{6}Li reactions.

The other state at 9.6 MeV, seen in all the spectra but with lower intensity, is sometimes quoted to be a 2^+ state. It should be noted that the "unnatural" parity states like 1^+, 3^+ and 4^-, expected at these energies by some calculations [15,16], cannot be observed in the decay.

Fig. 3b. shows a ^{10}Be excitation energy spectrum obtained from the triple-coincidence (d-α-^{6}He) data from our recent study of the ^{6}He+^{6}Li reactions at E_0=17 MeV [19]. Again strong contribution of the 10.2 MeV state is seen, which is in this case formed by the α-particle pick-up process. Also, from the double-coincidence (d-Be) data much larger cross sections are observed for formation of the ^{10}Be excited states around 6 and 7.5 MeV than of the ground state. All the results again support predictions that some states close to the break-up thresholds have the molecular type structure.

At the end instead of a summary and as an inspiration for future searches of a more complicated molecular structure in light nuclei we mention here an accidental finding from the ^{6}He+^{6}Li experiment. Namely, poorly known, weakly bound $(^4H^1H_2)^+$molecular ions were observed as an impurity in the $^6He^+$ beam [20].

References

1. J.B. Marion, *Phys. Lett.* B **14**, 315 (1965).
2. N. Godinović *et al.*, *Phys. Rev.* C **60**, 037601 (1999).
3. Y. Kanada-Enyo, H. Horiuchi and A. Ono, *Phys. Rev.* C **52**, 628 (1995).
4. K. Arai, Y. Ogawa, Y. Suzuki and K. Vargà, *Phys. Rev.* C **54**, 132 (1996).
5. N. Soić, Ph.D. thesis, University of Zagreb, 1999.
6. N. Soić *et al.*, *Europhys. Lett.* **41**, 489 (1998).
7. N. Soić, *et al.*, *Eur. Phys. J.* **3** (1998) pp. 303.
8. Roy Subinit *et al.*, *Phys. Rev.* C **52**, 1524 (1995) and references therein.
9. P.G. Roos *et al.*, *Phys. Rev.* C **15**, 69 (1977) and references therein.
10. A.A. Cowley *et al.*, *Phys. Rev.* C **50**, 2449 (1994) and references therein.
11. J. Takahashi *et al.*, *Phys. Rev. Lett.* **78** 30 (1997).
12. W.Z. von Oertzen, *Phys.* A **354**, 37 (1996); **357**, 355 (1997).
13. K. Fujiwara *et al.*, *Phys., Rev.* C **59**, 817 (1999).
14. Y. Ogawa *et al.*, RIKEN-AF-291 report (1998).
15. Y. Kanada-Enyo, H. Horiuchi and A. Doté, *Phys. Rev.* C, to be published.
16. N. Itagaki and S. Okabe, priv. comm.
17. S. Hamada *et al.*, *Phys. Rev.* C **49**, 3192 (1994).
18. N. Soić *et al.*, *Europhys. Lett.* **34**, 7 (1996).
19. M. Milin *et al.*, *Europhys. Lett.* to be published.
20. Đ. Miljanić *et al.*, *Nucl. Instr. Meth. Phys. Res.* A to be published.

^{6}HE+^{6}HE BREAKUP STATES IN ^{12}BE

MARTIN FREER - FOR THE CHARISSA-DEMON COLLABORATION
School of Physics and Astronomy, University of Birmingham, Birmingham, B15 2TT, UK
E-mail: mf@np.ph.bham.ac.uk

The breakup of ^{12}Be into ^{6}He+^{6}He and ^{4}He+^{8}He has been studied using a 378 MeV ^{12}Be beam inelastically excited by ^{12}C and $(CH_2)_n$ targets. The measurements indicate that breakup occurs from rotational states in the 10 to 25 MeV excitation energy interval, with spins in the range of 4 to 8 $\hbar$. The inferred moment of inertia is consistent with the cluster decay of an exotic molecular structure in ^{12}Be, which may be associated with an α-4n-α cluster configuration.

1 Introduction

The evolution of the energies of single-particle levels with increasing neutron and proton numbers have important implications for nuclear structure. The appearance of nuclear halos is just one example. However, this reorganization might be expected to signal the disappearance of a structural mode observed in nuclei with equal numbers of protons and neutrons, namely clustering. Clustering is most widely associated with light nuclear systems composed of even numbers of protons and neutrons. The association of pairs of neutrons and protons with common shell model orbits, and hence in a relative s-state, permits a description of the nuclear properties in terms of collections of α-particles. The introduction of, for example, valence neutrons to form neutron-rich isotopes might be expected to disrupt the underlying cluster structure. However, recent calculations using the Antisymmetrized Molecular Dynamics description (AMD)[1,2,3,4] suggests that rather than clustering vanishing, it is still present in non A=4n nuclei and is actually enhanced for nuclei close to the neutron drip-line. This latter feature is consistent with the ideas developed by Ikeda[5] that clustering should be important for weakly bound systems. Interestingly, the AMD description has *a priori* no cluster structure built into the model as the nuclear wave-function is constructed from a Slater determinant of Gaussian wave-functions representing the individual nucleons. Nevertheless, this model predicts a series of pronounced cluster structures for isotopic chains for beryllium, boron and carbon isotopes[1,2,3,4]. These cluster structures echo strongly of the underlying α-cluster symmetries, indicating that the α-cluster also plays an important role in neutron rich nuclei. In fact, these hybrid α-cluster structures are formed in a region of isospin where the single-particle levels are perturbed, offer the possibility for the study of new nuclear phenomena.

Nuclear-molecular-analogue states bear a striking resemblance to the structure and properties of simple atomic molecules such as H_2 and O_2, which are covalently bound by virtue of the exchange interaction involving the valence electrons. For example the case of ^{9}Be. In this instance, the considerable binding energy of the ^{4}He nucleus results in the 2-^{4}He system, ^{8}Be, being unbound. But, the addition of a single neutron to the unstable ^{8}Be produces the stable ^{9}Be nucleus. In this system, the nuclear potential that the valance neutron experiences results from the underpinning $\alpha+\alpha$ cluster structure and should be two-centred in form. Indeed, the Antisymmetrized Molecular Dynamics (AMD) calculations of the structure of ^{9}Be also reveal the prevailing two-centred structure. An examination of the low-lying states of this nucleus shows rotational bands with moments of inertia consistent with the underlying two-centred potential. These nuclear orbits are the analogues of the atomic π- and σ-molecular bonds, and are revealed in the densities associated with the single-particle wave-functions. The introduction of a further neutron into these molecular type orbits produces states in ^{10}Be again molecular in nature, with large deformation. However, these states are not related to the ground state but lie at an excitation energy of 6 MeV. The systematics of these molecular structures have been studied by von Oertzen[7,8,9], who found evidence for such configurations in systems with both valence protons and neutrons. The present contribution describes an experimental study of the α-4n-α system ^{12}Be[6].

2 Breakup of ^{12}Be

A 378 MeV ^{12}Be secondary beam, with an energy spread of 18 MeV and intensity 2×10^4 particles per second, was produced from a 63 MeV per nucleon primary ^{18}O beam of intensity 2 μA, provided by the GANIL coupled cyclotrons, incident on a 200 mgcm^{-2} thick beryllium target. Beam purification ($\sim$95% ^{12}Be) was achieved using the LISE3 spectrometer. The beam was tracked onto 10 mgcm^{-2} ^{12}C and 20 mgcm^{-2} $(CH_2)_n$ target foils, using two x-y position sensitive parallel plate avalanche counters. The breakup products from the (^{12}Be, ^{6}He^{6}He) and (^{12}Be, ^{4}He^{8}He) reactions, on proton and ^{12}C targets, were detected in an array of 10 Si-CsI telescopes. The Si elements were 500 μm thick, 5×5 cm^2 two-dimensional position sensitive detectors. These were backed by 4.0 cm thick CsI detectors. The detectors were arranged in a symmetric fashion around the beam axis to pick up breakup fragments in opposing telescopes. The ^{4}He, ^{6}He and ^{8}He reaction products were identified

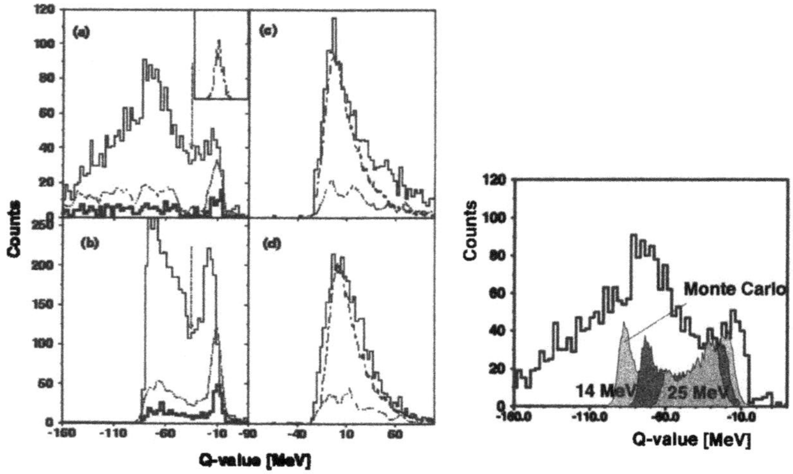

Figure 1. a) Q-value spectra for the reactions (a) $^{12}C(^{12}Be, ^{6}He^{6}He)$, (b) $^{12}C(^{12}Be, ^{4}He^{8}He)$, (c) $p(^{12}Be, ^{6}He^{6}He)$ and (d) $p(^{12}Be, ^{4}He^{8}He)$ from the $(CH_2)_n$ target. The arrows indicate the Q-value used to discriminate between the two target components. The bold histograms in (a) and (b) are for the reactions from the $^{12}C_{nat}$ target. The dotted curves in (a) to (d) correspond to the $^{12}C_{nat}$ target data scaled to account for differences in beam exposure and thickness of the two targets, and in (c) and (d) indicates the level of background in these spectra. The dot-dash curves in (c) and (d) represent the results of Monte Carlo simulations of the reactions from the proton target. The inset in (a) shows the result of the simulation for the $^{12}C(^{12}Be, ^{6}He^{6}He)$ reaction for the $(CH_2)_n$ target. The right hand spectrum provides a comparison of spectrum (a) with Monte Carlo simulations of the $p(^{12}Be, ^{6}He^{6}He)$ reaction with ^{12}Be excitation energies 14, 20 and 25 MeV, reconstructed assuming a ^{12}C recoil.

from their characteristic energy-loss in the Si-CsI telescopes. Using the measurements of the mass, energy and emission angle of the fragments, the energy of the undetected recoil was calculated via momentum conservation, and the reaction Q-value obtained. In the case of reactions on the $(CH_2)_n$ target, the ambiguity in the target nucleus led to an uncertainty in the mass of the recoil. Fig. 1a shows the Q-value spectrum reconstructed for $^{6}He+^{6}He$ coincidences assuming the reaction $^{12}C(^{12}Be, ^{6}He^{6}He)$ (Q=-10 MeV). The bold histogram corresponds to reactions from the natural carbon target (denoted $^{12}C_{nat}$), and the narrow-lined histogram corresponds to reactions from the $(CH_2)_n$ target producing two ^{6}He nuclei, assuming a ^{12}C recoil. The peaks in these spectra at Q=-10 MeV indicate that the $^{12}C(^{12}Be, ^{6}He^{6}He)$ reaction has been correctly identified. The inset displays Monte Carlo simulations of the reaction process in the case of the $(CH_2)_n$ target. These simulations include the effects of the position and energy resolutions of the detectors, energy and angular spread of the beam and energy and angular straggling of the beam and reaction products in the target, and suggest a Q-value resolution of 3.5 MeV for the $^{12}C_{nat}$ target and 9 MeV for the $(CH_2)_n$ target, in agreement with the data. The dotted-line is an indication of the background contribution to the $(CH_2)_n$ target data. This clearly demonstrates that there are additional ^{6}He+^{6}He coincidences produced from the ^{1}H component of the $(CH_2)_n$ target. The simulations indicate that the proton recoil data extend into the region under the peak identified with the ^{12}C(^{12}Be, ^{6}He^{6}He) reaction. This indicates that the reactions from the proton and ^{12}C components of the $(CH_2)_n$ target are not completely resolved. Figure 1 shows the Q-value spectrum for the $(CH_2)_n$ target (from Fig. 1a) compared with Monte Carlo calculations of the p(^{12}Be, ^{6}He^{6}He) reaction reconstructed assuming a ^{12} recoil. This comparison clearly demonstrates that the broad bump in the spectrum at Q$\simeq$-70 MeV is associated with reactions in which the recoil protons are emitted at large centre-of-mass angles.

Figure 2a shows the excitation energy spectrum for the ^{6}He+^{6}He coincidences (the ^{8}He+^{4}He breakup channel is less structured as both odd and even spin states can contribute). There appears to be a series of states between 12 and 25 MeV with widths of the order of 700 keV (the experimental resolution). Tentative spins have been assigned to these states using angular correlation techniques. The resulting excitation energy, spin systematics are shown in Fig. 2b. The two lines correspond to a fit to the experimental data points and the extrapolated rotational band of the ^{12}Be ground state, assuming that the first 2^+ state is rotational. The latter is consistent with a ^{12}Be nucleus which possesses only a small deformation whilst the former suggests a structure with a deformation which may be approximated by two touching ^{6}He nuclei. The large moment of inertia, and an extrapolated bandhead (10.8 MeV) that is close to the ^{6}He decay threshold, may indicate that these data correspond to a ^{6}He+^{6}He cluster structure in this nucleus.

3 Discussion

The proposed bandhead energy is, however, in disagreement with he description of nuclear dimers developed by von Oertzen[7,8], which would place a molecular α-4n-α cluster structure at $\sim$6 MeV. We believe that this difference may be related to uncertainties in the underlying ^{12}Be single-particle structure. Von Oertzen's model suggests that the ^{10}Be molecular states are

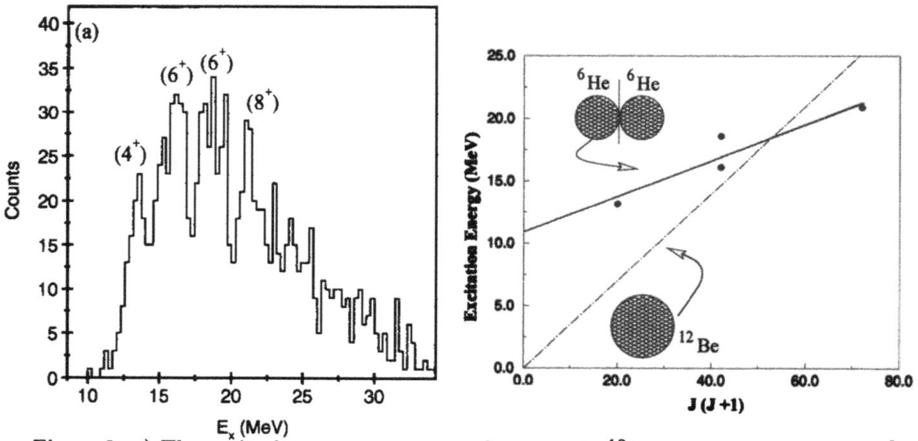

Figure 2. a) The excitation energy spectrum for states in ^{12}Be that breakup into two ^{6}He nuclei. b) The energy, spin systematics of the ^{6}He breakup states.

0^+ (6.26 MeV), 2^+ (5.958 MeV), 1^- (5.960 MeV) and 2^- (6.263 MeV). The associated single-particle configurations are formed by combinations of neutrons in two centre molecular orbits originating from the spherical $p_{1/2}$, $p_{3/2}$ and $d_{5/2}$ orbits, or $[\sigma_{1/2^-}, u]$, $[\pi_{3/2^-}, g]$ and $[\sigma_{1/2^+}, u]$ in the molecular orbital notation[7,8]. In this description the ^{10}Be ground state corresponds predominantly to two neutrons in the $[\pi_{3/2^-}, g]$ molecular orbit coupled to J=0 ($[\pi_{3/2^-}, g]^2_{0^+}$). Further, it was suggested that the most likely ^{12}Be ground state configuration involves 2 neutrons in each of the molecular orbitals originating from the $p_{3/2}$ and the $d_{5/2}$ shell model orbitals with both pairs coupled to a total angular momentum of zero, i.e. $[\pi_{3/2^-}, g]^2_{0^+} \otimes [\sigma_{1/2^+}, u]^2_{0^+}$. Then in analogy with the ^{10}Be 5.958 MeV 2^+ molecular state ($[\pi_{3/2^-}, g]^2_{2^+}$), the ^{12}Be molecular band was thus predicted to be at ~6 MeV, much lower than the present observations.

However, the presence of the N=8 shell closure should favour a less deformed ^{12}Be ground state with two neutrons in each of the spherical $p_{3/2}$ and $p_{1/2}$ orbitals ($[\pi_{3/2^-}, g]^2_{0^+} \otimes [\sigma_{1/2^-}, g]^2_{0^+}$). AMD calculations also indicate that the positive parity ^{12}Be ground state is associated with a spherical neutron closure[3,4]. This would then imply a molecular type configuration for the states in the present work based on a 2p-2h excitation from either the $p_{3/2}$ or $p_{1/2}$ levels to the $d_{5/2}$ originating orbit ($[\sigma_{1/2^+}, u]$) which would lie higher in energy, as in the present study, and also in agreement with the predictions of Seya *et al.*[10]. However, the large log(ft) value for the β-decay of ^{12}Be implies that the neutrons do not form a perfect closed shell[11], and that the interaction of the prolate deformed protons with the neutrons causes there to be a mixed $[\sigma_{1/2^+}, u]$-$[\pi_{3/2^-}, g]$ configuration for the 4 valence neutrons.

A study of the p(^{12}Be,^{11}Be)d and d(^{11}Be,^{12}Be)p reactions could thus provide important insights into the structure of the ^{12}Be ground state and the location of the molecular structures, and indicate if the present ^{6}He+^{6}He breakup states are linked with the α-4n-α molecular structure.

References

1. Y. Kanada-En'yo, H. Horiuchi and A. Doté, *J. Phys.* G, **24**, 1499 (1998).
2. A. Doté, H. Horiuchi, and Y. Kanada-En'yo, *Phys. Rev.* C **56**, 1844 (1997).
3. Y. Kanada-En'yo and H. Horiuchi, *Phys. Rev.* C **52**, 628 (1995).
4. H. Horiuchi and Y. Kanada-En'yo, *Nucl. Phys.* **A616**, 394c (1997).
5. K. Ikeda, *Suppl. Prog. Theor. Physics (Japan) Extra Numbers*, 464 (1968).
6. M. Freer *et al.*, *Phys. Rev. Lett.* 82, 1383 (1999).
7. W. von Oertzen, *Z. Phys.* **A354**, 37 (1996).
8. W. von Oertzen, *Z. Phys.* **A357**, 355 (1997).
9. W. von Oertzen, *Nuovo Cimento*, **110**, 895 (1997).
10. M. Seya, M. Kohno, and S. Nagata,*Prog. Theor. Phys.* **65**, 204 (1981).
11. N. Itagaki, *Contribution to the Cluster '99 conference*, Rab.

INVESTIGATION OF ^{6}HE AND ^{8}HE CLUSTER STRUCTURES BY TRANSFER REACTIONS AT BEAM ENERGY OF 25 MEV/N

R. WOLSKI
Flerov Laboratory of Nuclear Reactions, JINR, Dubna, 141980 Russia
Institute of Nuclear Physics, Cracow,Poland

A.S. FOMICHEV, A.M. RODIN, S.I. SIDORCHUK, S.V. STEPANTSOV, G.M. TER-AKOPIAN, AND YU.TS. OGANESSIAN
Flerov Laboratory of Nuclear Reactions, JINR, Dubna, 141980 Russia

P. ROUSSEL-CHOMAZ AND W. MITTIG
GANIL BP5027, 14021 Caen, France

I. DAVID
Institute of Atomic Physics, P.O.B. MG6 Bucharest, Romania

Angular distributions of the 2-neutron transfer reaction for ^{6}He+^{4}He and ^{6}He+^{1}H systems at 25 MeV/nucleon beam energy are presented and relevant analysis are reviewed. The results point out the importance of the "di-neutron" ^{6}He configuration for the investigated process and suggest a lower triton spectroscopic amplitude for this nucleus as compared to the shell model calculation. A preliminary result of search for 4-neutron transfer in the ^{8}He+^{4}He system is given

1 Introduction

Studies of the structure of light exotic nuclei and nuclear systems close to the neutron drip line are of principal meaning for nuclear physics. Some of these nuclei (^{6}He, ^{8}He, ^{11}Li, ^{11}Be, ^{14}Be, ect.) have neutron halos - a recently discovered phenomenon. Two neutrons make the halo in the ^{6}He nucleus, whereas four neutrons are in the ^{8}He skin. Light exotic nuclei close to (and even beyond) the neutron or proton drip line make one of the main objects of interest in the contemporary physics of nuclear structure[1]. The extreme conditions of large charge asymmetry and week binding of valence nucleons in these nuclei provide stringent tests of the concepts of nuclear structure physics formulated on the basis of data obtained for well bound and almost charge symmetric systems. Theory makes sometimes detailed predictions about halo nuclei. For the ^{6}He nucleus involving three practically structureless constituents (one α particle and two neutrons) the solution of the three-body problem (see Ref.[2]) gives the two most prominent spatial configurations - one resembling a "di-neutron" coupled in the field of α core, and another one, which is named the "cigar"-like configuration. According to shell model calculations[3], large spectroscopic amplitudes for α+2n and α+d as well as t+t and ^{3}He+t configurations characterize, respectively, ^{6}He and ^{6}Li. Experimental data e.g. Ref.[4-6] confirm the significance of the ^{3}He+t - cluster configuration in the ^{6}Li nucleus. Analogously, one could expect a high probability of the t+t clustering of ^{6}He. Experimental verifications of the two configurations of this nucleus and searches for manifestations of its t+t clustering are necessary. For the ^{8}He nucleus, theory predicts the co-existence of the three configurations where the halo-neutrons form, respectively, a "tetra-neutron", a pair of "di-neutrons" and a more or less homogeneous formation surrounding the α core[7]. Transfer reactions are one of the most adequate and powerful methods of the study of light exotic nuclei. In the past, these reactions with light and heavy ion beams have been a major source of spectroscopic information in the nuclei lying close to the β-stability line[8]. Beams of exotic nuclei, such as ^{6}He, ^{8}He, and heavier ones, provide excellent conditions for such study in the region of drip-line nuclei. For obvious reasons, data on transfer reactions with radioactive beams are scarce. The quality of presently available radioactive beams does not match the usual requirements for measuring of angular distributions for transfer processes. To our knowledge the first measurement was done by Kurchatov Institute group[9]. The authors observed the alpha particle energy spectrum from p (^{6}He,α) t reaction at a low incident beam energy of 20 MeV.

A program for investigation of structural properties of ^{6}He and ^{8}He nuclei by means of transfer reactions has been launched at FLNR.

2 Experimental results and discussion

The ACCULINNA separator was designed and built for the realization of a physics program dedicated to the direct reaction study with radioactive ion beams. The details of this separator, one can find in Ref.[10]. With primary beams of (32-34)A MeV ^{7}Li and ^{11}B ions delivered by the U-400M cyclotron of the Flerov Laboratory of Nuclear Reactions (FLNR, JINR, Dubna) we obtain (20-30)A MeV secondary beams of ^{6}He and ^{8}He nuclei focused on a physics target of - 0.8 cm^2 in its area. These secondary beams have intensities of 3×10^5 and 2×10^4 respectively, when the routinely produced primary beam intensity on the Be target makes 4 pμA. The energy resolution is about 5%, but due to the fact that all excited states of the nuclei investigated so far are particle unstable, the identification of both participants for each binary reaction allowed to determine the reaction channel uniquely. We concentrated our attention on the transfer reactions occurring between ^{6}He and ^{8}He projectiles and helium and hydrogen target nuclei. In the case of a ^{4}He target, these reactions could be the two- and four-neutron transfers, respectively, for ^{6}He and ^{8}He, i.e. an exchange effect which should be observed in the center-of-mass frame as elastic scattering in the backward direction. The choice of the projectile energy is essential in these experiments. In the case of low-energy collisions, potential scattering makes a sizeable contribution to the elastic cross section in the whole angular range, i.e. it interferes with the searched exchange effect. For the collision energy approaching (or even exceeding) the nuclear Fermi energy the probability of the cluster neutron exchange becomes too small. Thus, for a clear manifestation of the exchange effect, the most favorable energy range for the ^{6}He (^{8}He) projectile seems to be 20-30A MeV.

The binary reaction transfer products were obtained in coincidences. Pairs of $\Delta E \times \Delta E \times E$ detector telescopes were used for the identification of both reaction products. Silicon strip detectors were used in these telescopes to know the entrance positions of nuclei in the telescopes. This resulted in a precision of 5-10 mrad in the measurements of laboratory system angles. The resulting angular dependence of the differential cross section for the elastic scattering of 151 MeV ^{6}He nuclei from ^{4}He[11] is shown in Fig.1.

The forward angle points were obtained from inclusive ^{6}He energy spectrum, backward ones from the correlated coincident energy spectrum. A one-step, two-neutron exchange mechanism could account for the obtained distinct rise in the cross section at the backward center-of-mass (CM) angles. DWBA calculations were performed[11] with the use of a two-body model wave function for ^{6}He, which was a crude simulation of the exact three-body problem[2]. The wave function assuming the presence of two prominent configurations - the "di-neutron" and "cigar"-like ones - showed a good agreement with the experimental data obtained for the backward CM angles. This result was interpreted as an experimental evidence in support of the theoretical wave function[2]. The calculations also showed that the "cigar"-like configuration alone could contribute only negligibly to the obtained two-neutron exchange, which, thus, is by about 100% due to the "di-neutron" configuration, see Fig.1a). Recently the data were reanalysed[12] using the full 3-body internal wave function of ^{6}He and a more proper technique for discrimination of these configuration. The new result is close to the earlier one though the contributions from the two different configurations are now less distinctive.

The data of the elastic scattering ^{6}He on ^{4}He was also a subject to analysis in terms of the coupled channels (CC) approach[13]. The authors have assumed a simple two-body α + "di-neutron" model for the ^{6}He structure. In the calculations the coupling to the 2^+ resonance and 16 non-resonant continuum

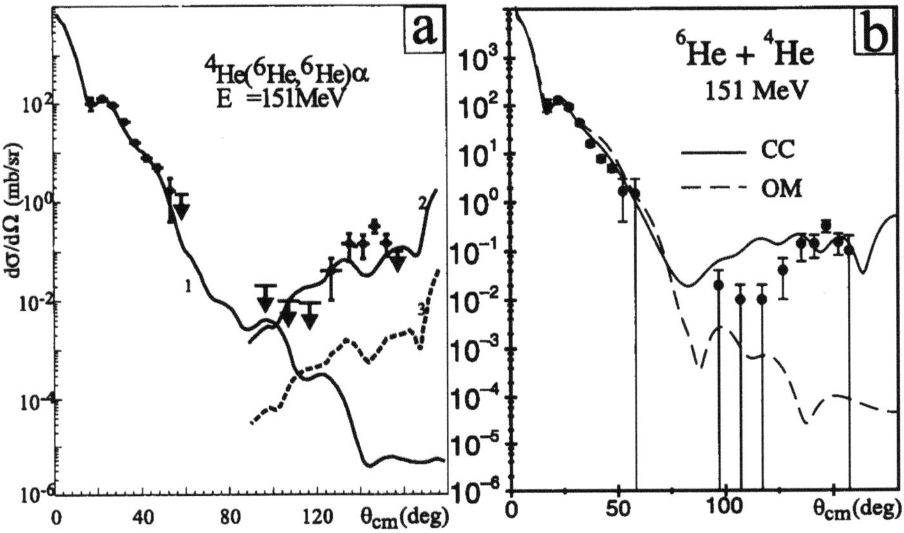

Figure 1. a)and b) Differential cross section for elastic scattering $^6He+^4He$, Ref.[11] Curves: 1a) 2n transfer analysis; Optical Model (1), "di-neutron" (2) and "cigar-like (3) contributions to the cross-section, 1b) Couple channels analysis, see text

energy bins were taken into account. All the central and coupling potential used in the calculations were derived from α - ^4He and "di-neutron" - ^4He potentials by means of the single folding method. It should be noted that the 2n- transfer process was not included explicitly in this treatment nevertheless the experimental data are reproduced by the calculations, see Fig.1b)

To further look into the validity of the tree-body approach to the structure of ^6He and to see whether or not the manifestations of the t+t ^6He configuration predicted by the shell model are present in experiments, we investigated reactions of 150 MeV ^6He with hydrogen target nuclei[14]. Differential cross section data for the reaction $^6He+p\to \alpha+t$ are shown in Fig.2 together with analogous data for the reaction $p+^6Li\to \alpha+^3He$[15].

A simple inspection of the data reveals significant differences between the two data sets. The two measured oscillations for the ^6He+p reaction are more pronounced, and their relative magnitudes are reversed with respect to the $p+^6Li$ data.

In a schematic view of direct reaction one could expect that 2n- transfer process contributes to the cross-section at the forward angles, whereas t- transfer process is responsible for the rise f the cross section at the back-

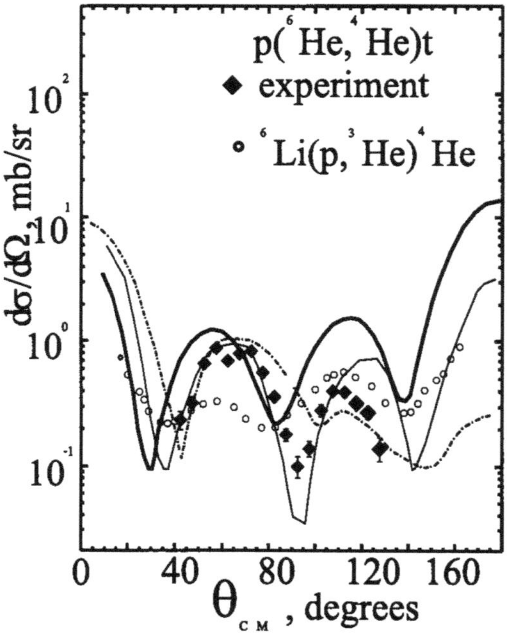

Figure 2. Experimental data (symbls) and calculations (lines) of differential cross section for ^{6}He+p→ α+t reaction; thick solid curve is the result of the two-body DWBA calculation for 2n and t transfer using the shell-model spectroscopic amplitudes (SA) for α+2n and t+t clustering of ^{6}He, the thin solid and doted curves show the results obtained, respectively, with two-triton SA reduced by a factor of 2 and with no t+t clustering.

ward angles. The experimental points shown in Fig.2 suggest thus a lower probability of finding the t+t-clustering in ^{6}He than the ^{3}He+t clustering in ^{6}Li. The calculations of relevant spectroscopic amplitudes (SA) according to translation invariant shell model (TISM)[3] predict just opposite, see Table 1.

A one-step, single particle transfer DWBA analysis incorporating the TISM spectroscopic amplitudes for the 2n+α and t+t partitions of ^{6}He nucleus overestimated the ^{6}He+p→ α+t cross section in the region of large CM angles. A better agreement with the experimental data was obtained when the two-triton amplitude of ^{6}He was reduced by a factor of 2 (see relevant curve in Fig.2). For completeness the effect of removing the t-transfer is shown in Fig.2 by the dotted line.

Prior to the publication of the original data for ^{6}He+p→ α+t reaction, the

Table 1. Spectroscopic amplitudes (SA) for ^{6}Li and ^{6}He nuclei [3]

Nucleus	Configuration	nlj	SA
^{6}Li	α+d	$2S_1$	-1.0564
^{6}Li	α+d	$1D_{3/2}$	-0.017
^{6}Li	^{3}He+t	$2S_{1/2}$	0.939
^{6}He	α+2n	$2S_0$	1.0606
^{6}He	t+t	$2S_{1/2}$	-1.297

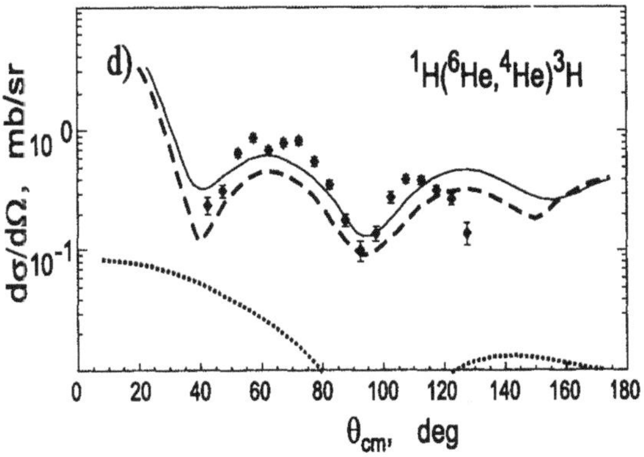

Figure 3. The 4-body analysis of 2n transfer process for the data shown in Fig.2. Curves show contributions of "di-neutron" (dashed), "cigar"-like (doted) components and the total cross-section (thin solid)

authors of Ref.[12] made an attempt to describe the angular distribution using one-step, DWBA calculation of 2n transfer with the 3-body wave functions for both ^{6}He and triton systems. The results of the analysis are shown in Fig.3. The agreement with the data in the angular range measured seems to be acceptable, although the t-transfer is neglected a priori in this calculation. It it can be seen in Fig.3 that the "di-neutron" component of 3-body α+n+n configuration in the ^{6}He nucleus dominates in the 2n-transfer reaction.

The both analysis, 3-body and 4-body, seem to suggest the apparent damping of the ^{6}He (t+t) clustering respect to TISM prediction.

In order to make a quantitative estimation of the strength of the t+t configuration in ^{6}He and compare it with the analogous configuration in ^{6}Li a more involved analysis of both reactions should be performed with inclusion of two-step transitions and CC effects. It is highly desirable too to extend the experimental angular distribution to the scattering angles smaller than 40°

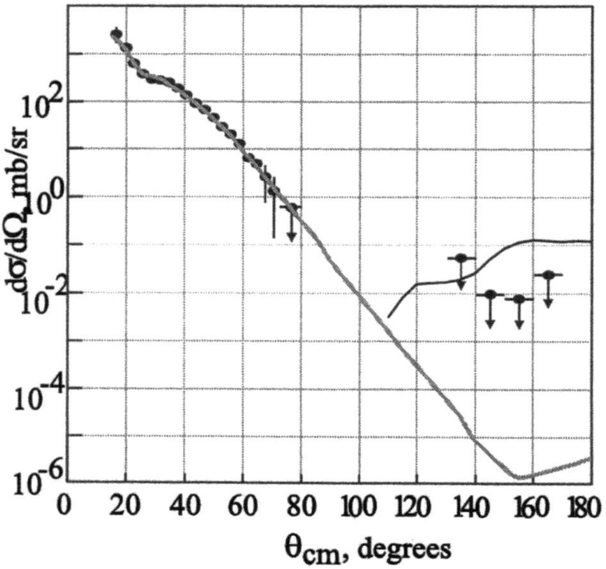

Figure 4. Experimental data of elastic scattering of ^{8}He on ^{4}He. Symbols with a downward arrow mark upper limits. Curves for the calculations: OM and 4n transfer process

and/or to the angles larger than 135°, where the forward-backward ratio is expected to be strongly correlated to the two-triton configuration.

Recently, we carried out experiments aimed at the observation the 4n-exchange effect in the elastic scattering of 210 MeV nuclei of 8He from 4He target nuclei. The data shown Fig.4 indicate that, in spite of rather low cross section limits achieved for the CM angles lying between 140° and 165°, the 4n-transfer effect is not seen. A simple, one-step, single particle, DWBA analysis of this process shows that the expected cross-section values for the 4n-transfer is close to and exceed the achieved sensitivity limits in this angular range. Therefore, it could be preliminary concluded, that the 4n- transfer hardy occurs with a large spectroscopic factor.

Our intention is to improve the sensitivity limit to go down to a cross-section of 0.1μb/sr. In a new series of experiments we will investigate 2n configuration in the ^{8}He nucleus using 2n transfer reactions for this system.

3 Acknowledgments

Experiments presented in this paper were carried out in Flerov Laboratory of Nuclear Reactions of JINR. The partial support of the work by Russian Basic Research Foundation (grant No. 96-02-17382) and INTAS grant No. 97-1668 is acknowledged. The fruitful discussions with Drs. J. Vaagen, Yu.M. Tchuvil'ski, A.T. Rudchik, N. Timofeyuk and N. Alamanos are very much appreciated.

References

1. I. Tanihata, *J. Phys.* G **22**, 157 (1996). See also Proceedings of International Conferences on Radioactive Nuclear Beams, of which the last, the Fourth one, was published in *Nucl. Phys.* A **616** (1997) Nos 1,2.
2. M.V. Zhukov *at al*, *Phys. Rep.* **231**, 151 (1993).
3. Yu.F. Smirnov and Yu.M. Tchuvil'sky, *Phys. Rev.* C **15**, 84 (1977); A.T. Rudchik and Yu.M. Tchuvil'sky, *Ukr. Phys.* J. **30**, 819 (1985); A.N. Boyarkina, *Structure of 1p-shell nuclei*, MSU, Moscow, 1973.
4. Yu.G. Mashkarov *at al*, *Izv. RAN, Ser. Fiz.* **59**, 176 (1995).
5. M.F. Werby *at al*, *Phys. Rev.* C **8**, 106 (1973).
6. S.I. Higuchi *at al*, *Nucl. Phys.* A **384**, 51 (1982).
7. M.V. Zhukov, A.A. Korsheninnikov, and M.H. Smedberg, *Phys. Rev.* C **50**, R1 (1994).
8. G.R. Satchler, *Direct Nuclear Reactions*, Clarendon Press, Oxford, New York, 1983.
9. D.V. Aleksandrov, E.Yu. Nikolsky, B.G. Novatsky and D.N. Stepanov, *Pisma* JETF **59**, 301 (1994).
10. A.M. Rodin *at al*, *Nucl. Inst. Meth.* B **126**, 236 (1997).
11. G.M. Ter-Akopian *at al*, *Phys. Lett.* B **426**, 251 (1998).
12. Yu.Ts. Oganessian *at al*, *Phys. Rev. Lett.* **82**, 4996 (1999).
13. K. Rusek, privat information, K. Rusek and K.W. Kemper, to be published.
14. R. Wolski *at al*, JINR communication E-15-98-284, Dubna, 1998, to be published in *Phys. Lett.* B.
15. K. Schenk *at al*, *Phys. Lett.* B **52**, 36 (1974).

PAIRING CORRELATIONS IN LIGHT NEUTRON-RICH NUCLEI

K. KATŌ[1], S. AOYAMA[2], T. MYO[1], T. YAMADA[3] AND K. IKEDA[4]
[1] *Graduate School of Science, Hokkaido University, Sapporo 060-0810, Japan*
[2] *Information Processing Center, Kitami Institute of Technology, Kitami 090-8507, Japan*
[3] *Laboratory of Physics, Kanto Gakuin University, Yokohama 236-8501, Japan*
[4] *The Institute of Physical and Chemical Research (RIKEN), Wako 351-0198, Japan*

The structure and excitations of neutron-rich nuclei are studied by using the core-plus-neutrons model. Pairing correlations between valence neutrons and core-neutrons are shown to play an important role in the binding and excitation mechanism of neutron-rich nuclei.

1 Introduction

The discovery[1] of a neutron halo in ^{11}Li provided us with a very challenging problem of exotic nuclear structure and excitation. Although, considerable studies have been done so far and our knowledge have been developed, we are still far from the satisfactory understanding. We have learned that the halo structure indicates that the nucleus consists of different structural parts of a strongly-coupled part and a weakly-coupled part. They are studied by using the core-plus-neutrons model[2] where the strongly-coupled part and the weakly-coupled part are described by a core cluster and valence neutrons, respectively. In many calculations, the excitation of the core cluster and its coupling to the motion of valence neutrons are neglected. Under such an assumption, the weakly-binding properties of valence neutrons have been extensively studied, where the meaning of the weak binding is twofold: one is the binding between core cluster and valence neutrons and another is the binding of valence neutrons. These couplings may have the same order of the weakness, and then must be treated carefully in the same footing.

In the present contribution, we discuss these features of the core+n+n systems via systematic studies of light neutron-rich nuclei, especially He- and Li-isotopes. The lightest neutron-halo nucleus, ^{6}He, is investigated by using a semi-microscopic ^{4}He+n+n model where the ^{4}He-n and n-n interactions and the model space describing the motion of two valence neutrons are examined in detail.[3] The ^{4}He+n+n model for ^{6}He is shown to describe well the halo character of the ground state and excited three-body resonance states. After confirming the reliability of the core+n+n model in ^{6}He, we discuss the characteristic property of ^{11}Li applying a ^{9}Li+n+n model. A serious problem in the ^{9}Li+n+n model, however, is that the calculated binding energy is smaller than the observed value of ^{11}Li. This problem is discussed from two points of view: one is the pairing correlations of valence neutrons, and another is the ^{9}Li-n interaction.

We reconstructed a folding potential based on the recent experiments of Bohlen et al.[4] This new interaction[5] is more attractive than the previous one which was fitted to the old data[6] on ^{10}Li. However, the calculated binding energy obtained by applying the new attractive ^{9}Li potential is still small. Then, since the employed basis space is large enough to describe the pairing correlation between valence neutrons, we return to a problem of the ^{9}Li interaction. We try to take into account the $J^\pi = 0^+$ pairing correlation of neutrons both in the ^{9}Li core and between the valence neutrons. This pairing correlation in ^{9}Li is expected to be changed by the addition of a valence neutron in ^{10}Li due to the Pauli-blocking. Including such a dynamical effect, we investigate the state-dependence of the ^{9}Li-n interaction, which provides a repulsive effect on the p-orbit neutron because of the Pauli-blocking of pairing correlation in the ^{9}Li core.

The obtained ^{9}Li interaction causes the inversion of s- and p-orbital neutron energies. However, in ^{10}Li, the s-orbital neutron is predicted as a virtual resonance which is difficult to observe directly. Therefore, we propose to observe the s-orbital neutron in ^{10}N as a resonant state.[9] The s- and p-orbital states in a pair of mirror nuclei are experimentally observed in ^{11}Be and ^{11}N.[10] We point out the existence of the anomalous Thomas-Ehrman shift[10,11] in weakly-bound and resonant s- and p-states.

2 Model and method

The semi-microscopic core-plus-neutrons model is used due to the following reasons: (1) It is necessary to investigate the weak binding around the thresholds, and to reproduce the binding energies (thresholds) of constituent clusters. (2) It is very interesting to see that the correlations play an important role in the weak-binding. In order to describe the resonant states, we employ the complex scaling method.[12] The core-plus-neutrons model for drip-line nuclei has no bound states between the core and a neutron. We investigate the core-neutron interaction through the analyses of the resonant states for which experimental data are available.

First, we apply the ^{4}He+n+n model[3] To ^{6}He, and solve the three-body problem in a variational framework using the hybrid-TV model[13] of T-type cluster-orbital-shell-model (COSM)[14] and V-type extended cluster model (ECM)[13] configurations. Next, we apply a ^{9}Li+n+n model to ^{11}Li in the same framework as the ^{4}He+n+n model.

3 He-isotopes — ^{6}He —

In the ^{4}He+n+n calculations, we employ the KKN-potential[15] for ^{4}He-n and the Minnesota potential[16] for n-n, which well describe the observed properties of the two-body systems. The result of the binding energy calculations is shown in Fig.1, where the convergences of COSM and hybrid-TV model calculations are compared. In COSM, the convergence is very slow for the channel number $[n]=[(l_n j_n)^2]_{J^\pi=0^+}$. On the other hand, when we take $[l=0,\ L=0]_{J^\pi=0^+}$ in ECM of the hybrid-TV model, we obtain a very rapid convergence. This results indicates that the pairing correlation between valence neutrons is very important in the binding of ^{4}He+n+n. Applying the complex scaling method to this model, we obtain the excited states of ^{6}He as shown in Fig. 2. The important result is that all the possible configurations of the $(p)^2$-shell model are obtained as resonant states, though their widths are rather large. Recently, the observation of the second 2^+ state was reported.[17]

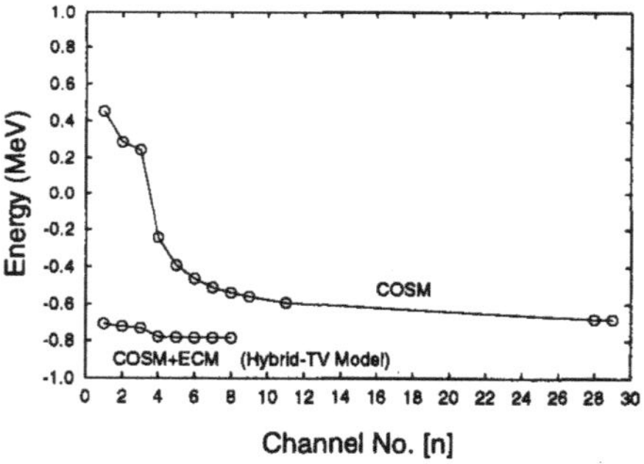

Figure 1: Bindind energy convergences of COSM and hybrid-TV model calculations of the ^{4}He+n+n system.

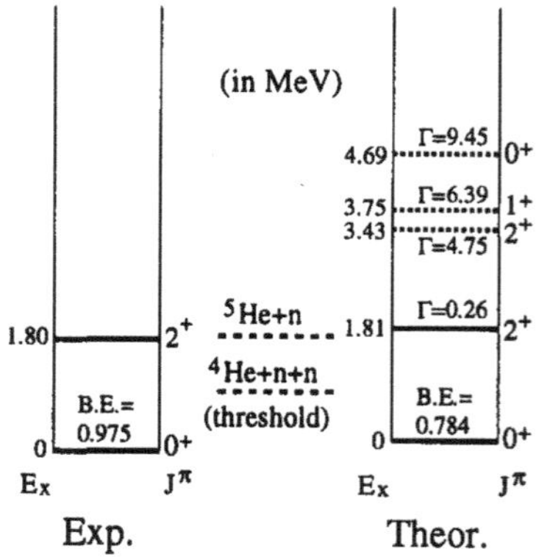

Figure 2: Energy levels of ^{6}He.

4 Li-isotopes — ^{11}Li —

In 1990, when Tosaka et al.[2] pioneeringly calculated the ^{9}Li+n+n model for the typical halo nucleus of ^{11}Li, only the data of Wilcox et al. [6] were available as information on the ^{9}Li-n interaction. Their calculation was performed using COSM, and presented too small binding energy in comparison with the observed one. Although the hybrid-TV model calculation was done,[18] the theoretical result could not explain the binding energy either. In 1993, Bohlen et al.[4] reported a new experimental result of the binding energy and two excited states of ^{10}Li. Their result suggested a more attractive ^{9}Li-n interaction.

We reconstructed the ^{9}Li-n folding potential based on these new data.[5] Using the obtained ^{9}Li-n interaction, we calculated the binding energy of ^{9}Li+n+n, and the result is shown in Fig. 3,[7] which is very similar to that of ^{4}He+n+n. It is seen that the pairing correlation is very important in obtaining a good convergence. But the observed ^{11}Li binding energy is still not explained. The calculated one is smaller by about 1 MeV.

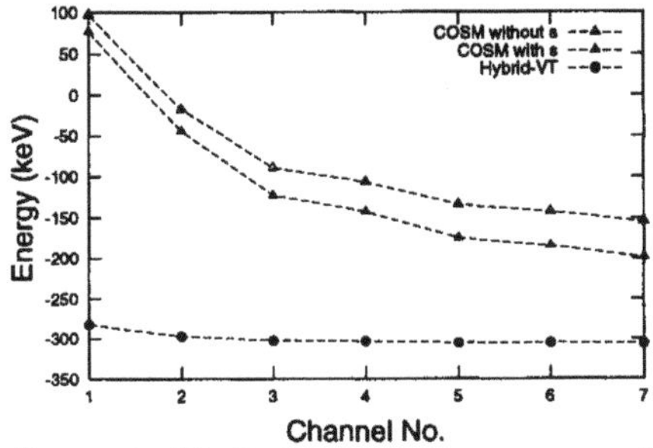

Figure 3: Bindind energy convergences of COSM and hybrid-TV model calculations of the ^{9}Li+n+n system.

To explain the halo structure of ^{11}Li, a possibility of the s-orbit configuration of valence neutrons has been discussed.[19] However, there is no reasonable explanation of low-lying s-orbital configuration in Li-isotopes so far. Here, we propose a new explanation of the origin of the low-lying s-orbit. The point of our idea is based on the Pauli-blocking of the pairing correlation in the ground state of ^{9}Li. To see this effect, we performed the coupled-channel calculation of $[^9\text{Li}((p_{3/2})^4_\nu)+\text{n}]+[^9\text{Li}((p_{3/2})^2_\nu(p_{1/2})^2_\nu)+\text{n}]+[^9\text{Li}((p_{3/2})^2_\nu(s_{1/2})^2_\nu)+\text{n}]$.[8] The Pauli-blocking provides a strong repulsive potential for the p-wave neutron but a small effects for other partial waves as shown in Fig. 4. The s-wave neutron has also a very small effect due to the small overlap with the core-

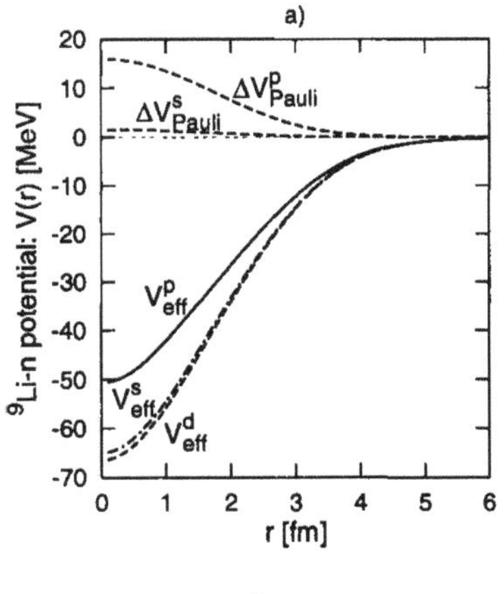

Figure 4: Effective ^{9}Li-n potentials and the Pauli-blocking potentials for s-, p- and d-wave neutrons.

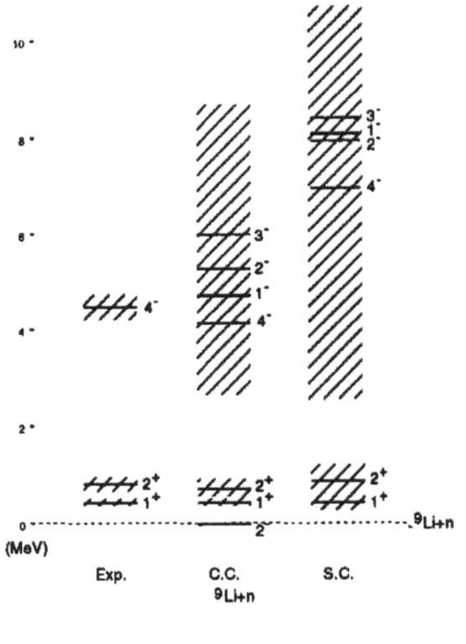

Figure 5: Energy levels of ^{10}Li.

neutron wave functions. Therefore, the $p_{1/2}$-orbit neutron is described by a shallower potential with the ^{9}Li core, but the $s_{1/2}$-orbit gives a deep one. As a result, the $s_{1/2}$-neutron shows a virtual resonance behavior. We can see a considerable improvement of the resonance spectrum in ^{10}Li in comparison with the experiments and with a single-channel calculation as shown in Fig. 5.

5 Summary

The halo nuclei have two different binding portions. A core part has a well-known binding property of normal stable nuclei, but the weakly binding of a halo part must be explained by effects from various kinds of correlation. Here, we investigated the important role of the pairing correlations in the weakly bound core+n+n systems by taking into account not only the correlation between valence neutrons but also coupling with the correlation in the core.

References

1. I. Tanihata *et al*, *Phys. Lett.* B **160**, 380 (1985).
2. Y. Tosaka and Y. Suzuki, *Nucl. Phys.* A **512**, 46 (1990).
3. S. Aoyama, S. Mukai, K. Katō and K. Ikeda, *Prog. Theor. Phys.* **93**, 99 (1995); **94**, 343 (1995).
4. H. G. Bohlen *et al*, *Z. Phys.* A **344**, 381 (1993).
5. K. Katō and K. Ikeda, *Prog. Theor. Phys.* **89**, 623 (1993).
6. K. H. Wilcox *et al*, *Phys. Lett.* B **59**, 142 (1975).
7. S. Mukai, S. Aoyama, K. Katō and K. Ikeda, *Prog. Theor. Phys.* **99**, 381 (1998).
8. K. Katō T. Yamada and K. Ikeda, *Prog. Theor. Phys.* **101**, 119 (1999).
9. S. Aoyama, K. Katō and K. Ikeda, *Phys. Lett.* B **414**, 13 (1997).
10. S. Aoyama, N. Itagaki, K. Katō and K. Ikeda, *Phys. Rev.* C **57**, 975 (1998).
11. S. Aoyama, K. Katō and K. Ikeda, *Prog. Theor. Phys.* **99**, 623 (1998).
12. J. Aguilar and J.M. Combes, *Commun. Math. Phys.* **22**, 269 (1971). E. Balslev and J.M.Combes, *Commun. Math. Phys.* **22**, 280 (1971).
13. K. Ikeda, *Nucl. Phys.* A **538**, 355c (1992).
14. Y. Suzuki and K. Ikeda, *Phys. Rev.* C **38**, 410 (1988).
15. H. Kanada, T. Kaneko, S. Nagata and M. Nomoto, *Prog. Theor. Phys.* **61**, 1327 (1979).
16. D. R. Thompson, M. LeMere and Y. C. Tang, *Nucl. Phys.* A **286**, 53 (1977).
17. J. Jänecke *et al*, *Phys. Rev.* C **54**, 1070 (1996).
18. Y. Tosaka, Y. Suzuki and K. Ikeda, *Prog. Theor. Phys.* **83**, 623 (1990).
19. I. J. Thompson and M. V. Zhukov, *Phys. Rev.* C **49**, 1904 (1994).

STRUCTURE OF THE A=6 NUCLEI WITH AN EXTENDED THREE-CLUSTER MODEL

K. ARAI, Y. SUZUKI
Department of Physics, Niigata University, Niigata 950-2181, Japan
E-mail: arai@nt.sc.niigata-u.ac.jp

R. G. LOVAS
Institute of Nuclear Research of the Hungarian Academy of Sciences, Debrecen, P. O. Box 51, H-4001, Hungary

The core breakup or distortion effects in the neutron-halo nucleus ^{6}He are studied with an extended microscopic $\alpha+n+n$ three-cluster model, in which the core particle, i.e., the α-cluster is described as a $3N+N$ two-cluster system. The allowance for the core breakup in ^{6}He enhances not only the binding energy but also the probability of the $t+t$ component. This model gives a binding energy and $t+t$ probability that are similar to the $\{\alpha+n+n;\ t+t\}$ model, and both can reproduce the very small binding energy of ^{6}He. Our results show that, for a weakly bound system like a halo nucleus, the core internal motion must be treated realistically since the tail behavior of the core affects the binding mechanism of the halo nucleons.

One of the most intriguing phenomena found in unstable nuclei is that some light nuclei near the neutron drip-line, e.g., ^{11}Li, ^{11}Be and ^{14}Be, have "neutron halos" [1]. The neutron-rich nucleus ^{6}He is also known to have a neutron-halo structure. The ground state of ^{6}He is expected to have a good $\alpha+n+n$ three-cluster structure with a binding energy of 1 MeV and the α-cluster is expected to be a rigid core. This $\alpha+n+n$ three-cluster structure has been studied with many microscopic and macroscopic models. These models can reproduce its halo structure well, but underbind the $\alpha+n+n$ system at least by 0.2 MeV. It seems that the reproduction of the correct binding requires the explicit inclusion of the $t+t$ channel [2], which is peculiar because the threshold of this channel is much higher than the ground state of ^{6}He. The $(0s)^4$ harmonic oscillator function which is used in the three-cluster model overestimates the excitation energy of the second 0^+ state and underestimates the tail part of the α-particle wave function. These defects of the core internal function are expected to affect the binding of the halo nucleus ^{6}He because the halo neutrons have spatially extensive orbits around the core.

We present an extended $\alpha+n+n$ three-cluster model, in which the α-particle core is described in a $3N+N$ cluster model, to study the binding mechanism and the $t+t$ component in the neutron-halo nucleus ^{6}He [3]. The two-cluster description improves the tail behavior of the α-particle with respect to the standard $0s$ harmonic-oscillator shell-model description, and makes the first excited state of the α-particle realistic[4,5].

The model is a microscopic cluster model of the type of the resonating group method (RGM), in which the Pauli principle is exactly observed for all nucleons. The effective nucleon-nucleon interaction used is the Minnesota potential.[6] The "stochastic variational method"[7,8] is used to set up a nearly exhaustive yet not excessively large basis.

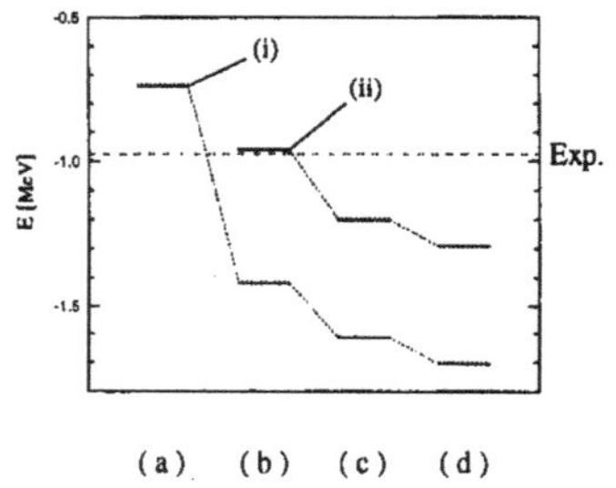

Figure 1: The binding energy of ^{6}He measured from the α+n+n threshold in models (a), (b), (c) and (d), with the two effective interactions, (i) and (ii).

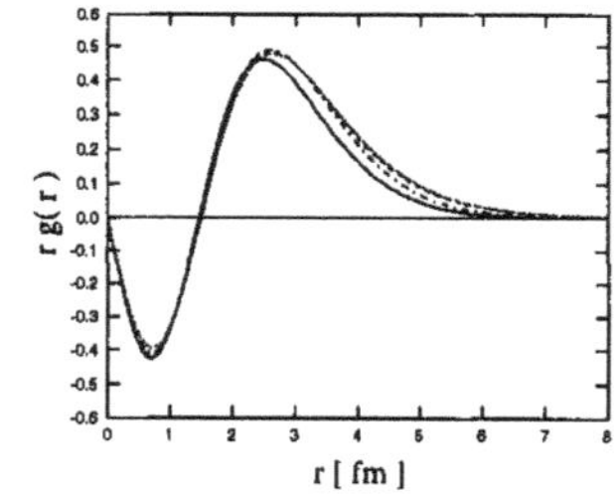

Figure 2: The t+t reduced width amplitudes for models (a) (solid line) (b) (dotted line), (c) (dot-dashed line) and (d) (dashed line).

Fig. 1 shows the resulting binding energies of ^{6}He with respect to the α+n+n threshold in four different models, viz. (a) $\{\alpha+n+n\}$, (b) $\{\alpha+n+n;\ t+t\}$, (c) $\{[3N+N]+n+n\}$ and (d) $\{[3N+N]+n+n;\ t+t\}$. The tabs (i) and (ii) in Fig. 1 denote two different parameter sets of the effective nucleon-nucleon interaction[2,3]. Model (a) with interaction (i), which is appropriate for the two-cluster subsystems, underbinds ^{6}He by about 0.24 MeV, but model (b) with interaction (ii), which is appropriate for the subsystems in a state space consistent with (b), reproduces the experimental binding energy correctly. In spite of the tight binding of the α-particle, the improvements caused by the inclusion of the t+t channel [model (b)] and of the α-breakup [model (c)] are substantial and of similar magnitudes. The α breakup seems not only more significant than the t+t contribution, but also seems to contain the latter; this is shown by the fact that the contribution of the t+t channel on top of the α breakup is only 0.09 MeV. Thus, by including the t+t channel, one gains energy not by forming another t cluster but by breaking up the α-cluster.

To analyse the role of the t+t channel, we have calculated "the amount of t+t clustering", which is the weight of the t+t subspace in the wave function[9]. The calculated amount of t+t clustering is 0.49 for model (a), 0.54 for model (b), 0.55 for model (c) and 0.57 for model (d). These results show that model

(a) underestimates the probability of the $t+t$ channel. This component can be made realistic either by explicitly including the $t+t$ channel as in model (b), or by including the $3N+N$ component of the α-cluster. Once one of them is included, the inclusion of the other causes hardly any change in the $t+t$ clustering. To see this more closely, we have calculated the $t+t$ reduced width amplitude and show it in Fig. 2 for all four models. For small intertriton distances ($r \leq 2$ fm), all amplitudes are very much alike, but for $r \geq 2$ fm, the pure three-cluster model (a) underestimates the amplitude, because the core internal function cannot reproduce correctly the tail part of the wave function.

The extended model was also applied to the ground state of ^{6}Li for comparison with the $\{\alpha+p+n\}$ and $\{\alpha+p+n;\ t+h\}$ models. This state has larger binding energy, $E=-3.7$ MeV, with respect to the $\alpha+p+n$ threshold. In this calculation only $\{L=0,\ S=0\}$ components were included. The results are similar to those for ^{6}He: both the $t+h$ channel and the core breakup effect enhances the binding energy significantly with respect to the pure $\alpha+p+n$ model and, when applied separately, their contributions are similar. When, however, the $t+h$ channel is added to the extended three-cluster model, the binding energy is only increased by 0.06 MeV. So the $t+h$ channel has much the same effect as the core breakup in ^{6}Li, just like the $t+t$ channel in ^{6}He. In fact, its contribution here is somewhat smaller than for ^{6}He. Since the ground state of ^{6}Li has a tighter binding and the two valence nucleons are closer to the core, the core breakup or distortion effect is weaker than in the neutron-halo nucleus ^{6}He.

Our results show that, for a weakly bound system like a halo nucleus, the core internal motion must be treated realistically since the tail behavior of the core affects the binding mechanism of the halo nucleons.

References

1. I. Tanihata, *J. Phys.* G **22**, 157 (1996).
2. A. Csótó, *Phys. Rev.* C **48**, 165 (1993).
3. K. Arai, Y. Suzuki and R. G. Lovas, *Phys. Rev.* C **59**, 1436 (1999).
4. H. Kanada, T. Kaneko, Y. C. Tang, *Phys. Rev.* C **34**, 22 (1986); H. Kanada, T. Kaneko, Y. C. Tang, *ibid.* **43**, 371 (1991).
5. A. Csótó and G. H. Hale, *Phys. Rev.* C **55**, 2366 (1997).
6. D. R. Thompson, M. LeMere, and Y. C. Tang, *Nucl. Phys.* A **286**, 53 (1977).
7. V. I. Kukulin and V. M. Krasnopol'sky, *J. Phys.* G **3**, 795 (1977).
8. K. Varga, Y. Suzuki, and R. G. Lovas, *Nucl. Phys.* A **571**, 447 (1994).
9. R. Beck, F. Dickman and R. G. Lovas, *Ann. Phys. (N. Y.)* **173**, 1 (1987).

^{12}BE IS NOT A MAGIC NUMBER NUCLEUS

N. ITAGAKI
RI beam science laboratory, RIKEN, Wako, Saitama 351-0198, JAPAN
E-mail: itagaki@postman.riken.go.jp

S. OKABE
Center for Information and Multimedia Studies, Hokkaido University, Sapporo 060-0810, JAPAN
E-mail: okabe@ec.hokudai.ac.jp

^{10}Be and ^{12}Be are investigated by a microscopic $\alpha+\alpha+n+n\cdots$ model based on the Molecular Orbit (MO) model. The low-lying states are characterized by several configurations of valence neutrons which are constructed as combinations of three basic orbits: $K^\pi=3/2^-$, $1/2^-$ and $1/2^+$. In ^{10}Be, all the observed positive parity bands and negative parity bands are described within the model, and the second 0^+ state has large $\alpha-\alpha$ structure. The configuration is $(1/2^+)^2$ corresponding to the sd-shell, and two valence neutrons along the $\alpha-\alpha$ axis enhance the $\alpha-\alpha$ distance. In ^{12}Be, this configuration strongly mixes in the ground state, and due to this, the observed slow β-decay to ^{12}B is explained. This is the breaking of $N=8$ magic number in ^{12}Be.

Numerous experiments with the use of unstable nuclear beams have succeeded in extending the observation of the neutron drip-line, and various features of β-unstable nuclei have been revealed. One of the characteristic phenomena is the anomaly of the shell structure. The experimental results show that the numbers of neutrons in the drip-line isotopes of p-shell nuclei Be and B are beyond magic number $N=8$ (^{14}Be, $N=10$ and ^{19}B, $N=14$). Moreover, recently, contributions of such higher shells are analyzed in $N=8$ nuclei. A calculation based on the shell model has shown that the slow β-decay of ^{12}Be to ^{12}B can be explained due to admixture of the sd-shell in ^{12}Be ($N=8$) where the single particle energy is artificially modified[1]. This show that the concept of magic number is vague in ^{12}Be.

There have been pioneering studies for Be isotopes based on the Molecular Orbit (MO) model[2,3,4], however, the ground state of ^{12}Be have been analyzed to be dominantly the closed p-shell configuration for the neutrons. Here, we propose a new $\alpha+\alpha+n+n+\cdots$ model based on MO where the model space for the valence neutrons is extended vastly. The total wave function is fully antisymmetrized and expressed by a superposition of terms with various configurations of valence neutrons. The α-α motion surrounded by the valence neutrons is also solved.

In MO, the orbits for valence neutrons are expressed as linear combination of orbits around each α-cluster. When the linear combination of two p-orbits around the left-α-cluster and the right-α-cluster is summed up by the same sign, the resultant MO also has negative parity and one node. In this case, the MO is restricted to spread along an axis perpendicular to the $\alpha - \alpha$ (z) axis, that is so-called π-orbit. It cannot spread along the z-axis where two α-clusters are already located. If the linear combination of p-orbit is summed up by the different sign, the resultant MO has two nodes and positive parity. This MO can spread to all directions, and the optimal direction becomes z-direction. This is so-called σ-orbit. In this way, three basic orbits $K^\pi = 3/2^-, 1/2^-$ and $1/2^+$ for the valence neutrons are introduced. The $1/2^+$ state is the σ-orbit, and the spin-orbit interaction splits the π-orbit to $3/2^-$ and $1/2^-$. These orbits are described by the superposition of local Gaussians whose positions are determined variationally for each $\alpha - \alpha$ distance before the angular momentum projection.

As effective two body $N - N$ interactions, we introduce Volkov No.2 potential with $M = 0.6, B = H = 0.125$ for the central part, and G3RS potential with the strength parameter V=2000 MeV for the spin-orbit part. These parameters are determined from the analyses of $\alpha - \alpha$, $\alpha - N$ and $N - N$ scattering phase shifts. First, we apply the model to ^{10}Be (α+α+n+n). The ground 0^+ state (-61.1 MeV) has dominantly the $(3/2^-)^2$ component for the two valence neutrons, and the 0_2^+ state at 8.1 MeV (experimentally 6.263 MeV) has dominantly the $(1/2^+)^2$ component. Then the level inversion of p-shell and sd-shell occurs in the second 0^+ state of ^{10}Be. The second 0^+ state has much larger charge radius (2.93 fm, proton radius 0.813 fm is used) than the ground state (2.51 fm), and the state is a band head of large clustering states. The two valence neutrons occupy higher nodal orbit along the $\alpha - \alpha$ axis and increase the distance between two α-clusters to reduce the kinetic energy of the system.

Next, we show that this $(1/2^+)^2$ component strongly mixes in the ground state of ^{12}Be. The levels of ^{12}Be are shown in FIG.1 with respect to the $\alpha - \alpha$ distance. Here, four valence neutrons rotate around two α-clusters, and two of them occupy $(3/2^-)^2$. In ^{12}Be, there appear two 0^+ states in low-lying energy region. When $\alpha - \alpha$ distance is small for example 2fm, the state with $(1/2^-)^2$ for the last two valence neutrons dominates the lowest 0^+ state that corresponds to the p-shell closed configuration at the limit of $\alpha - \alpha$ distance zero. However, the state with $(1/2^+)^2$ for the last two valence neutrons becomes lower as increase of the $\alpha - \alpha$ distance. At the $\alpha - \alpha$ distance of 3 fm which is the optimal distance of ^{12}Be, the ground 0^+ state has squared overlap with $(1/2^-)^2$ by 0.41 and with $(1/2^+)^2$ by 0.64. Therefore, in the ground state of ^{12}Be, two different configurations contributes, and the neutron magic number $N = 8$ disappears.

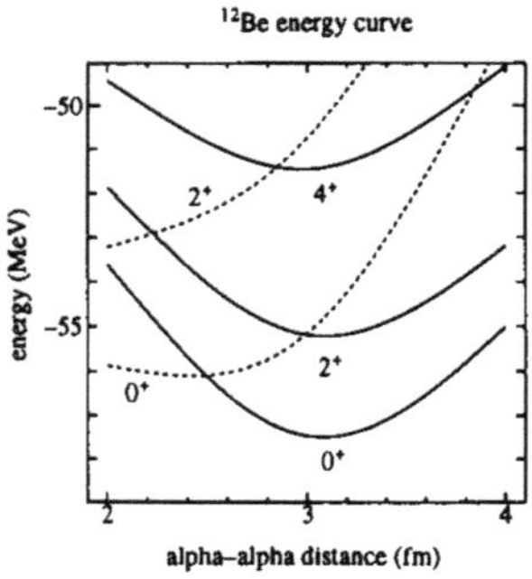

Figure 1. The energy curve with respect to the $\alpha - \alpha$ distance in ^{12}Be. The solid line has dominantly the $(1/2^+)^2$ configuration for the valence two neutrons.

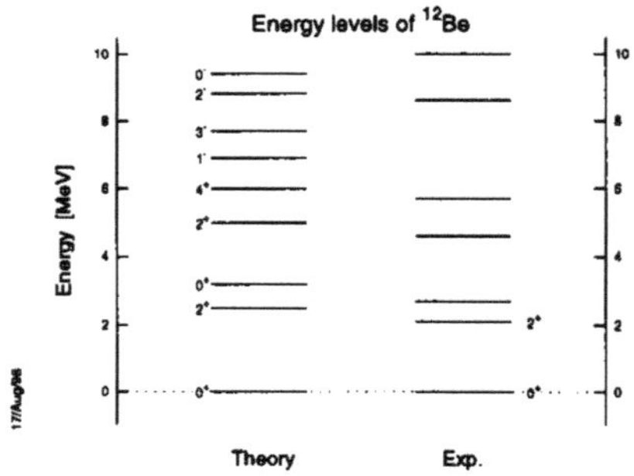

Figure 2. The calculated and experimental energy levels of ^{12}Be.

The energy levels of ^{12}Be are shown in FIG.2. The ground state has a rotation band structure and the 4^+ state is predicted at 6.0MeV. The 0_2^+ state is predicted at 3.1MeV whose candidate is experimentally observed at 2.7MeV. The electric transition B(E2) from 0_1^+ to 2_1^+ is predicted to be 50.3 $e^2\text{fm}^4$, and transition to 2_2^+ is about 1/4 (12.8 $e^2\text{fm}^4$).

One evidence of the $(1/2^+)^2$ contribution in the ground 0^+ state of ^{12}Be is the β-decay probability from ^{12}Be to ^{12}B. Using the present model, the $\log(ft)$ value of $^{12}\text{Be}(0^+)\rightarrow{}^{12}\text{B}(1^+)$ is calculated to be 3.81, and this value agrees with the experimental value of 3.8. However, if we restrict the configuration of the last two valence neutrons to $(1/2^-)^2$, the $\log(ft)$ value becomes 3.51. Therefore, the mixing of $(1/2^+)^2$ component suppresses this transition and increases the $\log(ft)$ value. The coincidence of the experimental and the theoretical $\log(ft)$ values suggests that the breaking of $N = 8$ magic number is realized in ^{12}Be.

References

1. T. Suzuki and T. Otsuka *Phys. Rev.* C **56**, 847 (1997).
2. Y. Abe, J. Hiura and H. Tanaka, *Prog. Theor. Phys.* **49**, 800 (1973).
3. H. Furutani, H. Kanada, T. Kaneko, S. Nagata, H. Nishioka, S. Okabe, S. Saito, T. Sakuda and M. Seya, *Prog. Theor. Phys. Suppl.* **68**, 193 (1980).
4. M. Seya, M. Kohno and S. Nagata, *Prog. Theor. Phys.* **65**, 204 (1981).

THE ALGEBRAIC THEORY OF THE THREE BODY PROBLEM APPLIED TO THE ^{12}C NUCLEUS

A. TUMINO
Dipartimento di Fisica, Università di Catania and Istituto Nazionale di Fisica Nucleare, L.N.S., Catania
ITALY
E-mail: Tumino@lns.infn.it

R. BIJKER
Instituto de Ciencias Nucleares, U.N.A.M., A.P. 70-543, 04510 México, D.F., MÉXICO
E-mail: Bijker@sphinx.nuclecu.unam.mx

F. IACHELLO
Center for Theoretical Physics, Sloane Laboratory, Yale University, New Haven, CONNECTICUT 06511
E-mail: Peggy@mirage.physics.yale.edu

The U(7) Spectrum Generating Algebra was used to study the ^{12}C nucleus. This formulation allows one to calculate in a straightforward way energy spectrum, transition rates and form factors. The results of the calculations together with the comparison with experiment will be discussed.

1 Introduction

The molecule-like structure is one of the fundamental forms of existence of nuclear systems together with shell structure. In the region of light nuclei, many of the properties of the self-conjugate 4n nuclei can be explained only by assuming the existence of this kind of structured levels; as it has been known for many years, the α-Cluster Model predicts a great amount of possible configurations for these nuclei. We have re-addressed the question of the extent to which it can provide a description of part of the spectrum comparable or superior to that of models based on the assumption of stable average field (shell model). To this aim, we have used a treatment of the n-body problem in terms of the U(3n-2) Spectrum Generating Algebra [ref. 1]. This formulation allows one to perform simple but detailed calculations of the spectrum, transition rates and form factors. In this work we focus on ^{12}C. The first assumption is that the lowest configuration is the one predicted by the α-Cluster Model with three α–particles at the corners of an equilateral triangle. The relevant degrees of freedom of this configuration are the two Jacobi coordinates, ρ and λ, related to the distance between two α-particles and to the distance between the third α-particle and the center of mass of the first two, respectively.

2 Construction of Algebra and operators of interest

We introduce a dipole boson with $L=1^-$ for each independent relative coordinate, and an auxiliary scalar boson with $L = 0^+$:

$$p^+_{\rho,m}\,,\; p^+_{\lambda,m}\,,\; s^+ \qquad\qquad m = -1,0,1$$

the scalar boson makes it possible to investigate the full dynamics of the two dipole ones, and is added under the restriction that the total number of bosons $N = n_\rho + n_\lambda + n_s$ is conserved. This procedure leads to the construction of U(7), whose 49 generators are the bilinear products:

$$b^+_{i,m}\, b_{j,m'}\,,\; b^+_{i,m}\, s\,,\; s^+\, b_{i,m}\,,\; s^+ s \qquad\qquad m, m' = -1,0,1 \qquad i, j = \rho, \lambda$$

We calculate the energy spectrum, form factors, radius and B(EL) values for ^{12}C, and compare these with the experimental results. There are two constraints for the Hamiltonian:

- the problem is rotationally invariant
- the three α-particles are identical objects.

Instead of writing down the full expression of the algebraic Hamiltonian [ref. 2], we will refer to a more intuitive description of its structure by looking at the geometric space attached to U(7). In order to do so, we make use of the coherent states and we assume to neglect any interaction between vibrational and rotational degrees of freedom. Mean field techniques give the equilibrium shape corresponding to an equilateral triangle; the corresponding energy spectrum expression is given by:

$$E = \xi_1 v_1 + \xi_2 v_2 + k\, L(L+1) + k' K^2$$

v_1 and v_2 are referred to the symmetric stretching vibration and the bending one respectively; L is the angular momentum and K its projection on the symmetry axis. The form factors of this configuration are given by the product of the matrix element of the transition operator $U(q\beta)$, with a folding term, $e^{-\alpha q^2}$, which represents the form factor of the α-particle. They depend on two parameters α and β, which are related to the shape under consideration [ref. 3]. Their values are chosen to better reproduce the behaviour of the experimental elastic form factor:

$$\alpha = (\alpha\text{-particle radius})^2 / 6 = 0.40 \pm 0.01\ \text{fm}^2$$
$$\beta = (\text{distance } \alpha\text{-}\alpha) / 3^{1/2} = 1.75 \pm 0.01\ \text{fm}$$

The B(EL) values correspond to the long wavelength limit of the form factor, and, hence, are proportional to some power of the β parameter.

3 Results and conclusions

The rotational band of the lowest configuration has the following members: 0^+(ground state), 2^+(4.43MeV), 3^-(9.64MeV), 4^+(10.84MeV). Our calculations of the energies for these states and form factors provide strong evidence for this interpretation even if the calculated BE(3) (70 fm^6) is somewhat smaller than the experimental value (107 fm^6); this disagreement could be reduced by making use of a value of the β parameter for the 3^- state different from that for the ground state, so taking into account the stretching effect on the former, caused by the centrifugal force. In figure 1 the calculated elastic form factor and the $0^+ \rightarrow 2^+$ one, are superimposed onto the experimental results, obtained from electron scattering [ref. 4÷9]:

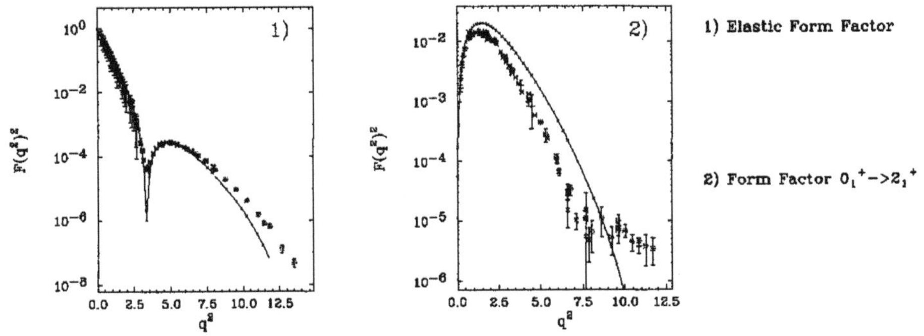

Figure 1

The experimental elastic form factor is very well reproduced by our calculations; for the second one it would be interesting to repeat the calculations by using a different β, or by folding in some other way, in order to improve the agreement.
Next, we have considered the 0^+ state at 7.65MeV. The assumption that it represents the stretching vibrational state of the triangular configuration gives a reasonable agreement with experiment. Figure 2 shows the comparison between the calculated form factor for the transition from the ground state to this state, and the experimental one; as one sees, the shape is well reproduced by the calculation, but not the absolute value. The present experimental data, available up to now, do not allow to exclude that this state could be the band head of another configuration as the one consisting of three α-particles in a line.

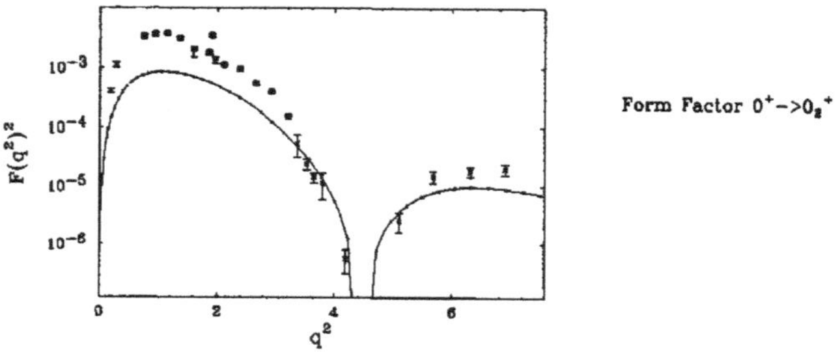

Figure 2

Moreover the 1^- state at 10.84MeV is assumed to represent the bending vibration, and this interpretation appears to be the only one consistent with the data. The symmetry of triangular configuration allows the negative parity states to be present, unlike the one of the α-chain. We also derived an estimation of the charge radius for ^{12}C, which comes directly from elastic form factor, as known. Our value is:

$$< r^2 >^{1/2}_{calc.} = 2.38 \pm 0.01 \text{ fm}$$

that is really close to the one reported in literature [7,8]:

$$< r^2 >^{1/2}_{exp.} = 2.46 \pm 0.02 \text{ fm}$$

In conclusion, the U(7) Spectrum Generating Algebra was used to study ^{12}C; this treatment has provided a significant agreement with experiment. The application to ^{12}C, for which many experimental and theoretical informations are available, would represent an inspection that can allow to proceed in describing more complex systems.

References

1. F. Iachello in "*Lie Algebras, Cohomologies and New Applications of Quantum Mechanics*", Contemporary Mathematics, AMS, N. Kamran and P. Olver eds., Vol. **160** (1994) p.151-171.
2. R. Bijker, F. Iachello and A. Leviatan, Ann. Phys. (N.Y.) **236** (1994) 69.
3. E.V. Inopin and B.I. Tishchenko, Soviet Phys. Jept **11 4** (1960) 840.
4. H.L. Crannel and Griffy, Phys. Rev. **136** (1994) B1580.
5. H.L. Crannel, Phys. Rev. **148** (1996) 1107.
6. A. Nakada et al., Phys. Rev. Lett. **27** (1971) 745.
7. W. Reuter et al., Phys. Rev. C **26** (1982) 606.
8. I. Sick and J.S.McCarthy, Nucl. Phys. **A150** (1970) 631.
9. Y. Torizuka et al., Phys. Rev. Lett. **22** (1969).

^{40}Ca+α+α CLUSTER MODEL OF ^{48}Cr

TOSHIMI SAKUDA AND YUKIYA IMAYOSHI

Department of Physics, Miyazaki University, Miyazaki 889-2192, Japan
E-mail: e04101u@cc.miyazaki-u.ac.jp

SHIGEO OHKUBO

Department of Applied Science and Environment, Kochi Women's University, Kochi 780-8515, Japan
E-mail: ohkubo@yukawa.kyoto-u.ac.jp

The microscopic calculation of ^{40}Ca+α+α cluster model is performed with the orthogonality condition model. Energy spectra of ^{48}Cr are obtained, and the structure of the levels are discussed throuh the wave functions. Low-lying levels are classified into several bands.

Recently, we have performed microscopic α-cluster model calculations with the orthogonality condition model (OCM) for ^{40}Ca$\sim$^{44}Ti nuclei.[1] It has been shown that the α-cluster model is successful in accounting for the energy spectra, electromagnetic transitions and the α-spectroscopic factors. The persistency of α-cluster states in this region is well established now. It also has been suggested that 2α cluster states may be present in these nuclei. It is very interesting to verify the 2α cluster states, which could show more variety of cluster structure. We will apply the ^{40}Ca+α+α OCM for the ^{48}Cr nucleus. The multi-cluster model spans a large model space which includes many shell-model states and cluster states and the dynamics of various kinds of cluster configurations. It is known that the nucleus ^{48}Cr has a rotational level structure and some measured $E2$-transitions of the yrast band are strongly enhanced.[2] Furthermore it seems difficult for full fp-shell calculations to reproduce the enhancement without using large effective charges. We will analyze these features of ^{48}Cr from the viewpoint of 2α-cluster structure.

The model space is described by a set of the wave functions

$$\Phi_J = \sqrt{\frac{4!4!40!}{48!}}\mathcal{A}\ \{\phi(\alpha)\ \phi(\alpha)\ \phi(\mathrm{Ca})\ [\mathrm{U}_{\mathrm{N}_{23},\mathrm{l}_{23}}(\mathbf{r}), \mathrm{U}_{\mathrm{N}_1,\mathrm{l}_1}(\mathbf{R})\ ]_\mathrm{J}\ \}\ , \qquad (1)$$

where ϕ's are the antisymmetrized internal wave functions and $U_{N_{23},l_{23}}(\mathbf{r})$ and $U_{N_1,l_1}(\mathbf{R})$ are harmonic oscillator functions with N_{23} and N_1 quanta for the relative motions. The model wave function is generated as a direct product of the two relative wave functions: $(N_{23}, 0) \times (N_1, 0)$. The Pauli-allowed states are obtained by diagonalizing the norm kernel. The resulting allowed

Table 1. The Pauli-allowed states of ^{40}Ca+α+α system. They are classified by the SU(3) label (λ, μ) with the multiplicity n.

N_T	$(\lambda, \mu)^n$
24	(16,4) (12,6) (8,8) (4,10) (0,12)
25	(19,3) (17,4) (15,5) (13,6) (11,7) (9,8) (7,9) (5,10) (3,11) (1,12)
26	$(22,2)(20,3)(18,4)^2(16,5)(14,6)^2(12,7)\ (10,8)^2\ (8,9)\ (6,10)^2\ (4,11)\ (2,12)^2$
27	$(25,1)\ (23,2)\ (21,3)^2(19,4)^2(17,5)^2(15,6)^2(13,7)^2(11,8)^2\ (9,9)^2\ (7,10)^2(5,11)^2(3,12)^2$ (1,13)
28	$(28,0)\ (26,1)\ (24,2)^2(22,3)^2(20,4)^3(18,5)^2(16,6)^3\ (14,7)^2(12,8)^3(10,9)^2(8,10)^3(6,11)^2$ $(4,12)^3(2,13)\ (0,14)$
29	$(29,0)\ (27,1)^2(25,2)^2(23,3)^3(21,4)^3(19,5)^3(17,6)^3\ (15,7)^3(13,8)^3(11,9)^3(9,10)^3(7,11)^3$ $(5,12)^3(3,13)^2(1,14)$
30	$(30,0)^2(28,1)^2(26,2)^3(24,3)^3(22,4)^4(20,5)^3(18,6)^4\ (16,7)^3(14,8)^4(12,9)^3(10,10)^4(8,11)^3$ $(6,12)^4(4,13)^2(2,14)^2$
31	$(31,0)^2(29,1)^3(27,2)^3(25,3)^4(23,4)^4(21,5)^4(19,6)^4\ (17,7)^4(15,8)^4(13,9)^4(11,10)^4(9,11)^4$ $(7,12)^4(5,13)^3(2,14)^2(1,15)$

states classified by SU(3) symmetry are listed in the Table 1. The states with total quanta $N_T \leq 23$ are not allowed as a matter of course. We can see that $N_T = 24$ space contains important $(fp)^8$ shell-model states and the configurations with a larger value of N_T have a capacity for presenting the 2α and also α cluster states. As for the interactions between clusters, we use the folding potentials with the effective two-nucleon interaction HNY. The depth parameter of HNY, Δ=65 MeV for ^{40}Ca-α and 18 MeV for α-α are chosen.

The OCM equation is solved for the positive-parity states up to $J^\pi = 16^+$ and the negative-parity states up to $J^\pi = 11^-$, by taking the total oscillator quanta $N_T = 24 \sim 31$ for the relative motions. The calculated energy spectra are compared with the experiments in Figs. 1 and 2. The energy scale is measured from the ^{40}Ca+α+α threshold (E_{th}=12.82 MeV). The ground state band is very well reproduced. These states are dominated by the $(\lambda,\mu)=(16,4)K^\pi=0^+$ configuration. The second band built on the second 2^+ state is also predicted. The main configuration of the band is $(16,4)K^\pi=2^+$. However, there are considerable mixing between these two bands at high angular momentum states ($J \geq 10^+$). This is probably consistent with a back-bending occuring at $J^\pi=10^+$ in the ground band. At higer energies than the two bands it is difficult to classify the states into bands. In the negative-parity states, the lowest and the second bands are largely $(17,4)K^\pi=0^-$ and $(17,4)K^\pi=2^-$, respectively. Some difference of ordering of the levels of the lowest band might be resolved, if we take account of the spin-orbit splittings of single-particle states.

As a whole, our result seems to be in good agreement and is very encouraging to the usefulness of the model.

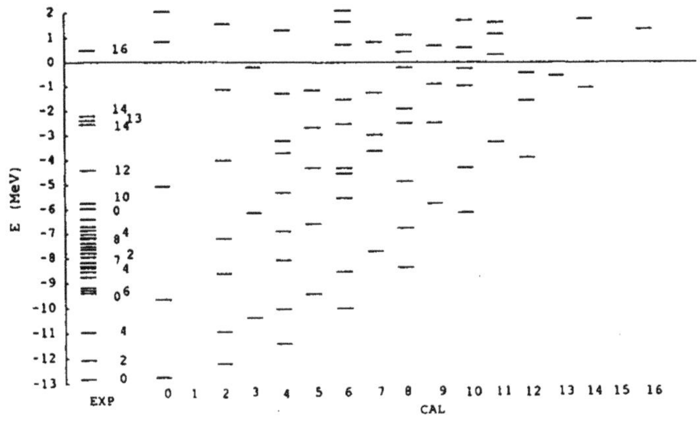

Figure 1. Calculated and experimental positive-parity states in ^{48}Cr.

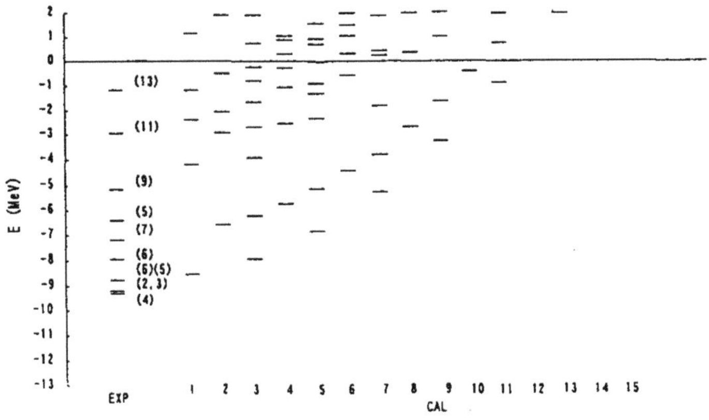

Figure 2. Calculated and experimental negative-parity states in ^{48}Cr.

References

1. T. Sakuda and S. Ohkubo, *Phys. Rev.* C **49**, 149 (1994); C **51**, 586 (1995); C **57**, 1184 (1998); *Prog. Theor. Phys. Suppl. No.* **132**, 103 (1998).
2. F. Brandolini *et al*, *Nucl. Phys.* A **642**, 387 (1998); J.A. Cameron *et al*, Phys. Lett. B **235**, 139 (1990); B **319**, *Phys. Rev.* C **58**, 808 (1998).

CLUSTER FORMATION IN NUCLEAR DYNAMICS BELOW 100 MEV/U

M. F. RIVET
Institut de Physique Nucléaire, IN2P3/CNRS, F-91406 Orsay Cedex France
E-mail: rivet@ipno.in2p3.fr

Different mechanisms of cluster production in heavy ion reactions below 100 MeV/u are reviewed. General aspects of the nuclear reactions are given, followed by a global separation between equilibrated and non-equilibrated emissions of clusters. Two mechanisms of production in which dynamical effects are particularly visible are presented in more details.

1 Introduction

"Clusters" in heavy-ion collisions include all reaction products but nucleons. Generally H and He isotopes (lcp) are distinguished from fragments (Z>2), because they are produced at every stage of the reaction. The production of fragments, assumed to occur in a restricted time interval, is particularly interesting to study in the incident energy range below 100 MeV/u: in this energy domain one observes a saturation of the number of fragments, for instance for Xe+Sn the fragment number saturates above 39 MeV/u (fig. 1 left) while the lcp multiplicity keeps regularly increasing; more generally the maximum fragment multiplicity per incident nucleon (fig. 1 right) goes through a maximum around an excitation energy of 8-9 MeV/u [1].

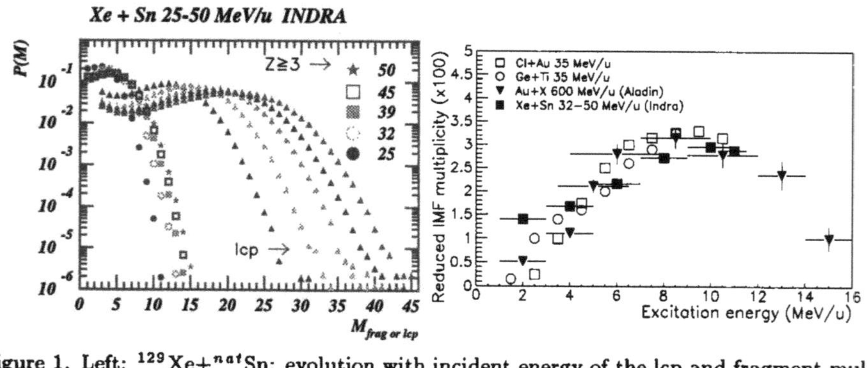

Figure 1. Left: ^{129}Xe+^{nat}Sn: evolution with incident energy of the lcp and fragment multiplicity distributions. Right: fragment multiplicity (normalised to the number of incident nucleons) as a function of the excitation energy. Extracted from [1].

From a fundamental point of view, nucleon-nucleon collisions progressively cease to be Pauli blocked, and are therefore expected to strongly compete with mean-field effects; this should allow to extract information such as relaxation time, nuclear viscosity, in-medium σ_{nn}... The relative time scales favor the occurrence of numerous phenomena: reaction times vary from 40 fm/c (peripheral collisions at 100 MeV/u) to 300 fm/c (central collisions at 30 MeV/u). The characteristic energy relaxation time is $\tau_{th} \sim$ 20-30 fm/c, independently of the incident energy [2]. The life-time of a nucleus heated at a temperature of 5 MeV is $\tau_n \sim 50$ fm/c [3]. These times show that before the end of the reaction, not only non-equilibrium but also both equilibrium emissions may occur. The more recent dynamical simulations have underlined the tremendous importance of fluctuations on the reaction mechanisms; indeed the reaction time is still larger than the characteristic times for the growth of fluctuations $\tau_{fluc} \sim 30$ fm/c [4], allowing the manifestation of shape, volume, surface ... instabilities.

A result now confirmed thanks to the results from high-quality 4π arrays is that fusion processes vanish ($< 5\% \ \sigma_R$) around 30 MeV/u, and practically all collisions are of binary nature independently of their violence, producing two hot entities (projectile-like and target-like sources PLS and TLS) and therefore ending with at least two fragments, residues of these sources. In a large fraction of the reactions these two residues are accompanied by other clusters (from deuterons to fragments) whose velocities lie between those of the main residues. Multifragment events are then dominant in intermediate energy heavy-ion reactions; For some rare events "multifragmentation" was characterized, in which all the fragments originate from a single equilibrated source ("fusion events") [5,6]. Multifragmentation of the PLS was also recently reported [7].

2 Global estimate of dynamical emission

If one defines dynamical emission as all products but those coming from the isotropic evaporation of *PLS/TLS having reached thermodynamical and shape equilibrium*, then a global estimate of the mass lost in dynamical emissions can be made.

In ref. [8,9,10] the PLS evaporation is subtracted from the measured products, and the remaining part is attributed to dynamical emission. The results, which strongly depend on the estimated PLS velocity (see refs [8,9,10] for details, and note that the actual methods for extracting V_{PLS} are different in [8] and [9,10]), are shown in fig. 2; the impact parameter scales are reversed for the two parts of the figure. Following fig. 2-right, for ^{129}Xe+^{nat}Sn dynamical

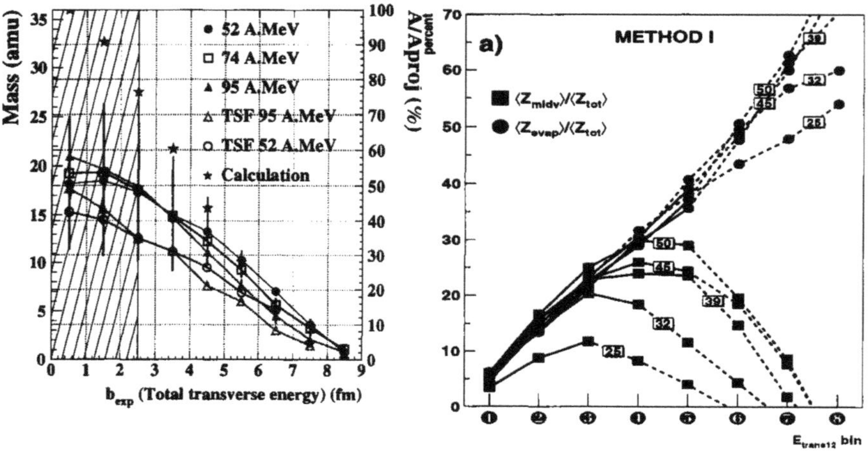

Figure 2. Evolution with energy of the mass of dynamical emissions as a function of the (experimental) impact parameter. Left: $^{36}Ar+^{58}Ni$: Symbols are experimental data, except the stars which display the calculated mass of "participant" nucleons. Extracted from [8]. Right: $^{129}Xe+^{nat}Sn$ (25-50 MeV/u): charge emitted through dynamical (squares) and evaporation (circles) processes, normalised to the total system charge; the charge of the heaviest fragment (PLS residue) excluded. the abscissa scale represents the impact parameter, bin 1 being the more peripheral; bins 1,2,3 cover the upper half of the total b range. Extracted from [10]. The dashed area (left) and lines (right) for central collisions indicate less accurate experimental results.

emission becomes sizeable around 20 MeV/u, and significantly increases with energy, and with the centrality of the collision; above 30 MeV/u, the mass lost in dynamical emission roughly equals the mass evaporated by the equilibrated PLS. In this picture most of the fragments are dynamically produced; this is in agreement with Quantum Molecular Dynamics simulations which depict early fragment formation from interparticle correlations built during the collision and well reproduce fragment properties in Xe+Sn reactions [10,11]. For the lighter system, $^{36}Ar+^{58}Ni$, the same evolution with centrality is observed, but no dependence on the energy, between 52 and 95 MeV/u. Note that for impact parameters larger than $0.5b_{max}$ the geometrical limit (the size of the overlap zone) is reached. This is also the case for Xe+Sn at 50 MeV/u [9]. These data evidence the influence of the geometry of the collision on the dynamical emission.

The "dynamical mass loss" so determined neglects any possibility of asymmetric statistical emission, due for instance to deformation of the emitter, or to the close presence of the TLS. Conversely the method used discards any dynamical emission forward of the PLS [12]. "Dynamical emission" includes charged products emitted at different times and through different mechanisms, from direct processes to neck rupture [13], oriented fission and even some statistical fission of the emitter. In the following sections, I will highlight some of these processes for which detailed data were obtained.

3 Formation of light cluster: Coalescence of direct nucleons

The origin of high energy light clusters in intermediate energy heavy ion collisions is not as straightforward to explain as that of nucleons. It was proposed that they come from the fast ejection of clusters preformed inside the nucleus [14], or from projectile break-up [15]; another long invoked process is the coalescence of direct nucleons [16]. Recently Danielewicz et al. have included coalescence in BUU calculations, formally reversing the particle breakup processes [17]. Calculations were carried out for the N+Sm system at 35 MeV/u, and the resulting high energy part of p,d,t spectra reasonably accounts for the experimental data [18]. Another new approach was tried in the INDRA

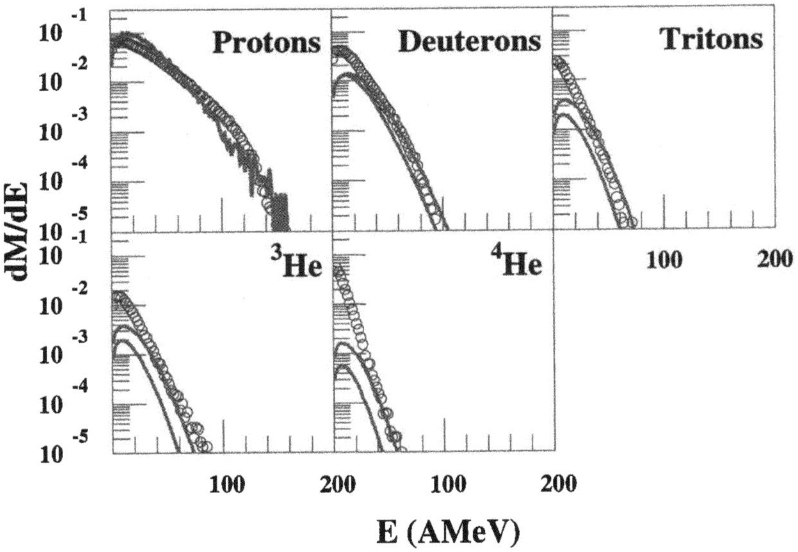

Figure 3. $^{36}Ar+^{58}Ni$ at 95 MeV/u: Comparison of the experimental (points) and calculated (lines) spectra of lcp emitted between 60 and 120°c.m. The double line for the calculation reflects uncertainties on the coalescence radius p_0. The experimental impact parameter bin is 5-6 fm, and there is no normalisation factor. Extracted from [20].

collaboration [20]. Direct nucleon densities in the momentum space given by an intranuclear cascade code (ISABEL [19]) were used as input of the coalescence model, instead of the measured nucleon spectra as commonly done [16] The subsequent evaporation stage is not taken into account. Comparison was done with the experimental spectra of lcp measured in the range 60-120°c.m. in the reaction $^{36}Ar+^{58}Ni$ at 95 MeV/u; in this angular range, high energy particles appear more separated from the lcp evaporated by the PLS and TLS. Fig. 3 shows, for mid-peripheral collisions, that such a process well accounts for the higher energy part of the measured spectra; the fraction of the lcp multiplicity attributed to this process decreases when the particle mass increase (inherent to coalescence), and with decreasing centrality, due to smaller direct nucleon multiplicities.

4 Fission-like processes

4.1 Oriented fission of the projectile-like source

It was predicted by several dynamical simulations that neck formation and rupture largely contribute to dynamical emission. Clusters so produced cover a large range of mass. Among these scenarios, it may happen that the neck sticks for a while to the PLS (TLS), creating a "nucleus" deformed beyond the fission saddle point, which will then be irremediably driven to fission. This phenomenon was found some years ago in the Mo+Mo reactions at 19 MeV/u [21], and more recently studied for various systems [22].

Two major differences with statistical fission emerge from the experimental observations: i) the break-up is dominantly very asymmetric, and the fission axis is preferentially "oriented" along the direction of separation of the primary PLS and TLS; the heavier fission fragment has the higher velocity (see fig. 4). ii) The relative velocity between the two fragments is higher than that predicted by the systematics of statistical fission. The privileged fission direction signs the memory of the dynamical part of the reaction, while an extra deformation velocity adds up to the "standard" (Coulomb + thermal) fission velocity. ~76% of the binary break-ups are oriented fissions for Xe-like fragments formed in the reactions Xe+Sn, against 24% statistical fissions, independently of the impact parameter and of the incident energy (25-50 MeV/u). Oriented fission of the PLS is negligible for heavy projectiles impinging on light targets (C, Al) and becomes dominant for heavy targets. In mid-central Xe+Sn collisions, fission (sum of both types) occurs for more than 20% of the well measured events

Such phenomena are therefore important contributors to "dynamical"

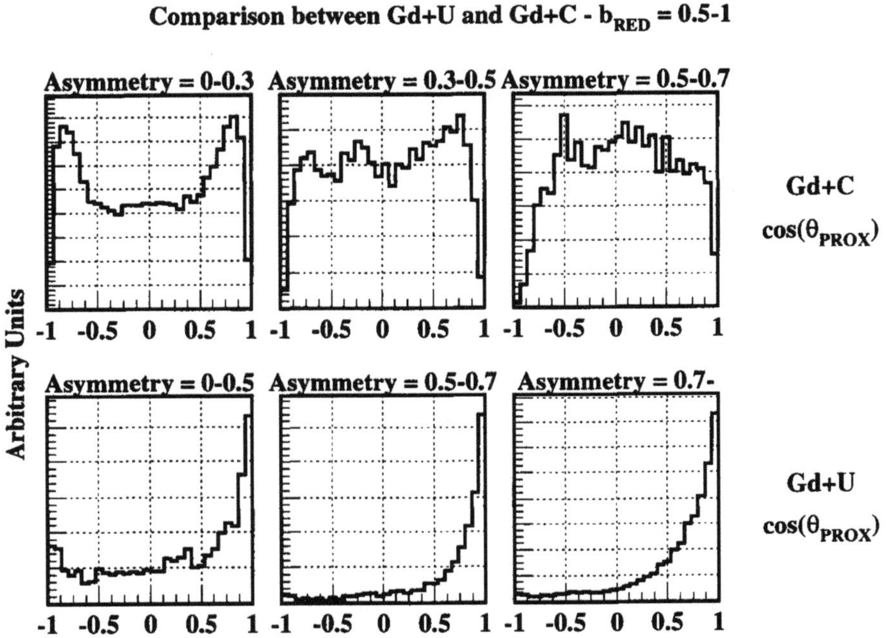

Figure 4. Angular distributions of the fission axis with respect to the PLS recoil direction, for the fission of 36 MeV/u Gd projectiles after reactions with C and U targets. Standard fission favors the reaction plane, and thus the PLS recoil direction (cos $\theta_{prox} = \pm 1$), as observed with the C target. Oriented fission dominates in the case of the U target; cos $\theta_{prox} = 1$ indicates that the heaviest fisson fragment is the fastest. Extracted from [22].

emissions. They are of primary interest in bringing information on the viscosity of nuclear matter in the intermediate energy domain.

4.2 Ternary fission

In the same kind of scenarios, one may place the ternary fission observed by the Indiana group in reactions between 90 MeV/u ^{3}He, 22 MeV/u ^{12}C and fissile (Th) targets [23]. Two types of light fragments (Z=3-12) are detected in coincident with the Th fission fragments: i) fragments evaporated by the composite system before fission; their characteristics can be reproduced in a statistical model framework. ii) fragments emitted at the scission time, focused perpendicularly to the scission axis. They are more abundantly produced in peripheral collisions, where the composite system should be more deformed and are expected to probe the nature of nuclear dissipation [23].

5 Multifragmentation

Multifragmentation, or simultaneous break-up of a well defined object has been reported first for "fused" heavy systems[24,5,6]. Models assuming that multifragmentation is initiated by dynamical volume instabilities, developing inside a system in thermal equilibrium (spinodal decomposition), allow to well account for experimental observables [6]. The final charge distributions indicate a full exploration of the phase space at some freeze-out time [5].

Multifragmentation of the PLS from 35 MeV/u Au+Au reactions was recently scrutinized over a large excitation range (1-8 MeV/u), and the presence of a peak in the energy fluctuations at a temperature of 3-5 MeV was claimed as the signal of a first order phase transition of finite nuclear matter [7].

6 Summary and open questions

Global attempts at separating dynamical and thermal emission of lcp/fragments were performed on various systems. The extreme constraints put on emitter properties to remove "equilibrated" emission should, in my opinion, maximize the dynamical component, as may be indicated by the identity between the mass of dynamical emission and that of the geometrical overlap region of the colliding nuclei. For specific processes dynamical effects are clearly visible: such lcp/fragments contribute for a sizeable part to the total mass attributed to dynamical emission. For separation between thermal emission and purely dynamical production the key factor is the time, which is essentially accessed through models (correlation experiments have however furnished some clues to relative emission times). In recent quantal mean-field approaches [25] emission time scales differ somewhat from those obtained in semi-classical simulations, where the quantal nature of the nuclei is not fully taken into account. While the detailed study of dynamical effects should give a deeper knowledge of the properties of nuclear matter, characterizing the features of sources in thermal equilibrium would open the way to the rich field of the thermodynamics of finite systems.

Acknowledgments

I hope that my friends from the INDRA collaboration will not feel too much betrayed by the lapidary comments supposed to account for all our discussions. Special thanks are due to B. Borderie, J. Colin, P. Pawlowski and E. Plagnol for permission to show data just submitted for publication.

References

1. D. Durand, *Nucl. Phys.* A **630**, 52c (1997).
2. B. Borderie et al., *Z. Phys.* A **357**, 7 (1997).
3. B. Borderie, *Ann. Phys. Fr.* **17**, 349 (1992).
4. D. Idier et al., *Ann. Phys. Fr.* **19**, 159 (1994).
5. N. Marie et al. (INDRA collaboration), *Phys. Lett.* B **391**, 15 (1997).
6. M.F. Rivet et al., (INDRA collaboration), *Phys. Lett.* B **430**, 217 (1998); J.D. Frankland, thèse, Université Paris XI Orsay, 1998, IPNO-T-98-06; B. Borderie et al., this conference.
7. M. D'Agostino et al, *Nucl. Phys.* A **650**, 329 (1999).
8. T. Lefort, D. Cussol et al. (INDRA collaboration), in *Proceedings of the XXXVI International Winter Meeting on Nuclear Physics*, Bormio, 1998, edited by I. Iori (University of Milan Press, Milan, 1998). and submitted to *Nucl. Phys.* A.
9. J. Lukasik et al. (INDRA collaboration), *Phys. Rev.* C **55**, 1906 (1997).
10. E. Plagnol et al. (INDRA collaboration) submitted to *Phys. Rev.* C.
11. O. Tirel, Thèse, Université de Caen (1998).
12. P. Eudes et al., *Phys. Rev.* C **56**, 2003 (1997).
13. C.P. Montoya et al., *Phys. Rev. Lett.* **73**, 3070 (1994).
14. Chinmay Basu and Sudip Ghosg, *Phys. Rev.* C **56**, 3248 (1997).
15. H. Fuchs and K. Möhring, *Rep. Prog. Phys.* **57**, 231 (1994).
16. T.C. Awes et al, *Phys. Rev.* C **24**, 89 (1981) and C **25**,2361 (1982).
17. P. Danielewicz et al., *Nucl. Phys.* A **553**, 712 (1991) and *Phys. Rev.* C **46**, 2002 (1992).
18. D. Prindle et al., *Phys. Rev.* C **57**, 1305 (1998).
19. Y. Yariv and Z. Fraenkel, *Phys. Rev.* C **20**, 2227 (1979) and *Phys. Rev.* C **24**, 488 (1981).
20. P. Pawlowski, B. Borderie et al. (INDRA coll.), unpublished data.
21. A. Stefanini et al., *Z. Phys.* A **351**, 167 (1995).
22. F. Bocage, J. Colin et al. (INDRA and Nautilus collaborations), in *Proceedings of the XXXVII International Winter Meeting on Nuclear Physics*, Bormio, 1999, edited by I. Iori (University of Milan Press, Milan, 1999) and submitted to *Nucl. Phys.* A.
23. S.L. Chen et al., *Phys. Rev.* C **54**, R2114 (1996); R. Yanez et al., *Phys. Rev. Lett.* **82**, 3585 (1999).
24. M. D'Agostino et al., *Phys. Lett.* B **368**, 259 (1996).
25. D. Lacroix and P. Chomaz, *Nucl. Phys.* A **636**, 85 (1998).

COMPREHENSIVE STUDY OF THE INTERACTION OF ^{12}C WITH $59 \leq A \leq 103$ NUCLEI

E. GADIOLI[1], M. CAVINATO[1], E. FABRICI[1], E. GADIOLI ERBA[1], C. BIRATTARI[1], D. LEONI[1], I. MICA[1], S. SOLIA[1], G. F. STEYN[2], S. V. FÖRTSCH[2], J. J. LAWRIE[2], F. M. NORTIER[2], T. G. STEVENS[3], S. H. CONNELL[3], J. P. F. SELLSCHOP[3], AND A. A. COWLEY[4]

[1]*Dipartimento di Fisica, Universitá di Milano and INFN, Sezione di Milano, Italia*

[2]*National Accelerator Centre, Faure, South Africa*

[3]*Schonland Research Centre for Nuclear Sciences, Wits, Johannesburg, South Africa*

[4]*Department of Physics, University of Stellenbosch, South Africa*

The results of a comprehensive study of the reactions induced by ^{12}C ions with energy up to about 45 MeV/nucleon on light and medium-mass nuclei are discussed. It is found that most of the ^{12}C energy is given to fast reaction products originating from projectile break-up and to pre-equilibrium ejectiles emitted during the thermalization of the excited composite nuclei created in the ^{12}C-nucleus interaction.

1 Purpose of this work

Experiments which provide comprehensive information on the reactions which occur when two nuclei interact and theories which consistently and comprehensively predict the values of the corresponding cross sections are of the greatest importance, not only for the basic understanding of the reaction mechanisms, but also for many applications of nuclear physics. Examples of such applied fields are the interaction of primary cosmic radiation with matter in space physics and the interaction of hadron beams with biological tissue in hadron therapy studies.

For many years integral and inclusive studies have been considered to be inferior to exclusive experiments where, by means of the choice of particular conditions (for instance by requiring particle-particle or particle-gamma coincidences), one tries to select a particular reaction channel. Without questioning the significance of such studies, one must say that they provide only partial information and it may be difficult, if not impossible, to unambiguously isolate a single reaction mechanism by the imposing of special conditions in the data taking. On the other hand, a complete set of inclusive measurements by no means excludes the possibility of clearly identifying each contributing reaction mechanism in a two-nucleus interaction, while it also provides complete information on all the reaction mechanisms which may occur. This becomes possible since the energy and the angular distributions of the reaction products originating from different reaction mechanisms may differ considerably, while their yields may also depend sensitively on the projectile energy and target mass.

We are currently carrying out such an investigation for the case of the interaction of ^{12}C ions with nuclei. The experimental data collected to date and additional data taken from the literature include a large number of excitation functions for production of radioactive residues in the interaction of ^{12}C with ^{103}Rh up to 33.5 MeV/nucleon (process A) [1–3]; radioactive residue yields in the interaction of ^{12}C with natural copper up to 45 MeV/nucleon (process B) [4–6]; recoil range distributions of residues from process A at 33.5 MeV/nucleon; average forward ranges of residues from process B; and inclusive spectra of the α particles emitted in the interaction of ^{12}C with ^{59}Co and ^{93}Nb at incident energies of 300 and 400 MeV [7,8].

2 Interpretation of the experimental data

The theoretical model which we use for analysing these data [1–3,8,9] is based on a set of assumptions which generalize the concepts at the basis of the theory of precompound emission in light-particle-induced reactions [10]. The first of these is the *random phase* approximation which leads to the assumption that the cross sections of the processes which contribute to a given reaction (the different *reaction paths*) add incoherently because their reaction amplitudes have a random phase. The second is the assumption that the mean-field interaction between the two nuclei, while they approach each other and come into contact, determines the reaction mechanisms (e.g., complete or incomplete fusion or any other process) and the initial configurations of the excited intermediate nuclei which result from this interaction. In the case of ^{12}C-induced reactions the main contributing reaction mechanisms are assumed to be, on the basis of a large number of previous experiments (see the works quoted in [2]), the complete fusion of the two nuclei, the incomplete fusion of a *participant* α-type fragment of ^{12}C (both a single α particle and two α particles weakly bound into a ^{8}Be) and the transfer of a nucleon from ^{12}C to the target nucleus. The cross sections of the complete and incomplete fusion processes are evaluated within the entrance channel *critical angular momentum* model: the two nuclei may fuse or a fragment of the projectile may fuse with the target only if, at some distance, the total potential that the fusing nucleus or the fusing fragment feels is attractive [11]. This occurs only if the sum of the repulsive centrifugal and Coulomb potentials and the attractive nuclear potential has a *pocket* of at least a few MeV deep. This leads to the assumption that

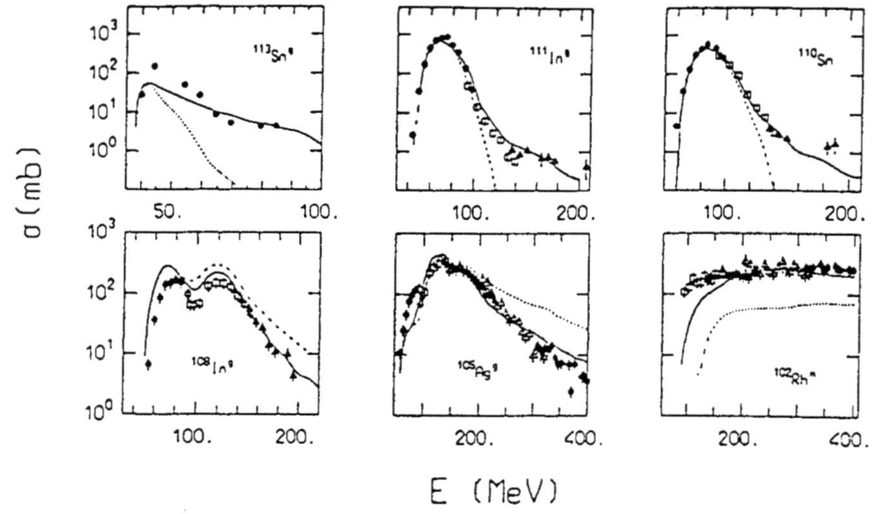

Figure 1. In the upper part of the figure are shown the excitation functions of complete fusion reactions which, in the interaction of ^{12}C with ^{103}Rh, lead to the production of $^{113}Sn^g$, $^{111}In^g$, ^{110}Sn. The full lines are the predictions of the Boltzmann Master Equation theory, the dashed line the predictions of a purely evaporative calculation. In the lower part of the figure are shown the excitation functions for formation of $^{108}In^g$ (mainly produced in a ^8Be incomplete fusion), $^{105}Ag^g$ (mainly produced in an α particle incomplete fusion) and $^{102}Rh^m$ (a target-like residue). The full lines are the results of calculations considering the re-emission of participant α fragments in the incomplete fusion reactions, while the dashed lines show the results obtained by ignoring such processes [3].

complete fusion may occur only if the two-nucleus relative angular momentum is less than a critical value L_f. Incomplete fusion and nucleon transfer reactions occur in windows of angular momentum values which increase with decreasing mass of the fusing fragment.

In complete fusion processes one produces a composite nucleus with fixed energy and *full transfer* linear momentum. In the case of incomplete fusion the excited nuclei which are created have a distribution of excitation energies and linear momenta which are evaluated by assuming that the break-up transition matrix is given by the Fourier transform of the wave function which describes the relative motion of an α particle and a ^8Be fragment in ^{12}C [7,8]. Once a composite excited nucleus is created, its subsequent *thermalization* and the emission of fast particles and clusters before reaching statistical equilibrium are assumed to result from a cascade of two-body nucleon interactions, which

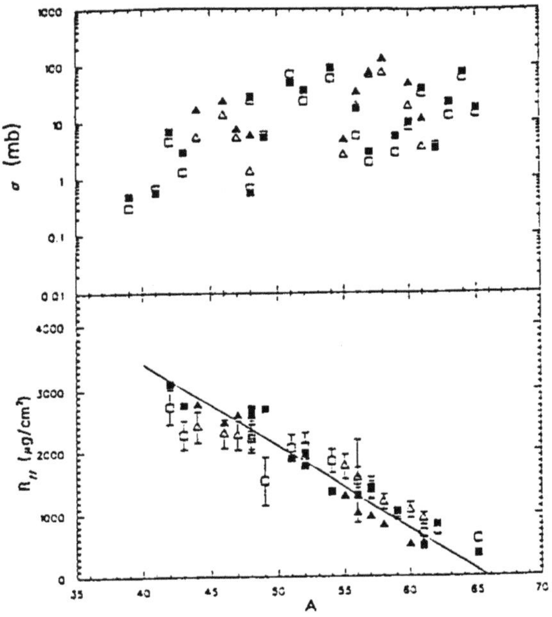

Figure 2. In the upper part of the figure are shown, as function of their mass, the cross sections for production of radioactive residues in the interaction of 45 MeV/nucleon ^{12}C ions with natural copper. The open and black symbols give, respectively, the experimental values [5] and the theoretical predictions. In the lower part of the figure are shown the average forward ranges of the residues. The symbols have the same meaning as before and the straight line is a χ^2 fit to the theoretical predictions.

is described by means of the Boltzmann Master Equation theory [12–16]. The equilibrated long-lived nuclei produced at the end of the thermalization further evaporate low energy particles as predicted by the statistical model [10].

The success of the Boltzmann Master Equation theory in accounting for the emission of pre-equilibrium particles may be appreciated, in the case of the interaction of ^{12}C with ^{103}Rh, by considering the cross sections for production of three residues ($^{113}Sn^g$, $^{111}In^g$ and ^{110}Sn) which can be produced only *via* a complete fusion reaction [1]. The results of these calculations are given by the full lines in the upper part of Fig. 1, which reproduce satisfactorily the excitation functions for production of these nuclei. The dashed lines show that the tails of the excitation functions cannot be reproduced if one does not account for the emission of pre-equilibrium particles. How-

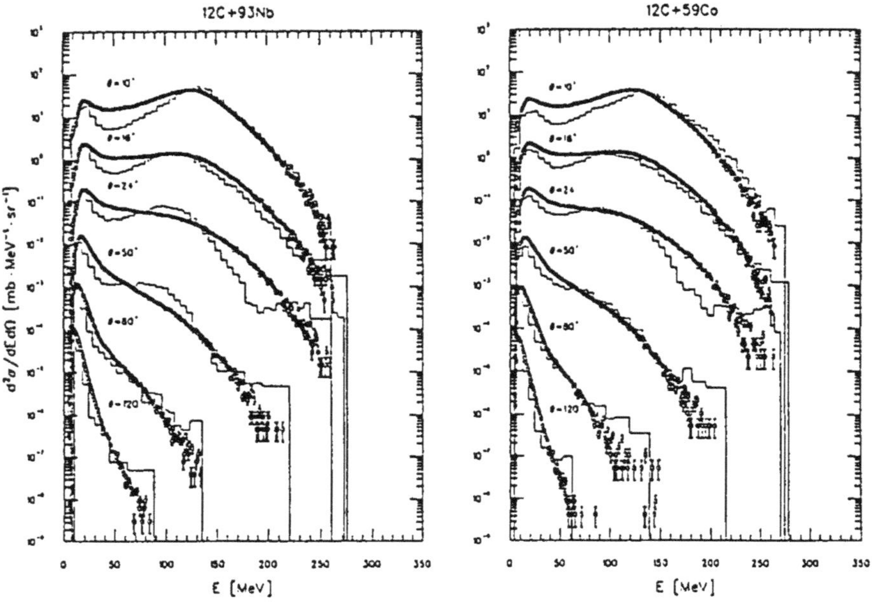

Figure 3. Spectra of α particles emitted at a few selected angles in the interaction of 400 MeV ^{12}C ions with ^{93}Nb (left part of the figure) and ^{59}Co (right part of the figure). The open squares give the experimental results, the histograms the theoretical predictions [8]. With increasing emission angle the spectra are progressively scaled down by a factor 10.

ever, by adopting incomplete fusion cross sections as predicted by the critical angular momentum model and evaluating the subsequent thermalization of the excited composite nuclei with the Boltzmann Master Equation theory, one cannot correctly predict the cross sections of incomplete fusion reactions, which are increasingly overestimated with increasing energy. This is shown in the lower part of Fig. 1, by the comparison of the results of the calculations (given by the dashed lines) with the experimental excitation functions for production of $^{108}\mathrm{In}^{g}$ and $^{105}\mathrm{Ag}^{g}$, which are mainly produced in the incomplete fusion of, respectively, a ^{8}Be and one α particle. At the same time the cross sections for production of nuclei with mass and charge almost equal to that of the target nucleus is greatly underestimated as is shown, in the case of ^{102}Rh in the same figure, by the dashed lines.

The discrepancy between the experimental and calculated cross sections for incomplete fusion reactions could be greatly reduced by adopting much smaller values for incomplete fusion cross-sections than those predicted by the critical angular momentum model. However, with this assumption complete and incomplete fusion reactions may account for only about one third of the reaction cross section, and one should assume in addition the existence of other less likely reaction mechanisms than those considered above. These supposed mechanisms should also greatly enhance the production of near target residues.

An alternative way of eliminating both discrepancies as mentioned above without reducing the incomplete fusion cross sections is revealed by certain studies of α-particle induced reactions. These studies show that an α-particle, after having been absorbed by a nucleus, has a high probability of being re-emitted with a reduced energy even after several α-nucleon interactions, reflecting the high stability of the α-particle [17–19]. In the case of α-particle induced reactions, this re-emission probability depends rather weakly on the mass of the target nucleus and, at incident energies higher than about 70 MeV, has a value of about 0.3 [18]. In the case of ^{12}C-induced reactions, the re-emission of α particles is expected to occur with a greater probability because both the incomplete fusions of ^{8}Be fragments and α particles occur in a very peripheral region of the target nucleus. Thus, one can expect that only those α-particles that, after their absorption, are directed toward the nuclear core will distribute their energy among the nucleons of the absorbing nucleus, while those which go through the low density nuclear surface region will have a rather large probability of escaping from the nucleus. As shown by the full lines in the lower part of Fig. 1, the measured excitation functions of both incomplete fusion reactions and reactions producing near-target residues can be satisfactorily reproduced by considering α-particle re-emission [2,3].

Incomplete fusion followed by prompt α-particle re-emission provides very little excitation energy and very low forward linear momentum to the intermediate excited nuclei, thus favouring the production of residues with mass and charge almost equal to those of the target nucleus (target-like residues), which therefore also have very low forward linear momentum. The production of these low momentum target-like residues has been observed in the case of the interaction of ^{12}C with ^{103}Rh [3]. In Fig. 2 it is shown that the production of low momentum target-like residues is also observed in the case of the interaction of ^{12}C with copper at an incident energy of 45 MeV/nucleon [5]. Both the yields of the residues observed in this reaction and their average forward momenta are satisfactorily reproduced by taking the re-emission of α particles into account. Obviously if this scenario is correct, one would expect to find a large yield of α-particle ejectiles (i.e. *spectator* α particles which are produced in the break-up of ^{12}C, *participant* α particles which undergo

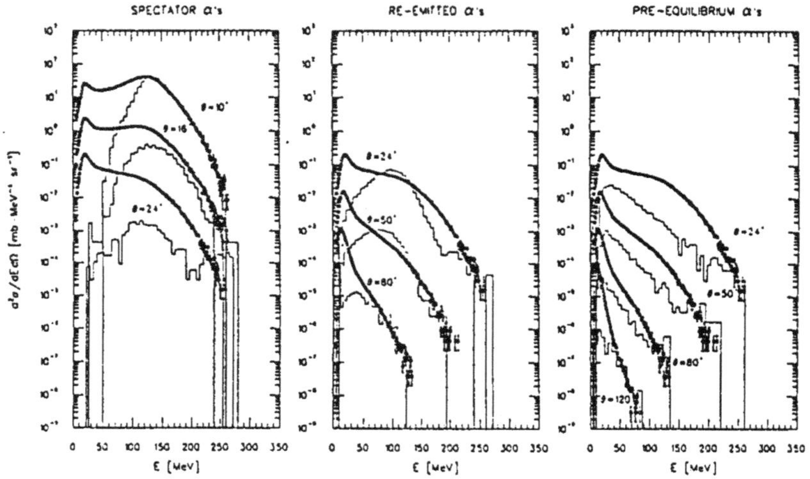

Figure 4. Contributions of the spectator α particles, the α particles re-emitted after an incomplete fusion and the pre-equilibrium α particles (histograms) to the measured double differential spectra (squares) of the α particles emitted in the interaction of 400 MeV ^{12}C ions with ^{93}Nb at selected scattering angles [8]. As in Fig. 3, with increasing emission angle the spectra are progressively scaled down by a factor 10.

an incomplete fusion but are subsequently re-emitted with a slightly reduced energy, *pre-equilibrium* and *evaporated* α particles). The measurements of inclusive α-particle spectra indeed show this conspicuous α-particle emission.

Fig. 3 shows that the inclusive spectra of the α particles emitted in the interaction of 400 MeV ^{12}C ions with ^{93}Nb and ^{59}Co are reasonably well reproduced by our theory. The non-evaporative contributions to the α-particle spectra are shown separately in Fig. 4 for the case of the interaction of ^{12}C with ^{93}Nb. The yields of the α particles produced in the different processes dominate in separate angular intervals, showing that a complete inclusive measurement may single out each contributing mechanism to a given reaction while at the same time also providing comprehensive information on all these mechanisms.

3 Conclusions

The analysis of a large set of data concerning the interaction of ^{12}C with $59 \leq A \leq 103$ nuclei over a large interval of incident energies leads to the conclusion that, at incident energies of a few hundred MeV, most of the kinetic energy of the incident ^{12}C is given to fragments produced in its break-up and to pre-equilibrium particles emitted by the excited intermediate systems which are produced in the projectile-target-nucleus interaction. These emissions considerably reduce the excitation energy of the equilibrated nuclei which are eventually produced at the end of the fast stage of the reaction. These nuclei further evaporate only a relatively small number of particles and, in a large number of cases, produce residues with mass and charge close to those of the target nucleus. A significant fraction of the high-energy ejectiles is constituted by α particles, produced in the ^{12}C break-up, which fused with the target nucleus (both as single units and as pairs weakly bound into a ^{8}Be) and were subsequently re-emitted with most of their energy after only a few two-body interactions with target nucleons. These processes, which are found to occur with large probability in the interaction of α particles with nuclei [17–19], are of even greater importance in ^{12}C-induced reactions because most of the interactions of the α-type fragments (produced in ^{12}C break-up) with target nucleons occur in the low density surface region of the target nucleus.

References

1. C. Birattari *et al*, *Phys. Rev.* C **54**, 3051 (1996).
2. E. Gadioli *et al*, *Phys. Lett.* B **394**, 29 (1997).
3. E. Gadioli *et al*, *Nucl. Phys.* A **641**, 271 (1998).
4. S.Y. Cho *et al*, *Phys. Rev.* C **36**, 2349 (1987).
5. S.Y. Cho, N.T. Porile and D.J. Morissey, *Phys. Rev.* C **39**, 2227 (1989).
6. J.P. Whitfield and N.T. Porile, *Phys. Rev.* C **47**, 1636 (1993).
7. E. Gadioli *et al*, in *The Nucleus: New Physics for the New Millennium*, Strand, South Africa, January 1999 (Plenum Press) in press.
8. E. Gadioli *et al*, *Nucl. Phys.* A in course of publication.
9. E. Gadioli *et al*, *Heavy Ion Phys.* **7**, 275 (1998).
10. E. Gadioli and P.E. Hodgson, *Pre-equilibrium Nuclear Reactions*, (Clarendon Press, Oxford, 1992).
11. K. Siwek-Wilczynska *et al*, *Nucl. Phys.* A **330**, 150 (1979).
12. E. Fabrici *et al*, *Phys. Rev.* C **40**, 2548 (1989).
13. E. Fabrici *et al*, *Z. Phys.* A **338**, 17 (1991).
14. M. Cavinato, *et al.*, *Z. Phys.* A **347**, 237 (1994).
15. C. Brusati *et al.*, *Z. Phys.* A **353**, 57 (1995).
16. M. Cavinato *et al.*, *Nucl. Phys.* A **643**, 15 (1998).
17. E. Gadioli and E. Gadioli Erba, *Z. Phys.* A **299**, 1 (1981).
18. E. Gadioli, E. Gadioli Erba and M. Luinetti, *Z. Phys.* A **321**, 107 (1985).
19. M. Chadwick *et al.*, *Heavy Ion Phys.* **2**, 347 (1995).

CLUSTER FORMATION DYNAMICS IN MEDIUM ENERGY NUCLEAR COLLISIONS

AKIRA ONO
Department of Physics, Tohoku University, Sendai 980-8578, Japan
E-mail: ono@nucl.phys.tohoku.ac.jp

Antisymmetrized molecular dynamics model has been developed and applied to the studies of cluster formation phenomena in medium energy heavy ion collisions, including the multifragmentation where many clusters are born in an expanding nuclear matter. Together with the introduction of the ideas of the model, the roles of the mean field, the two-body collisions, the wave packet diffusion, and the coalescence are discussed in various systems and energies.

1 Introduction

Various cluster formation mechanisms are possible in medium energy heavy-ion collisions. In many cases, a lot of new clusters are formed when excited nuclear matter cools down as it expands. Although multifragmentation is often regarded as a manifestation of the liquid-gas phase transition, microscopic dynamical understandings are necessary because of the nonequilibrium feature of the reactions. Furthermore, multifragmentation is a good touchstone for microscopic models because it includes a nontrivial mechanism, namely, the appearance of new cluster correlations with dynamical symmetry breaking from an almost uniform excited matter.

We have developed the antisymmetrized molecular dynamics (AMD) [1,2,3] as a time-dependent microscopic model for nuclear collisions. Contrary to the mean field models, AMD respects the fact that there are a huge number of reaction channels in the intermediate states and the final state. Each channel wave function is described by a Slater determinant of Gaussian wave packets, which is an extension of the Brink model when each nucleon is regarded as a cluster. The time evolution of each channel is determined by a deterministic equation of motion for the wave packet centroids. The effects which are not included in the AMD wave function are treated as successive stochastic quantum branching processes from an initial channel to many final channels. As the quantum branching processes, we consider the two-nucleon collision, the wave packet diffusion in the mean field, and the coalescence if necessary.

In this talk, the basic idea of each microscopic process in AMD is shown in parallel with the typical examples of cluster formation reactions where the process plays an essential role. Consequently, the variety of cluster formation mechanisms will be understood in relation to their microscopic backgrounds.

2 Equation of Motion and Two-Nucleon Collisions

We describe each channel by an AMD wave function [1],

$$\Phi(Z) = \det_{ij}\left[\exp\left\{-\nu\left(\mathbf{r}_j - \frac{\mathbf{Z}_i}{\sqrt{\nu}}\right)^2\right\}\chi_{\alpha_i}(j)\right], \tag{1}$$

where the complex variables $Z \equiv \{\mathbf{Z}_i\}_{i=1,\ldots,A}$ represent the centroids of the wave packets. The wave packet width is time-independent $\nu = 0.16\,\mathrm{fm}^{-2}$. The spin isospin states χ_{α_i} are also fixed.

The time evolution of Z is determined by the time dependent variational principle, from which one can derive the equation of motion for Z,

$$\frac{d}{dt}\mathbf{Z}_i = \{\mathbf{Z}_i, \mathscr{H}\} \tag{2}$$

where $\mathscr{H}$ is the expectation value of the Hamiltonian H after the subtraction of the spurious center-of-mass kinetic energies of fragments [1]. The quantum Hamiltonian H includes an effective two-body interaction such as the Gogny force. The expression for the Poisson brackets $\{\mathscr{F}, \mathscr{G}\}$ is rather complicated because Z are not canonical variables.

It is sometimes convenient to approximate the one-body Wigner function as a simple sum of Gaussian packets like in QMD [4]. The centroids of the packets are the physical coordinates $W = \{\mathbf{W}_i\}$ [1] which are defined as nonlinear functions of Z.

The equation of motion (2) describes the motion of the wave packet centroids in the mean field. However, in medium and high energy collisions, there should be the effect of the residual interaction which changes a Slater determinant to a superposition of many Slater determinants. This effect is introduced as the stochastic two-nucleon collision process. In AMD, two-nucleon collisions are introduced stochastically in the same way as in QMD [4], but by the use of the physical coordinates W. The Pauli blocking is automatically introduced because of the existence of the Pauli-forbidden region in the physical coordinate space [1].

AMD with two-nucleon collisions can describe the fragmentation reaction where the fragments are produced by breaking the correlations of the initial nuclei. In order to see the roles played by the mean field and the two-nucleon collisions, we show our investigations [1,5] of ^{12}C fragmentation in the energy region of $E/A \sim 30$ MeV. As shown in Fig. 1, the isotope distribution of the produced fragments is reproduced well by AMD. Especially the large production cross section of α particles is very well reproduced. The solid line of the lower part of Fig. 2 shows the distribution of the excitation energy of ^{12}C projectile after the interaction with the target [5]. The high excitation energy of $E^* \gtrsim 30$ MeV is possible due to the two nucleon collisions. On the other hand, The sharp peak at $E^* \sim 10$ MeV is essentially due to the mean field effect from the target. The α clustering states of ^{12}C in this energy region are strongly excited, and they contribute to the large yield of α particles (as shown by

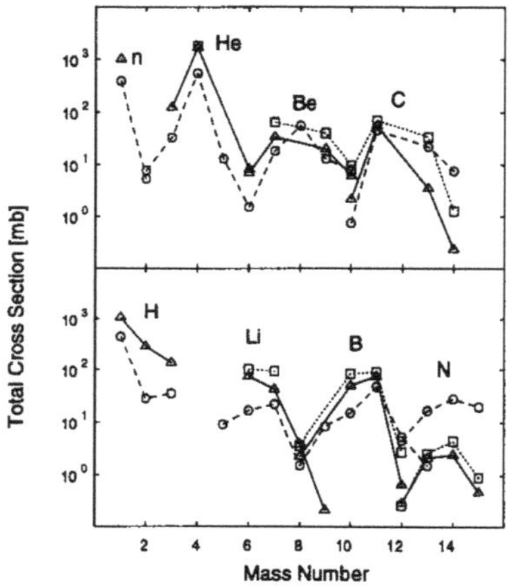

Figure 1. Isotope distribution in $^{12}C + ^{12}C$ reaction at 28.7 MeV/nucleon. Circles are the yields before the statistical decay, while triangles are those after the statistical decay. Squares are the data [6]. Lines connect isotopes.

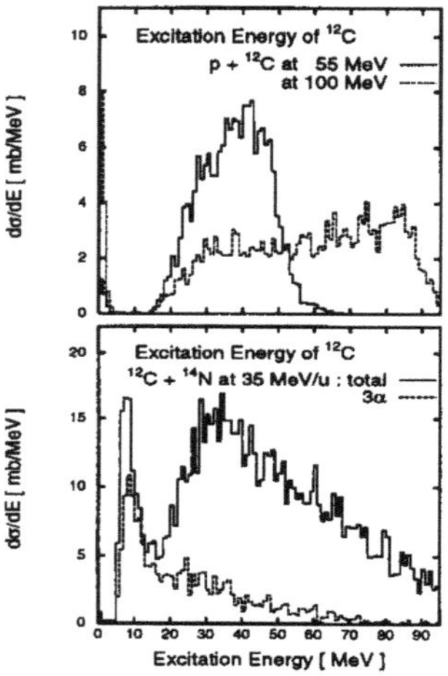

Figure 2. The distribution of excitation energy of ^{12}C projectile before its breakup. Upper figure is for $p + ^{12}C$ collisions, while the lower figure is for $^{12}C + ^{14}N$ collision at 35 MeV/nucleon.

dashed line). On the other hand, in the proton induced reactions, the excitation of ^{12}C is mainly of single particle nature as shown in the upper part of Fig. 2.

3 Wave Packet Diffusion in the Mean Field

The wave packet shape is not allowed to change in AMD, which is a disadvantage when compared with TDHF. However, we should not extend AMD to TDHF because TDHF has the problem of the spurious channel coupling. In Refs. [2,3], we introduced the precise single particle dynamics into AMD as a new stochastic branching process. The improved model is called AMD-V since it utilizes the Vlasov equation.

Each nucleon k in an AMD wave function $\Phi(Z(t))$ is represented approximately by $f_k(x,t) = 8\exp\left[-2\sum_{a=1}^{6}(x_a - X_{ka}(t))^2\right]$ in phase space, where we have introduced the phase space coordinate $\{x_a\}_{a=1,\dots,6} = \{\sqrt{\nu}\mathbf{r},\ \mathbf{p}/2\hbar\sqrt{\nu}\}$. The centroid $\{X_{ka}\}$ stands for $\mathbf{W}_k$. In the usual AMD, only X_{ka} can move following the equation of motion, while the shape of the wave packet is fixed. However, more reliable time evolution of f_k is given by the Vlasov equation,

$$\frac{\partial f_k}{\partial t} + \frac{\partial h}{\partial \mathbf{p}} \cdot \frac{\partial f_k}{\partial \mathbf{r}} - \frac{\partial h}{\partial \mathbf{r}} \cdot \frac{\partial f_k}{\partial \mathbf{p}} = 0. \tag{3}$$

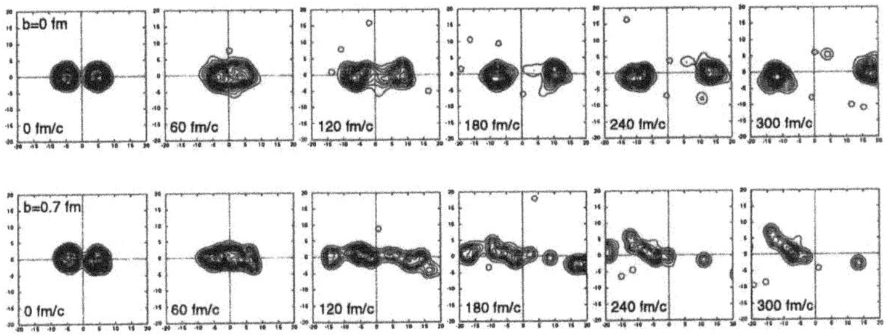

Figure 3. Time evolution of the density projected onto the reaction plane for central $^{40}Ca + ^{40}Ca$ collisions at 35 MeV/u. The upper part is the result of the usual AMD, and the lower part is the result of AMD-V.

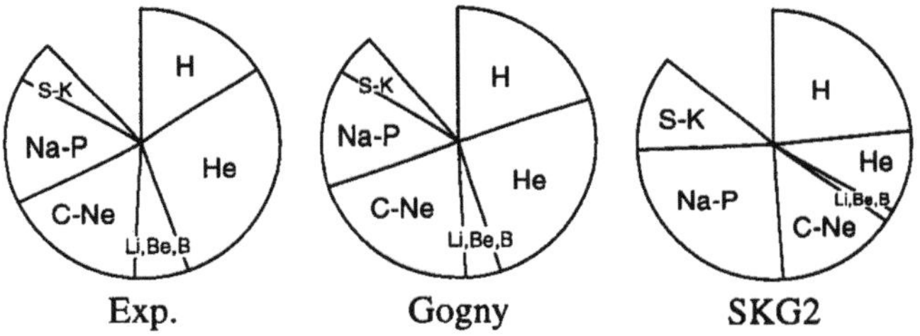

Figure 4. The partition of the total charges of the system into final products for $^{40}Ca + ^{40}Ca$ at 35 MeV/u. The calculated results of AMD-V with Gogny force and SKG2 force are compared with the data [7].

Based on this equation, we can evaluate $\dot{\sigma}^2_{kab}(t) \equiv \frac{d}{dt}\sigma^2_{kab}(t)$, where σ^2_{kab} are the second moments of f_k.

Instead of changing the shape of f_k, we now give fluctuation $\Xi_{ka}(t)$ to the centroid $X_{ka}(t)$ in order to introduce the wave packet diffusion effect $\dot{\sigma}^2_k$. If we assume the white noise for the fluctuation, it should satisfy

$$\overline{\Xi_{ka}(t)} = 0, \qquad \overline{\Xi_{ka}(t)\Xi_{kb}(t')} = [\dot{\sigma}^2_k]_{ab}(t)\delta(t-t'). \tag{4}$$

Let us introduce the expectation value $\mathscr{O}(Z,t)$ of a one-body operator [3] that generate the fluctuation $\Xi_{ka}(t)$. Then the equation of motion of AMD-V is

$$\frac{d}{dt}\mathbf{Z}_i = \{\mathbf{Z}_i, \mathscr{H}\} + \{\mathbf{Z}_i, \mathscr{O}\} + \mu\Big(\mathbf{Z}_i, \mathscr{H} + \sum_m \beta_m \mathscr{Q}_m\Big). \tag{5}$$

The third term is the cooling (or heating) term to ensure the energy conservation by canceling the energy violation by $\{Z, \mathscr{O}\}$. Conserved quantities and several one-body quantities $\{\mathscr{Q}_m\}$ are kept constant by the cooling term.

The effect of the stochastic incorporation of the wave packet diffusion can be seen dramatically in the cluster production in $^{40}Ca + ^{40}Ca$ at 35 MeV/u. As shown in

Figure 5. Time evolution of the density projected onto the reaction plane in a ^{197}Au + ^{197}Au central collision at 150 MeV/u. This is the result of AMD-V without coalescence.

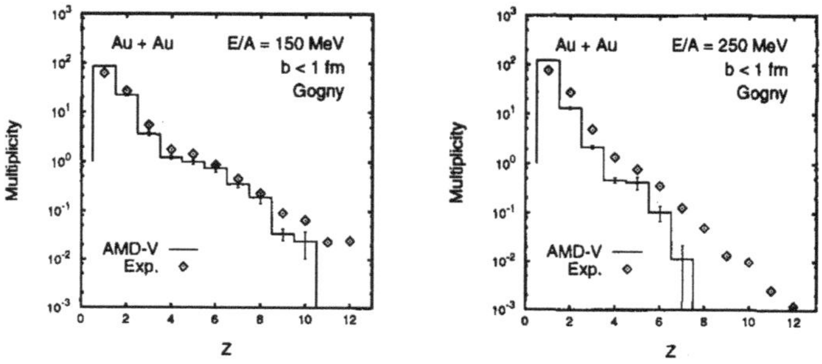

Figure 6. Charge distribution of the fragments in central ^{197}Au + ^{197}Au collisions at 150 MeV/u (left) and 250 MeV/u (right). The results of AMD-V without coalescence are compared with the data [8].

the lower part of Fig. 3, a neck region is often produced connecting the projectile and the target, and several clusters are produced from there in AMD-V calculation. The neck is formed by mixing the nucleons from both nuclei, which is enabled by the flexible single particle dynamics in AMD-V. On the other hand, in the usual AMD without the wave packet diffusion process, the neck is not created as shown in the upper part of Fig. 3. The decay of the nuclei is quite slow even though the excitation energy of the nuclei is as large as 5 MeV/u, while the created clusters are already cool in AMD-V. This is related to the improvement of the statistical properties in AMD-V such as the nucleon evaporation.

The AMD-V calculation reproduces the data quite well as in Fig. 4. The experimental data [7] shows that only 20 % of protons are emitted as isolated protons, and as many as 30 % of protons are contained in α particles. These features are quite well reproduced by AMD-V with the Gogny force, which corresponds to a soft EOS ($K = 228$ MeV) and the momentum-dependent mean field. On the other hand, another effective interaction (SKG2 force) which corresponds to a stiff EOS ($K = 373$ MeV) and the momentum-independent mean field, is inconsistent to the data.

In heavier systems such as ^{197}Au + ^{197}Au central collisions, the compression and the strong radial expansion are observed in calculation. At $E \gtrsim 100$ MeV/u, the thermal excitation is so high that the initial cluster correlations are completely lost. As the matter expands and cools down, new clusters are copiously formed, which is distinguished from the simple breakup of initial nuclei.

AMD-V reproduces the data [8] rather well for 150 MeV/u as shown in Fig. 6, while the deviation from the data increases for higher energy. AMD-V tends to fail when the dominant final products are light particles rather than the nuclei which can be described by the mean field, as discussed in the next section.

It will be interesting to discuss here the difference between AMD-V and the mean field models. In the mean field dynamics, as the compressed system expands, not only each single particle packet moves outwards but also it spreads out. If one continues the mean field calculation, the broadened smooth mean field is too weak to produce fragments, as demonstrated in Ref. [9] by a molecular dynamics with variable widths. On the other hand, in AMD-V, this expanding state is regarded as a superposition of many AMD Slater determinants of compact wave packets. In this case, the mean field is different from channel to channel, and each channel naturally evolves to a different multifragment final channel.

4 Coalescence to Form Light Clusters

As mentioned in the previous section, an important thing for the description of the new cluster formation is to avoid spurious channel correlations. It was done by restricting the channel wave functions to AMD states. However, it may be still possible that several channels are included in a single AMD state which is a superposition of energy eigenstates. Although it will not cause problems in the case of a compound nucleus where the levels have similar characters, there should be serious problems in the case of a light nucleus which has few bound states. If continuum unbound components are included in an AMD state, the bound state probability will decrease as time progresses due to the effect of the unbound components, which is an unphysical situation.

In order to remove the spurious channel couplings, we introduce a quantum stochastic branching process again. For each subsystem with a proton and a neutron, this coalescence process moves the centroids $\mathbf{Z}_a$ and $\mathbf{Z}_b$ to the same point $\frac{1}{2}(\mathbf{Z}_a+\mathbf{Z}_b)$. If the coalescence takes place, the deuteron probability will increase from $P_d = |\langle d|\mathbf{Z}_a-\mathbf{Z}_b\rangle|^2$ to 1. The probability or the rate of the coalescence α should be chosen so as to keep the probability to find the system in $|d\rangle$,

$$\frac{d}{dt}P_d + (1-P_d)\alpha = 0. \tag{6}$$

This modification for the coalescence is necessary not only for a pair of a proton and a neutron but also for pairs of light clusters when they form a single loosely bound intrinsic state. The considered coalescence processes are $p+n \to d$, $p+n+n \to t$, $p+p+n \to {}^3\text{He}$, $\alpha+d \to {}^6\text{Li}$, $\alpha+t \to {}^7\text{Li}$, $\alpha+{}^3\text{He} \to {}^7\text{Be}$, and $\alpha+\alpha \to {}^8\text{Be}$. The coalescence rates are obtained in a similar way to the deuteron case so that the ground state probability of each subsystem is kept constant. The coalescence

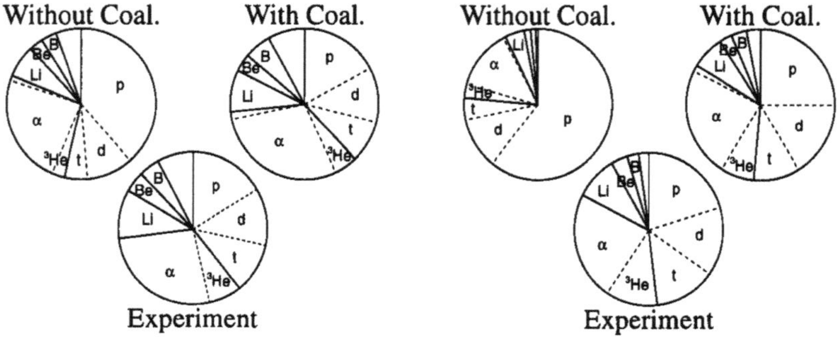

Figure 7. Partitioning of total charges into final products in ^{197}Au + ^{197}Au at 150 MeV/u (left) and 250 MeV/u (right). The results of AMD-V with and without coalescence are compared with the data [8].

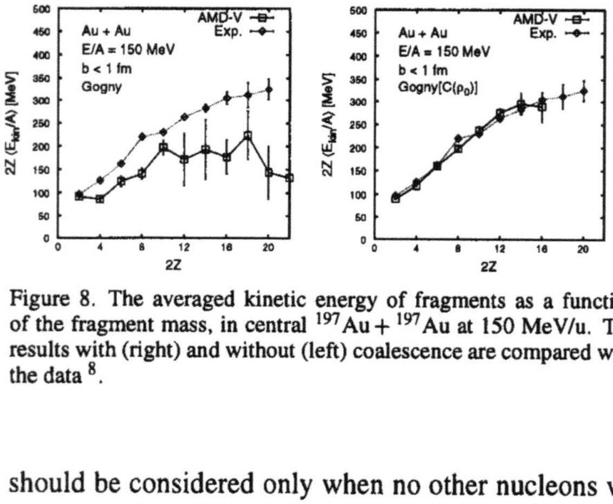

Figure 8. The averaged kinetic energy of fragments as a function of the fragment mass, in central ^{197}Au + ^{197}Au at 150 MeV/u. The results with (right) and without (left) coalescence are compared with the data [8].

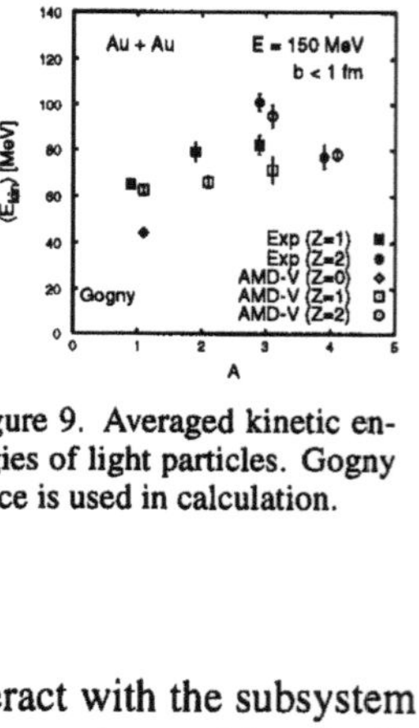

Figure 9. Averaged kinetic energies of light particles. Gogny force is used in calculation.

should be considered only when no other nucleons will interact with the subsystem any more. In order to take account of this point, the coalescence is considered only for the subsystems in the low phase space density region of $f < 0.7$.

As can be expected, the coalescence process has a drastic effect for the reactions when the dominant final products are light particles rather than heavier nuclei. As shown in Fig. 7, in central ^{197}Au + ^{197}Au collisions, the coalescence process reduces the proton yield drastically and increases the yields of light clusters, so that the calculated results are now consistent to the data. The coalescence process also has an effect on the kinetic energies of clusters. Figure 8 shows the kinetic energy of clusters as a function of the cluster mass. The almost linear relation suggests the strong collective expansion. In the calculation without the coalescence process, the heavier clusters are only produced in the inner region of the expanding system with low velocities, so that their kinetic energies are underestimated. On the other hand, with the coalescence process, the heavier clusters can be produced also with relatively large velocities from the outer region, which reproduces the data.

In Fig. 9, the calculated averaged kinetic energies of light particles are compared with the data. The difference between ^{3}He and t is well reproduced by the Gogny force which corresponds to a very soft EOS for neutron rich matter. On the other hand, with another force which has a density-independent symmetry coefficient, the difference is underestimated. Thus the difference may carry an information of the EOS of asymmetric nuclear matter.

5 Summary

We have shown various fragment or cluster formation mechanisms in heavy ion reactions in parallel with the basic ideas of AMD. A simple fragmentation takes place with a light projectile. Fragments are produced by breaking the initial nucleus, where the mean field (in the equation motion) sometimes plays an interesting role as well as the stochastic two-nucleon collisions. Next, we have shown the cluster production from the neck region, and the copious new cluster formation from an expanding system. Such reactions can be explained by introducing the wave packet diffusion effect into AMD in a stochastic manner, as far as the dominant products are not light particles. On the other hand, in higher energies where dominant products are the light particles with a single bound state, the stochastic incorporation of the coalescence is important.

Acknowledgments

Most of the presented results are based on the collaboration with H. Horiuchi. The work of Fig. 2 by H. Takemoto is also acknowledged.

References

1. A. Ono, H. Horiuchi, Toshiki Maruyama and A. Ohnishi, Phys. Rev. Lett. **68**, 2898 (1992); Prog. Theor. Phys. **87**, 1185 (1992).
2. A. Ono and H. Horiuchi, Phys. Rev. **C53**, 2958 (1996).
3. A. Ono, Phys. Rev. **C59**, 853 (1999).
4. J. Aichelin, Phys. Rep. **202**, 233 (1991).
5. H. Takemoto, H. Horiuchi and A. Ono, Phys. Rev. **C54** 266 (1996).
6. J. Czudek et al., Phys. Rev. C43 1248 (1991).
7. K. Hagel et al., Phys. Rev. **C50**, 2017 (1994).
8. W. Reisdorf et al., Nucl. Phys. **A612**, 493 (1997).
9. D. Kiderlen and P. Danielewicz, Nucl. Phys. **A620**, 346 (1997).

STATISTICAL EXPLORATION OF FRAGMENTATION PHASE SPACE IN NUCLEAR MULTIFRAGMENTATION

L. G. MORETTO, L. BEAULIEU, L. PHAIR, AND G. J. WOZNIAK
Nuclear Science Division, Lawrence Berkeley National Laboratory, Berkeley, California 94720

1 Introduction

Is nuclear multifragmentation a statistical or a dynamical process? Let us consider a fragmentation space defining the number and masses/charges of the produced fragments. The thermal population of each cell can be expressed in terms of suitable Boltzmann factors. The thermal nature of this population can be tested in a variety of ways. A very visual way is the Arrhenius plot where the log of the population is plotted vs. the reciprocal temperature $1/T$.

The question however, remains whether, apart from the Boltzmann factor, all the cells of the fragmentation space are uniformly explored. A test of such a uniform filling implies the knowledge of the size (mass/charge) of the source. A new empirical feature observed in many reactions has led us to a way to verify the uniform filling of fragmentation space and to determine simultaneously the source size.

It has been shown [1] that the intermediate mass fragment (IMF) multiplicity distribution P_n at any given transverse energy E_t is *empirically* given by a binomial distribution

$$P_n = \frac{m!}{n!(m-n)!} p^n (1-p)^{m-n}. \tag{1}$$

This implies that fragments are emitted nearly independently of each other, so that the probability P_n of observing n fragments can be written by combining a single one-fragment emission probability p according to Eq. (1). The parameter m (the total number of throws) represents the maximum possible number of fragments, which is immediately related to the source size.

The simplest statistical equilibrium model of multifragmentation has exactly the structure of Eq. (1). Let us assume that the source is made up of m fragments. The "outside" fragments have energy ϵ_2, and those "inside" have energy ϵ_1. A generic partition of n fragments outside and $m-n$ fragments inside has the probability:

$$P_n = \frac{m!}{n!(m-n)!} \frac{e^{-(n\epsilon_2 + (m-n)\epsilon_1)/T}}{(e^{-\epsilon_1/T} + e^{-\epsilon_2/T})^m} \tag{2}$$

which leads to Eq. (1) when

$$p = \frac{e^{-\epsilon_2/T}}{e^{-\epsilon_1/T} + e^{-\epsilon_2/T}}. \tag{3}$$

Thus, a simple way to obtain the size of the source is to multiply m by the fragment size.

When the definition of IMF covers a range of atomic numbers (IMF: $Z_{th} \leq Z \leq 20$, with Z_{th} equal to 3), one should multiply m by a "suitably" averaged Z. In fact, a dependence of m on the lower threshold Z_{th} has been found such that $m(Z_{th}) \times Z_{th} \approx constant$. [1] The natural next step is to restrict the fragment definition to a single atomic number Z.

A straightforward generalization of Eq. (1) to the production of fragments with charges 1, 2, ...Z_0 is given by the multinomial distribution

$$P = \frac{Z_0!}{n_1!n_2!...n_{Z_0}!} p_1^{n_1} p_2^{n_2} ... p_{Z_0}^{n_{Z_0}} \tag{4}$$

with

$$Z_0 = \sum_Z Z n_Z. \tag{5}$$

Here there is no single quantity m as in Eq. (1), since the constraint is now on the total charge rather than on the total number of fragments. However, this leads immediately to a simple scaling that must be obeyed if the fragmentation phase space is to be completely explored.

Let us first consider, as a familiar example, the fragmentation space spanned by the Euler number partitions. This case is particularly simple because all partitions are unbiased, or equally probable. This would correspond, for the different model defined in Eqs. (4) and (5), to setting p_i=constant.

Let us indicate with $W(Z_0)$ the statistical weight (number of partitions) associated with the integer or "charge" Z_0. If we now select the partitions containing at least n_Z integers of size Z, their number is given by the number of partitions of the residue

$$P_{n_Z} = W(Z_0 - n_Z Z) \tag{6}$$

by definition. Since the "charge" constraint is applied "minimally," what counts is only the product Zn_Z, rather than the individual Z values. Consequently, the weight of the residue does not change if we substitute $n_Z Z$ with $n_{Z'} Z'$ provided that $n_Z Z = n_{Z'} Z'$. In other words, $P_{n_Z, Z} = P_{n_{Z'}, Z'}$ if $n_Z Z = n_{Z'} Z'$.

This gives immediately the scaling laws

$$n_Z / n_{Z'} = Z'/Z \tag{7}$$

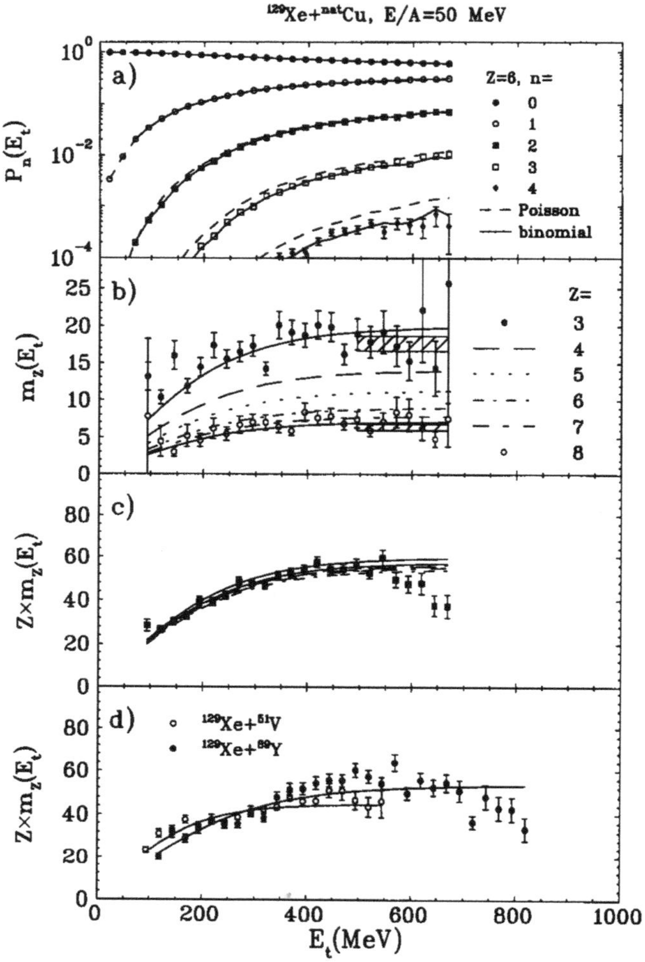

Figure 1. Panel a): the n-fold probability distribution (symbols) and Poisson (dashed) and binomial (solid) fits are plotted as a function of transverse energy for carbon fragments emitted from the reaction ^{129}Xe+^{nat}Cu at 50 AMeV. Panel b): the extracted binomial parameter m_Z (the number of "throws") is plotted as a function of transverse energy for lithium (solid circles) and oxygen (open circles) emission. The solid lines are hyperbolic tangent fits to the indicated data. The other lines are fits to data not shown. The two hatched regions represent weighted averages of the top 5% most central collisions (based on the integrated E_t spectrum) of the m_Z values for lithium and oxygen. Panel c): The fits from panel b) are scaled by the atomic number Z of the emitted particle. The square symbols represent an average source size. Panel d): The symbols represent an average source size for ^{129}Xe+^{51}V and ^{129}Xe+^{89}Y. The lines are hyperbolic tangent fits to the data.

or for the extreme value of $n_Z = m_Z$,

$$m_Z = Z_0/Z. \tag{8}$$

This result follows directly from the uniform exploration of the fragmentation phase space. It amounts to a *complete decoupling* of all the fragmentation paths. Once the system attains the charge loss of $q = nZ$, it is indifferent to how this was attained. *Any* path $q = \sum n_i Z_i$ is *equivalent* and *substitutable.* Similarly, for the remaining charge $Z_0 - q$.

This argument works when all partitions are unbiased, as for instance in the Euler number partition mentioned above. However, in the binomial/multinomial distributions of Eqs. (1) and (4), each partition is weighted by the probabilities p_i (reflecting Q-value effects) which may or may not be Boltzmann factors.

Yet, the binomial/multinomial analysis elegantly and automatically separates out the Q-value dependent probability p from the parameter m_Z which now contains direct information on the accessible fragmentation space and must follow the scaling given by Eq. (8). Thus the $1/Z$ scaling is the counterpart of the Arrhenius plot which verifies the thermal (Canonical) population of each fragmentation configuration.

2 Excitation functions for fragments with a given Z

When the restriction to *individual* Z values is made experimentally, [2] the multiplicity distributions are found to be nearly Poissonian, namely $mp << m$. This introduces interesting simplifications in the analysis and interpretation of the data, but at the cost of a loss of scale. In the Poisson limit the average multiplicity $\langle n \rangle = mp$ is the only accessible parameter, and the decomposition into m and p becomes impossible. The recovery of scale for an individual Z is highly desirable in view of the possibility that the number of throws m_Z (for a single species Z) might obey the simple scaling of Eq. (8).

Thus, we have attempted binomial fits of the multiplicity distributions for *individual* Z values in an effort to extract m_Z. Fortunately, a number of reactions (^{36}Ar+^{197}Au at 35 to 110 AMeV, [3] ^{129}Xe+^{27}Al,^{51}V,$^{\rm nat}$Cu, ^{89}Y, at 50 AMeV, [4] and ^{129}Xe+^{197}Au at 50 to 60 AMeV [5]) have been studied with good Z resolution and high statistics. We first consider the asymmetric, intermediate-energy reactions in reverse kinematics exemplified by ^{129}Xe+$^{\rm nat}$Cu at 50 AMeV, for which we can expect a single dominant fragment source.

Examples of both binomial and Poisson fits to the carbon yield from this reaction are shown in panel a) of Fig. 1. An improvement of the fit by using the binomial expression is observed for large fold numbers. A similar improvement is observed for each Z in all reactions listed in this letter.

The E_t dependence of the parameters m_Z from the binomial fits to the multiplicity distributions associated with each fragment atomic number leads to several observations. For each Z value, m_Z increases to a near constant value with increasing E_t. We approximate this behavior with the form $m_Z = m_Z^0 \tanh f E_t$.

The parameter m_Z^0 represents the saturation value of m_Z for large E_t and f controls the rise of m_Z with increasing E_t. The solid lines in panel b) of Fig. 1 are the empirical fits to m_Z values extracted for lithium and oxygen emission from the reaction ^{129}Xe+^{nat}Cu at 50 AMeV. The other discontinuous lines are fits to data not shown (Z=4-7).

3 $1/Z$ scaling of m_Z

At all E_t values there is an overall decrease of m_Z with increasing fragment Z value in agreement with the expected scaling $Zm_Z = Z_0$. This remarkable dependence is exemplified in panel c) of Fig. 1 and in Fig. 2. By applying the expected scaling (Zm_Z), all of the fits to the ^{129}Xe+^{nat}Cu data collapse together, resulting in the approximate source "size" as a function of E_t. A weighted average ($\langle Z_0 \rangle$) of the data over different exit channels is shown by the symbols in panel c) of Fig. 1. A similar behavior is observed in two additional asymmetric reactions ^{129}Xe+^{51}V and ^{89}Y (see Fig. 1d).

The E_t dependence of the source size is tantalizing. The source size increases quickly to a saturation value. The fact that E_t is related to impact parameter as well as to the total excitation energy may explain the observed features. In the highly asymmetric reverse kinematic reactions one quickly achieves sufficient overlap to produce a dominant Xe-like source as one moves from peripheral to central collisions.

As a special case of the $1/Z$ scaling, the "saturation" m_Z values from central collisions of the reverse kinematics reactions (the top 5% of the E_t scale, shown by the hatched regions in panel b) of Fig. 1), are shown in Fig. 2. The open symbols represent the scaled quantity $Zm_Z = Z_0$. The solid lines are weighted averages for the different reactions. The same data in the form m_Z/Z_0 vs. Z are shown by the solid symbols, where the $1/Z$ dependence is manifested by the good agreement of the data with the values of $1/Z$ (solid line). This striking $1/Z$ dependence of the parameter m_Z is observed for all asymmetric systems measured.

These overall results for asymmetric reactions suggest the dominance of a single source, *strongly support the hypothesis of uniform (statistical) ex-*

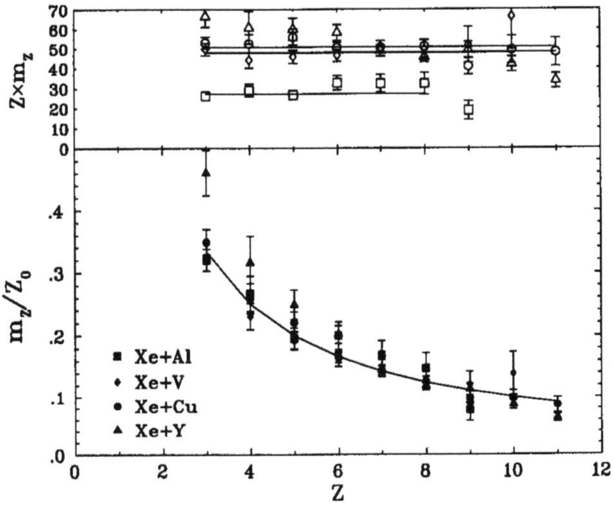

Figure 2. Top panel: The m_Z values scaled by Z as a function of Z for the indicated reactions (open symbols) at 50 AMeV extracted for the 5% most central collisions. Bottom panel: The values of m_Z divided by the value of the extracted source charge Z_0 for the indicated reactions (solid symbols). The solid line is $1/Z$.

ploration of the fragmentation phase space, and lead to the interpretation of $Z_0 = Zm_Z$ as the source "size." The $1/Z$ scaling is general and should be observed in models used to describe multifragmentation.

4 Source size extraction

In the less asymmetric reactions ^{129}Xe+^{197}Au at 50 and 60 AMeV for which at least two sources are plausible, we shall refer directly to $Z_0 = Zm_Z$ as the source size, although we shall see that now it depends on the fragment Z value as well as on E_t. The weak decrease of the source size with increasing fragment size Z, already observable in ^{129}Xe+^{89}Y (Fig. 2), becomes more visible in the case of the ^{197}Au target. At low fragment Z values, the source size Z_0 is ≈ 70 and it decreases monotonically with increasing fragment size Z to to a source size Z_0 of approximately 40-50. This fragment size dependence seems to suggest that for the ^{89}Y target and, most of all, for the ^{197}Au target there may be a distribution of sizes, the higher Z fragments being emitted preferentially by the smaller source(s).

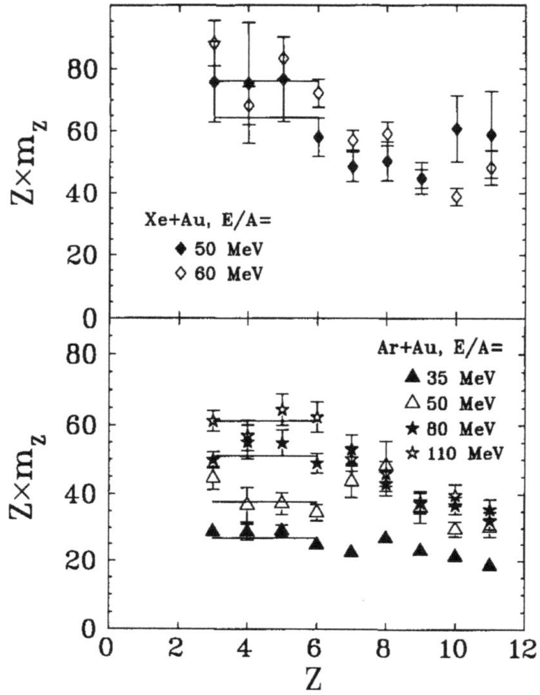

Figure 3. The scaled values $Z \times m_Z$ are plotted for the indicated reactions of the 5% most central collisions. The solid lines represent weighted averages from $Z = 3$ to 6.

The reactions ^{36}Ar+^{197}Au at 35, 50, 80, 110 AMeV give a picture intermediate between the ^{129}Xe+^{197}Au and the ^{129}Xe induced reverse kinematics reactions. They also give information of the source size dependence on bombarding energy. The source size at low fragment Z increases from $Z_0 \approx 30$ to $Z_0 \approx 60$ as the bombarding energy increases, AMeV, while at higher fragment Z the source size increases from $Z_0 \approx 20$ to $Z_0 \approx 40$.

In previous work, [6] it was empirically shown that the observed charge distributions resulting from nuclear multifragmentation obey the following invariant form:

$$P_n(Z) \propto \exp\left(-\frac{B(Z)}{\sqrt{E_t}} - cnZ\right) \tag{9}$$

where n is the total intermediate mass fragment multiplicity of the event; E_t is the total transverse energy; and $B(Z)$ is the "barrier" distribution.

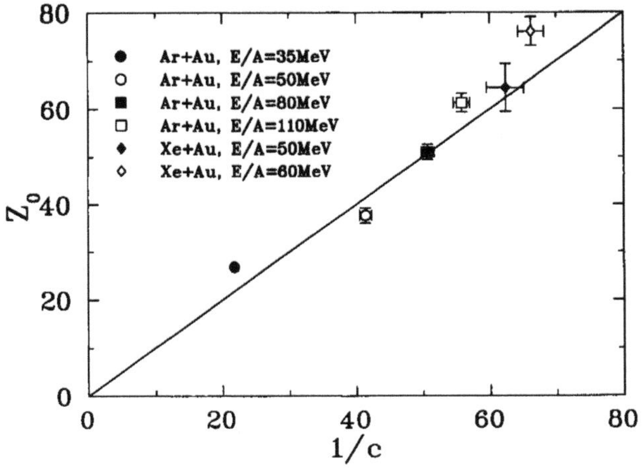

Figure 4. The source size Z_0 extracted from the binomial analysis as a function of the source size $(1/c)$ determined as per ref. [6] from central collision of the indicated reactions.

From thermodynamic considerations and percolation simulations, it was shown that c in Eq. (9) vanishes when the gas of IMFs is in equilibrium with a liquid (residue source, or percolating cluster), and assumes a value $\propto 1/Z_0$ (Z_0 being the source size) when the source is wholly vaporized. [1,6,7]

Experimentally, the parameter c undergoes an evolution with (transverse) energy from approximately zero to a positive constant. [7] Thus the source size evolves from near infinity (an "infinite" reservoir of fragments) to the actual size of the source. With the exception of the reactions $^{129}Xe+^{27}Al,^{51}V$, Cu, ^{89}Y, this limit is attained. Thus, it is possible to plot the value of Z_0 determined from m_Z against that obtained from the c parameter (both for the top 5% most central collisions in E_t).

Such a plot is shown in Fig. 4. The result is striking. Not only are the two quantities well correlated, but they also agree quite well in absolute value. This good agreement gives confidence that we have gained direct access to the source size. This source (sources) is specifically the entity that generates the fragments through "chemical equilibrium". It does not contain the pre-equilibrium part which is often incorporated in other source reconstruction methods.

In conclusion, we have shown that: a) the binomial (nearly Poissonian) multiplicity distributions for individual fragment atomic numbers permit the extraction of the parameter m_Z, the number of throws; b) m_Z for reactions where a single source is clearly dominant has the form $m_Z = Z_0/Z$; c) the $1/Z$ dependence is dramatic proof that the fragmentation phase space is statistically explored; d) source size(s) can be extracted and should reflect the region(s) where chemical (as opposed to physical) equilibrium is achieved; e) these source sizes agree with the sizes obtained from the analysis of multiplicity selected charge distributions in the E_t range where a single gas phase, or thermodynamic bivariance, prevails.

Acknowledgments

This work was supported by the Nuclear Physics Division of the US Department of Energy, under contract DE-AC03-76SF00098.

References

1. L. G. Moretto *et al.*, Phys. Rep. **287**, 249 (1997).
2. L. Beaulieu *et al.*, Phys. Rev. Lett. **81**, 770 (1998).
3. R. T. Desouza *et al.*, Phys. Lett. B **268**, 6 (1991).
4. D. R. Bowman *et al.*, Phys. Rev. C **46**, 1834 (1992).
5. K. Tso *et al.*, Phys. Lett. B **361**, 25 (1995).
6. L. Phair *et al.*, Phys. Rev. Lett. **75**, 213 (1995).
7. L. G. Moretto *et al.*, Phys. Rev. Lett. **76**, 372 (1996).

DESCRIBING FRAGMENTS WITH LV AND DYWAN MODELS

V. DE LA MOTA, S. PRIAULT AND F. SÉBILLE
SUBATECH, Université de NantesIN2P3 (CNRS)Ecole des Mines de Nantes, Nantes Cedex 03, France
E-mail:delamota@subatech.in2p3.fr

We study the disassembly of highly excited nuclear systems into small fragments in two different ways. The onset and evolution of bounded structures in phase-space in HIC at intermediate energies is explored, within a semiclassical description, the LV transport model, supplemented with a cluster recognition algorithm. Secondly, we present the basis of a new dynamical approach, the DYWAN model, which takes into account more completely the underlying microscopic properties of the colliding system and is therefore better suited for the description of cluster dynamics.

1 Introduction

Among the different theoretical approaches describing the dynamics of heavy-ions collisions at intermediate and high energies are those based on some semi-classical transport equation which describe the reaction dynamics through the evolution of the one body distribution function in phase space. They have been successful in the description of global features of these reactions but the possibility of extracting information about fragment formation has been quite an elusive point. In the first part of this work I will discuss what kind of information about fragmentation can we obtain from the semi-classical Landau-Vlasov (LV) aproach supplemented with a cluster recognition algorithm. According with the quality of the experimental data, theoretical models have to be improved in order to take into account more completely the underlying microscopic properties of nuclei which can play an important role when the system disassembles. In this spirit, we developped a new dynamical approach which combines the mathematical theory of wavelets and the projection methods of statistical physics, the so-called DYnamical WAvelets in Nuclei model (DYWAN). I will show why this model is better suited than semiclassical models to the description of cluster formation, since it is able to take into account the structure of wave functions on order to preserve as precisely as possible the relevant quantum properties of the system

2 Cluster analysis with the semiclassical LV model

The LV equation gives the time evolution of the semiclassical one-body distribution function $f(\vec{r},\vec{p})$ of a system of A interacting particles in phase space:

$$\frac{\partial f(\vec{r},\vec{p})}{\partial t} + \{f(\vec{r},\vec{p}), H(f)\} = I(f) \tag{1}$$

where { , } stands for the Poisson bracket, $H(f)$ is the one-body Hamiltonian and $I(f)$ is a Boltzmann-like collision kernel [1]. The funcion $f(\vec{r},\vec{p})$ represents at each time the mean occupation in an elementary volume $d^6\Gamma = d^3r\ d^3p$ in phase space Γ and therefore is the mean number of particles in $d^6\Gamma$ when a number M of experiments have been performed. In this sens the LV distribution function can be viewed as a microcanonical ensemble average of microscopic "A-body" configurations giving the same one-body description. Taking advantage of this "statistical" interpretation of f we can generate M configurations of A particles by choosing phase space coordinates according to the LV distribution function. Such a mapping respects: a) conservation of a given set of relevant observables with respect to the mean behaviour, and b) Pauli exclusion principle. The most general ensemble of A-body states, satisfying Eq. (1) together with conditions a) and b) contains informations about the dissipative behaviour of the system generated by the LV model. In order to obtain reliable information on observables typically measured in experiments, this process is performed for a wide range of impact parameters. Whithin the ECRA recognition procedure [2], a cluster ζ is defined as that set of particles i satisfying the condition $\forall i \in \zeta,\quad e_i = T_{iCM} + \sum_j V_{ij} < 0$, where T_{iCM} is the kinetic energy of particle i calculated in the C.M. of the cluster ζ and V_{ij} is the interaction energy between particles i and j, both belonging to the same cluster ζ. This cluster definition is free from arbitrary "clusterization parameters" but is also a very complicated task, because the problem is highly self consistent. In order to compare with experimental results, a simplified version of an experimental filter has been applied in order to simulate the detector acceptances (angular and energy detection thresholds). In Fig. 1 we show the evolution of the charged product multiplicity for complete events with the incident energy in the case of the ^{36}Ar + ^{58}Ni reaction at 32, 52 and 95 MeV/nucleon performed at GANIL. In the upper panel the experimental data [3] are shown, on the lower one we show the results of this calculation. In spite of the low statistics in the calculation, the main characteristics of the bell-shaped distributions are reproduced. The theoretical histogram is too wide at 95 MeV/nucleon, with an important contribution at high multiplicities. This is coherent with the fact that multiparticle correlations are absent

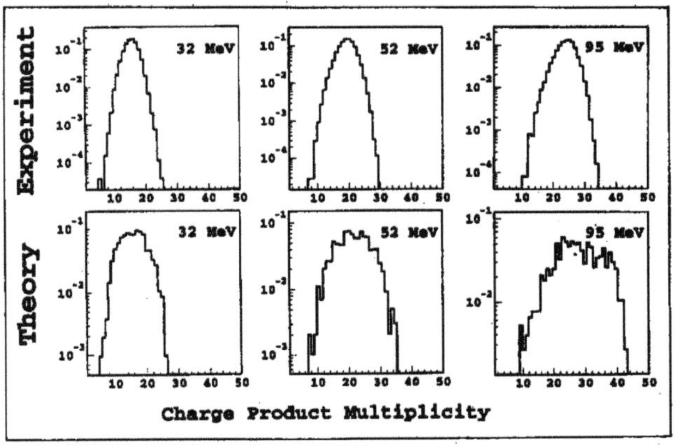

Figure 1. Evolution of the charged product multiplicity for complete events as a function of the incident energy.

and, that the implemented effective interaction is the Zamick force. Even if this force describes rather well the mean field properties, it does not account for surface terms nor quantum effects, necessary to generate bound fragments of few nucleons.

3 The DYnamical Wavelets in Nuclei model

The quantal counterpart of the semiclassical kinetic Eq. (1):

$$i\dot{\rho} = [h(\rho), \rho] + iI(\rho) \tag{2}$$

has been derived by different approaches. ¿From the projection methods (PM) point of view [4], it is assumed that the essential information for the description of the system is is given in terms of relevant variables. They define a subspace in state space which is orthogonal to the one concerning the irrelevant degrees of freedom. The characteristic scales in both subspaces are very different, nevertheless, their time evolution are coupled. Eq. (2)is derived from PM in the case where the relevant variables are one-body observables.

The projection onto subspaces concerning two, three, ... N-body observables will give equations of motion for multiparticle densities, and generate, in this way, a hierarchy of equations concerning different levels of description. In the case of Eq. (2) the reference level of the description is the mean field. In our approach, the equations of motion in those subspaces are solved by using a strongly compressed representation given by a basis of wavelets. The density matrix is expanded on a set of orthogonal wavelets $\alpha_{j,k}(x)$, generated by dilation and translation of the generating function $\alpha_(x)$ (the mother wavelet). A multiresolution analysis is a scheme for the decomposition of the information, by which a given signal $f(x)$ can be projected onto orthogonal, embedded, subspaces V_j corresponding to the different scales j:....$V_2 \subset V_1 \subset V_0 \subset V_{-1} \subset V_{-2} \subset$ ..... The approximation of $f(x)$, at the j-level, is given by the complete set of coefficients $C_{j,k} =< f, \alpha_{j,k} >$. In this way, the derivation and resolution of the equations of motion are consistent.

In our model the initial state of nuclei is prepared as follows. A multiresolution analys is applied to the description of the ground state of a system of nucleons in a self-consistent potential. Starting from a set of trial s.p. wave functions, an iterative procedure is implemented in which, the one-body density matrix and the self-consistent mean-field are constructed from the analyzed wave functions. The dynamical evolution, ruled by Eq. (2), can be transcribed in terms of equations of motion for the non-stationnary wavelet basis. It can be shown [5] that this gives a set of coupled differential equations for the first and second moments in position and in momentum coordinates and that the populations of s.p. states obey to a master equation. By construction, the orthogonality of the elements of the basis is conserved in time during the dynamics and this fact is essential to insure Pauli blocking.

The lowest (one-body) order of the mentionned hierarchy of approximations, is an ETDHF-like description, which describes the mean behaviour of the system, including dissipation. We would like now to find the least-biased many-body wave function compatible with one-and-two-body information contained in Eq. (2). We proceed then as follows: at any time we represent our many-body system by a linear combination of Slater determinants ψ_l of single particle wave functions $\{\varphi_i\}$ such that the one-body density matrix $\rho = \sum_i \eta_i \mid \varphi_i >< \varphi_i \mid$ is solution of the ETDHF problem Eq. (2). According with our resolution scheme, the s.p. wave functions are projected on a non-stationary basis of wavelets:

$$\mid \varphi_i(t) >= \sum_n c_n^i \mid \alpha_n^i(t) > \qquad i = 1, ..., N \tag{3}$$

The $\{\varphi_i\}$ are orthonormal and since the set of wavelets $\{\mid \alpha_n^i >\}$ constitutes a basis, the coefficients c_n^i satisfy the clausure relation: $\sum_n \mid c_n^i \mid^2 = 1$. Due to the orthonormality properties of wavelets, any Slater determinant of s.p. wave functions can be written as a linear combination of Slater determinants of wavelets and then the most general N-body wave function takes the following form:

$$| \Psi(t) >= \sum_{l} \sum_{n_1,...,n_N} \Omega_l(t) C_n^l \mid \alpha_{n_1}^{i_1} ... \alpha_{n_N}^{i_N} > \tag{4}$$

Here l represents a set of s.p. indices $\{i_1, i_2, ...i_N\}$, in the same way as n denotes the set $\{n_1, n_2, ...n_N\}$ corresponding to the associated wavelets indices. In this equation, $C_n^l = c_{n_1}^{i_1}...c_{n_N}^{i_N}$, $\mid \alpha_{n_1}^{i_1}...\alpha_{n_N}^{i_N} >$ are determinantal wave functions of wavelets and the $\Omega_l(t)$ are connected with the time evolution of the s.p. populations. In this sens, Eq. (4) represents the least biased N-body wave function compatible with the one-and-two-body information contained in Eq. (2). It does not contain informations concerning higher order multi-particle correlations, except the one concerning the particle statistics. Then it is possible to construct the complete hierarchy of reduced density matrix describing many-body clusters:

$$\rho_s = \aleph \; Tr_{\{s+1,...,N\}} \mid \Psi >< \Psi \mid, \qquad s = 1, ...N$$

where $\mid \Psi >$ is given by Eq. (4) , $\aleph$ is a normalization constant and $Tr_{\{s+1,...,N\}}$ denotes the trace over the degreees of freedom of particles $s+1$,...,N. The result is a statistical mixture of s-body Slater determinants of non-stationnary wavelets. In our numerical simulation we work in a truncated N-body space, starting from a well defined initial state, characterized by the coefficients C_n^l. The subsequent evolution, at the N-body level, is given by Eq. (4), which is treated by means a Montecalo procedure, in consistentcy with the simulation of the master equation. A clusterization algorithm like ECRA can be applyed to these Slater determinants of non stationnary states at any time during the evolution. Compared with the semiclassical description in Sec. 2, a better agreement with experimental results is expected since, even at the lowest level of the approximation hierarchy, the dynamical description in terms of wavelets takes care of the underlying quantum properties of the system, which are not contained in semi-classical models. This work is now in progress.

References

1. C. Grégoire *et al*, *Nucl. Phys.* A **585**, 317 (1987).
2. A. Strachan and C.O. Dorso, *Phys. Rev.* C **56**, 995 (1997).
3. T. Reposeur, V. de la Mota, F. Sébille and C.O. Dorso, *Z. Phys.* A **357**, 79 (1997)
4. R. Balian, Y. Alhassid and H. Reinhardt, *Phys. Rev.* **131**, 1 (1986).
5. B. Jouault, F. Sébille and V. de la Mota, *Nucl. Phys.* A **628**, 119 (1998).

EVIDENCE FOR NUCLEAR MATTER TRANSPARENCY IN CENTRAL HEAVY-ION REACTIONS FOR LIGHT SYMMETRIC SYSTEMS AROUND THE FERMI ENERGY

Z. BASRAK

Ruđer Bošković Institute, HR-10 001 Zagreb, Croatia

E-mail: basrak@irb.hr

P. EUDES

Laboratoire SUBATECH, Université de Nantes, École des Mines de Nantes, IN2P3/CNRS, F-44 072 Nantes Cedex 03, France

E-mail: eudes@subatech.in2p3.fr

Reaction mechanisms in central heavy-ion collisions around the Fermi energy are investigated within the Landau-Vlasov semiclassical transport model. It is shown that the reaction dynamics precludes the formation of a fusion nucleus: In head-on collisions, a transition from fusion process to binary mechanisms occurs. A pure nuclear transparency is underlined for light symmetric systems around the Fermi energy. For medium-heavy systems, the memory of the entrance channel still exists but, in addition, dynamical multifragment production is observed. These predictions are qualitatively consistent with recent experimental observations.

1 Introduction

In the Fermi energy domain, the main mechanism observed in heavy-ion reactions around the Fermi energy is of binary nature. [1] However, the classical fusion process still contributes by a small part of the reaction cross section and it is restricted to the most central collisions. [2] By classical fusion we refer to a process leading to the formation of a single cold residue at the end of the reaction. At low incident energies (below 15 MeV/nucleon), the fusion mechanism is complete and fusion nucleus decays by the standard evaporation process. With increasing incident energy, the fusion becomes more and more incomplete owing to pre-equilibrium emission. Moreover, the classical fusion cross section decreases and vanishes around 50 MeV/nucleon. This trend has been observed in various experimental studies [2] and confirmed by very recent data for the symmetric Ar+KCl system at incident energies ranging from 32 to 74 MeV/nucleon. [3,4] It has been found that the classical fusion cross section is about 40 mb at 32 MeV/nucleon and decreases rapidly to zero for incident energies higher than 50 MeV/nucleon. Similar results have been obtained for the slightly asymmetric system Ar+Ni. [5]

Two interpretations are propound to explain the vanishing of classical fusion: *i*) Opening of new decay channels for the fusion nucleus, such as the evaporation of intermediate mass fragments (IMF) or the so-called multifragmentation process; the fusion process still exists and the vanishing of fusion is an exit channel effect. *ii*) The reaction dynamics precludes the formation of the fusion nucleus; the fusion disappears owing to an entrance channel effect. In this work we are going to show that in the frame of the Landau-Vlasov (LV) dynamical model, [6] a transition from the classical fusion process to the binary mechanism (BM) is predicted in head-on collisions around the Fermi energy. Obviously, in such a scenario, the fusion process really disappears.

The LV model [6] is based on the quantum Boltzmann equation which is solved for a nonlocal effective force (Gogny D1-G1) [7] by projecting the one-body nuclear phase space onto a continuous basis of thousands (here 60 to 100 per nucleon) of coherent states taken as elementary Gaussian functions. One of the main ingredients of the model is the residual nucleon-nucleon (NN) cross section $\sigma_{\rm NN}$ that acts through the collision term. In the simulations presented in this work, the free nucleon-nucleon cross section with its usual energy and isospin dependence was chosen. In-medium effects are studied in [8]. Several symmetric systems with a total mass ranging from 32 up to 250 were investigated for incident energies ranging from 10 MeV/nucleon up to 50 MeV/nucleon. In this paper we focus mainly on the light Ar+Ar system as it has recently been studied by the INDRA collaboration. Preliminary results are also shown for the medium-heavy systems Ag+Ag and Xe+Sn.

2 Transition from fusion to the binary mechanism: the case of light symmetric systems

Figure 1 shows the simulation results for the Ar+Ar head-on collisions at two incident energies: 29 MeV/nucleon (left column) and 30 MeV/nucleon (right column). This figure allows to clearly illustrate the transition from fusion to BM predicted by the model. Indeed, at 29 MeV/nucleon, a single fusion residue is present at the end of the reaction. However, a large deformation of dinuclear type persists for a long period which finally recontracts itself. This deformation has already been observed in calculations carried out at slightly lower incident energies, but it develops with increasing incident energy up to the rupture which occurs at 30 MeV/nucleon (see the bottom panel of the right column). In this case, two spherical nuclei emerge from the reaction around 150 fm/c. Moreover, the grey scale shows a weak mixing between the nucleons of the projectile and the target, proving a strong memory of the entrance channel.

This mechanism implies an important degree of nuclear matter trans-

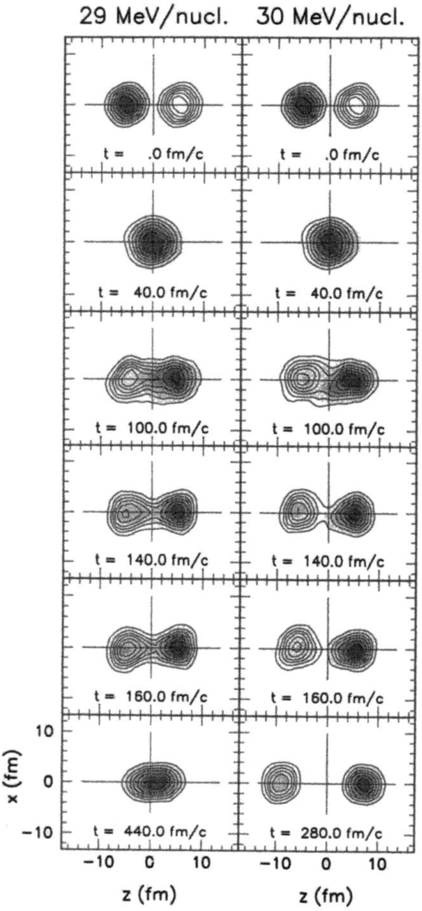

Figure 1. Equidistant density-profile contours in the configuration space at several times of the reaction projected onto the reaction plane for Ar+Ar head-on collisions at 29 MeV/nucleon (left column) and at 30 MeV/nucleon (right column). Curves refer to the whole system, whereas the scale of grey shows the projectile matter density. The z axis is along the projectile direction.

parency, a scenario which has already been mentioned in several works [2] in order to qualitatively explain experimental observations. However, neither experimental quantitative probes nor careful theoretical studies have been performed on this subject yet. A remark concerning the memory of the entrance channel can also be made at 29 MeV/nucleon. Indeed, as shown in the left column of figure 1, there is a slow shape relaxation and a few nucleon mixing over a large period of spatial rearrangement lasting several hundreds of fm/c. Thus, the fusion nucleus, detected at the end of the reaction, can be rigorously considered as a fully thermalized compound nucleus only beyond this period. Such dynamical effects are generally not taken into account,

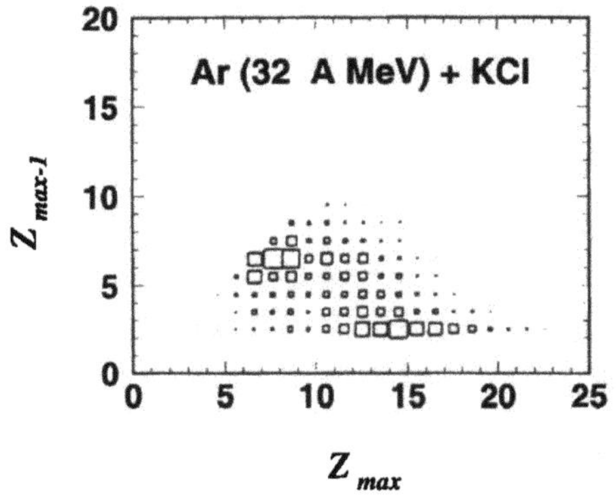

Figure 2. Charge of the largest fragment Z_{max} versus the charge of the second largest fragment $Z_{\mathrm{max}-1}$ for the Ar+KCl system at 32 MeV/nucleon. The analysis is carried out only for the so-called complete events [4] and restricted to the most central collisions which are selected by global variable analysis.

as the deexcitation of fusion nucleus is described using statistical models in which the equilibration of all degrees of freedom is assumed from the very early instants of a collision. [9] Study of the heavier Ni+Ni system displays the same trend, but the fusion/BM transition occurs at a slightly higher incident energy (between 31 and 32 MeV/nucleon). That is not so surprising since the nuclear matter which has to be passed through is larger in the case of Ni+Ni in comparison with the Ar+Ar system.

The value of the transition energy deserves two comments: *i*) The transition energy is strongly dependent on the nucleon-nucleon cross section. [8] Thus, it could be a right observable to infer on in-medium effects. *ii*) There are no fluctuations in the code; the model is able to provide only an average value for the transition energy between fusion and BM. Obviously, any sharp transition will never be found experimentally: The fusion process is present up to higher incident energies, although with a sharply decreasing weight [3,4] which cannot be predicted within the frame of deterministic calculations such as provided by the LV model.

3 Experimental indications

The recent INDRA experimental results on the Ar+KCl reaction could be in agreement with the above scenario. Figure 2 displays the yields of two largest fragments as a function of fragment charge for the subset of the most central collision events. [4] In this plot one distinguishes two components. The component with the large values of Z_{max} and the small values of $Z_{\mathrm{max}-1}$ corresponds to events in which a single fusion residue is detected and can be attributed to classical fusion. The other component is characterized by two main fragments with approximately the same charge ($Z_{\mathrm{max}} \approx Z_{\mathrm{max}-1}$) and could be attributed to BM such as observed in the model. The latter component was interpreted in [4] in terms of dynamical fission of the fusion residue, a rather surprising mechanism for such a light system. Let us add that at 32 MeV/nucleon, the experimental cross section associated to each component is found to be approximately the same: 43 and 44 mb, respectively. Moreover, from the same plots at 40, 52, and 74 MeV/nucleon one infers that the classical fusion component disappears progressively, while at the same time the BM component is always present but with a diminishing average values of Z_{max} and $Z_{\mathrm{max}-1}$ [4] (probably due to an increasing contribution of dynamical emission [10]). Both the presence of binary processes after selecting very central collisions at 32 MeV/nucleon and the disappearance of classical fusion events with increasing incident energy qualitatively endorses our simulation predictions.

4 Transition from fusion to binary mechanism: the case of medium-heavy systems

It is intriguing to extend this study to heavier systems. Figure 3 shows the simulation results for the Ag+Ag (top) and the Xe+Sn (bottom) systems in head-on collisions. At 45 MeV/nucleon (left top panel), there is still a fusion residue (the comments concerning the slow shape relaxation and the few nucleon mixing remaining valid). At 50 MeV/nucleon (right top panel), there are two main fragments in the exit channel showing that transparency effects still exist for this heavier system (see also the grey scale). However, there is a major difference in comparison with light systems: a multifragment emission occurs in-between the two main fragments. This multifragment production is much clearer for the Xe+Sn system at 50 MeV/nucleon as can be seen in the right bottom panel of the figure. Since our study of heavy systems is at the beginning, one cannot draw any definitive conclusion. However, let us make two comments: *i*) It seems that there is a limit in the system size if the pure transparency with no IMF in-between the two outgoing nuclei is to be observed. *ii*) From the above figures of central collisions, the multifragment production appears as an induced dynamical process rather than a process following the formation of a fully equilibrated fusion nucleus. Let us mention that the dynamical production of fragments in the Xe+Sn system at 50 MeV/nucleon has recently been investigated by the INDRA collaboration. [11]

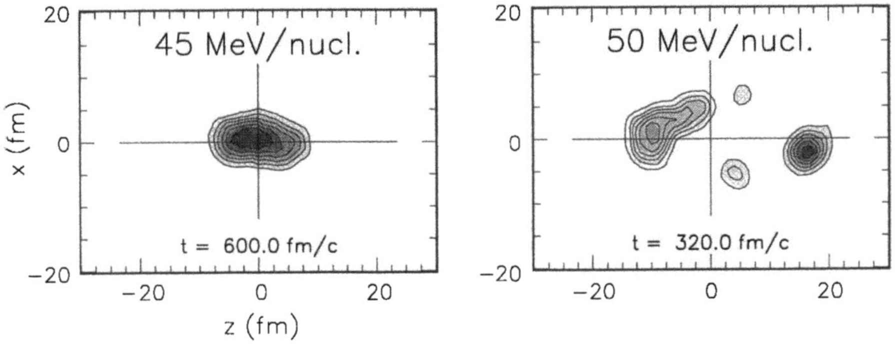

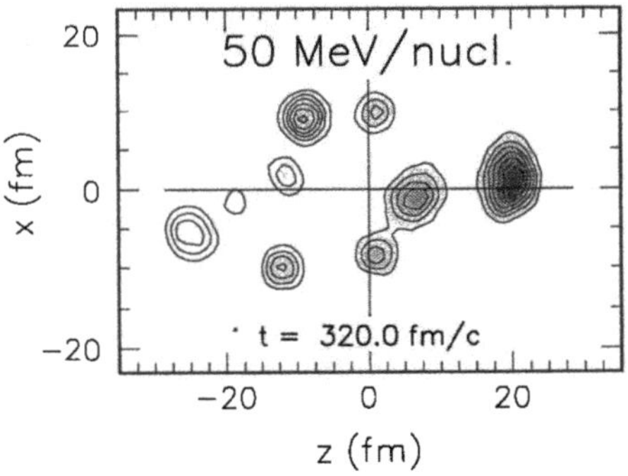

Figure 3. Equidistant density-profile contours projected onto the reaction plane for the Ag+Ag (top) and the Xe+Sn (bottom) head-on collisions at 45 MeV/nucleon (left) and 50 MeV/nucleon (right column). For more details see caption to Fig. 1.

5 Conclusion

We have found that the vanishing of fusion is due to dynamical effects: the binary mechanism becomes the dominant process up to the most central collisions. For light symmetric systems, we observe an effect of pure nuclear transparency which leads to the formation only of two massive fragments in the exit channel of the reaction, a quasiprojectile and a quasitarget. For heavier systems, the transparency is also observed in the sense that two remnants of the projectile and of the target are present. In addition, a number of fragments are also produced in the neck region between the two remnants. This effect is of dynamical origin and the dynamical multifragment production seems to be related to the size of the system. The system-size limit for the onset of the multifragment production has to be precised and a comparison with experiment has to be performed. It will also be interesting to study the same phenomena in asymmetric systems. Let us underline that the above predictions are qualitatively consistent with experimental observations, particularly with the recent results of the INDRA collaboration on the Ar+KCl and Xe+Sn systems. Nevertheless, more quantitative comparisons are needed, especially to experimentally probe the existence of pure nuclear transparency for light systems.

6 Acknowledgments

It is a pleasure to thank Prof. F. Sébille for supplying us with the Gogny version of the Landau-Vlasov computer code.

References

1. B. Borderie et al., *Phys. Lett.* B **205**, 26 (1988); B. Lott et al., *Phys. Rev. Lett.* **68**, 3141 (1992); E. Vient et al., *Nucl. Phys.* A **571**, 588 (1994); J. Yee et al., *Phys. Lett.* B **356**, 191 (1995); M. F. Rivet et al., *Phys. Lett.* B **388**, 219 (1996); B. Borderie et al., *Phys. Lett.* B **388**, 224 (1996); J. Péter et al., *Nucl. Phys.* A **593**, 95 (1995); J. C. Steckmeyer et al., *Phys. Rev. Lett.* **76**, 4895 (1996); L. Beaulieu et al., *Phys. Rev. Lett.* **77**, 462 (1996); Y. Larochelle et al., *Phys. Rev.* C **55**, 1869 (1997); E. Colin et al., *Phys. Rev.* C **57**, R1032 (1998).
2. E. Suraud, C. Grégoire, and B. Tamain, *Prog. Nucl. Part. Sci.* **23**, 357 (1989).
3. V. Métivier, *Ph.D. Thesis*, Univ. of Caen, 1995.
4. E. Bisquer, *Ph.D. Thesis*, Univ. of Lyon, 1996.
5. L. Nalpas and the INDRA collaboration, *Private communication.*
6. C. Grégoire et al., *Nucl. Phys.* A **436**, 365 (1985); C. Grégoire et al., *J. Phys.*, Paris **48**, C2 (1987).
7. B. Remaud et al., *Nucl. Phys.* A **447**, 555c (1985); F. Sébille et al., *Nucl. Phys.* A **501**, 137 (1989).
8. Z. Basrak and Ph. Eudes, *Proceedings of the XXXVI International Winter Meeting on Nuclear Physics,* Bormio, 1999, edited by I. Iori (University of Milan Press, Milan, 1999), p. 284 and to be published.
9. F. Gulminelli and D. Durand, *Nucl. Phys.* A **615**, 117 (1997).
10. Ph. Eudes, Z. Basrak, and F. Sébille, *Phys. Rev.* C **56**, 2003 (1997); in *Proceedings of the XXXVI International Winter Meeting on Nuclear Physics,* Bormio, 1998, edited by I. Iori (University of Milan Press, Milan, 1998), p. 277.
11. J. Łukasik et al, *Phys. Rev.* C **55**, 1906 (1997); E. Plagnol et al, *Proceedings of the International Nuclear Physics Conference,* Paris, 1998, Nucl. Phys. A **654**, 815c (1999).

CLUSTERS AND NUCLEAR FLOW

W. REISDORF
GSI, Planckstr. 1
64291 Darmstadt, Germany
E-mail: W.Reisdorf@gsi.de

FOPI Collaboration

We discuss clusterization in central heavy ion collisions in the energy range between 90 and 400 A MeV. Pronounced system-size dependences are observed for participant matter. Possible correlations with size-dependences of sideflow, pion multiplicity and momentum space topology are pointed out. A mixing experiment is described that demonstrates the importance of non-equilibrium dynamics.

1 Introduction

One of the surprising phenomena recently observed [1] in *central* Au on Au collisions at intermediate energies, is the large probability of emitted protons to appear within 'clusters' rather than as single nucleons. Fig.1 shows that the probability of participant protons to attach themselves to at least one other nucleon at 'freezeout' time, is as high as 80% at 150A MeV and still roughly 50% at 1A GeV.

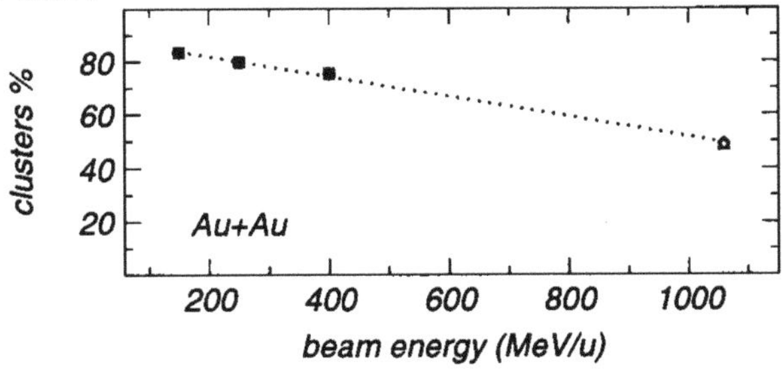

Figure 1: Excitation function for the proton clusterization probability in central Au+Au collisions

In the energy range primarily discussed here (90-400A MeV) the available c.o.m. energy varies from 22 to 95A MeV, well beyond the known binding energy of 8A MeV for stable nuclei. The survival of so many clusters in this energy regime calls for models that go beyond naive global statistical equilibration concepts. The observation [2,3,1] of pronounced mass-dependences of the kinetic energies was interpreted as evidence for strong radial flow caused by an expansion of the hot and possibly initially compressed fireball. Such scenarios could be used to understand[3,1] the kinetic energy spectra of the emitted clusters, as illustrated in Fig.2. The energy taken up by such an expansive flow was estimated to be as high as 60%. At 250A MeV incident energy, to be specific, this still meant that about 20 MeV per nucleon (the Q-values are about −5A MeV) were available for 'local use'. It is therefore not surprising that statistical multifragmentation calculations, allowing for the available *local* energy, still failed to reproduce the measured fragment yields as illustrated in Fig.3. To obtain fair agreement with the data one has to assume that the primary fragments are cold, in obvious contradiction to statistical expectations.

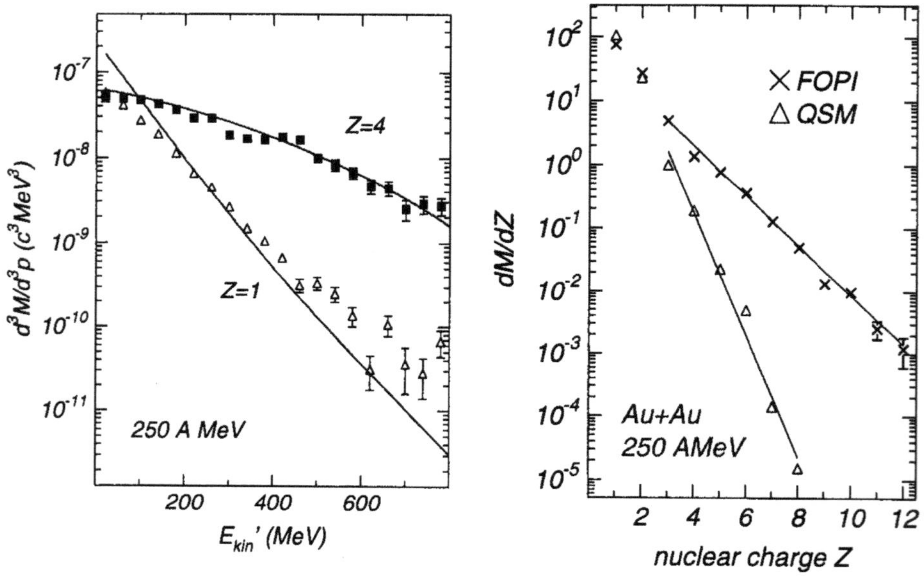

Figure 2: Left panel: kinetic energy spectra of Z=1 and Z=4 fragments emitted in central Au on Au collisions near 45° (c.m.). The smooth curves are blast model fits. Right panel: Fragment charge distribution. Comparison of the experimental data (× FOPI) with constrained statistical model predictions (△).

In the following we discuss new, partially preliminary results obtained by the FOPI Collaboration, that shed additional light onto this complex subject. A theoretical approach is presented by A. Ono[4].

2 Observation of system-size dependences

Our Collaboration uses a 4π detector[5] which is located at the SIS/ESR accelerator facility of GSI, Darmstadt, and which allows to identify Light Charged Particles (LCP), Intermediate-Mass Particles (IMF), as well as produced particles, such as pions and kaons, using a combined magnetic rigidity, time-of-flight and energy-loss analysis

It is of interest to study the system-size dependence of multifragmentation and its possible correlations with other observables. In Fig. 3 we compare the IMF multiplicity in the system Au+Au with that for Ni+Ni.

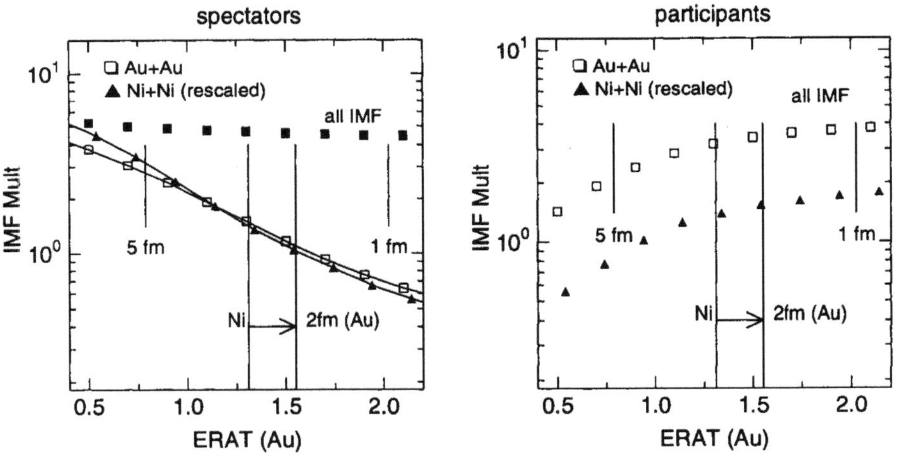

Figure 3: IMF multiplicities in spectator (left) and participant (right) matter for Au+Au and Ni+Ni as a function of ERAT, see text.

To remove trivial size dependences we have scaled the Ni data by 79/28, the ratio of projectile charges in the two symmetric systems. We vary the centrality using as global event selector ERAT, the observed ratio of total transverse to total longitudinal kinetic energy. For the Au+Au system the geometrical interpretation of the cross sections (in fm) is indicated in the figures. In the left hand panel we compare 'spectator' sources, somewhat arbitrarily defined as originating from scaled rapididities $|y^{(0)} \pm 0.4|$. For projectile (target) $y^{(0)} = 1$ (-1). We find that the (scaled) IMF multiplicities of the Ni system are very close to those observed in the Au system, independent on the event selection. Note that the abscissa (ERAT) has been shifted for the Ni system, as shown in the figure in order to overlay data belonging to the same reduced impact parameter b/b_{max}, b_{max} being the maximum (grazing) value of b. This size-independence is in line with the universality of spectator fragmentation observed by the ALADIN collaboration [6].

The right panel of Fig. 3 shows that outside the spectator rapidities, universality is lost: at comparable reduced impact parameter about twice as many IMF's *per nucleon* are created in the heavier system, again very weakly dependent on ERAT, i.e. the centrality. For participant matter we are here in an energy region where the IMF multiplicity is expected to decrease with increasing available excitation energy: hence, applying thermal language, we can say that the heavier participant system appears to be colder.

Is there a correlation with other observables? It turns out that size dependences are also seen for sideflow, Figs. 4,5, the scaled pion multiplicity, Fig. 6, and the momentum-space topology of events with the same centrality, Fig. 7.

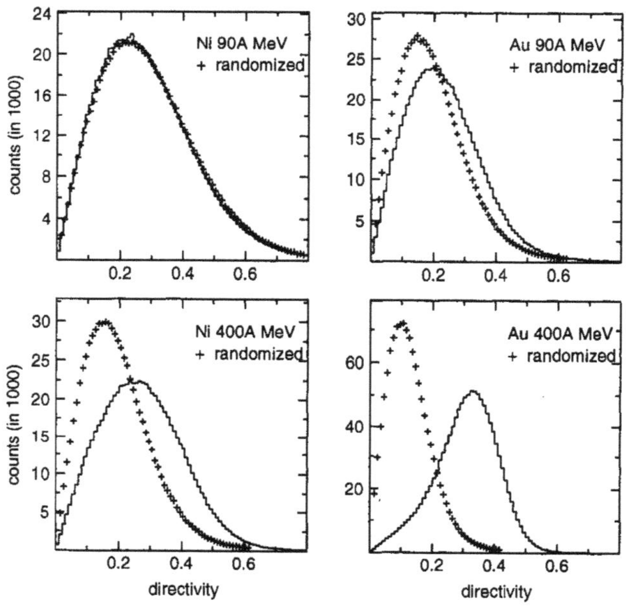

Figure 4. Directivity distributions (histograms) for various systems and energies indicated in the panels. Randomized distribution (crosses) are shown for comparison.

Briefly, in Fig. 4, we present distributions of the directivity D defined by

$$D = \frac{\sum sign(y)\, \vec{p_t}}{\sum |p_t|} \tag{1}$$

where $sign(y)$ is $+1$ in the forward hemisphere and -1 else and the sums are taken over all observed particles. The directivity is a good measure for side-flow. The data shown are integrated over a broad range of impact parameters, $b < 0.5\, b_{max}$. As all observables are subject to finite-size fluctuations, it is important to remove this effect before drawing conclusions on non-trivial system-size dependences. The random walk distributions of D (crosses), obtained by randomizing the azimuthal emission angles in each event, are compared with the actual distributions (histograms).

One finds that the random distributions of this scale-invariant observable are broader for lighter than heavier systems (i.e. the *relative* fluctuations are larger), see the top two panels in Fig. 4, and narrow down at higher energies because of increasing particle multiplicities (compare the top with the bottom panels). Comparing with the actual distributions one sees that

1) At 90A MeV there is no net effect for the Ni system, while a statistically significant signal is seen for Au.

2) At 400A MeV sideflow has increased substantially in both systems.

3) The sideflow excitation function, Fig. 5, after correction for the finite number fluctuation, shows a dramatic system-size dependence.

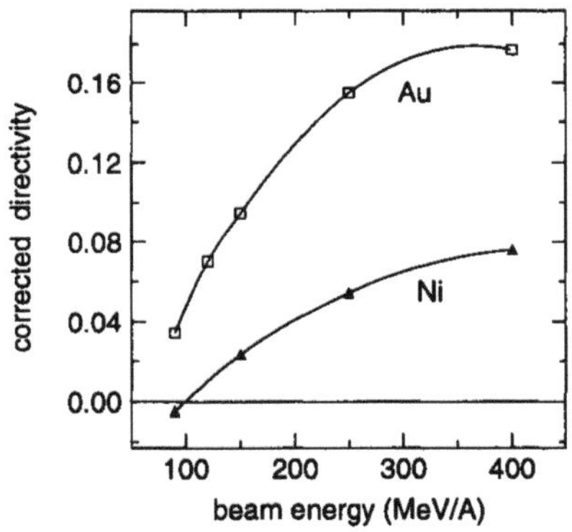

Figure 5: Excitation function of the finite-size corrected directivity for Au+Au and Ni+Ni

While flow is naturally understood using hydrodynamical considerations, this observation is incompatible with the expectations from ideal one-fluid dynamics which relies on complete local equilibration.

Mid-rapidity pion transverse 4-velocity spectra (scaled with the projectile 4-velocity) are shown in Fig.6 for both systems, Ni and Au. While the ratio of π^+/π^- multiplicities is higher for the Ni system due to a higher proton/neutron ratio in this system, it is clear that the sum of pions emitted by the lighter system is larger: again, the Ni data have been scaled with 79/28 and normalized to the same number of events.

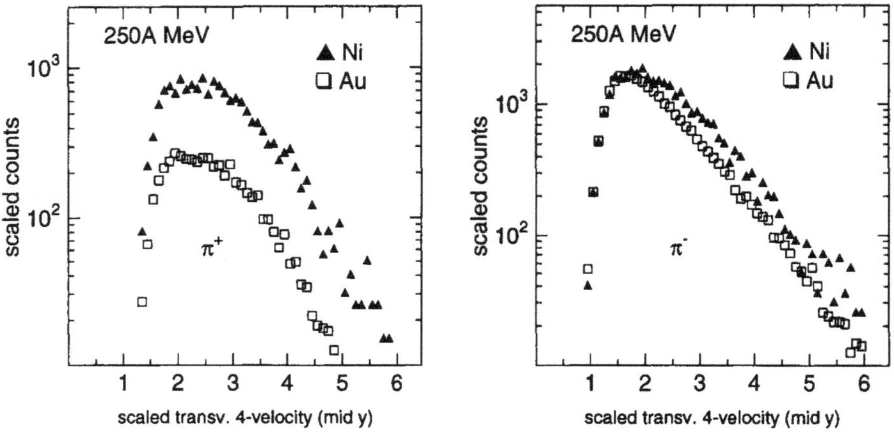

Figure 6: Mid-rapidity transverse 4-velocity spectra of pions (π^+ left, π^- right) for Ni+Ni and Au+Au at 250A MeV.

Finally, from Fig. 7, we read off that the momentum-space topology is more elongated in the lighter Ni system. The fact that transverse momenta are larger in the chemically 'colder' system (Au+Au) is remarkable. It suggests again that a sizeable amount of these momenta are in *ordered* motion, or radial flow. But in addition the size dependence tells us that the development of such flow requires a minimum amount of ***bulk*** material, not available in light systems, where surface effects dominate. But since we have seen earlier that a *locally* equilibrated fluid dynamics scenario seems to fail quantitatively, we are still left with the question of how to assess non-equilibrium effects.

3 Observation of incomplete mixing

We have performed an experiment that allows to detect non-equilibrium effects more directly. At 400A MeV we studied the isospin-asymmetric, but mass-symmetric system ^{96}Zr + ^{96}Ru. The results are summarized in Fig.8. The left panel shows for highly central collisions the observed yields of protons plus deuterons as a function of the center-of-mass angle θ_{cm}, the right panel displays the observed t/^{3}He ratios as a function of centrality (increasing from left to right), in a fixed rapidity bin ($y^{(0)} = 0.25 - 0.75$). As ^{96}Zr and ^{96}Ru differ only by 4 protons (or 10%), one has to make some additional effort to judge quantitatively if complete mixing of projectile and target nucleons has taken place.

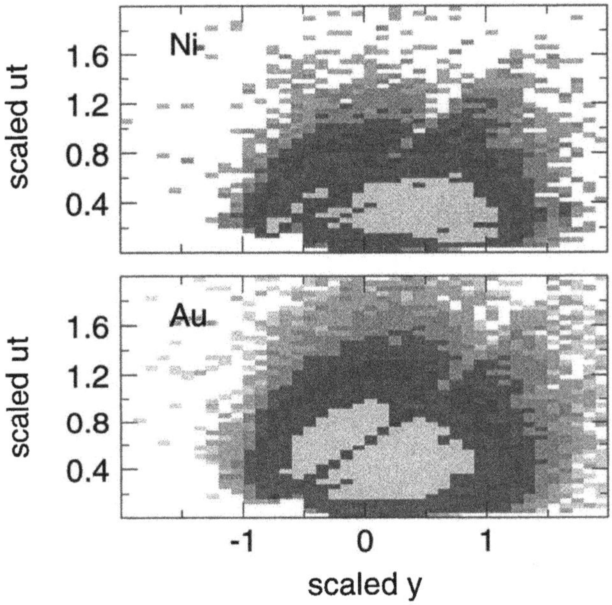

Figure 7: Momentum space distributions of Li fragments in central Ni+Ni (top) and Au+Au (bottom) collisions at 250A MeV

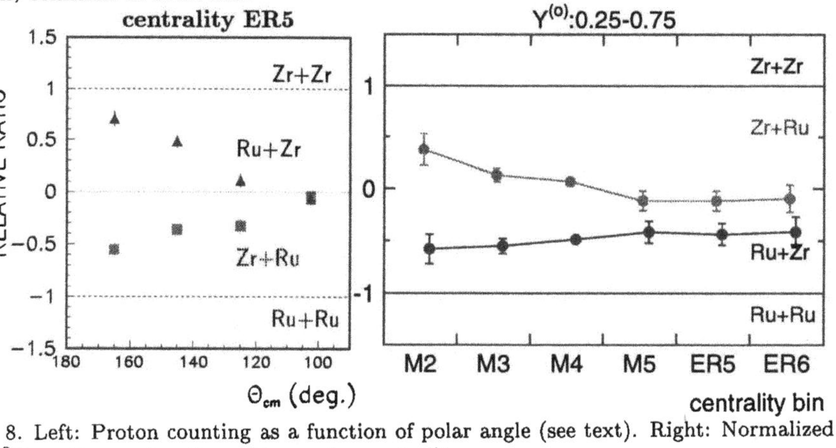

Figure 8. Left: Proton counting as a function of polar angle (see text). Right: Normalized triton/^{3}He ratio as a function of increasing centrality.

One repeats the same measurements for the completely symmetric systems ^{96}Zr + ^{96}Zr, respectively ^{96}Ru + ^{96}Ru to obtain reference values of the studied observables, which for convenience were normalized to +1 (Zr), resp. (-1) (Ru)

in the figures. Furthermore, to overcome apparatus asymmetries one inverts also target and beam in the asymmetric combination.

In Fig. 8 the mixed systems should meet at some constant common intermediate value if complete mixing had been achieved in phase space. They do so only at 90° in the c.o.m. where, *by principle*, incomplete mixing cannot be tested. This experiment has therefore shown unambiguously that mixing equilibrium has not been achieved. For more details see refs. [8].

4 Conclusion

To summarize, the heavier system apears to be be 'chemically colder': more clusterization, less pions; it has larger sideflow and higher transverse momenta. If one interpretes the latter as indicative of higher *radial* flow, one is tempted to conclude that flow is cooling the system, as one would expect qualitatively from an adiabatic expansion scenario. On the other hand this fluid dynamics scenario does not lead to a simultaneous and quantitative understanding of clusterization, momentum-space population and size dependences. Since, furthermore, nonequilibrium effects were shown directly to be important in a mixing experiment, it seems clear that transport calculations are necessary in order to understand the dynamics and eventually draw more 'fundamental' conclusions about the acting nuclear forces and the equation of state.

References

1. W. Reisdorf *et al*, *Nucl. Phys.* A **612**, 493 (1997).
2. S.C. Jeong *et al*, *Phys. Rev. Lett.* **72**, 3468 (1994).
3. M. Petrovici *et al*, *Phys. Rev. Lett.* **74**, 5001 (1995).
4. A. Ono , contribution this conference.
5. A. Gobbi *et al*, *Nucl. Instrum. Methods* A **324**, 156 (1993), J. Ritman *et al*, *Nucl. Phys.* B **44**, 708 (1995).
6. A. Schüttauf *et al*, *Nucl. Phys.* A **607**, 457 (1996).
7. F. Rami, contribution this conference.
8. B. de Schauenburg, Ph. D. thesis, Strasbourg (1999), F. Rami, Y. Leifels, B. de Schauenburg, A. Gobbi *et al*, to be published.

MULTIFRAGMENTATION AND SPINODAL DECOMPOSITION OF FINITE NUCLEAR SYSTEMS

B. BORDERIE[1], J.D. FRANKLAND[1], M.F RIVET[1], CH.O. BACRI[1], M. PÂRLOG[2], M. SQUALLI[1], G. TĂBĂCARU[2], M. COLONNA[4],A. GUARNERA[3] AND P. CHOMAZ[3]
G. AUGER[3], N. BELLAIZE[5], F. BOCAGE[5] R. BOUGAULT[5], R. BROU[5], P. BUCHET[6], A. CHBIHI[3], J. COLIN[5], R. DAYRAS[6], A. DEMEYER[7], D. DORÉ[6], D. DURAND[5], E. GALICHET[7], E. GENOUIN-DUHAMEL[5], E. GERLIC[7], D. GUINET[7], P. LAUTESSE[7], J.L. LAVILLE[3], J.F. LECOLLEY[5], T. LEFORT[5], R. LEGRAIN[6], N. LE NEINDRE[5], O. LOPEZ[5], M. LOUVEL[5], A.M. MASKAY[7], L. NALPAS[6], A.D. NGUYEN[5], J. PÉTER[5], E. PLAGNOL[1], E. ROSATO[8], S. SALOU[3], F. SAINT-LAURENT[3], J.C. STECKMEYER[5], M. STERN[7], B. TAMAIN[5], O. TIREL[3], L. TASSAN-GOT[1], E. VIENT[5], C. VOLANT[6] J.P. WIELECZKO[3]

INDRA COLLABORATION

[1] *Institut de Physique Nucléaire, IN2P3-CNRS, F-91406 Orsay Cedex, France.*
[2] *National Institute for Physics and Nuclear Engineering, RO-76900 Bucharest-Măgurele, Romania.*
[3] *GANIL, CEA et IN2P3-CNRS, B.P. 5027, F-14076 Caen Cedex, France.*
[4] *Laboratorio Nazionale del Sud, Viale Andrea Doria, I-95129 Catania, Italy.*
[5] *LPC, IN2P3-CNRS, ISMRA et Université, F-14050 Caen Cedex, France.*
[6] *DAPNIA/SPhN, CEA/Saclay, F-91191 Gif sur Yvette Cedex, France.*
[7] *Institut de Physique Nucléaire, IN2P3-CNRS et Université, F-69622 Villeurbanne Cedex, France.*
[8] *Dipartimento di Scienze Fisiche e Sezione INFN, Universit di Napoli "Federico II", I80126 Napoli, Italy.*

A bulk effect in multifragmentation of very heavy fused systems has been experimentally evidenced. Such an effect is very well reproduced by a complete dynamical simulation of collisions in which multifragmentation occurs through the growth of density fluctuations in the unstable spinodal region of the phase diagram of nuclear matter.

1 Introduction

The decay of highly excited nuclear systems through multifragmentation (emission of several fragments in a short time scale) is, at present time, a subject of great interest in nucleus-nucleus collisions. If this process has been observed from many years, its experimental knowledge in the Fermi energy domain was strongly improved only recently with the advent of powerful 4π devices. Well defined systems or subsystems which undergo multifragmentation have been carefully selected. Moreover central collisions between heavy nuclei have revealed the importance of a compression phase followed by expansion to cause multifragmentation [1].

We report here on studies of multifragmentation of very heavy fused systems formed at the same available excitation energy : $^{129}Xe +^{nat} Sn$ at 32 MeV/u and $^{155}Gd +^{nat} U$ at 36 MeV/u. They can be identified to well defined pieces of nuclear matter and reveal bulk properties at low density to be compared to models in which bulk or volume instabilities are present.

Many theories have been produced to explain multifragmentation (see for example ref. [2] for a general review of models). One can come in particular to the concept of multifragmentation by considering a liquid-gas phase transition in excited nuclear matter. It is commonly believed that, during a collision, a wide zone of the nuclear matter phase diagram might be explored and that the nuclear system may enter the liquid-gas phase coexistence region (at low density) and even more precisely the unstable spinodal region (domain of negative incompressibility). Thus, a possible origin of multifragmentation may be found through the growth of density fluctuations in this unstable region.

Among the models some are related to statistical approaches whereas other, more ambitious, try to describe the dynamical evolution of systems resulting from collisions between two nuclei taking into account the dynamics of the phase transition. Thus theoretical dynamical scenarios to be compared with experimental data are simulated (with approximations) via molecular dynamics or stochastic mean field approaches.

In this last type of approach, that we shall compare to our experimental data, a powerful approximate tool is provided by the Brownian One-Body (BOB) dynamics [3,4], which consists in employing a Brownian force in the kinetic equations.

2 Experimental evidence for bulk effect in multifragmentation

Using the INDRA detector, multifragmenting fused systems have been carefully selected for two reactions leading to the same available excitation energy per nucleon ($\sim 7MeV$) : $^{129}Xe +^{nat} Sn$ at 32 MeV/u and $^{155}Gd +^{nat} U$ at 36 MeV/u. The selection was performed by examining the evolution of fragments' kinematics as a function of dissipated energy and loss of memory of the colliding nuclei [5].

Fig. 1 shows that, for the two fused systems, we observe the same Z distribution for fragments while, as a consequence, the fragment multiplicities scale as the size of the total systems. This independence of the Z distribution, experimentally observed for the first time, can be considered as a strong experimental evidence for a bulk effect.

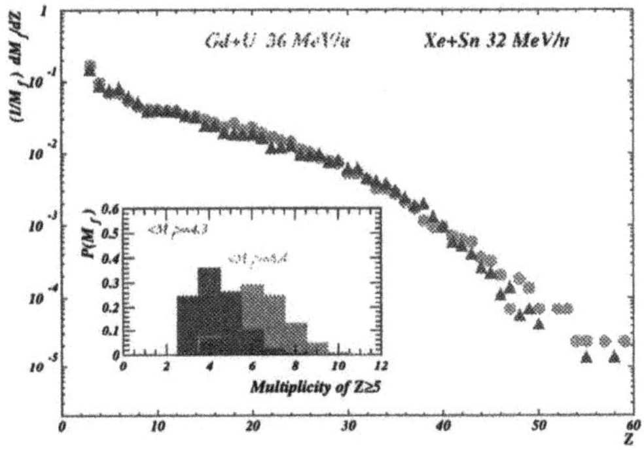

Figure 1. Experimental fragment multiplicity distributions and differential charge multiplicities distributions for the 32 MeV/u Xe+Sn (black histogram and triangles) and 36 MeV/u Gd+U (grey histogram and circles).

The scaling observed also results from some statistical models [6] for the heavy systems considered ($\sim$ 250 and 400 nucleons). However in this case the dynamical phase of the reaction is ignored and parameters such as the mass and charge of the multifragmenting system, its excitation energy, its volume have to be backtraced to the experimental data. Neither do these models account for the fragment kinetic energy, unless an extra expansion energy is added.

3 Spinodal instabilities and stochastic mean field simulations

On the other hand this experimental observation may also sign an effect of bulk instabilities and it was worth comparing the data with a complete simulation of collisions, including fluctuations in the spinodal region to generate the formation of fragments [7,8]. In reference [9], it was shown that the multiplicity and charge distributions for both systems - and thus the scaling- were correctly reproduced, but that the calculated fragment kinetic energies were too low. Since then two major improvements have been implemented in the simulations. Firstly, in deterministic or stochastic simulations, the collision term is assumed to be local in both space and time; this simplification is no more justified when fast unstable modes are present as in the spinodal region. Then quantal fluctuations connected with collisional memory effects are now taken into account with the determinant result of about doubling the overall amplitude of fluctuations [10]. Secondly the previous simulation method, which consisted in simply injecting the correct (classical) fluctuations when the system enters the spinodal region (Stochastic Initialisation method) led to a damping of these fluctuations. The fragment formation time was thus artificially (and incorrectly) increased, leading to smaller kinetic energies due to the decrease with time of the radial expansion of the system. A better simulation method was proposed [3,4] which takes into account the time dependence of the fluctuation source. Density fluctuations are created by the introduction of a Brownian force with the correct magnitude in kinetic equations, which is maintained while the system is unstable. This method, analogous to the Langevin treatment of Brownian motion, is called Brownian One-Body dynamics (BoB).

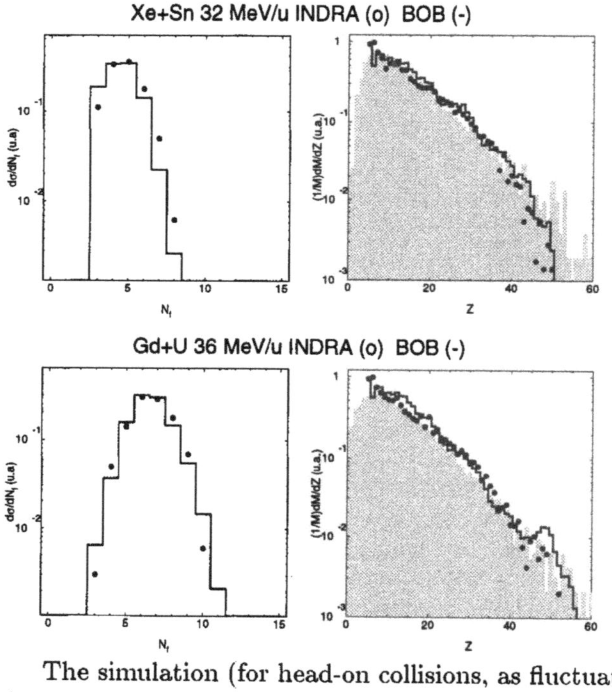

Figure 2. Experimental (points) and simulated (lines) fragment multiplicity and charge distributions for fused multifragmenting systems formed in 32 MeV/u Xe+Sn and 36 MeV/u Gd+U collisions. The grey histograms refer to the hot primary fragments.

The simulation (for head-on collisions, as fluctuations calculated for infinite matter can be rather safely extrapolated only to systems having a spherical symmetry) thus proceeds as follows [5]: the dynamics is followed through a BNV/BoB simulation, up to the time where the fragments are formed and well separated. The fluctuating force starts acting at the time of maximum compression in order to get the correct amplitude at the entrance in the spinodal region. The maximum density is 25-30% higher than normal density, corresponding to a moderate compression. Slightly after entering the spinodal

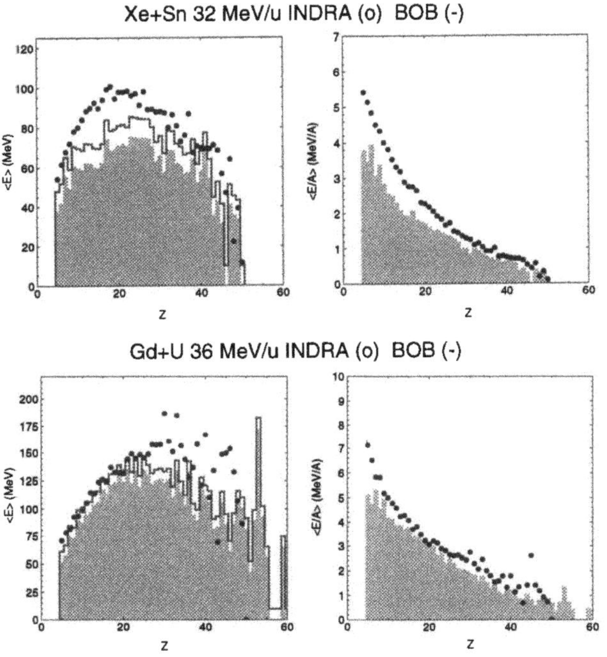

Figure 3. Experimental (points) and simulated (filled histograms) average fragment kinetic energies. The lines are the calculated values when adding the thermal energy. The results are for total energies on the left, and in MeV/u on the right.

region, the systems are in thermal equilibrium at low density, with a temperature of 4 MeV. The radial velocity at the surface is rather large ~0.1c. When the fragments are well separated, they still bear an excitation energy of ~3.2 MeV/u. Then the de-excitation of the hot primary fragments, whose mass accounts for 80% of the total system mass is followed through the evaporation part of the SIMON code. Finally the results are filtered to take into account the experimental set-up.

4 Comparison to the data

Figures 2 and 3 show the comparison between experimental data and simulations. Calculated multiplicity and charge distributions of fragments well match the experimental ones. More detailed comparisons on the charge distributions of the three largest fragments display the same excellent agreement [5]. Shape variables, such as the isotropy ratio, with its multiplicity dependence, are also well accounted for [5]. Finally the most crucial test is performed on the fragment kinetic energies.

The better simulations now correctly reproduce the fragment energies for the Gd+U system, especially if the average thermal energy (3/2T) of fragments is added; indeed in simulations this energy is strongly minimized because of the number of test particles. For Xe+Sn, the calculated energies fall within 20% of the measured values, which is satisfactory if one remembers that there were no adjustable parameters in the simulation.

5 Summary and future prospects

We have experimentally observed that two heavy multifragmenting systems with the same available energy produce the same fragment charge distributions, while fragment multiplicities scale with the total masses. These experimental observations naturally emerge in a full dynamical calculation including fluctuations , where multifragmentation arises from the spinodal decomposition of finite systems. The origin of multifragmentation may then be traced back to this phenemenon, at least for central collisions around 30-35 MeV/u, owing to the excellent agreement between static and kinetic variables observed in the experiments and in the calculations.

To put ultimate constraints on calculations, studies on correlations in events are in progress. They concern size correlations for fragments and fragment-fragment correlations (reduced velocity) to trace back the fragment topology at freeze-out.

References

1. B. Borderie, Int. Symposium on Large-Scale collective motions of Atomic Nuclei, Brolo Italy (1996) (eds G. Giardina, G. Fazio, M. Lattuada) World Scientific (1997) p.1.
2. L. G. Moretto and G. J. Wozniak, Ann. Rev. of Nuclear and Particle Science Vol 43 (1993) 379
3. Ph. Chomaz et al., *Phys. Rev. Lett.* **73**, 3512 (1994).
4. A. Guarnera et al., *Phys. Lett.* B **403**, 191 (1997).
5. J.D. Frankland, thèse, Université Paris XI Orsay, 1998, IPNO-T-98-06.
6. A. Le Fèvre, thèse, Université Paris 7-Denis Diderot (1997), GANIL T 97 03.
7. M. Colonna and Ph. Chomaz, *Phys. Rev.* C **49**, 1908 (1994)
8. A. Guarnera et al., *Phys. Lett.* B **373**, 267 (1996).
9. M.F. Rivet et al. (INDRA collaboration), *Phys. Lett.* B **430**, 217 (1998).
10. S. Ayik and J. Randrup, *Phys. Rev.* C **50**, 2947 (1994).

CLUSTERS AS SENSITIVE PROBES FOR A DEEPER UNDERSTANDING OF BARYON FLOW OBSERVABLES

M. PETROVICI

National Institute for Physics and Nuclear Engineering,
P.O. - MG6, 76900 Bucharest, ROMANIA
E-mail: mpetro@ifin.nipne.ro

FOPI COLLABORATION

Details on the azimuthal- and polar angle- dependence of the flow phenomena in semi-central and respectively central heavy ion collisions, using reaction products composed of few nucleons, are presented.

1 Introduction

As the first predictions on heavy ion collisions at relativistic energies have been based on hydrodynamical models it is obvious that part of the terminology has been adopted from hydrodynamics. This is the case also with the flow term. In this contribution we will concentrate on some aspects related to the flow in terms of correlated or ordered motion in which many particles are involved. In the moving frame the particles are not in rest, usually being characterized by a thermal motion. Concequently the associated thermal velocities are smaller for heavier objects. This is the most obvious argument which recommends the clusters as cleaner probes for detailed quantitative estimates on any type of flow.

The collective expansion of hot and dense nucler matter produced in heavy ion central collisions is well confirmed by now. The mass dependence of the fragments kinetic energy and their abundancy give strong support for the existence of an important collective expansion which cools down the disassembling matter so that heavy clusters can be produced. Although the central collisions appear to deliver the cleanest signal on the collective expansion, two aspects are worth to be mentioned: i) if the axial symmetry for such geometry holds, the spherical one has to be carefully studied, preequilibrium emission and tranparency effects influence the conclusion on the spheroidal shape of the expanding object, ii) the nuclear matter can escape freely in any direction perpendicular to the collision axis starting from the very first moments of the collision. The confinement is realized only by the remaining regions of the nuclei not yet entered in the reaction zone along the collision axis.

At lower centrality one has to deal with the complicated situation of rotating expanding objects in the presence of the spectator matter. However, there are some advantages of such studies: i) rotation and shadowing can be used as internal clocks for deeper information on the expansion dynamics, ii) the centrality can be used to control the shape and content of the fireball and shadowing matter, iii) for a given centrality the passage time of the shadowing objects can be controled varying the incident energy, iv) the confinement of the fireball by the spectators is more compact in the reaction plane.

Based on symmetry arguments these two type of phenomena are called azimuthaly symmetric and elliptic flows, respectively.

Systematic studies of different aspects of flow evidenced in the first experiments on relativistic heavy ion collisions performed with Phase I of the FOPI detector at GSI Darmstadt [1] have been followed in a series of experiments on Ni+Ni, Xe+CsI and Au+Au at energies between 90 and 400 A·MeV with complete configuration of the FOPI setup. The results presented here are based on this set of data. Details on the experimental configuration and analysis technique can be found elsewhere [2,3,4,5,6].

Section 2 is dedicated to the results of a multidimensional analysis of the incident energy where a transition from in-plane to out-of-plane azimuthal enhancement takes place, as a function of the mass of emitted particles, their transverse momentum and collision geometry. The excitation functions of the mean kinetic energy, flow and temperature for Z=2 reaction products in- and out-off-reaction plane for mid central collisions, are presented in Section 3. The polar angle- and A_{part}-dependence of the azimuthally symmetric correlated motion in highly central collisions is the subject of Section 4. The conclusions are presented in Section 5.

1.1 Transition from in-plane to out-of-plane emission for mid-central heavy ion collisions

Almost a decade ago, two types of azimuthal anisotropies have been evidenced in mid-central heavy ion collisions [7,9,10,11]. Following their shape relative to the reaction plane and the main mechanism behind them they have been baptized as in-plane and out-of-plane enhancement or rotational-like [7] and squeeze-out [8,9,10,11] phenomena, respectively. How the transition from one to the other scenario takes place as a function of incident energy, for different collision geometries and type of the reaction products was a question addressed few years later for light [12] and heavy [13,14,15,5,16,17,6] systems. Relative to the Phase I FOPI studies [13,16,17], the present data make possible the azimuthal distribution studies in the rotated reference frame, with no limitation on transverse momentum and as a function of mass of light charged reaction products. The procedure used to extract the transition energy (E_{tran}) can be followed in Fig.1. For a given centrality range, transverse momentum and mass of the analyzed reaction products, the squeeze-out ratio - $R_N=(1\text{-}a_2)/(1+a_2)$ (a_2 being the second coefficient of a Fourier expansion in azimuth), is represented as a function of incident energy (E_{inc}).

Squeeze-out signal corresponds to $R_N > 1$ while in-plane enhancement of the azimuthal distribution is characterized by $R_N < 1$. The continuous lines correspond to the result of fits using a polinomial function. The intersections of these lines with the one corresponding to $R_N = 1$ value are defined as E_{tran}. Fig.1 shows a clear mass dependence of the transition energy, lower values corresponding to heavier fragments for a given centrality and a given range in the scaled transverse momentum $p_t^{(0)} = (p_t/A)/(p_P^{cm}/A_P)$.

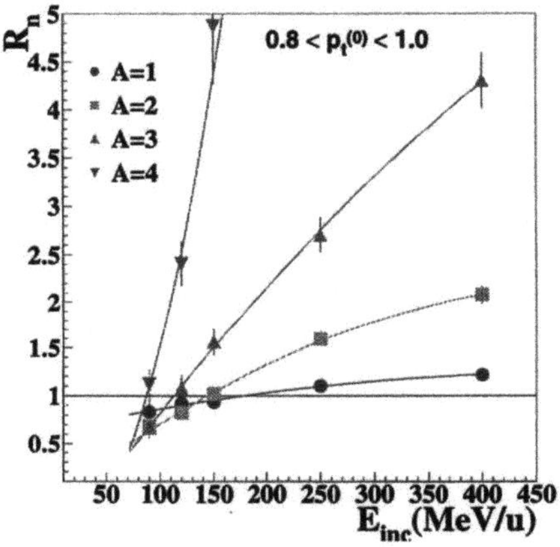

Figure 1. Squeeze-out ratio as a function of incident energies for Au + Au, CM3 centrality and A=1,2,3 and 4.

For a rotating emitting source one would expect a larger in-plane alignment for heavier fragments [14]. This effect alone can not explain the mass dependence of E_{tran}. Therefore a dynamical effect has to be considered besides the pure geometrical one of shadowing. Following the same recipe for different regions of $p_t^{(0)}$, a two dimensional dependence of the E_{tran} values as function of mass of the analyzed reaction products and transverse momentum can be obtained. The result is represented in Fig.2. A continuous decrease of the E_{tran} as a function of $p_t^{(0)}$ is evidenced for all analyzed particles and the difference in E_{tran} values for different particles is decreasing towards larger values of $p_t^{(0)}$. At large $p_t^{(0)}$ the contribution comes from larger expansion velocities, earlier expansion phase of the fireball [18,4], higher shadowing. At lower values of $p_t^{(0)}$, the expansion zones are less localized specially for light fragments due to larger contribution of the thermal velocities relative to the collective ones, the shadowing being less effective on the in-plane yields. These studies can be done for different centralities. The result of such analysis is represented in Fig.3 for Z=1,2 and 3 integrating on $p_t^{(0)}$. The shown centrality bins, ER5, ER4, CM3 and CM2, correspond in a geometrical "sharp cut-off" approximation to the impact parameter intervals 0-2, 2-4, 4-6 and 6-8 fm, respectively. At lower impact parameters the amount of energy which can be transferred in a rotational-like motion becomes smaller. In the same time the volume of the spectator matter causing the shadowing is reduced, while the number of baryons involved in the fireball region increases, contributing to larger expansion [19,20,21]. All these effects play a role in the observed experimental azimuthal distributions.

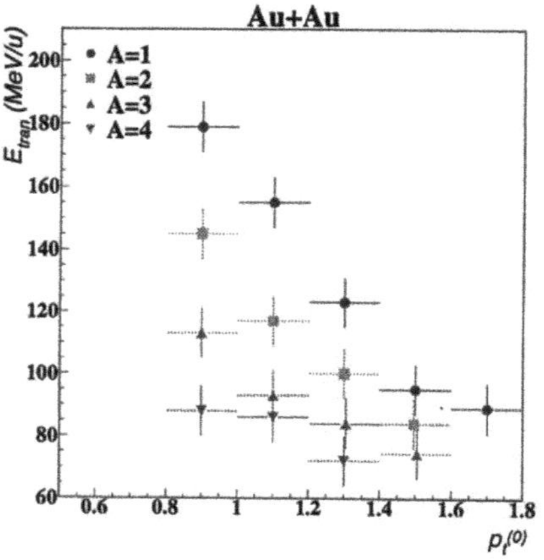

Figure 2. E_{tran} values as a function of $p_t^{(0)}$ for CM3 centrality and A=1,2,3 and 4.

However, the centrality dependence of the E_{tran} suggests that the reduction of the rotational motion of the fireball is the main source of the observed effect. Preliminary comparison with the IQMD (Isospin Quantum Molecular Dynamics) [22] indicates that both versions, HM (hard EoS - K=380 MeV, momentum dependent nucleon interaction) and SM (soft EoS - K=200 MeV, momentum dependent nucleon interaction) show the same qualitative behaviour of E_{tran} as a function of centrality; quantitatively the HM version predicting higher E_{tran} values than SM one. We have to mention here that the results presented above, peformed in the system of coordinate rotated with the sidwards flow angle relative to the beam direction, are based on corrected squeeze-out signals using a new method, recently proposed by Ollitrault [23],

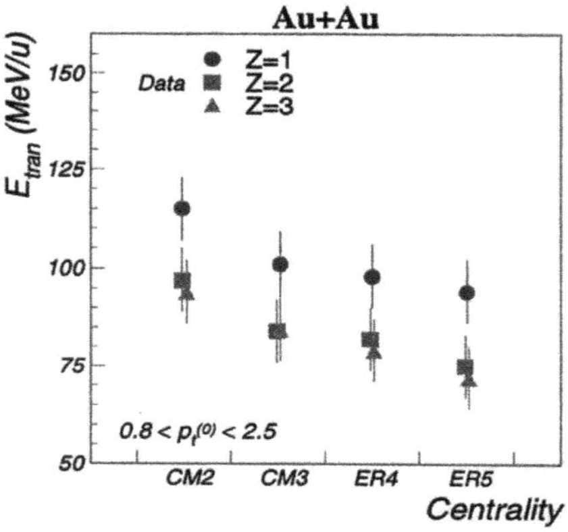

Figure 3. E_{tran} - centrality dependence for Z=1,2 and 3 reaction products.

deviced for the laboratory system. A consistent correction procedure is in progress. While the correction plays an important role on the absolute values of the squeeze-out signal, the preliminary results show that the effect on E_{tran} values is within the error bars.

2 Incident energy dependence of the mean kinetic energy, flow and temperature azimuthal distributions

The squeeze-out phenomena initially predicted by hydrodynamical calculations [8] has been extensively studied experimentally [9,10,11,24]. A unified representation of the observed trends can be done in terms of the azimuthal distribution of the total mass [25] or kinetic energy [4]. Previously reported comprehensive description of the squeeze-out phenomena in the range 0.25 - 1.15 A·GeV [25] was based on the parametrization of the total mass spectra with an expresion characteristic for radially symmetric shell expansion. As far as such a situation can hardly be encountered in heavy ion collisions and is not specific at all for mid-central geometries, we prefer to present also the experimental information free of any model which can be used to extract it.

Due to this reasons we report also the results on azimuthal distribution of the mean kinetic energy ($< E^{cm}_{kin} >$) besides the ones corresponding to β and T values. The azimuthal distributions of $< E^{cm}_{kin} >$, β and T can be fit with the following expresion: $(< E^{cm}_{kin} >, \beta, T) = (< E^{cm}_{kin} >, \beta, T)_0 - (\Delta < E^{cm}_{kin} >, \Delta\beta, \Delta T) \cdot \cos 2\Phi$. The excitation functions of $(< E^{cm}_{kin} >, \beta, T)_{max}$ and $(< E^{cm}_{kin} >, \beta, T)_{min}$ from 90 A·MeV to 400 A·MeV are represented in Fig.4. The elipsoidal character of the flow with the major axis perpendicular to the reaction plane constantly enhances from 90 to 400 A·MeV.

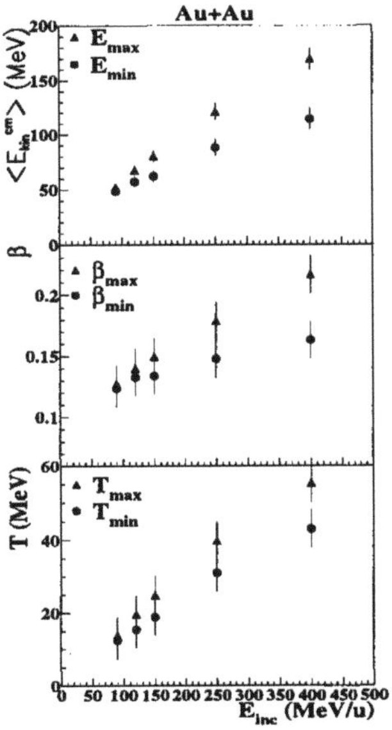

Figure 4. Excitation function for out-off-plane and in-plane $< E^{cm}_{kin} >$, β and T values for ER4 centrality and Z=2 fragments.

The lower slope in $(E, \beta, T)_{min}$ is due to the fact that the violence of the expansion increases faster as a function of incident energy than the decrease of the passage time. At higher incident energy the relativistic contraction and the gradient of the thermal pressure in the reaction plane due to the confinement created by the spectator matter are increasing. The two slopes reverse and a crossing point is expected. It would correspond to an elliptic flow with the major axis aligned in the reaction plane [26,27].

A larger modulation in the mean kinetic energy or flow relative to temperature can be explained by the larger variation of the expansion relative to the temperature as a function of break-up time [18,28].

3 Polar angle- and A_{part}- dependence of the correlated motion in heavy ion central collisions

Highly central collisions (1% cross section) have been slected using $E_{rat} = \sum_i E_{\perp,i} / \sum_i E_{||,i}$ observable, where i reffers to all particles identified by our detector. The experimental average kinetic energies represented as a function of mass of the corresponding reaction products show a linear dependence. The extracted slopes from such representations are presented in Fig.5 as a function of incident energy for the three symmetric systems and for two polar regions, $\theta_{cm} \in (25°\text{-}45°)$ (with a condition in azimuth of $\phi \in (90°(270°) \pm 45°)$) and $\theta_{cm} \in (80°\text{-}100°)$. One could easily follow the difference between the extracted slopes, representing the amount of the correlated motion per nucleon, in the forward and transversal direction relative to the collision axis. While in the forward region ($\theta_{cm} \in (25°\text{-}45°)$) no dependence of the extracted slopes on the mass of the colliding systems is observed, in the transversal direction ($\theta_{cm} \in (80°\text{-}100°)$) the dependence on the the barionic content of the fireball is quite clear.

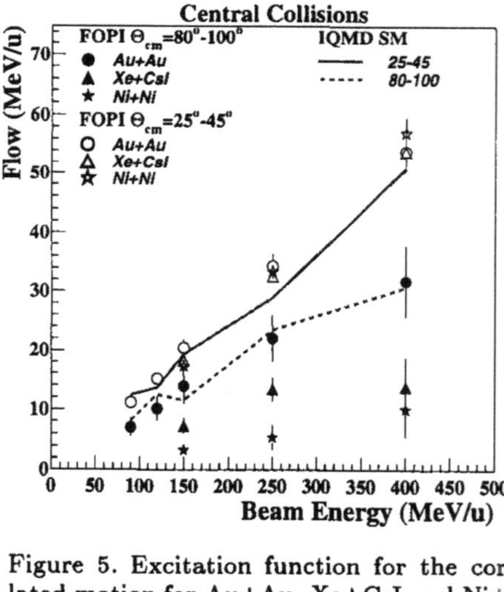

Figure 5. Excitation function for the correlated motion for Au+Au, Xe+CsI and Ni+Ni in two polar regions and comparison with the IQMD-SM model predictions.

The continuous and dashed lines represent the results of IQMD - SM estimates of the corresponding slopes for Au + Au in the corresponding polar regions. The agreement with the experimental data is quite satisfactory. IQMD simulations show a dependence of the average number of collisions per nucleon as a function of polar angle. In the average 2 times more colisions are suffered by the nucleons emitted in the transversal direction relative to the forward ones. The experimental trends corroborated with the results of the microscopic approach bring to the conclusion that the higher values of the slopes and their independency on the barionic content of the fireball in the forward direction is mainly due to transparency effect. A better estimate of the amount of expansion and its dependence as a function of the participating number of nucleons is visible in the transverse direction relative to the collision axis where the contamination from other effects has the lowest value.

4 Conclusions

Using a new generation of data collected with the complete FOPI experimental device we performed a systematic study of the azimuthal- and polar angle-dependence of the flow phenomena in semi-central and respectively central heavy ion collisions based on reaction products composed of few nucleons. A clear transition from in-plane (rotational-like) to out-of-plane (squeeze-out) was evidenced. The incident energy at which this transition takes place - E_{tran}, was studied as a function of mass of reaction products, their scaled transverse momentum ($p_t^{(0)}$) and centrality. The excitation functions of in- and out-off-plane $< E_{kin} >^{cm}$, β and T show that the elipsoidal character of the flow with the major axis perpendicular to the reaction plane constantly enhances from 90 to 400 A·MeV. A clear polar angle- and A_{part}-dependence of the correlated motion in highly central collisions was evidenced. Microscopic transport model (IQMD) calculations predict the same qualitative behaviour of E_{tran} as a function of centrality. Soft EoS explains better the trends observed in highly central collisions.

Acknowledgments

Special credit for the data presented here is given to A. Andronic and G. Stoicea. This work has been supported in part by the German BMBF under contract RUM-005-95, Deutsche Forschungsgemeinschaft (DFG) under projects 436 RUM-113/10/0 and contract B7/1999 financed by the Romanian Agency for Science, Technology and Inovation.

References

1. A. Gobbi, FOPI Collaboration, *Nucl. Instr. and Meth. in Phys. Res. A* **324** (1993) 156
2. J. Ritman, FOPI Collaboration, *Nucl. Phys. B (Proc. Suppl.)* **44** (1995) 708
3. D. Pelte, FOPI Collaboration, *Z. Phys. A* **357** (1997) 215
4. M. Petrovici, FOPI Collaboration, *Heavy Ion Physics at Low, Intermediate and Relativistic Energies using 4π Detectors*, ed. M. Petrovici et al., World Scientific, 1997, p. 216
5. A. Andronic, FOPI Collaboration, *Heavy Ion Physics at Low, Intermediate and Relativistic Energies using 4π Detectors*, ed. M. Petrovici et al., World Scientific, 1997, p. 209 and PhD Thesis - Bucharest, 1998
6. A. Andronic, FOPI Collaboration, will be published
7. W.K. Wilson et al., *Phys. Rev. C* **41** (1990) R1881
8. H. Stöcker et al., *Phys. Rev. C* **25** (1982) 1873
9. H.H. Gutbrod et al., *Phys. Lett. B* **216** (1989) 267
10. M. Demoulins et al., *Phys. Lett. B* **241** (1990) 476
11. H.H. Gutbrod et al., *Phys. Rev. C* **42** (1990) 640
12. R. Popescu et al., *Phys. Lett. B* **331** (1994) 285
13. A. Buţă et al., GSI Scientific Report 1994, 95-1 (1995) 57
14. W.K. Wilson et al., *Phys. Rev. C* **51** (1995) 3136
15. M.B. Tsang et al., *Phys. Rev. C* **53** (1996) 1959
16. N. Bastid, FOPI Collaboration, *Nucl. Phys. A* **622** (1997) 573
17. P. Crochet, FOPI Collaboration, *Nucl. Phys. A* **624** (1997) 755
18. M. Petrovici, FOPI Collaboration, *Phys. Rev. Lett.* **74** (1995) 5001
19. P. Danielewicz and Q. Pan, *Phys. Rev. C* **46** (1992) 2002
20. M.A. Lisa et al., *Phys. Rev. Lett.* **75** (1995) 2662
21. R. Pak et al., *Phys. Rev. C* **54** (1996) 1681
22. C. Hartnack et al., *Phys. Lett. B* **336** (1994) 131, *Mod. Phys. Lett. A* **9** (1994) 1151
23. J.-Y. Ollitrault, nucl-ex/**9711003** (1997)
24. D. Brill et al., *Phys. Rev. Lett.* **71** (1993) 336, Y. Leifels et al., *Phys. Rev. Lett.* **71** (1993) 963, D. Lambrecht et al., *Z. Phys. A* **350** (1994) 115, N. Bastid et al., *Nucl. Phys. A* **622** (1997) 573, Y. Shin et al., *Phys. Rev. Lett.* **81** (1998) 1576
25. S. Wang et al., *Phys. Rev. Lett.* **76** (1996) 3911
26. P. Danielewicz, *Phys. Rev. Lett.* **81** (1998) 2438
27. C.Pinkenburg et al., nucl-ex/**9903010** (1999)
28. P. Danielewicz, *Phys. Rev. C* **51** (1995) 716

SYSTEM EXPANSION AND CLUSTERS

M. KOROLIJA
Ruđer Bošković Institute, P.O.B. 1016, HR-10 001 Zagreb, Croatia
E-mail: korolija@faust.irb.hr

FOPI COLLABORATION

The kinetic energy spectra of light charged particles produced in the most central Au + Au collisions have been measured at the SIS/GSI accelerator facility using the *FOPI* detector. The spectra of the measured particles have been described by a model that includes thermal and collective degrees of freedom. By incorporating the conservation laws of total charge, baryon number, and energy into the analysis, we have found that there is more energy in the collective degrees of freedom than in the thermal ones. With the inclusion of velocity flow profile effects, the kinetic energy spectra of the measured particles could all be described with a relatively low freeze-out temperature.

1 Introduction

Heavy-ion collisions at relativistic energies offer a unique possibility of studying the properties of nuclear matter far from its ground state. The hydrodynamic [1,2], cascade [3], as well as models based on the quantum molecular dynamic picture [4,5], all predict that very hot and compressed nuclear matter can be formed as a transient state in high-energy collisions of medium and heavy nuclei. In particular, at a beam energy of $\approx$ 1 AGeV, the models predict that, for the most central collisions, nuclear matter is compressed by a factor of $2-3$ in comparison with its ground state. These head-on collisions are associated with high excitation energies in the central cell of the collision zone and, therefore, are followed by a large number of emitted particles.

The gross features of the single-particle inclusive spectra are well described by the existing theoretical models [6]. However, it has recently been observed that among these emitted particles there is a relatively large amount of light composite particles, i.e., clusters [7]. This is quite surprising since it is well known that light clusters are weakly bound particles. Therefore, they can only survive in a rather dilute system where strong interactions with the environment are rare. On the other hand, the density of the system can not be too dilute. As an example, let us consider the formation of the lightest composite particle, the deuteron. In this case, at least three nucleons must be involved: two to make the deuteron, and the third one to fulfil the conservation laws of energy and momentum.

2 Outline of the experiment and event selection criteria

The experiment was performed with the UNILAC/SIS accelerator facility at GSI, Darmstadt. The ^{197}Au + ^{197}Au reaction was measured with the *FOPI* detector at an incident energy of 1.06 AGeV. The *FOPI* detection system [8,9] consists of two major subsystems. The first subsystem is a forward plastic-wall detector (*PLA*) covering polar angles from 7° to 30°. The second subsystem is a superconducting solenoid which houses the central drift chamber. The target is positioned at the center of the solenoid.

Central events were selected by applying the cuts on multiplicity, and multiplicity plus directivity [10]. The charged-particle multiplicity distribution measured with the *PLA* detector was divided into 5 bins. The highest multiplicity bin (PM5) was defined in such a way that its lower multiplicity limit equaled half the plateau value of the multiplicity distribution. The other four (PM1-4) multiplicity bins were then divided into equal-width multiplicity bins. The cross sections $\sigma = 119\ mb$ ($\sigma = 47\ mb$) correspond to the highest, so called PM5 (PM5D1) multiplicity bin analyzed in this work. For a more thorough discussion on centrality criterions, we refer to Ref. [11]. Unless otherwise stated, the PM5 multiplicity bin will be used to select central events.

3 The model

The model used in this paper combines the early success of purely thermal models with the collective degrees of freedom describing the reexpansion of the initially compressed nuclear matter [12,13]. From the thermal point of view, the model relies on the assumption of a local thermal equilibrium, i.e., in the fluid rest frame, all particle spectra have the Maxwell-Boltzmann shape. From a hydrodynamic point of view, the model assumes a symmetric radial expansion, i.e., instead of solving the full hydrodynamic equations of motion, the solution is approximated to a symmetric radial flow of emitted particles.

The Lorentz-invariant shape of a Maxwell-Boltzmann spectrum having temperature T, and boosted radially with velocity β_f, is given as [13]

$$E\frac{d^3N_i}{dp^3} = \frac{d^2N_i}{p\ dE\ d\Omega} = C_i{\cdot}E{\cdot}e^{-\frac{\gamma_f E}{T}}\cdot\left[\frac{sinha}{a}\left(\gamma_f + \frac{T}{E}\right) - \frac{T}{E}cosha\right], \quad (1)$$

where N_i, E, and T are, respectively, the total number, the energy, and the temperature of the particle species i. The symbol C_i is an overall normalization for a given particle species i. The flow parameter a is given as $a = p\beta_f\gamma_f/T$, where p is the particle momentum. The parameters γ_f and β_f are the usual relativistic parameters γ and β given as ${\gamma_f}^2 = 1/(1-{\beta_f}^2)$.

To describe the spectra of emitted particles, Eq. (1) was used with C_i, β_f, and T as fitting parameters. In the fitting procedure the measured proton, deuteron, and triton/^{3}He kinetic energy spectra were simultaneously fitted at a given slice in the θ_{cm} angle. The values of the fitting parameters were determined by taking into account the conservation laws of charge and energy.

4 Results

As an example, Fig. 1 shows the measured (open symbols) and extrapolated (lines) triton/^{3}He kinetic energy spectra. The summed spectra of tritons and ^{3}He isotopes were used throughout the analysis in view of the fact that no difference was observed in the kinetic energy spectra of midrapidity tritons and ^{3}He particles [14]. As one sees from Fig. 1, owing to the geometrical limitations, the measured data lack the low-energy part of the spectra. Note, however, that the contributions from regions unaccessible to the detector are strictly bounded by the conservation laws. The kinetic energy spectra shown in Fig. 1 are obtained by applying the PM5 multiplicity cut. Almost identical spectra are obtained using the PM5D1 cut.

The open symbols in Fig. 1 refer to the *FOPI* data, whereas full symbols refer to the data measured by the *EOS* collaboration [14]. For comparison, the dashed lines in Fig. 1 refer to extrapolations into the full solid angle using the values $\beta_f = 0.32$ c and $T = 81$ MeV from Ref. [14]. A glance at Fig. 1 reveals that the solid lines do not describe the measured (symbols) triton/^{3}He spectra.

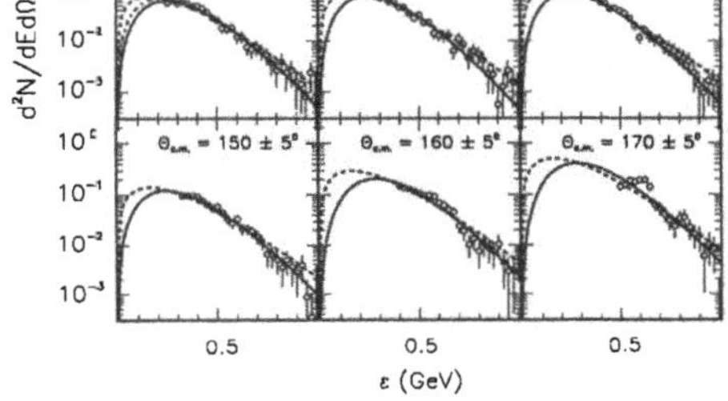

Figure 1. The measured triton/^{3}He kinetic energy spectra (open symbols) for different slices in the θ_{cm} angle. The full symbols refer to the results of Ref. [14], while the lines refer to two different extrapolations into the full solid angle. The dashed lines correspond to the values of the parameters β_f and T from Ref. [14], while the solid ones refer to the values of these parameters as deduced from the simultaneous fits. The dash-dotted lines at $\theta_{cm} = 110^0$ and 120^0 are used to obtain smooth extrapolation between the dashed lines, at $\theta_{cm} = 90^0$ and 100^0, and the solid lines at the most backward angles.

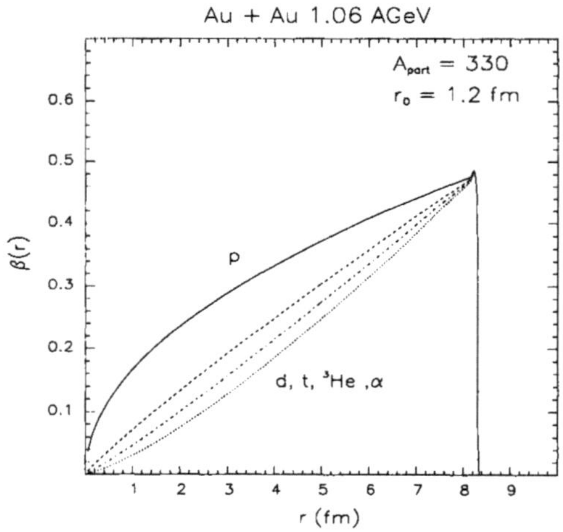

Figure 2. The extracted velocity flow profiles of protons (full line), deuterons/^{3}He (dashed line), tritons (dotted line), and alpha particles (dot-dashed line) obtained using the *FOPI* and the data from Ref. [14]. The freeze-out radius R was calculated as $R = r_0 \ A_{part}{}^{(1/3)}$, with $r_0 = 1.2$ fm, and the number of participants (A_{part}) is equal to 330.

However, by introducing velocity flow profile effects [15,16], it becomes possible to simultaneously fit the kinetic energy spectra of all measured baryons (p, d, t, ^3He, and α particles), with a common freeze-out temperature for all particle species.

The dependence of the radial flow velocity β_f on the radius r is parametrized as $\beta_r(\mathrm{r}) = \beta_0 \ (r/R)^n$, where β_0 and n specify the amount of radial flow and the shape of the velocity flow profile, respectively. With this dependence of $\beta_r(\mathrm{r})$ on r, the mean radial velocity $< \beta_r >$ is given as

$$< \beta_r > = \int_0^R \beta_r(r) r^2 dr \Big/ \int_0^R r^2 dr = \beta_0 \ \frac{3}{n+3} \ . \qquad (2)$$

As one sees from Fig. 2, the values of n for heavier fragments are scattered around a linear velocity flow profile ($n = 1$). On the other hand, protons with $n = 0.5$ (not shown) are less influenced by velocity flow profile effects. In fact, in Ref. [17] it was already observed that all particle species do not participate equally in the process of collective expansion. Owing to the assumed proportionality of β_r and r (see Eq. (2)), the increase of n with the particle mass m also implies that protons are the last particle species to freeze out, i.e., the heavier fragments freeze out somewhat earlier than protons. Such behavior points into the direction of a rather complex freeze-out surface. By introducing velocity flow profile effects, one finds relatively low freeze-out temperatures, i.e., ~ 46 MeV.

Regarding the division of the available kinetic energy into the collective and thermal degrees of freedom, we found that, using a single shell-expansion scenario, $\sim$ 43 % of the available kinetic energy is in the collective degrees of freedom, while only $\sim$ 32 % is in the thermal motion. The collectivity of $\sim$ 43 % is on its minimal side since velocity flow profile effects increase the collectivity of heavier fragments.

5 Conclusions

Light charged particles produced in the most central Au + Au collisions have been measured with the *FOPI* detector at SIS/GSI at a beam energy of 1.06 AGeV. The kinetic energy spectra of these particles have been described by a model that includes thermal and collective degrees of freedom. Supplementing this model with the conservation laws of total charge, baryon number, and energy, the possibilities of sharing the available energy between thermal and collective modes are limited. As a result, we have found more energy in the collective than in the thermal motion of participating nucleons. Introducing velocity flow profile effects, the kinetic energy spectra of light charged particles could all be described as having a common, and relatively low, freeze-out temperature.

References

1. W. Scheid *et al*, *Phys. Rev. Lett.* **32**, 741 (1974).
2. H. Stöcker *et al*, *Phys. Rev. Lett.* **44**, 725 (1980).
3. J. Cugnon *et al*, *Nucl. Phys.* A **352**, 505 (1981).
4. J. Aichelin and H. Stöcker, *Phys. Lett.* B **176**, 14 (1986).
5. J. Aichelin, *Phys. Rep.* **202**, 233 (1991).
6. S.A. Bass *et al*, *Prog. Part. Nucl. Phys.* **41**, 255 (1998), and references therein.
7. W. Reisdorf *et al*, *Nucl. Phys.* A **612**, 493 (1997).
8. A. Gobbi *et al*, Nucl. Instrum. & Methods **A324**, 156 (1993).
9. J. Ritman, Nucl. Phys. B (Proc. Suppl.) **44**, 798 (1995).
10. J.P. Alard *et al*, Phys. Rev. Lett. **69**, 889 (1992).
11. D. Pelte *et al*, Z. Phys. **A 357**, 215 (1997).
12. J.P. Bondorf *et al*, *Nucl. Phys.* A **296**, 320 (1978).
13. P.J. Siemens and J.O. Rasmunssen, Phys. Rev. Lett. **42**, 880 (1979).
14. M.A. Lisa *et al*, *Phys. Rev. Lett.* **75**, 2662 (1995).
15. K.S. Lee, U.Heinz, and E. Schnedermann, *Z. Phys.* C **48**, 525 (1990).
16. E. Schnedermann and U.Heinz, *Phys. Rev. Lett.* **69**, 2908 (1992).
17. W.C. Hsi *et al*, *Phys. Rev. Lett.* **73**, 3367 (1994).

EQUILIBRATION AND PHASE TRANSITION IN SPECTATOR REACTIONS

W. TRAUTMANN FOR THE ALADIN COLLABORATION
Gesellschaft für Schwerionenforschung mbH, Planckstr. 1, D-64291 Darmstadt, Germany

New results are reported from two recent experiments with the ALADIN spectrometer at SIS in which spectator decays following ^{197}Au on ^{197}Au collisions at bombarding energies up to 1 GeV per nucleon were studied. The evidence, obtained from these data, for the existence of an equilibrated breakup-state at low density is discussed. Particular attention is given to the kinetic degrees of freedom, and it is suggested that the slope temperatures derived from the fragment energy spectra may reflect the Fermi motion within the equilibrated spectator system prior to its decay.

1 Introduction

The observation of multifragmentation has given rise to the hope that signals of the predicted liquid-gas phase transition in nuclear matter may be identified in reactions of finite nuclei. Excited spectator systems, produced in reactions at relativistic bombarding energies, are of particular interest here as they seem to pass through highly equilibrated states of high temperature and low density before they decay. These conditions are similar to those expected for the coexistence region in the nuclear-matter phase diagram.

In this talk, the experimental evidence for equilibration at low density in spectator reactions will be reviewed. Temperatures obtained with different methods will be compared and tested for their compatibility with equilibrium at breakup. Particular attention will be given to the longstanding problem presented by the fragment kinetic energies. Their properties indicate equilibrium, however with corresponding temperatures that are much higher than those obtained from observables related to the chemical or thermal degrees of freedom. It will be shown that a consistent interpretation of these observations can be achieved if it is assumed that the fragment kinetic energies reflect the Fermi motion within the equilibrated spectator system prior to its decay.

The data were obtained in two recent experiments with the ALADIN spectrometer at SIS in which reactions of ^{197}Au on ^{197}Au in the regime of relativistic energies up to 1 GeV per nucleon were studied. In the first experiment, the ALADIN spectrometer was used to detect and identify the products of the projectile-spectator decay. The Large-Area Neutron Detector (LAND) was used to measure coincident free neutrons emitted by the projectile source.

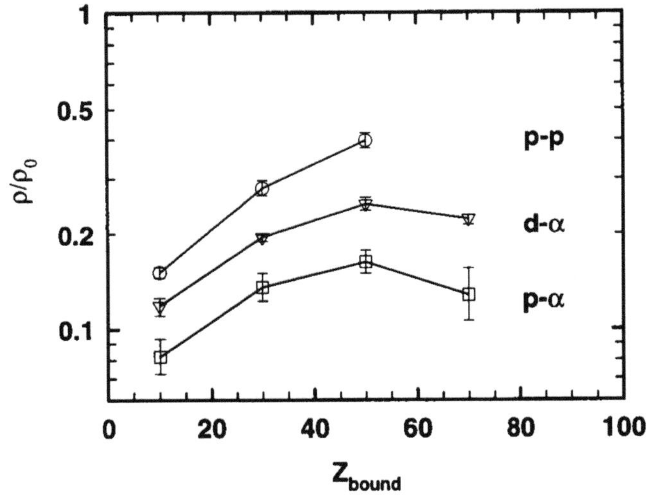

Figure 1. Breakup density ρ/ρ_0 as deduced from p-p, p-^{4}He, and d-^{4}He correlation functions for ^{197}Au + ^{197}Au collisions at 1000 MeV per nucleon. The error bars represent the statistical uncertainty.

In the second experiment, three multi-detector hodoscopes, comprising a total of 216 Si-CsI(Tl) telescopes, and three high-resolution telescopes, each consisting of three Si detectors followed by a 6-cm long CsI(Tl) scintillator, were positioned at backward angles and used to measure the yields and correlations of isotopically resolved light fragments from the target-spectator decay.

2 Density at Breakup

A breakup configuration with low density is one of the basic assumptions made in statistical multifragmentation models. [1,2] It seems supported by the fact that these models successfully reproduce the main features of multi-fragment decays. A direct experimental confirmation has now been obtained from the correlation functions constructed from proton-proton coincidences and from coincidences of protons and deuterons with α particles, measured for ^{197}Au + ^{197}Au collisions at 1000 MeV per nucleon with the multi-detector hodoscopes. [3] The analysis of the p-p correlation functions was performed with the Koonin-Pratt formalism, [4] while for the p-α and d-α correlations the numerical results of Boal and Shillcock were used. [5] Radii were deduced in the limit of zero-lifetime for the source, and densities were calculated by dividing the number of spectator constituents (from Refs. [6,7]) by the source volume. The results are shown in Fig. 1, plotted as a function of Z_{bound}. The sorting variable Z_{bound} is defined as the sum of the atomic numbers Z_i of all projectile fragments with $Z_i \geq 2$. Z_{bound} is close to the total charge of the original spectator system and is correlated with the impact parameter of the reaction.

The radii were found to be approximately constant, with values slightly larger than the radius of a gold nucleus, but the derived densities vary considerably with centrality. This is caused by the variation of the spectator masses which, in excellent agreement with the prediction of the geometric participant-spectator model, decrease with increasing centrality nearly in proportion to Z_{bound}. In the p-p case, the mean relative densities decrease from $\rho/\rho_0 = 0.4$ for the near-peripheral to below $\rho/\rho_0 = 0.2$ for the most central collisions. These values compare well with the densities assumed in the statistical multifragmentation models. [1,2] The differences between the p-p results and those from the resonances involving α particles are statistically significant and may be partially due to the different formalisms used. At present, their origin is not yet fully clear.[3]

3 Evidence for Equilibration

The universal features of the spectator decay, as apparent in the observed Z_{bound} scaling of the measured charge correlations, were the first and perhaps most striking indications for equilibrium. [8,9] The invariance of the fragmentation patterns with respect to the parameters of the entrance channel suggests that the memory of the dynamics governing the primary interaction of the colliding nuclei is lost.

Invariance with respect to the bombarding energy has also been established for the breakup temperature T_{HeLi} deduced from the measured ^{3,4}He and ^{6,7}Li isotopic yields. [10] In Fig. 2 the results from three measurements for ^{197}Au on ^{197}Au collisions are shown as a function of Z_{bound}. The acceptance of the ALADIN spectrometer provides full coverage for the projectile decay, and integrated yields were measured. The temperatures obtained at 600 and 1000 MeV per nucleon are found to be the same within the errors. For the target spectator, isotopic yields were measured at $\Theta_{lab} = 150°$ with a four-element telescope while Z_{bound} was determined for the coincident projectile decay with the ALADIN time-of-flight wall. The agreement with the temperatures obtained for the projectile decay at the same energy is expected from the symmetry of the collision and mainly illustrates the accuracy achieved in the experiments.

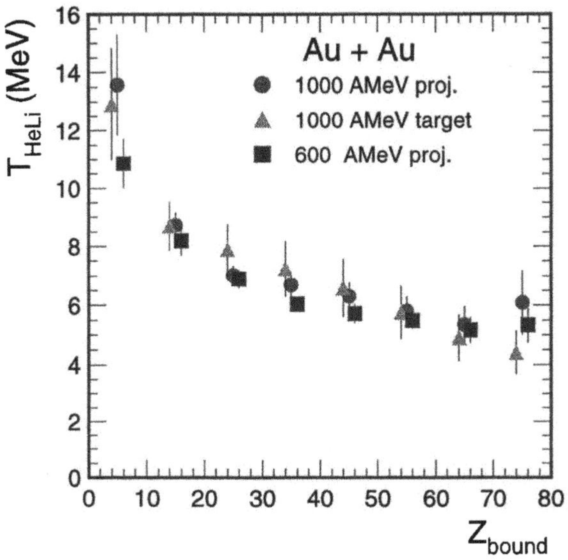

Figure 2. Temperatures $T_{\rm HeLi}$ for target (E/A = 1000 MeV) and projectile spectators ($E/A = 600$ and 1000 MeV) as a function of Z_{bound}. The data symbols represent averages over bins of 10-units width.

The success of the statistical multifragmentation models in describing the observed population of the partition space may be seen as a further argument for equilibration. Here the main task consists of finding an appropriate ensemble of excited nuclei to be subjected to the multi-fragment decay according to the model prescription. With empirically derived ensembles the measured correlations have been reproduced remarkably well, including their dispersions around the mean behaviour. [10,11] In addition, the equilibrium temperature obtained in this way from the model description coincides with the experimentally determined breakup temperature $T_{\rm HeLi}$. [10]

The emerging picture of a statistical decay of equilibrated breakup configurations appears, at first sight, not to be supported by other temperature observables. The largest discrepancies exist with the slope temperatures, obtained from fitting Maxwellians to the kinetic-energy spectra. They are much higher than the breakup temperatures discussed above. [9,12] This is a long-standing problem and will be addressed, in the light of the new data, in the following sections. Excited-state temperatures, derived from the yields of

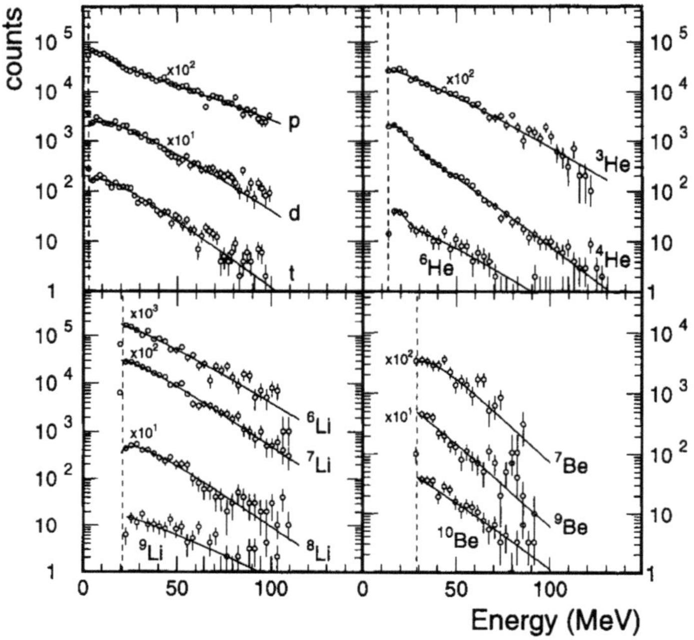

Figure 3. Energy spectra of light charged particles and fragments with $Z \leq 4$ for ^{197}Au + ^{197}Au at 1000 MeV per nucleon, measured at $\theta_{lab} = 150°$ and integrated over $20 \leq Z_{bound} \leq 60$. The full lines represent the fit results, the dashed lines indicate the identification thresholds of the telescope.

particle-unstable resonances, [13,14] are similar to the isotope temperatures for moderate excitations but then saturate at $T \approx 5$ MeV with further increasing excitation energy while the isotope temperatures continue to rise. This has been observed for several heavy-ion collisions, [15,16] including spectator reactions in ^{197}Au on ^{197}Au at 1000 MeV per nucleon. [7] Several mechanisms have been proposed as explanations (see [15] and references therein), and it is worth noticing that none of them questions the basic assumption of equilibrium at breakup. In the case of volume emission, it has to be considered that the resonances used for the temperature evaluation, even after breakup, may still weakly interact with the surrounding nuclear medium. They may not appear with their asymptotic energies and widths before they are released into vacuum. Excited-state temperatures may thus represent the internal fragment temperatures at the very final stages of the fragmentation process.

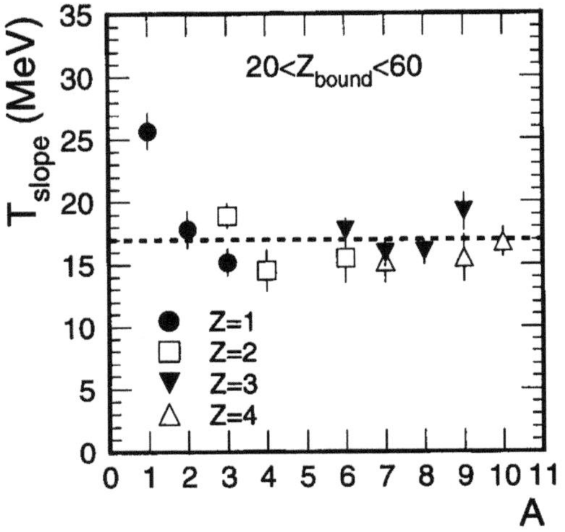

Figure 4. Slope temperatures for isotopically resolved charged particles and fragments with $Z \leq 4$ from the reaction ^{197}Au + ^{197}Au at E/A = 1000 MeV, measured at $\Theta_{lab} = 150°$ and integrated over $20 \leq Z_{bound} \leq 60$, as a function of the mass number A. For protons and α particles the value of the high-temperature component is shown.

4 Fragment Kinetic Energies

Energy spectra of light charged particles and fragments were measured with the high resolution telescopes at backward angles. Examples, obtained at $\theta_{lab} = 150°$ and integrated over $20 \leq Z_{bound} \leq 60$, are shown in Fig. 3. Most spectra exhibit Maxwellian shapes to a good approximation. For protons and α particles, two individual components have been identified of which the low-temperature component has a temperature of $T \approx 5$ MeV and an intensity correlated with the production rate of heavy residues, as expected for evaporation.

The slope parameters of the high-temperature components are shown in Fig. 4. The high value observed for protons of $T \approx 26$ MeV is consistent, after Coulomb corrections, with the results obtained with LAND for neutrons from the projectile decay in the same reaction. [7] It is believed to have its origin partly in the initial cascading stages of the collision. For the other species with $A \geq 2$, the spectra exhibit temperatures that are scattering closely around a mean value of 17 MeV (dashed line).

A constant temperature and the corresponding mass invariance of the mean kinetic energies is a very general indication of equilibrium. In the present case, the effects of the mutual Coulomb repulsion after breakup and of the small but finite motion of the target spectator have to be taken into account. This was done in a model study, based on realistic partitions and including many-body Coulomb trajectory calculations, which showed that the two effects are relatively small and, at backward angles, should cancel each other to a very good approximation. [17] The measured slope parameters of $T \approx 17$ MeV (Fig. 4) may thus be directly taken as representing the thermal components of the fragment kinetic energies.

5 Modes of Fragment Formation

It was suggested many years ago by Goldhaber that the large kinetic energies of fast-fragmentation products may have their origin in the nucleonic Fermi motion within the colliding nuclei. [18] He also pointed out that the resulting behavior is indistinguishable from that of a thermalized system with rather high temperature. For the Fermi momentum $p_F \approx 265$ MeV/c of heavier nuclei the corresponding temperature is $T = 15$ MeV. This is very close to the measured slope temperatures $T = 17$ MeV.

Goldhaber's idea, initially formulated for cold nuclei, has been extended to the case of expanded fermionic systems at finite temperatures by Bauer. [19] To use the prescription given there the temperature and density of the system prior to its decay have to be specified. For the temperatures, it was assumed that they are given by the measured breakup temperatures T_{HeLi} (Fig. 2). For the breakup density two values $\rho/\rho_0 = 1.0$ and 0.3 were used which correspond to two significantly different scenarii for the process of fragment production. The higher value is expected for a fast abrasion stage that will induce fragmentation of the remaining spectators but not affect much the nucleon motion within them. The lower value is in the range usually assumed for multi-fragment breakups of expanded systems. Here the development of instabilities, as the system enters the spinodal region, may cause a rapid fragmentation, whereby the Fermi motion of the nucleons may contribute to the fragment kinetic energies. For a homogeneous system at a lower density the Fermi motion is reduced and its effect will be smaller.

The resulting slope temperatures, represented by the lines in Fig. 5, are in the range of the experimental values. They reproduce the rise with decreasing Z_{bound} which follows as a consequence of the rising breakup temperature T_{HeLi}. It also seems that a more quantitative comparison will favor the prediction for $\rho/\rho_0 = 1.0$ and a fast breakup rather than that for $\rho/\rho_0 = 0.3$ and the scenario of homogeneous expansion.

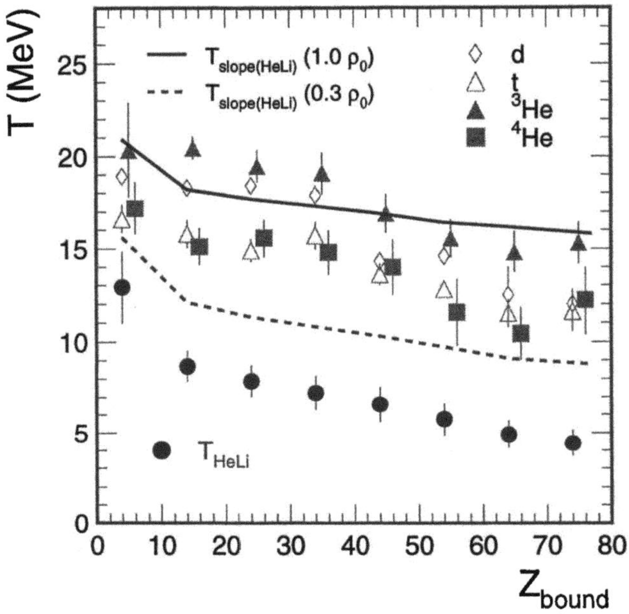

Figure 5. Slope temperatures for light charged particles of mass $2 \leq A \leq 4$ (squares, triangles and diamonds) and the isotope temperature T_{HeLi} (dots) as a function of Z_{bound}. The full and dashed lines are the slope temperatures obtained for the fast breakup of a Fermi gas with finite temperature T_{HeLi} with densities $\rho/\rho_0 = 1.0$ and 0.3.

6 Concluding Remarks

There is strong evidence in the data for the spectator systems' passing through equilibrated breakup states. The corresponding phase space determines the partitioning into fragments and light particles. The mean density is low as assumed in the statistical multifragmentation models which describe this phase space well. The measured breakup temperature T_{HeLi} is in good agreement with the equilibrium temperature of the model.

Corrections to the idealized breakup scenario result from the comparison of the different temperature observables. The temporal evolution of the system is emphasized by the saturation of the excited-state temperatures. The saturation temperatures $T \approx 5$ MeV is similar to the asymptotic temperatures obtained in dynamical models describing the reaction process. [21] The partitioning may be determined at an earlier time, as suggested by quantum or classical molecular dynamics models. [12,20]

The high values of the slope temperatures are quantitatively understood as a result of the Fermi motion of the constituent nucleons in the equilibrated system before its decay, as assumed in the Goldhaber model. This is consistent with a partitioning according to phase space but contradicts the equilibration of the kinetic degrees of freedom after breakup that is assumed in the statistical multifragmentation models.

The present data for the spectator decay in ^{197}Au on ^{197}Au collisions seem to favor a fast fragmentation over an intermediate stage of expansion of a homogeneous system. This may be different for other types of reactions. A comparison of fragment kinetic energies has shown that they may be smaller in reactions induced by light ions of high energy. [22] This question should be further investigated as it is related to the hierarchy of time scales for equilibration and disintegration of highly excited nuclear systems. [23]

References

1. D.H.E. Gross, *Rep. Prog. Phys.* **53**, 605 (1990).
2. J.P. Bondorf *et al.*, *Phys. Rep.* **257**, 133 (1995).
3. S. Fritz *et al.*, preprint GSI-99-17, nucl-ex/9906001.
4. S. Pratt and M.B. Tsang, *Phys. Rev.* C **36**, 2390 (1987).
5. D.H. Boal and J.C. Shillcock, *Phys. Rev.* C **33**, 549 (1986).
6. J. Pochodzalla *et al.*, *Phys. Rev. Lett.* **75**, 1040 (1995).
7. C. Groß, PhD thesis, Universität Frankfurt, 1998.
8. P. Kreutz *et al.*, *Nucl. Phys.* A **556**, 672 (1993).
9. A. Schüttauf *et al.*, *Nucl. Phys.* A **607**, 457 (1996).
10. Hongfei Xi *et al.*, *Z. Phys.* A **359**, 397 (1997).
11. A.S. Botvina *et al.*, *Nucl. Phys.* A **584**, 737 (1995).
12. P.B. Gossiaux *et al.*, *Nucl. Phys.* A **619**, 379 (1997).
13. J. Pochodzalla *et al.*, *Phys. Rev.* C **35**, 1695 (1987).
14. D.J. Morrissey *et al.*, *Ann. Rev. Nucl. Part. Science* **44**, 65 (1994).
15. V. Serfling *et al.*, *Phys. Rev. Lett.* **80**, 3928 (1998).
16. Hongfei Xi *et al.*, *Phys. Rev.* C **58**, R2636 (1998).
17. T. Odeh, PhD thesis, Universität Frankfurt, 1999.
18. A.S. Goldhaber, *Phys. Lett.* **53B**, 306 (1974).
19. W. Bauer, *Phys. Rev.* C **51**, 803 (1995).
20. H.W. Barz *et al.*, *Phys. Lett.* B **382**, 343 (1996).
21. C. Fuchs *et al.*, *Nucl. Phys.* A **626**, 987 (1997).
22. V.A. Karnaukhov *et al.*, *Phys. of Atomic Nuclei* **62**, 237 (1999).
23. M. Colonna *et al.*, *Phys. Rev.* C **55**, 1404 (1997).

SYSTEM SIZE EFFECTS IN CLUSTER PRODUCTION AT SIS ENERGIES

F. RAMI, B. DE SCHAUENBURG, P. WAGNER
(FOPI COLLABORATION)
Institut de Recherches Subatomiques, IN2P3-CNRS, Université Louis Pasteur, F-67037 Strasbourg Cedex 2, France
E-mail: fouad.rami@ires.in2p3.fr

Intermediate mass fragment production has been experimentally investigated in central collisions of Ni+Ni, Zr+Zr, Xe+CsI and Au+Au reactions at bombarding energies going from 90 to 400A MeV. The number of intermediate mass fragments produced per unit of incoming charge is found to be larger in heavier reactions. This system-size effect takes place with the occurence of the radial flow and its magnitude increases as the expansion velocity increases. The observed features indicate that the production mechanism of intermediate mass fragments in central collisions is strongly influenced by the expansion dynamics which seems to favour clusterization.

1 Introduction

The origin of the multifragmentation process is one of the central questions in heavy-ion collisions. A good characterization of this process is of crucial importance for the delimitation of the nuclear equation of state and the possible liquid-gas phase transition in nuclei.

At incident energies of a few hundreds MeV per nucleon, fragmenting nuclear systems can be produced through two types of heavy-ion reactions: peripheral collisions leading to the formation of highly excited spectator nuclei and central collisions in symmetric systems producing a nuclear fireball. Multifragmentation of spectator matter has been extensively investigated over the last decade, in particular by the ALADIN collaboration [1]. It was shown that the decay properties of spectator nuclei exhibit a universal behaviour (beam energy and projectile/target size independence) consistent with a purely thermal process where fragments are emitted from an equilibrated source. In the present work, we study the multi-fragment production in central reactions. Here, in contrast to peripheral collisions, large collective expansion velocities were observed [2,3]. The collective expansion energy was found to represent a significant fraction of the energy available in the center-of-mass (c.m.) frame. The question then arises whether such collective expansion effects may influence the fragment formation process. To explore this question, we compare the fragment yields in different symmetric systems with masses ranging from

$A_P = A_T = 58$ to 197 (A_P and A_T are the masses of, respectively, the projectile and the target nuclei): one expects an increasing effect of the compression-decompression process with increasing size of the interacting system.

2 Experimental conditions and event selection

The data presented in this contribution were taken in a series of experiments performed recently using the phase II setup of the FOPI detector [4,5] at the SIS accelerator of GSI-Darmstadt. Four symmetric reactions, Ni+Ni, Zr+Zr, Xe+CsI and Au+Au, were measured at beam energies going from 90A MeV to 400A MeV. Clusters or intermediate mass fragments, IMF's ($3 \leq Z \leq 15$), were detected in the forward wall [4] covering laboratory polar angles below 30°, corresponding to the coverage of a very large fraction of the forward c.m. hemisphere.

Central collisions were selected using the E_{RAT} criterium [2,3]. E_{RAT} is determined event wise by including all particles detected by the forward wall: $E_{RAT} = \sum_{i=1}^{M} E_t^i / \sum_{i=1}^{M} E_l^i$, where E_t^i and E_l^i are the transverse and longitudinal energies of the particle i and M is the event multiplicity. The E_{RAT} cut used to select central collisions corresponds to about 2% of the total reaction cross section and to a maximum geometrical impact parameter $b \leq 0.1 \times b_{max}$ (b_{max} being the maximum impact parameter) in a sharp-cut-off approximation.

3 Influence of system size effects on IMF production

Multi-fragment emission in central Au+Au reactions was the object of detailed studies in previous experiments [3,6,7]. It was shown that IMF's are copiously emitted in these reactions. The maximum production was observed at an incident energy close to 100A MeV. At this energy, about 13 IMF's on the average are produced per event and a large fraction, about 40%, of the initial charge is emitted in the form of IMF's.

These studies were extended in the present work to several symmetric reactions: Ni+Ni, Zr+Zr and Xe+CsI in addition to Au+Au. This allowed us to vary the size of the interacting system over a broad range; the number of interacting nucleons increases by a factor of 3.4 from Ni+Ni to Au+Au. IMF multiplicities were extracted for highly central collisions under the same condition on the scaled impact parameter ($b/b_{max} \leq 0.1$), allowing therefore direct comparisons between the four measured reactions. Fig.1 shows examples of charge distributions, dM/dZ, measured at two incident energies of 150 and 400A MeV. As can be seen, the same typical exponential fall is observed

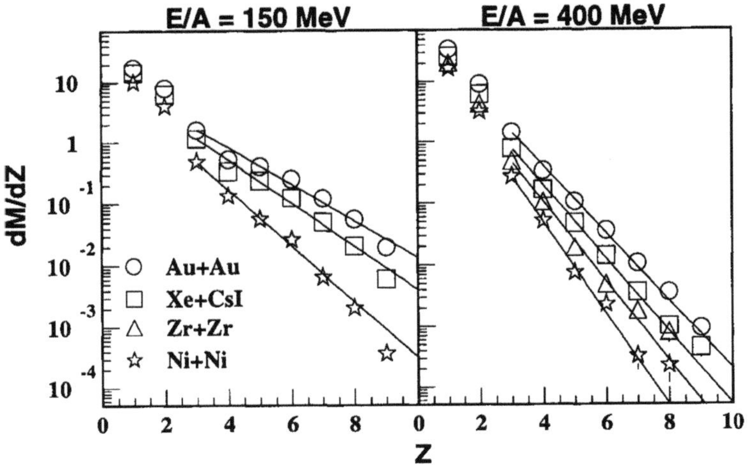

Figure 1. Measured multiplicities in the forward c.m. hemisphere as a function of the particle charge. The data are shown for central collisions of Au+Au, Xe+CsI, Zr+Zr and Ni+Ni at two bombarding energies : 150 AMeV (left) and 400 AMeV (right). The straight lines are exponential fits to the $Z \geq 3$ data points. Statistical uncertainties are smaller than the size of the symbols.

for all reactions. This trend is very different from the power-law dependence characterizing the multifragmentation of spectator nuclei. Another important feature of these dM/dZ-distributions is that their slope becomes steeper as the size of the system decreases: one produces relatively less heavy fragments in smaller systems. This observation is more quantitatively visible in Fig.2 where the results are expressed in terms of the number of IMF's per unit of incoming charge, i.e. IMF multiplicity scaled to the size of the interacting system. It is clearly seen that the production of IMF's raises rapidly, almost linearly, with increasing number of interacting protons. At 400A MeV, the number of IMF's per unit of incoming charge produced in Au+Au collisions is larger by a factor of ~ 2.1 than in Ni+Ni. This striking observation indicates that multifragmentation of the fireball in central collisions does not exhibit the universal behaviour observed for the spectator decay [1]. We will see in the next section that these differences in the properties of multifragmentation between central and peripheral reactions might be attributed to the role of the expansion dynamics.

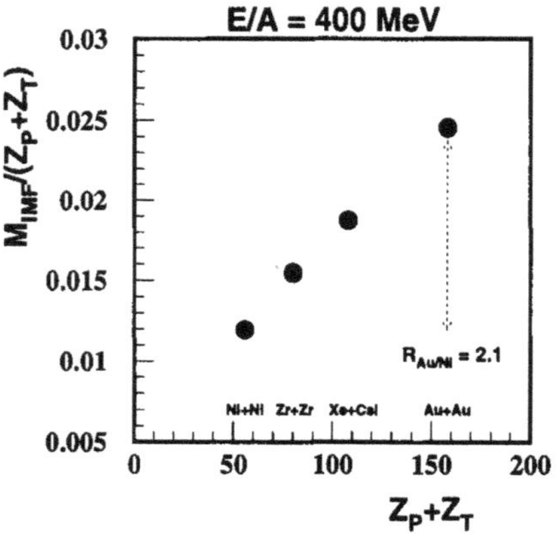

Figure 2. Scaled average IMF multiplicity as a function of the total projectile+target charge. M_{IMF} were measured in the forward c.m. hemisphere and then multiplied by a factor of 2. Data are shown for central collisions at the incident energy of 400 AMeV. Statistical uncertainties are smaller than the size of the symbols.

4 Interplay between IMF formation and the radial flow

The influence of system-size effects in IMF production becomes more and more pronounced as the incident energy raises from 90 to 400 AMeV. This is shown in the upper panel of Fig.3 where the quantity $R_{Au/Ni}$, i.e. the ratio of the scaled IMF multiplicities for Au+Au over Ni+Ni, is plotted as a function of the energy available in the c.m. frame. As is seen in this figure, size effects seem to take place at an available c.m. energy of $E_{cm} \sim 10$ AMeV. The latter value is close to the threshold energy for expansion extracted from the excitation function of the mean radial flow velocity (see lower panel of Fig.3).

Fig.3 shows a remarkable correlation between system-size effects and the collective flow motion. Size effects, characterized by the ratio $R_{Au/Ni}$, take place with the occurence of expansion and their magnitude increases as the expansion velocity increases. This correlation indicates that here, in contrast to the spectator decay, the production mechanism of IMF's is intimately connected to the expansion dynamics of the nuclear system. In the absence of expansion (below the threshold), one expects a system-size independent partition of the nuclear system; this would correspond to a ratio $R_{Au/Ni} = 1$ in Fig.3. Recent results from the INDRA experiment [8], at an incident energy where the expansion effect is negligible, show indeed a system-size independent multifragmentation.

The large production of IMF's in Au+Au collisions, as compared to lighter reactions, can be attributed to the fact that heavy systems are subjected to a larger collective expansion motion which seems to favour clusterization. This is supported by recent preliminary results [9] where the collective flow energy was found to be larger in heavier systems. Within a hydrodynamical picture,

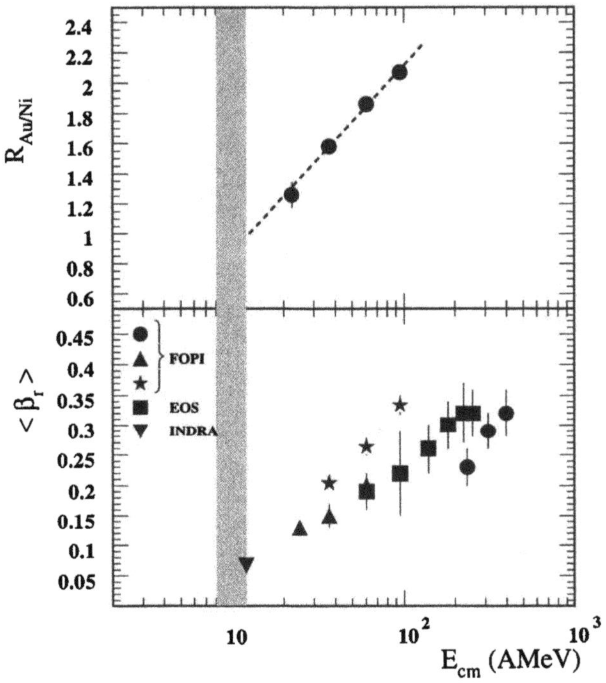

Figure 3. Upper panel: Ratio $R_{Au/Ni}$ of scaled IMF multiplicities measured in Au+Au to those measured in Ni+Ni as a function of the energy available in the c.m. frame. Lower panel: Compilation of radial flow velocities obtained in various experiments as a function of c.m. available energy. Circles, triangles and stars are taken from the FOPI experiment [3,10,11], squares are from the EOS experiment [12] and inverted triangles are from the INDRA experiment [13]. The vertical shadowed area indicates the E_{cm} region where the collective radial expansion sets in.

it is tempting to attribute the observed size effects to the cooling action of bulk expansion which might be responsible for the higher condensation in larger systems [3,14].

5 Conclusion

In this contribution, we have presented a detailed experimental study of the multifragmentation process in central heavy-ion collisions at beam energies from 90 to 400*A* MeV. The data show evidence for system-size effects: the probability to produce IMF's is larger in heavier systems. These size effects take place with the occurence of expansion and their magnitude increases as the expansion velocity increases. These features depart significantly from the universal behaviour observed in the multifragmentation of spectator nuclei. They indicate that the production mechanism of IMF's in central collisions is intimately connected to the expansion dynamics which seems to favour clusterization.

Acknowledgments

This work was supported in part by the French-German agreement between GSI and IN2P3/CEA.

References

1. A. Schüttauf *et al*, ALADIN Collaboration, *Nucl. Phys.* A **607**, 457 (1996).
2. S.C Jeong *et al*, FOPI Collaboration, *Phys. Rev. Lett.* **72**, 3468 (1994).
3. W. Reisdorf *et al*, FOPI Collaboration, *Nucl. Phys.* A **612**, 493 (1997).
4. A. Gobbi *et al*, FOPI Collaboration, *Nucl. Instrum. Methods* A **324**, 156 (1993).
5. J. Ritman *et al*, FOPI Collaboration, *Nucl. Phys.* B (Proc. Suppl.) **44**, 708 (1995).
6. J.P. Alard *et al*, FOPI Collaboration, *Phys. Rev. Lett.* **69**, 889 (1992).
7. C. Kuhn *et al*, FOPI Collaboration, *Phys. Rev.* C **48**, 1232 (1993).
8. M. F. Rivet *et al*, *Phys. Lett.* B **430**, 217 (1998).
9. B. de Schauenburg, Ph-D thesis, Strasbourg, France, IReS 99-06 (1999).
10. B. Hong *et al.*, FOPI Collaboration, *Phys. Rev.* C **57**, 244 (1998).
11. G. Poggi *et al.*, FOPI Collaboration, *Nucl. Phys.* A **586**, 755 (1995).
12. M. Lisa *et al.*, *Phys. Rev. Lett.* **75**, 2662 (1995).
13. R. Bougault *et al.*, XXXV International Winter Meeting on Nuclear Physics, Bormio (1997).
14. W. Reisdorf, Contribution to this Conference.

GENERALIZED APPROACH TO ENTROPY DETERMINATION IN HEAVY-ION COLLISIONS

M. DŽELALIJA

Department of Physics, University of Split, Teslina 12, 21000 Split, Croatia
E-mail: Mile.Dzelalija@pmfst.hr

N. CINDRO, Z. BASRAK, R. ČAPLAR, AND M. KOROLIJA

Ruđer Bošković Institute, Bijenička 54, 10000 Zagreb, Croatia

Assuming the uniformly distributed nucleons in configuration space at the time of freeze out, we propose the way for a unique cell-size determination in momentum space. Then, using the nucleon occupation probability function in momentum space, we determine the values of generalized entropy for the Au + Au reaction at incident energies of 150, 250, and 400 A MeV. The data have been obtained with the FOPI facility at SIS/GSI, Darmstadt. Central collisions have been selected by a proper cut of the ERAT observable defined as the ratio of the transverse to the longitudinal kinetic energies of the fragments in the center-of-mass forward hemisphere. The extracted values of entropy per nucleon are in good agreement with the values obtained in earlier FOPI Collaboration studies.

1 Introduction

One of the most interesting goals in relativistic nuclear physics is to study the behavior of hot and dense nuclear matter in terms of different physical quantities. It is generally assumed that the quantity that remains almost unchanged through the final stages of the interaction is entropy. [1] Thus, the measurement of baryonic entropy yields information on the state of hot and dense nuclear matter. However, the entropy information on nuclear matter produced in heavy-ion collisions is not easily accessible.

In a number of experimental papers the value of entropy has been determined for various nuclear systems and projectile energies using many computer models and different global observables. [2,3,4] These different observables have been compared with those calculated using different models. Such a determination of entropy is obviously model dependent. In the present paper we propose to study the amount of entropy generated at the time of freeze-out without recurring to simulation models. For this purpose, we use the general approach to entropy extraction in fermionic systems. Historically, this connection was introduced by Boltzmann and provides the statistical basis of thermodynamics. In general, entropy is just proportional to the logarithm of the number of the possible microscopic states in phase space at given excitation energy,

$$S = k \ln N(E) = -k \ln P(E), \tag{1}$$

where $k = 1$ is the Boltzmann constant and $P(E) = N^{-1}(E)$ is the probability assumed to have the same value for each one of the microstates.

For a nonequilibrated system of independent fermions, the generalized entropy can be defined as

$$S = -k\frac{g}{h^3}\int\int \{f(\vec{r},\vec{p})\ln f(\vec{r},\vec{p}) + [1 - f(\vec{r},\vec{p})]\ln[1 - f(\vec{r},\vec{p})]\}\, d\vec{r}\, d\vec{p}, \tag{2}$$

where g is the spin-isospin degeneracy factor ($g = 4$ in our case), h is the Planck constant and $f(\vec{r},\vec{p})$ is the single-particle occupation probability of phase space. The explicit dependence of $f(\vec{r},\vec{p})$ on time is dropped since our study is restricted to the freeze-out conditions. This definition leads to the usual thermodynamic entropy in the case of global thermodynamic equilibrium.

This paper is organized as follows. Section 2 briefly introduces the idea of a unique entropy determination at the time of freeze-out. The values of generalized entropy per nucleon in Au + Au collisions are presented and discussed in Sec. 3. A summary is given in Sec. 4.

2 Entropy Determination

It has to be stressed that the definition of generalized entropy is not unique and depends on the cell size. [5] Thus, according to the Heisenberg uncertainty principle, we divided 6-dimensional single-particle phase space into several cubic cells with the size

$$(\Delta\Omega_x)^3 = (\Delta x)^3(\Delta p_x)^3 = h^3. \tag{3}$$

After that the generalized entropy from Eq. (2) becomes

$$S = -g\sum_{\vec{r}}\sum_{\vec{p}} \{f(\vec{r},\vec{p})\ln f(\vec{r},\vec{p}) + [1 - f(\vec{r},\vec{p})]\ln[1 - f(\vec{r},\vec{p})]\}\,. \tag{4}$$

The distribution function $f(\vec{r},\vec{p})$ depends on six variables and it would not be practical to make a brute-force division into cells in all six dimensions. Since from heavy-ion data we can have the occupation probability only in momentum space, we have to reduce the dependence of the distribution function to its dependence on momentum variables. The spatial density of nucleons is related to the occupation probability $f(\vec{r},\vec{p})$ by

$$\rho = \frac{g}{h^3} \int f(\vec{r},\vec{p})\, d\vec{p}. \tag{5}$$

After dividing the momentum space into cubic cells with the size $(\Delta p_x)^3$, for the spatial density one obtains

$$\begin{aligned} \rho &= \frac{(\Delta p_x)^3}{h^3} g \sum_{\vec{p}} f(\vec{r},\vec{p}) \\ &= \frac{(\Delta p_x)^3}{h^3} A, \end{aligned} \tag{6}$$

where A is the total number of nucleons. Using the above expression and Eq. (3) one obtains that the cell size in configuration space equals the total volume of the system

$$(\Delta x)^3 = \frac{A}{\rho} = V. \tag{7}$$

Now, we can suppose that nucleons are uniformly distributed in configuration space at the time of freeze-out and, hence, the entropy becomes

$$S = -g \sum_{\vec{p}} \{ f(\vec{p}) \ln f(\vec{p}) + [1 - f(\vec{p})] \ln[1 - f(\vec{p})] \}\,. \tag{8}$$

From Eq. (6) it is easy to calculate the size of cubic cells in momentum space as a function of freeze-out density ρ as

$$(\Delta p_x)^3 = h^3 \frac{\rho}{A}. \tag{9}$$

For the Au + Au reaction ($A = 394$) at the assumed freeze-out density $\rho = 0.3\rho_0$, one infers that the value for the size of a cell is $\Delta p_x \approx 60$ MeV/c.

In this paper we have used the data collected at the SIS/ESR accelerator at the GSI, Darmstadt, using the 4π-detector system FOPI described in detail elsewhere. [6,7] The ratio of the transverse to the longitudinal kinetic energy in the center of mass [8]

$$ERAT = \left[\frac{\sum_i E_\perp^{(i)}}{\sum_i E_\parallel^{(i)}} \right]_{(y \geq 0)_{c.m.}} \geq 0.76 \tag{10}$$

is used as a criterion to select central events. To determine the experimental occupation-probability function, all detected light and heavier particles in the center-of-mass polar-angle range 25^0-45^0 are used and extrapolated to the whole system. Each reaction product is taken as a (sub)system of the corresponding number of nucleons. The determined entropy is discussed in the next section.

3 Discussions

For the chosen freeze-out density $\rho = 0.3\rho_0$, the present values of entropy per nucleon are shown in Fig. 1, together with the results of Doss *et al.* [4] and the results of the earlier analyses of the FOPI Collaboration data. [2,3,9] The results of calculations corresponding to a fireball source [10] and a hydrodynamic model [11] are also shown in the figure. The solid hydrodynamic curve

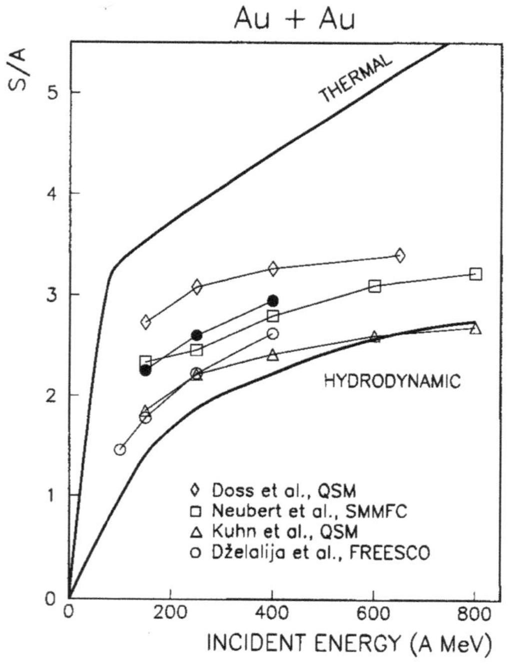

Figure 1. Baryonic entropy S/A plotted as a function of beam energy. The earlier results of Doss *et al.* are shown as diamonds, whereas the earlier results of the FOPI Collaboration (Kuhn *et al.*, Dželalija *et al.* and Neubert *et al.*) are displayed as indicated in the figure. The present results are shown as full circles. The solid curve labeled THERMAL is calculated from the fireball model and the other solid curve, labeled HYDRODYNAMIC, from a very hard equation of state. The thin solid lines connecting the extracted values of S/A are to guide the eye.

in the figure corresponds to a very hard equation of state ($K = 550$ MeV). The discrepancy between the values of entropy obtained in this paper and those calculated from the fireball model [10] (which assumed that all of the initial center-of-mass energy was converted into thermal energy) points out the importance of including the collective-motion component for a better understanding of the mechanism of relativistic heavy-ion collisions. The reasonable values obtained for entropy in this paper and the good agreement with the values obtained in earlier FOPI studies are an *a posteriori* justification for this approach.

The present approach, however, shows an inherent uncertainty in the obtained values of entropy. The principal uncertainty stems from the freeze-out density assumption. Also, this approach gives a single-particle entropy, while, in general, one should consider an N-body entropy, for which the distribution in the full 2N-dimensional phase space should be invoked.

4 Summary

We have used a generalized approach to the single-particle entropy determination in central nucleus-nucleus collisions. This approach has been applied to Au + Au collisions in the energy range from 150 to 400 A MeV. The approach is based on the unique cell-size determination in momentum space for a chosen freeze-out density. In the region of the overlap, the values of entropy obtained in this paper agree fairly well with the values of entropy reported in earlier FOPI Collaboration studies.

5 Acknowledgments

It is a pleasure to thank all members of the FOPI Collaboration for helpful and stimulating discussions. The authors also gratefully acknowledge the financial support of the Ministry of Science and Technology of Croatia.

References

1. I.M. Mishustin *et al*, *Phys. Lett.* B **95**, 361 (1980).
2. C. Kuhn *et al*, *Phys. Rev.* C **48**, 1232 (1993).
3. M. Dželalija *et al*, *Phys. Rev.* C **52**, 346 (1995).
4. K.G.R. Doss *et al*, *Phys. Rev.* C **37**, 163 (1988).
5. O. Penrose, *Rep. Prog. Phys.* **42**, 129 (1979).
6. J.P. Coffin, *Int. J. Mod. Phys.* E **1**, 739 (1992).
7. A. Gobbi *et al*, *Nucl. Inst. Meth.* A **324**, 156 (1993).
8. W. Reisdorf, *Proceedings of the XX International Workshop on Gross Properties of Nuclei and Nuclear Excitation*, ed. H. Feldmeier, Darmstadt, 1992, p. 38.
9. W. Neubert *et al*, to be published.
10. J. Gosset *et al*, *Phys. Rev.* C **16**, 629 (1977).
11. H. Stöcker *et al*, *Phys. Lett* **103** B, 269 (1981)

FRAGMENT PRODUCTION IN PROTON INTERACTIONS WITH LIGHT NUCLEI

L. N. ANDRONENKO, M. N. ANDRONENKO, D. M. SELIVERSTOV
St.Petersburg Nuclear Physics Institute, 188350 Gatchina, Russia
E-mail: land@rec03.pnpi.spb.ru

W. NEUBERT
Forschungszentrum Rossendorf e.V., 01314 Dresden, Germany
E-mail:neubert@fz-rossendorf.de

Isotopic yield ratios were derived from double-differential cross sections $d^2\sigma/dEd\Omega$ measured in $p+{}^9Be$ and $p+{}^{12}C$ collisions at 1 GeV. Enhanced ^6He yields were found compared to heavier target nuclei.

1 Experiment and Results

Inclusive measurements were performed at the external beam of the PNPI synchrocyclotron. The reaction products were registered by two spectrometers arranged at $\Theta_{lab} = 30^o$ and 126° with respect to the beam axis. In each branch, two Parallel Plate Avalanche Counters provided the TOF and the trigger signal for the following twin Bragg Ionization Chamber (see ref. [1]). The pulse shape measured by Flash ADC's allowed to derive the kinetic energy, energy loss, nuclear charge (Z) and range of the fragments. From these quantities we obtained isotopically resolved kinetic energy distributions to which a moving-source formula was fitted (ref. [2]). Integral yields obtained by restoring the missing part of the spectrum were converted into absolute cross sections using the monitor reaction $p+{}^{27}Al \rightarrow {}^7Be+x$. In order to minimize systematic uncertainties we consider isotope yield *ratios* R=yield(N_2, Z)/yield(N_1, Z) which were classified by the difference in the neutron number $\Delta N = (N_2, Z) - (N_1, Z)$.

isotope	σ (mb)	T_{slope} (MeV)	$\beta \cdot 10^{-3}$ (units of c)
^3He	27.8±0.9	6.8±0.2	6.7±0.3
^4He	149.0±1.6	5.8±0.3	7.6±0.2
^6He	0.71±0.06	6.4±0.2	9.2±0.4
^6Li	9.6±0.3	7.1±0.2	7.6±0.2
^7Li	10.4±0.3	8.4±0.2	8.2±0.2
^7Be	5.7±0.2	11.1±0.2	12.5±0.3
$^{9,10}Be$	3.3±0.3	12.7±0.4	11.8±0.3
$^{10,11}B$	8.2±0.3	17.9±0.4	11.5±0.3

Table 1: Fragment yields from $p+{}^{12}C$ at 1 GeV, ref. [3]. Temperature T_{slope} and velocity β: moving-source fit parameters to kinetic energy spectra, the given errors consider only statistical fluctuations. Afterwards the Be and B isotopes at the bottom have been resolved by Gaussian decomposition of the corresponding mass distributions.

isotope	relative yield	T_{slope} (MeV)	$\beta \cdot 10^{-3}$ (units of c)
^{3}He	$1.71 \cdot 10^{-1} \pm 1.0 \cdot 10^{-3}$	5.3±0.1	9.1±0.7
^{4}He	$1.00 \pm 1.0 \cdot 10^{-3}$	5.5±0.1	6.1±0.2
^{6}He	$1.05 \cdot 10^{-2} \pm 3.0 \cdot 10^{-4}$	7.3±0.3	9.5±0.7
^{6}Li	$3.97 \cdot 10^{-2} \pm 8.3 \cdot 10^{-4}$	8.3±0.1	7.6±0.3
^{7}Li	$5.53 \cdot 10^{-2} \pm 7.8 \cdot 10^{-4}$	11.2±0.1	9.6±0.3
^{8}Li	$2.53 \cdot 10^{-2} \pm 5.8 \cdot 10^{-4}$	20.0±0.2	9.0±0.5
^{7}Be	$1.22 \cdot 10^{-2} \pm 1.3 \cdot 10^{-3}$	16.3±0.6	12.6±0.8

Table 2: Fragment yields (related to ^{4}He) from 1 GeV proton interactions with ^{9}Be. The denotations are the same as in table 1.

2 Equilibration probed by isotopic yield ratios

The classification with reference to ΔN is closely connected to a statistical fragmentation scenario in the grand canonical approach. Supposed that the neutron chemical potential μ_n is a linear function of the neutron-to-proton ratio the yield ratio R of two isotopes can be expressed as

$$R \sim exp\left[\frac{\mu_n}{T} \cdot ((N_2, Z) - (N_1, Z))\frac{N_p + N_t}{Z_p + Z_t}\right]. \qquad (1)$$

The neutron-to-proton ratio $\frac{N_p+N_t}{Z_p+Z_t}$ of the combined target (t) and projectile (p) systems simplifies to $N_t/(Z_t + 1)$ in the case of a proton-nucleus interaction. T is the corresponding equilibrium temperature. In fact, the bulk of data obtained from medium-weight and heavy targets (black symbols in fig. 1) increases with $N_t/(Z_t+1)$ as expected from relation (1). The *mean slopes* obtained from exponential fits to the data were found to be (ΔN=1)=1.85±0.78, (ΔN=2)=4.49±1.20 and (ΔN=3)=7.25±0.93 which are close to the relation expected for these *slopes* (Δ N= 1, 2, 3) = 1 : 2 : 3 . The exponential fits were used to value our new data obtained in p+^{12}C and p+^{9}Be interactions which complete the data systematics in the region $0.85 \leq N_t/(Z_t + 1) \leq 1$.

Statistical model calculations for two scenarios *i)* cluster evaporation (EV) and *ii)* multifragmentation (SMM) from a single source of *fixed* excitation energies ranging from 3 MeV/u to 5 MeV/u) qualitatively verify the trend of the data but fail in the details. There are remarkable deviations, especially for the ratios ^{6}He/^{4}He, ^{8}Li/^{6}Li and ^{10}Be/^{7}Be obtained from p+^{12}C and p+^{9}Be interactions. In light nuclei, already moderate excitation energies can represent a substantial part of the binding energy and evaporation becomes meaningless if the cluster size reaches that of the emitting residual nucleus. The alternative scenario is the Fermi break-up which has been applied to fragmentation at the first time in ref. [4]. Implemented into the PEANUT code of the FLUKA program package [5] yields a perfect reproduction of the cross sections given

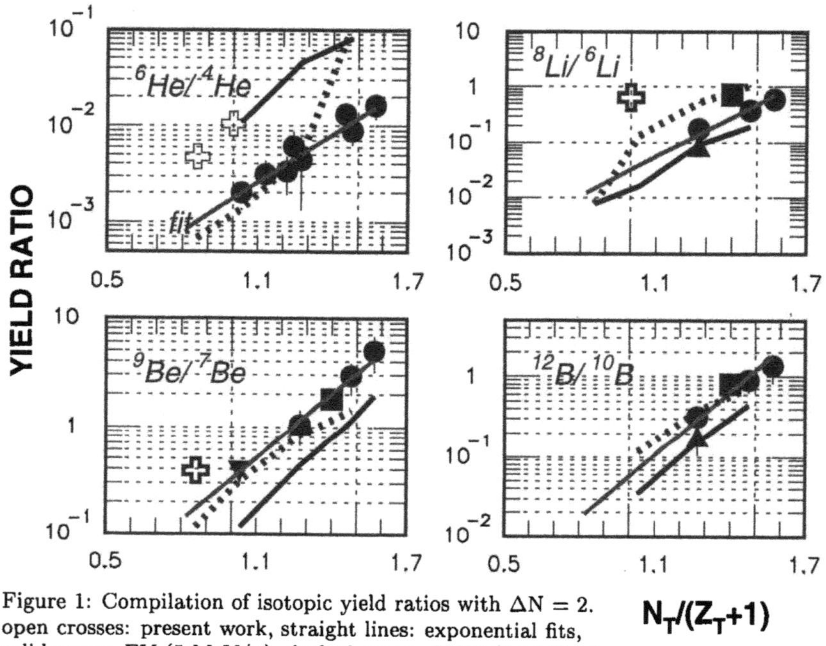

Figure 1: Compilation of isotopic yield ratios with $\Delta N = 2$. open crosses: present work, straight lines: exponential fits, solid curve: EV (5 MeV/u), dashed curve: SMM (5 MeV/u).

in table 1. Otherwise, the Intranuclear Cascade Model combined with the Statistical Multifragmentation Model reasonably reproduces the shapes of the fragment charge distribution and the kinetic energy spectra of the emitted fragments.

Acknowledgments

This work was supported by the Russian Foundation for Fundamental Research Grant No. 95-02-03671 and the German Scientific-Technical Collaboration project WTZ RUS-622-96.

References

1. L.N. Andronenko *et al*, *Nucl. Instrum. Methods* A **312**, 467 (1992).
2. G.D. Westfall *et al*, *Phys. Rev.* C **17**, 1368 (1978).
3. L.N. Andronenko *et al*, *Preprint PNPI* **NP-3**, 2217 (1998).
4. A.S. Botvina *et al*, *Preprint INR* **P-0657**, Moscow (1990).
5. A. Ferrari and P. Sala, *Proc. Int. Conf.* **MC93**, Tallahassee(FL) (1993).

CLUSTERING PHENOMENA AND ISOTOPIC EFFECTS IN PRE-EQUILIBRIUM REACTIONS

E. BĚTÁK

Institute of Physics, Slovak Acad. of Sciences, 84228 Bratislava, Slovakia,
Faculty of Philosophy and Sciences, Silesian University, 74601 Opava, Czech Rep.
E-mail: betak@savba.sk

R. ČAPLAR

Institute "R. Bošković, P.O.B. 1016, 10001 Zagreb, Croatia
E-mail: caplar@faust.irb.hr

Co-existing models of pre-equilibrium cluster (complex particle) emission are compared. The conclusion about the proper mechanism is not unique yet in the general case, however. As a special case, we show the model predictions compared to the data for the 14 MeV (n,α) isotopic cross sections.

1 Introduction

The pre-equilibrium model [1,2] is widely used to describe nuclear reactions below about 200 MeV. So far, reactions involving only nucleons are reasonably understood, whereas those of cluster (complex particle) and gamma emissions are still missing some of important facets of the problem.

Pre-equilibrium emission of light clusters (d to α) is treated in several distinctive ways. Already in the initial period, two opposite mechanisms have been suggested. The concept of pre-formed α particles [3] stresses that the α particle is a strongly coupled object, and assumes that it can be treated as a single (special) exciton. On the other hand, coalescence models initiated by Cline [4] and Ribanský and Obložinský [5] assume forming a cluster in the course of a reaction from excitons, or — later — also from already unexcited nucleons [10,9]. The coalescence models are of more general nature that the pre-formed one and they are currently applied to all types of light complex particles. Apart of these two groups of models with straightforward physical background, also phenomenological descriptions are popular [6].

2 Pre-equilibrium complex particle emission

The energy spectrum of the emitted particles and/or γ quanta in the spin-independent formulation of the model is $\mathrm{d}\sigma/\mathrm{d}\varepsilon = \sigma_{\mathrm{R}} \sum_n \tau_n \lambda_x^c(n, E, \varepsilon)$, where $\lambda_x^c(n, E, \varepsilon)$ is the particle (or γ) emission rate from an n-exciton state ($n =$

$p + h$) of excitation energy E to continuum, the energy of the ejectile of type x is ε, and τ_n and σ_R are the time spent in an n-exciton state and the cross section of creation of the composite system, respectively.

The particle emission rate (see, e.g., [2,5]) is proportional to $\omega(p - p_x, h, U)\omega(p_x, 0, \varepsilon + B)/\omega(p, h, E)R_x(p)\gamma_x$, where p_x is the exciton number associated with the emitted particle and the charge factor $R_x(p)$ takes into account the charge composition of the ejectile, and is not generally accepted [a]. The last factor, γ_x, is the formation probability [5] of the coalescence models, or the α pre-formation factor γ_α if we assume their existence as special entities within the nucleus [3]. A simple estimate of the formation probability of the coalescence model [8] yields $\gamma_x \approx {p_x}^3 (p_x/A)^{p_x - 1}$. However, the presence of formation probabilities is not strictly jusfiable from the detailed balance, and it is therefore rejected by some groups, even though it means worsening the quality of the agreement between theory and experiment .

The original simple coalescence model has been made more sophisticated as to allow the cluster to be formed not only of excitons, but also from so far unexcited nucleons below the Fermi level. This approach became popular as the Iwamoto-Harada model [9] (though it has been suggested and applied five years earlier [10]). Essentially, it implies replacing of the density product $\omega(p - p_x, h, U) \times \omega(p_x, 0, \varepsilon + B)$ by [10] $\sum_{p^*=1}^{p_x} \int_{\varepsilon+B}^{E} \omega(p - p^*, h, E - \varepsilon_1)\omega(p^*, 0, \varepsilon_1)\omega(0, p_x - p^*, \varepsilon_2)\, d\varepsilon_1$, where p^* is the number of excitons contributing to form the cluster, and the remaining nucleons are picked up from the Fermi sea. The cluster density is proportional to the forming probability, making this approach to be parameterless for the cluster emission [10].

Bisplinghoff suggested that not all nucleons be available for the cluster formation within the model, but only those close to the Fermi energy, and the energy width of the "band of availability" is determined by the binding energy of nucleons inside the cluster [12]. Therefore strongly bound entities, like α's, have large energy space available for their creation (what makes the approach close to the original Iwamoto-Harada ideas [9,10]), and loosely coupled objects (e.g. deuterons) practically get close to the standard coalescence model. Thus the model is very close to the coalescence one, both in its model assumptions and in predictions.

The coalescence model [4] often works well for deuteron emission, usually fails for α's, and there is a half-to-half chance of reasonable description for tritons and ^{3}He. The Iwamoto-Harada model does not contain any free parameter, as was the formation probability in the former case. The overall fit is of similar quality (though it may be significantly different for a spe-

[a] A recent discussion of various forms of the charge factor is in [7].

cific reaction) than in the coalescence model, but one cannot find any drastic discrepancies with respect to the data here, rather successful model at wide scale of reactions is the phenomenological one due to Kalbach [6] describing also pickup and stripping. Anyway, it contains many parameters without clear physical justification, just to fit the data. Therefore, it can be (and is) successfully applied to calculate cross sections and other quantities, but there is not much sense in discussing the underlying physics.

3 Isotopic effect

The (n,α) reactions at 14 MeV represent bulk amount of experimental data of various kind which cannot be found at any other (higher) energy. One of interesting trends observed here is the isotopic effect in the (activation) cross sections, i.e. the exponential decrease of the cross section with increasing $(N - Z)$ (see also [14], and the data in [15]).

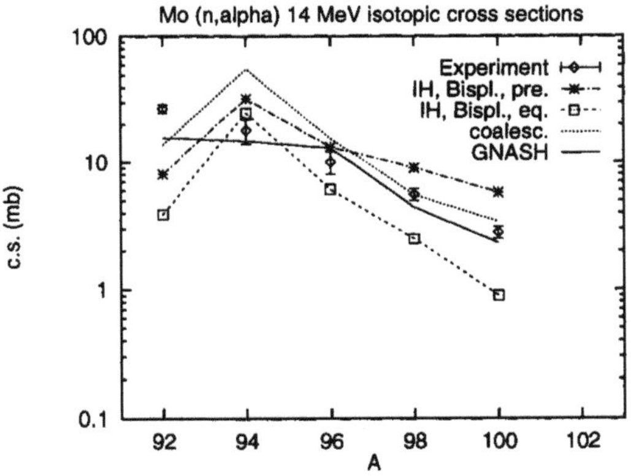

Figure 1. Isotopic effect of 14 MeV (n,α) on Mo. The data are from Refs. **REFS** the full curve is the GNASH calculation, dotted line is the coalescence model (code CLUEX) and the lines with points are the Iwamoto-Harada model with Bisplinghoff's suggestion in the pre-equilibrium case (stars) and its compiund nucleus limit (squares).

An example of the experimental data and the calculated dependences can be seen in Fig. 1. The calculations have been performed using codes GNASH [16] (α emission in accord with the phenomenological model [6]) and CLUEX [13] using both the standard coalescence model with $\gamma_\alpha = 0.0025$ and the Iwamoto-Harada one. The calculations describe the cross section decrease at higher A and yield reasonable absolute value. At ^{92}Mo, however, some discrepancy remains, probably caused by necessity to pick neutrons from a magic shell, in which case the theoretical description predicts significant lowering of the cross section.

4 Conclusions and outlook

Cluster emission can be expressed in several different ways, each of them having their pros and cons. Possible improvement of the Iwamoto-Harada model that only the nucleons near the Fermi level take active part in the complex particle formation (as suggested by Bisplinghoff) makes it close to the original coalescence one and is promising for future model developments.

Acknowledgments
The authors are grateful to M.B. Chadwick, J. Dobeš and P. Obložinský for discussions. The work has been supported by Grant No. 2/5122/99.

References

1. J.J. Griffin, *Phys. Rev. Lett.* **17**, 478 (1966).
2. E. Gadioli and P.E. Hodgson, *Pre-Equilibrium Nuclear Reactions* (Clarendon Press, Oxford, 1992); and the references quoted therein.
3. L. Milazzo-Colli and G.M. Braga-Marcazzan, *Phys. Lett.* B **36**, 447 (1971); *ibid.* B**38**, 155 (1972).
4. C.K. Cline, *Nucl. Phys.* **A193**, 417 (1972).
5. I. Ribanský and P. Obložinský, *Phys. Lett.* B **45**, 318 (1973).
6. C. Kalbach, *Z. Phys.* **A283**, 401 (1977).
7. E. Běták and P.E. Hodgson, *Rep. Prog. Phys.* **61**, 483 (1998).
8. E. Běták and P. Obložinský in *5th Int. Symp. Interacts. of Fast Neutrons with Nuclei, Gaussig 1975.* Report ZfK-324 (Rossendorf, 1976), p. 77.
9. A. Iwamoto and K. Harada, *Phys. Rev.* **C26**, 1821 (1982); K. Sato, A. Iwamoto and K. Harada, *Phys. Rev.* **C28**, 1527 (1983).
10. J. Dobeš and E. Běták in *Int. Conf. Reaction Models '77, Balatonfured 1977*, ed. L.P. Csernai, p. 195.
11. E. Běták in *Coll. Motion & Giant Res., Proc. 15th Mikołajki Summer Sch.*, eds. Z. Wilhelmi, M. Kicińska-Habior. (Warsaw, 1984), p. 132.
12. J. Bisplinghoff, *Phys. Rev.* **C50**, 1611 (1994).
13. E. Běták, Report INDC(CSR)-016/LJ (IAEA Vienna 1989) and its later developments (unpublished).
14. E. Běták, R. Čaplar and E. Rurarz, *Acta Phys. Pol.* **B30**, 1511 (1999).
15. K. Gul, INDC(PAK)-009 (IAEA Vienna 1995)
16. P.G. Young, E.D. Arthur, M.B. Chadwick, Report LA-12343-MS (UC-413) (Los Alamos 1992); *updated* 22 Sept., 1995.

DISCOVERING SUPERHEAVY ELEMENTS

G. MÜNZENBERG
Gesellschaft für Schwerionenforschung, GSI mbH, Planckstr. 1, 64291 Darmstadt and Johannes Gutenberg-University at Mainz

In recent years significant progress has been made approaching the superheavy elements. Element 112, the heaviest identified unambiguously at present, is already close to that region. The principal result of heavy-element research is the discovery of a new shell stabilised deformed region interconnecting the trans-uranium elements and the spherical superheavies. Recently evidence for the creation of elements 114, and 118 has been reported.

1 Introduction

The motivation of heavy-element research is, to which extent shell stabilisation can extend the number of elements beyond the macroscopic limit determined by fission. Element 112 [1], the heaviest element safely identified at present, is already far beyond and exists only by microscopic stabilisation. This is explained in the frame of macrosopic- microscopic models in terms of a hexadecapole deformation in the ground state [2,3]. The region of deformed nuclei centred at Z = 108 and N = 162 connects the trans-uranium elements to the predicted spherical superheavy nuclei.

All of the heaviest elements have been produced by complete fusion of heavy ions. The production cross sections range down to picobarns. The key to the discovery of new elements which are produced only in amounts of few atoms with half-lives of the order of milliseconds is in-flight separation and identification by their decay properties with the α-α parent-daughter correlation technique. The identification relies on single-atom decay sequences leading to known transitions and is thus independent from theoretical predictions - an essential prereqiesite for the exploration of far unknown regions. As an example fig. 1 displays one out of the two α- decay chains observed for element 112 [1].

2 The Production of Superheavy Elements

Two types of complete fusion reactions have been used so far successfully for heavy-element synthesis: the cold heavy-ion fusion with lead or bismuth targets and appropriate projectile beams of the most neutron rich isotopes available, such as ^{64}Ni, ^{70}Zn, or ^{86}Kr for the production of elements 110,111,112,

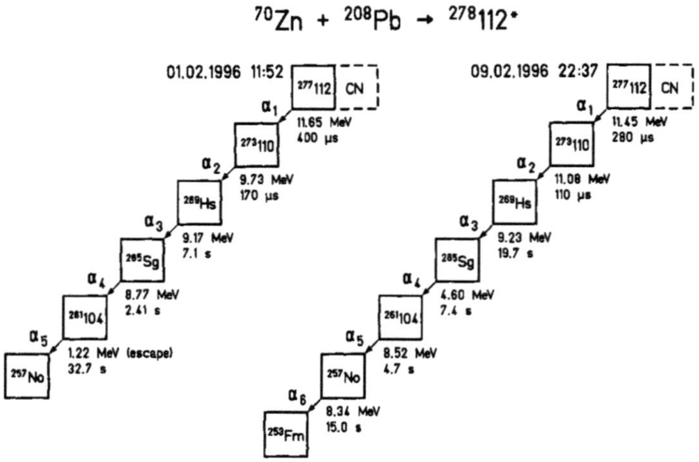

Figure 1. The two α-decay sequences observed for element 112.

or 118, respectively, and the hot fusion starting from actinide targets such as ^{235}U or ^{244}Pu and beams of ^{48}Ca to produce elements 112, and 114. The production cross-sections for the heaviest elements are of the order of $10^{-36}cm^{-2}$. Symmetric reactions have also been investigated in the early stage of heavy element research without success. These experiments however are not conclusive as their sensitivity was by far not comparable to the new generation of experiments discussed here.

Heavy-element production is dominated by the strong Coulomb forces of the massive nuclear system: In the entrance channel the Coulomb interaction of projectile and target hinders the fusion process, in the exit channel it reduces the survival probability of the Compound nucleus by strong fission competition in the de-excitation process of the excited system on its way to the ground-state.

Therefore the cold fusion of heavy ions has been considered to be most favourable for the production of the most heavy systems. For the fusion with lead or bismuth targets only one-neutron evaporation has been observed beyond element 108. The excitation energy of the compund system of about 10 - 12 MeV is already close to the neutron of binding energy, the natural lower limit for this reaction channel. The increasing fissility of the compound nucleus towards increasing proton number lowers the upper limit of the excitation function.

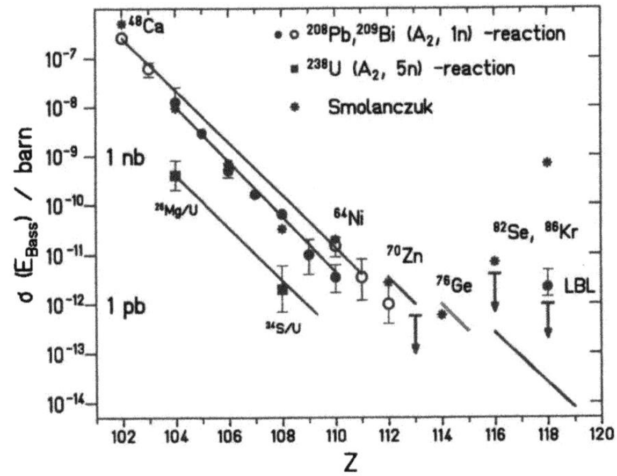

Figure 2. Production cross sections for the elements nobelium and beyond. The lines connect reactions with the same projectile isospin. The figure includes the prediction of Smolanczuk [5], asterisks, and the Berkeley cross section given for the production of element118, together with the upper limit of the GSI experiment.

As the consequence of these competing processes the excitation function for heavy element production is rather narrow an amounts to about 5 MeV. This limits the usable target thickness to 10^{18} atoms cm^{-2} to match with the energy loss of the projectile beam. Beam intensities of $10^{12}s^{-1}$ up to the $10^{14}s^{-1}$ expected for the new high-current accelerators yield luminosities f 10^{30} $s^{-1}cm^{-2}$ to $10^{32}s^{-1}cm^{-2}$, corresponding to production rates of 1 atom per 11 days for a cross section of 1 pb with exisiting, and 1 atom per 2 hours with future accelerators.

Fig. 2 displays the production cross-sections for the elements nobelium and beyond. Projectiles with the same isospin projection are connected by solid lines. The production cross section of element 110 profits from the isospin enhancement when using ^{64}Ni instead of ^{62}Ni, an effect which could not be confirmed for element 112. Here it should be noted, that for element 112 in contrast to element 110 no excitation function was measured, because of the low cross section. The isospin enhancement may still exist.

Theoretical models to predict the production cross sections for superheavy elements are under development, but up to now have only limited predictive power [4]. It is generally accepted that clusters, e. g. shell effects in target and projectile, as proven for the fusion with lead targets, enhance the fusion probability, whereas the stabilisation of shell effects against fission of the compound system has not been observed. A recent calculation [5] predicts an increase of the production cross section up to several hundreds of picobarns for the elements beyond Z = 114 (see Fig. 2), which could not be confirmed experimentally up to now. To our present knowledge the creation of new elements will be picobarn physics and remain the domain of intense beams of stable nuclei.

Radioactive beam facilities of the first generation will not deliver beam intensities competing with stable beams, production cross sections may however profit from isospin enhancement and neutron rich species in the deformed region near Z = 108 and N = 162 and even beyond will become accessible.

3 Experimental Developments

In-flight separation is the method commonly used at present in heavy-element research. Various types such as gas filled magnetic separators, energy-, and velocity filters are in use [6]. The separation characteristics of the various separator types are slightly different. While gas filled separators have the largest transmission and excellent suppression of scattered projectiles as compared to the other in-flight separators, they produce background from energetic projectile-scattered nuclei of the filling gas and do not separate from reaction products other than the neutron evaporation channels, such as α,xn and transfer products.

The separated nuclei are implanted into silicon detectors where their decay can be observed in situ. Characteristic for the decay of a heavy nucleus are long α decay chains as well as sequences of α decays terminated by spontaneous fission. The unambiguous identification of new elements is based on the connection of the observed decay sequences to known nuclides. In the example of the identification of element 112 the isotopes starting from element 106 and below were safely identified before (see Fig. 1).

Recently two exciting discoveries were reported. The Dubna laboratory reported the observation of α-fission sequences [7] in irradiations of $^{244,242}Pu$ with ^{48}Ca assigned to decays of the isotopes with masses 287, and 289 of the superheavy element with Z = 114. Berkeley announced the discovery of superheavy nuclei [8] in the reaction $^{86}Kr + {}^{208}Pb = {}^{298}118 + n$. The conclusive identification of the observed decays remains a problem, as the decay sequences end far in the unknown neutron rich trans-actinide region. Their assignment is primarily based on general considerations, consistency checks, and comparisons of the decay chains with theoretical models.

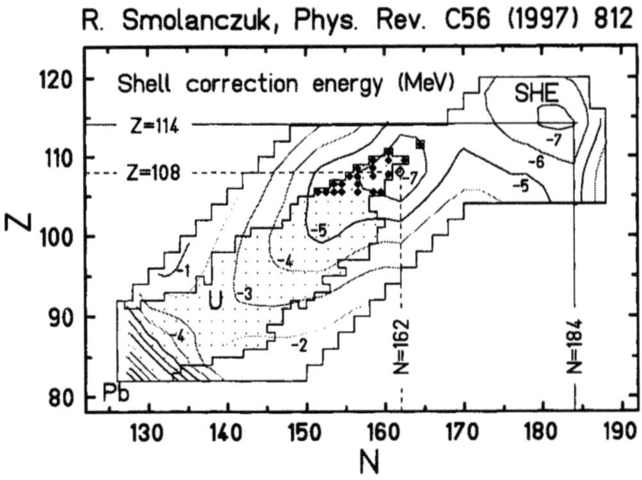

Figure 3. Shell correction energies form a macroscopic-microscopic model [9] for the region from lead to the spherical superheavy nuclei. The solid symbols mark the nuclides of the heaviest elements synthesised at GSI close to the deformed shell at Z=108 and N=162.

The irradiation of ^{208}Pb with ^{86}Kr to create element 118 was repeated at GSI with the projectile energy of the Berkeley experiment. The Berkeley result could not be confirmed down to a level of 1 pb (Note added after the presentation at the conference).

4 Ground-state Properties - Experiment and Predictions

The heaviest elements reveal two remarkable features:
- spontaneous fission dominates in the region around element 104. For the heavier elements beyond α decay prevails.
- half-lives increase towards 162 neutrons .

This was explained theoretically by a strong microscopic stabilisation and attributed to a hexadecapole deformation in the ground-state. Fig. 3 shows the shell-correction energies for the elements between the two shell closures at ^{208}Pb and the superheavy nucleus ^{298}114. In between there is the deformed region at Z = 108 and N = 162. Shell correction energies are as high as -7 MeV.

The calculated shell correction energies agree well with experimental data. The decay chains of element 112 pass close by the maximum of the shell correction and clearly show the step in decay energies and correlation times in the 110 to 108 region as a direct confirmation of the shell passing (Fig 1). The figure also reveals the character of the spherical superheavy shell which is much less pronounced as the much stronger lead shell.

The principal problems for the theoretical treatment of the heaviest nuclei are the high level density of the large nuclear shells with large angular momentum and the strong Coulomb field of the many protons which may influence the diffuseness of the nuclear proton surface. The spherical superheavy shell is by far not as strong as the lead shell (Fig. 3). Precise knowledge of spin-orbit coupling and the proton density at the nucleon surface are essential to calculate the location of the proton shell closure. Magic proton numbers of 114, 120, and 126 are currently predicted [10,11], depending on the models and forces. The neutron shell at N = 184 is fairly stable. Macroscopic-microscopic models tend to predict the magic proton number Z = 114, relativistic mean field models prefer Z = 120, whereas Skyrme-Hartree Fock models also predict Z = 126. A detailed comparative investigation of a great number of forces has been carried out by Bender et al. [10]. For this discussion see also the contribution of W. Greiner to this conference.

5 New Experimental Developments

With the elements seaborgium and beyond the predicted new species of shell nuclei has been discovered. They close the gap between the trans-uranium elements and the spherical superheavy shell closures. This transition region will offer a rich field of nuclear structure research as unexpected long half-lives have been observed experimentally. These will allow new types of experiments such as the extension of heavy-element chemistry beyond the present limit of seaborgium (Z = 106) and the application of ion traps for precision mass determination or the investigation of atomic properties of the heaviest elements.

The transition region between the deformed N = 162 shell and the spherical N = 184 shell is of specific interest for nuclear structure research. New features such as fast changes in e. g. their ground state deformation due to the absence of the smoothing macroscopic forces may be expected. The high level density and large angular momentum will contribute to fast changes in structure as high spin and low spin states are close together. Already in the known isotopes a number of isomeric states have been found.

A key problem is the understanding of the production of superheavy nuclei, especially the importance of nuclear structure in the entrance and exit channels. New types of reactions need exploration. The fragmentation model

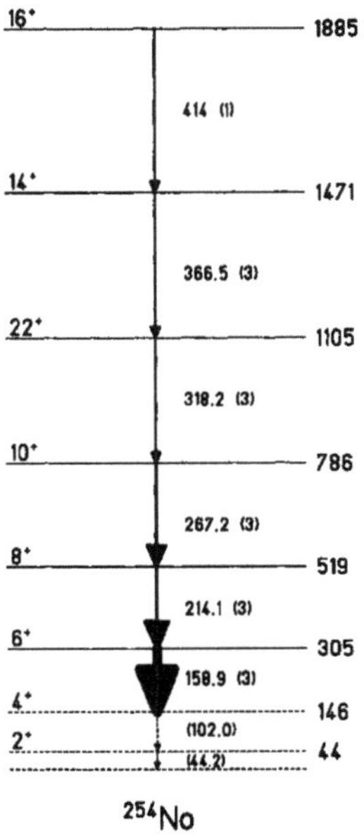

Figure 4. Level scheme [14] as measured with recoil-decay-tagging for ^{254}No

predicts three paths to superheavy elements [12]: the cold fusion with lead and bismuth targets, the hot fusion with actinides, and as the third path the symmetric fusion or reversed fission. Presently only the cold fusion has been explored systematically, for the hot, actinide based fusion also data are available, whereas the symmetric fusion just started to be explored at sufficiently high sensitivity [13].

Accelerator developments will allow to proceed into the femtobarn cross sections and to extend our knowledge towards still heavier elements into the up to now inaccessible region of neutron rich species. High-current accelerators and new ion sources will provide beam intensities of more than 10^{14} ions/s allowing to proceed towards Z = 114 and possibly to go beyond. Low-charge state accelerators will deliver still larger intensities. Increased beam intensities and new developments such as the new generation of efficient 4π-Ge-arrays will allow a detailed investigation of the structure of the trans-actinide elements. First interesting results have been obtained with recoil tagged in-beam-spectroscopy. The rotational band of ^{245}No was observed which allowed to extract the deformation [14] and in addition the limiting angular momentum for the fusion process. Fig. 4 displays the ground-state band. Please note that the lowest levels 2^+ and 4^+ are converted. Intense projectile beams will open up the possibility of decay spectroscopy to explore in detail the character of the new species of hell stabilised trans-actinides.

The new radioactive beam facilities providing intense exotic nuclear beams with energies above the Coulomb barrier will produce neutron rich beams for heavy-element research [15]. Existing and planned laboratories as SPIRAL of GANIL (France), EXCYT (Catania, Italy), RIKEN (Japan) and its upgrade, MSU and its upgrade (Michigan State, USA), and the new US - ISOL project deliver already or will provide beams with intensities up to 10^7 /s for the running, and up to 10^{10} /s for the projected facilities. They will allow to explore the transition region between the deformed neutron shell at N = 162 and the spherical N = 184 region. The still open problems of isospin enhanced production cross section and the reversed fission are ideally explored with radioactive beams.

The combination of recoil separators and ion traps will open up new possibilities such as high precision mass measurements, laser spectroscopy to determine for instance deformations or ionisation potentials, or for chemical investigations. Such a trap could also serve as ion source to inject the in-flight separated nuclei into the GSI accelerator system. SHIPTRAP, a trap attached to SHIP is being planned, with a broad scientific program is in progress.

6 Conclusions

In-flight separation is now the established method for heavy-element research. It permits the unambiguous identification of new elements and allows to obtain gross nuclear properties of a new nuclide from the decay of only single atoms. Recent heavy element research led to the discovery of deformed shell stabilised nuclei creating an extended region of enhanced stability which connects the trans-uranium region to the magic proton number 114, already closely approached with the discovery of element 112. First evidence for the superheavy elements 114 and 118 has been reported. New experimental developments will not only allow to extend the region of known elements but also give new access to their nuclear, atomic, and chemical nature. New high-current ion sources for stable beams will allow to produce and investigate new elements near element 114. The use of neutron rich beams of radioactive nuclei will open new perspectives for the investigation of heavy and superheavy elements in still inaccessible regions. Reaction studies to investigate the isospin dependence of the fusion cross sections will be the naturally first generation experiments with radioactive beams and give new insights in the production of superheavy nuclei. Trans-uranium elements up to the region around element 104 are expected to be accessible with secondary beams available from the presently planned facilities. The new region of neutron rich species near the deformed neutron shell and beyond, where the transition towards the spherical N = 184 shell is expected, is a rich and interesting field to explore. New insights in the shell stabilisation and the related existence of heavy elements in nature are expected. The long half-lives of these nuclei will allow the application of new experimental techniques, giving new information about nuclear, atomic, and chemical properties of the elements at the top of the nuclear table.

Acknowledgments

The experiments at GSI were carried out by S. Hofmann, F.P. Hessberger, V. Ninov, P. Armbruster, A.Yu. Lawrentew, M.E. Leino, G. Mñzenberg, A. G.Popeko, A. V. Eremin, and S. Saro. Fruitful discussions with A. Antonenko, W. Greiner, W. Nazarewicz, R. Smoloanczuk, and A. Sobiczewski are highly appreciated.

References

1. S. Hofmann *et al*, *Z. Phys.* A **354**, 229 (1996).
2. S. Cwiok *et al*, *Nucl. Phys.* A **410**, 254 (1983).
3. P. Möller *et al*, *Z. Phys.* A **323**, 41 (1986).
4. G. Adaminan, contribution to this conference.
5. R. Smolanczuk, *Phys. Rev.* C **59**, 2634 (1999).
6. G. Münzenberg in *Experimental Techniques in Nuclear Physics*, eds. D.N. Poenaru and W. Greiner, Walter de Gruyter, Berlin, 1997.
7. Yu. Ts. Oganessian *et al*, *Nature* **400**, 242 (1999).
8. V. Ninov *et al*, *Phys. Rev. Lett.* **83**, 1104 (1999).
9. R. Smolanczuket *et al*, in *Proc. EPS XV Nucl. Phys. Conf.*, St. Petersburg, World Scientific, Singapore (1995) 313, *Phys. Rev.* C **56**, 812 (1997).
10. K. Rutz *et al*, *Phys. Rev.* C **56**, 238 (1997).
11. S. Cwiok *et al*, *Nucl. Phys.* A **611**, 211 (1996).
12. A. Sandulescu and W. Greiner, *J. Phys.* G **3**, 189 (1977).
13. C. Stodel *et al*, GSI Annual Report 1996; GSI-97-1
14. P. Reiter *et al*, *Phys. Rev. Lett.* **82**, 509 (1999).
15. H. Geissel, G. Münzenberg and K. Riisager, *Ann. Rev. Nucl. Part. Sci.* **45**, 163 (1995).

SUPERSHORT MODE IN THE FISSION OF SUPERHEAVY NUCLEI IN SUBBARRIER FUSION REACTION

M. G. ITKIS, YU. TS. OGANESSIAN, E. M. KOZULIN, N. A. KONDRATIEV, L. KRUPA, I. V. POKROVSKI, A. N. POLYAKOV, V. A. PONOMARENKO, E. V. PROKHOROVA

Flerov Laboratory of Nuclear Research, JINR, 141980 Dubna, Russia

A. YA. RUSANOV

Institute of Nuclear Physics of the National Nuclear Center of Kazakhstan, 480082 Alma-Ata, Kazakhstan

1 Introduction

In the last year, considerable progress has been made in experiments on the synthesis of superheavy nuclei. Works [1,2] have reported that the heaviest and longest-lived isotopes of elements 112 and 114 have been synthesised, which suggests that there exists the "Island of Stability" predicted in theory.

However, the fission of excited superheavy nuclei is known to be the dominant channel of their decay. In reactions of the subbarier fusion-fission of heavy ions and targets, low excitation energies ($E^* < 30$ MeV) are fairly easily obtainable, and then the structural shell properties of the fissioning nucleus and the fragments produced should be expected to manifest themseves.

To research the structure of the mass-energy distributions (MED) of the fragments of superheavy nuclei at low E^*, we have studied the reactions ^{48}Ca $+^{208}$Pb $\rightarrow^{256}$No, ^{22}Ne $+^{248}$Cm $\rightarrow^{270}$Sg and ^{22}Ne $+^{249}$Cf $\rightarrow^{271}$Hs at the energies of incident ions above and deep below the Coulomb barrier.

The first reaction is a cold fusion reaction, in which the minimum E^* are obtainable. The second and third reactions are interesting owing to the fact that the compound nuclei have 164 neutrons each and their fission into equal parts results in fragments of the magic number $N = 82$.

2 Experiments

The experimental study of the fission of superheavy elements was performed on the U-400 accelerator at the Flerov Laboratory of Nuclear Reactions, Dubna. The targets were 100-200 μg/cm^2 thick layers of ^{208}Pb, ^{248}Cm and ^{249}Cf produced by deposition on carbon backings 50-60 μg/cm^2 thick. Time-coincident fission fragments were registered by the CORSET double-arm time-of-flight spectrometer [3]. Each arm of the spectrometer included a start detector built on the basis of microchannel plates with an electrostatic mirror and two stop position-sensitive (x-, y- sensitivity) detectors also built on the basis of microchannel plates 6 × 4 cm each. The start detectors were placed at a distance of 4 cm from the target. The minimal start-stop flight distance was 10 cm. So, the spectrometer consisted of 2 start detectors and 4 stop detectors, the solid angle being 360 msr. The position resolution of the stop detectors was $\pm 0.1^o$. The estimated mass resolution was $3-5$ a.m.u.

In the experiment carried out on the beam of ^{22}Ne and ^{48}Ca the arms of the spectrometer were set at such correlation angles as for it to register the energies and masses of fragments up to the ration 5. The data were processed by regular techniques on the basis that the fission was a two-body process. Special attention was paid to the angular folding correlations both in the plane and out of the plane of the reaction. Only those registered events were selected and then analysed that corresponded to a two-body process of total momentum transfer. Fission events associated with incomplete fusion reactions, sequential fission, were rejected. The date processing took account of the energy losses of fragments in the target itself, its backing and in the electron emitters of the start detectors. In addition, because the geometrical dimensions of the spectrometer did not allow registration of the whole set of the energy-mass ratios of fragments with equal probability, we made corrections for its geometrical efficiency.

3 Reaction ^{48}Ca $+^{208}$Pb

The products of the reaction ^{48}Ca $+^{208}$Pb were measured for six values of the energy of bombarding ^{48}Ca ions, $E_{\rm lab} = 211, 217, 220, 225, 232$ and 242 MeV. Fig. 1 shows the MED of the products of this reaction for the four largest values of the energy, which correspond to the values of the excitation energy of the ^{256}No compound nucleus $E^* = 25, 29, 35$ and 43 MeV. Presented from top to bottom are: the two-dimensional matrices (TKE-Mass), mass distributions (MD), dependence TKE(M) and dependence $\sigma^2_{\rm TKE}$(M). It is clearly seen that the characteristics of the MED of the fragments for the mass range $A/2 \pm 25$ are of the form typical for the fission of heated nuclei, where MD is a gaussian, TKE(M) and $\sigma^2_{\rm TKE}$(M) parabolas. In the case of more asymmetric partition in the region of light masses $65-80$ and additional to them, "shoulders" are observed whose origin we associate with a quasi-fisson component. This explanation follows from the results of works [4,5], where in the close reactions ^{40}Ar, ^{50}Ti $+^{208}$Pb for the large mass-asymmetry of the fragments their angular distributions were obtained characterised by clearly defined asymmetry

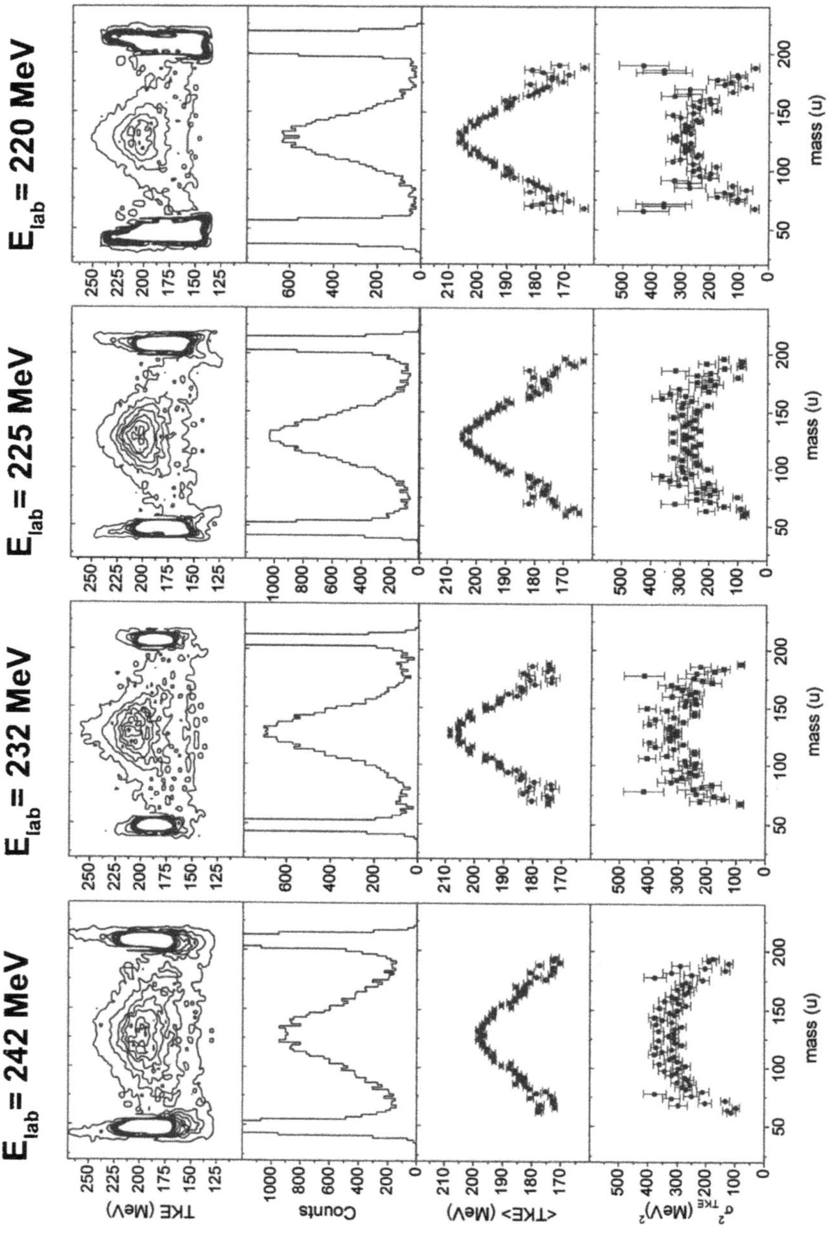

Figure 1: Mass-energy disrtibutions of the $^{48}Ca + ^{208}Pb$ reaction products. From top to bottom: two-dimensional matrices TKE-Mass, Mass distributions, <TKE> and σ^2_{TKE} dependences on the fragment mass.

Figure 1.

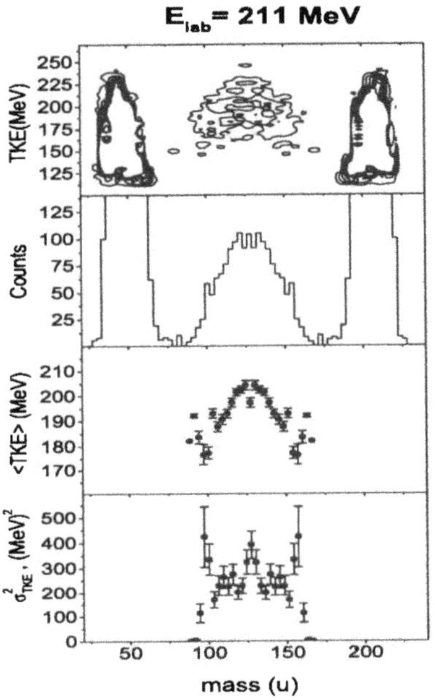

Figure 2. Mass-energy distributions of the products of the $^{48}Ca+^{208}Pb$ reaction for the ^{48}Ca ion energy $E_{\rm lab} = 211$ MeV.

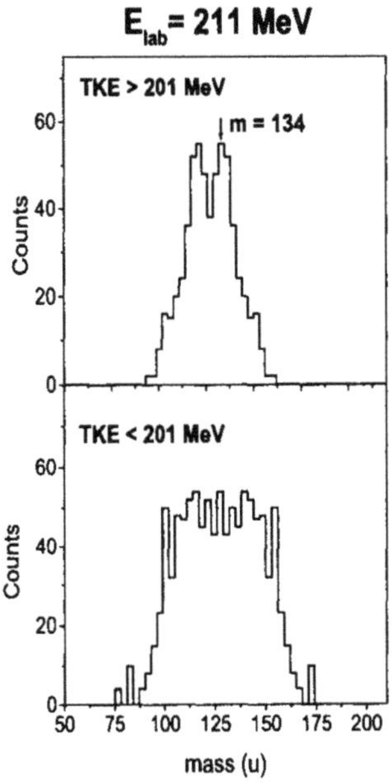

Figure 3. Differential distributions of fission fragment masses for ^{256}No at $E^* =$ 18 MeV for the TKE range of $>$ 201 MeV (top) and $<$ 201 MeV (bottom).

forward-back, which suggests that the process is nonequilibrium one. At the same time for masses $M \sim A/2$, according to works [4,5], the angular distributions of the fragments are symmetric relative to 90°. As seen from Fig. 1, TKE for the "quasi-fission shoulders" is also larger than that expected for normal fission in the region of these masses. This confirms the results of work [6], where, on the basis of a large data set, it has been shown that large TKE are, on the average, characteristic of quasi-fission.

Figure 2 shows the MED of the reaction products for $E_{\rm lab} = 211$ MeV ($E^* = 18$ MeV). The MD of the fragments of ^{256}No is of quite a different form compared with that for higher excitation energies E^*, it has patterns characteristic of low energy fission. Higher mass yields of heavy fragments $M = 132 - 134$ and $M = 140 - 150$ are observed. The dependence TKE(M) at $A/2$ is a much more inclined parabolic curve than that in Fig. 1.

Figure 3 presents the differential mass distributions of the fragments for $E^* = 18$ MeV: at the top for the region of TKE $>$ 201 MeV, at the bottom

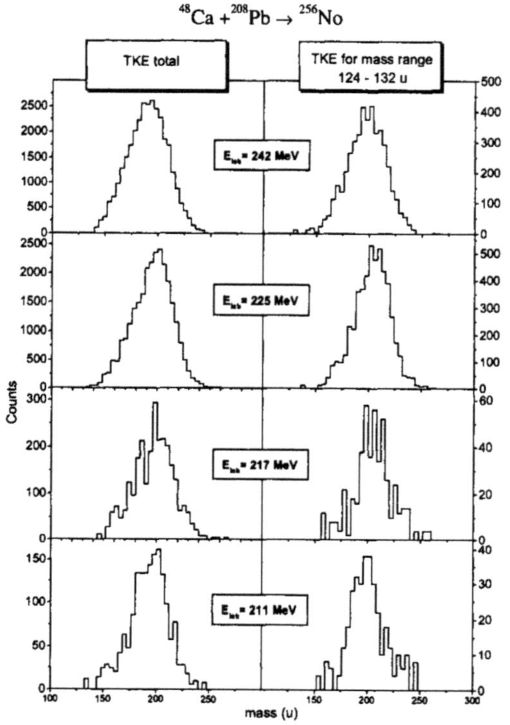

Figure 4. TKE distributions for all masses of the fission fragments (left column) and for symmetric masses in the ranges $M = 124 - 132$ (right column).

for TKE < 201 MeV. For high TKE, MD proves to be two-humped, the peak corresponding to the heavy fragment of $M = 132 - 133$. For low TKE, the MD peak is, on the average, flat, but it is clearly seen that this is due to the fragments of masses $M = 140 - 150$. Figure 4 shows the TKE spectra for four values of the energy of ^{48}Ca ions. On the left there are the total spectra for all the fragment mass, on the right for the symmetric mass region $M = 124 - 132$. In the right bottom figure ($E^* = 18$ MeV), as well as in MD, two components are observed: one with average $\langle \text{TKE} \rangle$ of ~ 200 MeV and the other having a high energy arm with $\langle \text{TKE} \rangle \sim 230$ MeV, which is a distinctive indication pointing to the fact that the fission of these nuclei [7] is bimodal. Thus, comparing Figs. 3 and 4, we come to the conclusion that

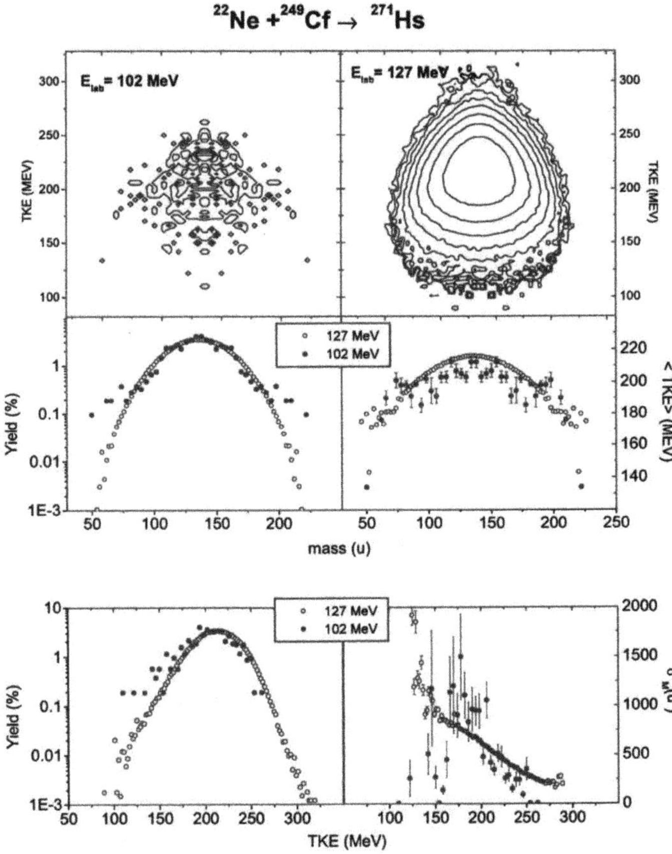

Figure 5. Mass-energy distributions of the ^{271}Hs fission fragment for the ^{22}Ne projectiles energy of 102 and 127 MeV.

in the low energy fission of ^{256}No, two modes with masses $M = 140 - 150$ and $M = 132$ are clearly seen. The high energy mode in terms of Brosa et al [8] is referred to as the super-short mode and is connected with the fact that both fragments are close to $Z = 50$ and $N = 82$, closed spherical shells. In the spontaneous fission of this nucleus, the super-short mode has not been observed [9], although it has been predicted in theory for this region of nuclei [10].

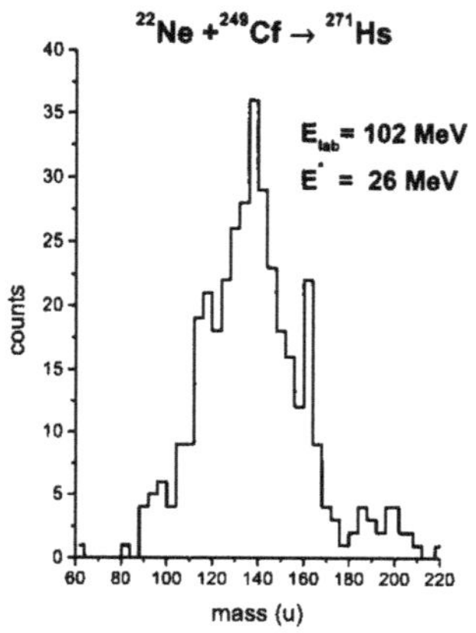

Figure 6. The ^{271}Hs fission fragment mass distribution for the ^{22}Ne energy of 102 MeV.

4 Fission of ^{270}Sg and ^{271}Hs

We have studied the fusion-fisson reactions ^{22}Ne $+^{248}$Cm , ^{249}Cf at the energy of bombarding ^{22}Ne E_{lab} = 102 and 127 MeV, which corresponds to the excitation energies of both compound nuclei of ^{270}Sg and ^{271}Hs $E^* = 27$ and 50 MeV.

We have reported the measured results for ^{270}Sg in work [11], so we are not going to cite them, but will observe that the super-short mode has been found to manifest itself in the case of this nucleus as well.

For ^{271}Hs, we are going to present preliminary results. Figure 5 shows the comparision of the following MED characteristics for ^{271}Hs for the two energy values of bombarding ^{22}Ne ions: two-dimensional matrices TKE-M, mass yields, energetic distributions $\langle$TKE$\rangle$(M) and $\sigma^2_{\mathrm{TKE}}(M)$, and also TKE yields and σ^2_M(TKE). Nevertheless, the statistics for E_{lab} = 102 MeV is essentially lower than for ^{270}Sg [11], the indicated characteristics repeat qualitatively the effects caused by the MED modality.

Figure 6 presents the data on the mass distributions of the fragments of ^{271}Hs at E_{lab} = 102 MeV. Only 365 pair fission events were collected in the experiment, but a sharp peak is clearly visible at M $\approx A/2$, because, as for neutrons, this nucleus is practically similar to ^{270}Sg. So, we probably also observe manifestation of the super short mode.

Thus, we have observed manifestation of the multimodal structure of the MED of the fission fragments of ^{270}Sg and ^{271}Hs at middle excitation energies. At E_{lab}(^{22}Ne)= 102 MeV, narrow symmetric structure was found to be present in the total mass distribution of the fragments. This structure is connected with manifestation in both fragments of the magic neutron number $N = 82$. The wide structure that forms the pediment of the distribution is likely to consist of two modes. So, it can be stated that the investigation of the deep subcoulomb fission of superheavy nuclei at rather low energies gives abundant information about the properties of fission.

Acknowledgments

The work has been supported by the Russian Foundation for Basic Research under the grant No 99-02-17891 and by INTAS under grant No97-11929.

References

1. Yu.Ts. Oganessian *et al*, *Eur. Phys. J.* A **5**, 63 (1999).
2. Yu.Ts. Oganessian *et al*, *Nature*, **400**, 242 (1999).
3. N.A. Kondratiev *et al*, *Fourth Int. Conf. on Dynamical Aspect of Nuclear Fission (DANF98)*, Časta-Papiernička, Slovak Republic, October 1998, (World Scientific, Singapore, 1999).
4. H. Keller *et al*, *Z. Phys.* A **326**, 313 (1987).
5. K. Lutzenkirchen *et al*, *Nucl. Phys.* A **452**, 351 (1986).
6. M.G. Itkis and A.Ya. Rusanov, *Phys. Part. Nucl.* **29**, 160 1998.
7. E.K. Hullet *et al*, *Phys. Rev.* C **40**, 770 (1989).
8. U. Brosa *et al*, *Phys. Rep.* **197**, 167 (1990).
9. E.K. Hullet *et al*, *Phys. At. Nucl.* **57**, 1099 (1994); M.R. Lane *et al*, *Phys. Rev.* C **53**, 2893 (1996).
10. P. Möller *et al*, *Nucl. Phys.* A **469**, 1 (1987); A **492**, 349 (1989).
11. M.G. Itkis *et al*, *Phys. Rev.* C **59**, 3172 (1999).

THEORY OF NUCLEAR REACTIONS LEADING TO SUPERHEAVY ELEMENTS

Y. ABE[1], K. OKAZAKI[2], Y. ARITOMO[3], T. WADA[2] AND M. OHTA[2]

[1] *Yukawa Institute for Theoretical Physics, Kyoto Univ., Kyoto 606-01, Japan*

[2] *Department of Physics, Konan Univ., Kobe 658, Japan*

[3] *Flerov Laboratory of Nuclear Reactions, JINR, Dubna 141980, Russia*

Dynamical reaction theory is presented for synthesis of superheavy elements. Characteristic features of formation and surviving are discussed, which combinedly determine final residue cross sections of superheavy elements. Preliminary results on Z=114 are also given.

1 Introduction

Superheavy elements around Z = 114(or 126) and N = 184 have been believed to exist according to theoretical prediction of stability given by the shell correction energy in addition to average nuclear binding energy.[1] This means that heavy atomic nuclei with fissility parameter $x \geq 1$ could be stabilized against fission by a huge barrier which is resulted in by the additional binding of the shelll correction energy around the spherical shape. In other words, if superheavy compound nuclei(C.N.) are formed in such high excitation that closed shell structure is mostly destroyed, they have no barrier against fission and thus are inferred to decay very quickly. Therefore, the point is how to reach the ground state of the superheavy nuclei, or how to make a soft-landing at them. In order to minimize fission decays of C.N. or maximize their survival probabilities, so-called cold fusion reactions have been used, which succeeded in synthesizing SHEs up to Z = 112.[2] They have the merit of large survival probabilities, but suffer from the demerit of small formation probabilities because of the sub-barrier fusion. On the other hand, so-called hot(warm) fusion reactions have the merit of expected large formation probabilities and the demerit of small survival probabilities due to relatively high excitation of C.N. formed. An optimum condition for large residue cross sections of SHEs is a balance or a compromise between formation and survival probabilities as a function of incident energy or excitation energy of C.N. formed.[3] Therefore, the whole reaction process has to be described, from the encounter of incident ions to the formation of compound nuclei, and then to fission decays with residues left of a small probability. An optimum path has to be searched for over all possible incident channels including secondary beams becoming available sooner or later. Whether the reaction is "cold" or "hot" fusion is automatically determined by choosing combinations of projectiles and targets. The theoretical framework, thus, has to accommodate both of them. As is described in section 2, the strong dissipation of energies of nuclear collective motions, which is well recognized by the fusion hindrance[4] and the long fission life,[5] has to be taken into account in formation process as well as in survival process.

2 Theoretical framework

In reactions of massive systems, it is not clear that the compound nucleus theory can apply, i.e., that the formation and the decay of compound nuclei are independent. But for simplicity, we assume that is valid, at least approximately. Then, the residue cross section is given by the following formula as usual,

$$\sigma_{SHE} = \frac{\pi}{k^2} \sum_J (2J+1) \cdot P^J_{for}(Ec.m.) \cdot P^J_{surv}(E^*) \tag{1}$$

,where P^J_{for} and P^J_{surv} denote the formation probability and the survival probability, respectively.

2.1 Formation probability

P^J_{for} is not equal to a simple transmission coefficient T_J(Ec.m.) of an optical potential, nor to a barrier penetration factor P_J(Ec.m.) of the combined potential of Coulomb repulsion and nuclear attraction. As is well known,[4] there is the fusion hindrance in massive systems, roughly those with $Z_1Z_2 \geq 1{,}800$, which could be interpreted by the overcome of so-called conditional saddle under the strong dissipation. The necessary additional energy is so-called extra-push energy. Therefore, P^J_{for} should take into account both the usual barrier penetration and the dissipative dynamics over the saddle. The former is approximated by the penetration factor for the parabolic barrier, while the latter is treated by Langevin equation, concerning the collective shape degrees of freedom of the compound systems starting with the contact configuration of the projectile and the target. In one-dimension, Langevin eq. is given as

$$m\frac{d^2q}{dt^2} + \frac{\partial V}{\partial q} - \gamma \cdot \frac{dq}{dt} + R(t) = 0 \tag{2}$$

The friction coefficient γ is taken to be equal to that of so-called one-body wall-and-window formula.[6] The last time-dependent term R(t) in Eq.(1) is a random force associated with the friction force. The last two terms are a phenomenological description of effects of the nucleonic degrees of freedom in excitation, considered as a heat both. Thus, we assume the fluctuation-dissipation theorem,

$$< R(t) \cdot R(t') > = \gamma \cdot T \cdot \delta(t - t') \tag{3}$$

where <> denotes an average over all the possible realizations of assumed gaussian noise of R and T does the temperature of the heat bath, i.e., the compound nucleus. The delta function in the r.h.s. is due to the Markovian assumption. Equivalently, one can use Kramers eq.

$$\frac{\partial f(q,p,t)}{\partial t} = \{-\frac{\partial}{\partial q}\frac{p}{m} + \frac{\partial}{\partial p}\frac{\partial V}{\partial q} + \frac{\partial}{\partial p}(\beta \cdot p + m\beta T\frac{\partial}{\partial p})\}f(q,p,t) \tag{4}$$

where p denotes the conjugate momentum to q, and β is the reduced friction coefficient γ/m. Eq.(4) describes a time-evolution1 of distribution function in the phase space, while Eq.(2) does trajectories of Brownian particle. Actually, Eq.(4) was proposed by Kramers[7] in order to interpret the fission width proposed by Bohr and Wheeler, from the dynamical view-point.

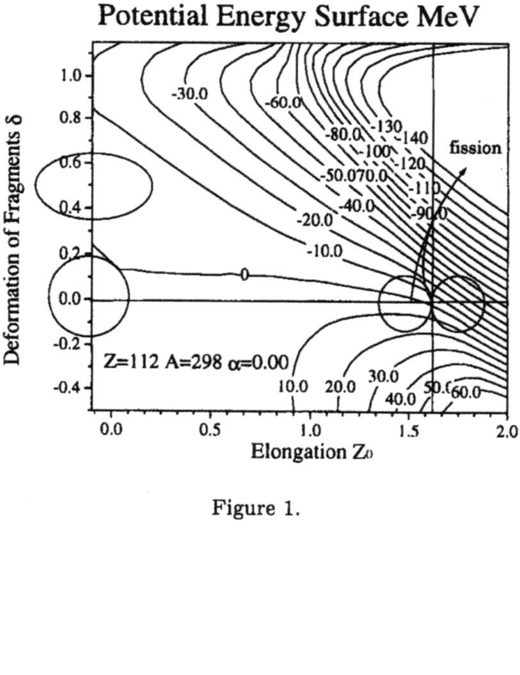

Figure 1.

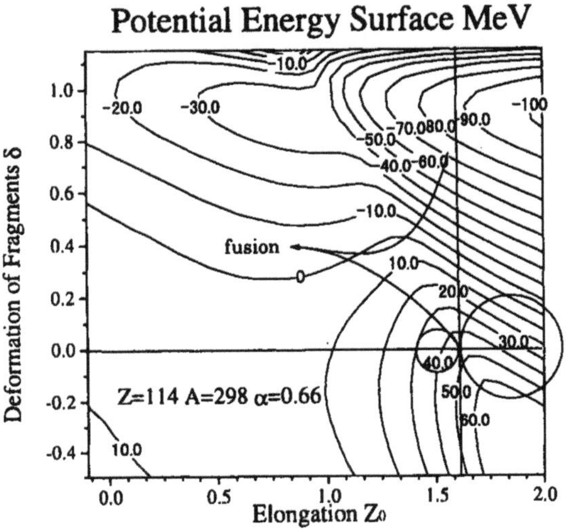

Figure 2.

In realistic situations, processes are in many dimensions including mass asymmetry degree of freedom etc. in addition to the elongation or the separation between fragments. An important case that we will discuss below is that the incident channel is with $Z_1 \cdot Z_2 \geq 1{,}800$ and the compound nucleus is with Z = 114. In Fig. 1, an example of contour maps of energy surfaces is given for the symmetric mass partition ($Z_1 Z_1 \geq 3{,}000$). The axes are the elongation (Zo), i.e., the relative distance between fragments and the deformation where fragment deformations are assumed to be proportional to their masses. It is seen that the contact point of two spheres is located below the conditional saddle point. The system has to climb up, governed by the fluctuation-dissipation dynamics (the multi-dimentional version of Eq.(2))in order to reach the spherical configuration of the total system beyond the saddle point. Apparently, an additional incident energy is required, which corresponds to the extra-push energy. On the other hand, if we think about a very mass-asymmetric incident channel, such as $^{48}Ca + ^{244}Pu$ ($Z_1 Z_2 \simeq 1{,}800$), we may not suffer from the extra energy, because the contact point of two incident nuclei can be located inside the conditional saddle point. Actually, the mass asymmetry parameter $\alpha = 0.66$ case is shown in Fig. 2, where the contact point is slightly inside and then the composite system evolves mainly over the flat region around the spherical configuration after touching each other. This aspect is surely favourable for formation probability. Anyhow, Langevin trajectories are calculated on the energy surfaces in order to obtain P^J_{for}.

2.2 Survival probability

The survival probability P_{surv} is usually given as follows,

$$P_{surv}(E^*) = \frac{\Gamma_n(E^*)}{\Gamma_n(E^*) + \Gamma_f(E^*)} \tag{5}$$

, assuming that fission and neutron emission are dominant decay modes. The neutron emission width Γ_n can be calculated by Weiskopf formula, while the fission width by Bohr-Wheeler one or Kramers' stationary limit.[5,7] If E^* is high enough for more than one neutron emission, then the expression at r.h.s. of Eq.(5) is repeatedly used for sequential E^* and factorized to obtain the final P_{surv}. Anyhow, this method is valid only for cases with the fission barrier height being larger than the temperature T. In SHE compound nuclei, especially in hot fusion reactions, there might be no fission barrier, because the barrier is given by the shell correction energy around the spherical configuration or somewhere in deformed region which is expected to melt in high excitation. The potential landscape changes depending on the temperature T, i.e., on neutron emission. Then, a dynamical treatment is called for. In other words, final residue cross sections of SHE are to be determined by a competition between time scale of fission and that of restoration of the shell correction energy due to cooling by neutron evaporation. Therefore, a speed of cooling and a temperature dependence of the shell correction energy are crucial. The latter is shown in Fig. 3, the energy being normalized by the maximum value, i.e., that of T = 0. The calculations are made by the use

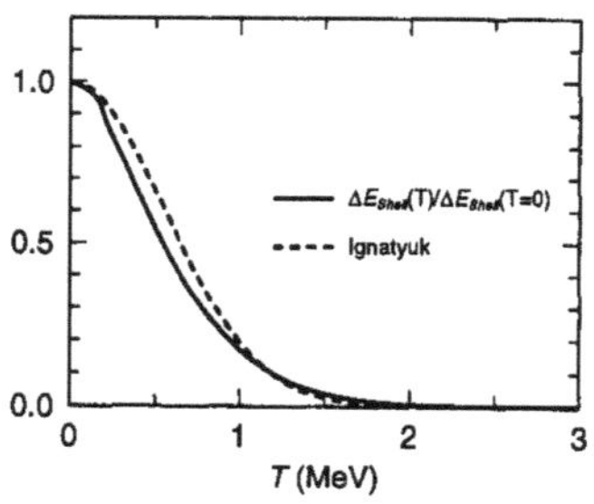

Figure 3.

of single particle levels in a Saxon-Wood finite potential well. The present results are in good agreement with the prescription proposed by Ignatyuk.[8] Using this temperature dependence and cooling obtained by the statistical code for evaporation, we obtain potential energy curves as a function of time. With this time-dependent potential, we calculate fission process with Eq.(2) or (4) or its simplified version and obtain a small fraction of probability captured by the pocket made by the restored shell correction energy, which gives a final survival probability P_{sur}.

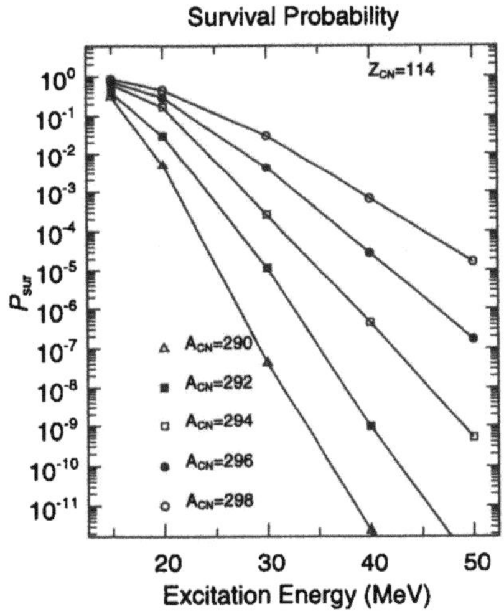

Figure 4.

In Fig. 4, P_{sur}'s with J=10 are shown for several isotopes of compound nuclei with Z=114, as a function of initial excitation energy. We readily see that P_{sur} depends strongly on excitation energy as well as on mass number, i.e., neutron number of initial compound nucleus. This is due to differences in cooling speed due to different neutron separation energies B_n's. Thus, we have to form a compound nucleus which has as small B_n as possible, i.e., as large neutron number N as possible in order to obtain large survival probabilities. In cases with small N such as GSI experiments, higher excitation is extremely unfavored, thus "cold" fusion process is used. Note that the results in Fig. 4 are at the time $2{,}000 \times 10^{-21}$ sec and therefore they should be reduced by a reduction factor due to the last neutron emission for the final survival probabilitites.

3 Preliminary results and discussion

In brief, using WKB approximation for tunneling and 3-dimensional Langevin equation for subsequent dynamical evolutions of the total system, we calculated formation probabilities, while one-dimensional Smoluchowski equation which is the approximation of Eq.(4) was used for fission decay or survival probabilities. Evaporations were taken into account in statistical model. Results are shown in Fig. 5 over possible incident channels leading to Z=114. Existences of the optimum enegies in various channels are clearly seen, which resulted in by the compromise of the two factors. Rapid increases in the left hand side are due to barrier penetration and dissipative dynamical evolutions of shape degrees of freedom, while decreases in the right hand side due to energy dependences of survival probabilities against fission. It is worth to notice that the recent Dubna experiment observed an event which would be a signature for Z=114 element, at the energy predicted here to be most favourable in the incident channel ^{48}Ca + ^{244}Pu.[9]

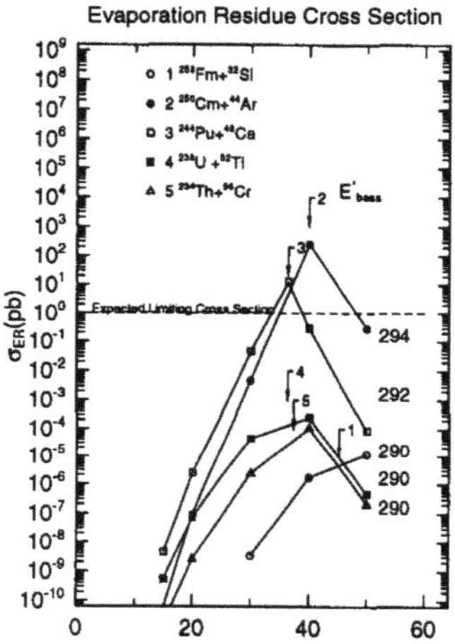

Figure 5.

As for a relation to so-called dinucleus system concept,[10] it would be helpful to have a look on contour map of energy surface in axes of the elongation (Zo), and their mass-asymmetry (α), which is shown in Fig. 6. There is a Businaro-Gallone peak at the upper right corner. Touching points with asymmetries $\alpha = 0.0$ (symmetric case) and $\alpha = 0.66$ (close to ^{48}Ca + ^{244}Pu case) are also shown, schematically, It should be noted here that the top horizontal line and the left vertical line both describe approximately the spherical configuration of the total system, thereby should converge to one point in a more realistic representation. The model based on the dinucleus system concept gives fusion probabilities over the Businaro-Gallone peak, so it evaluates evolutions along the vertical direction in Fig. 6, while the 3-D Langevin calculations take into account evolutions over all possible paths dynamically including the horizontal direction as well.

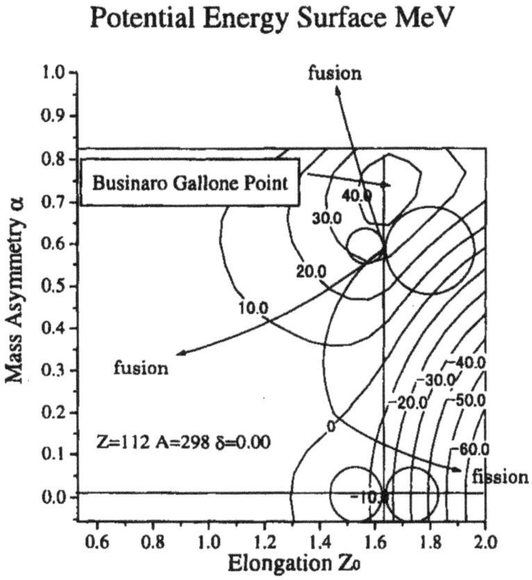

Figure 6.

References

1. see for example, P. Armbruster, Ann. Rev. Nucl. Sci. **35**, 135 (1985).
2. S. Hofmann et al., Z. Phys. **A354**, 229 (1996).
3. Y. Abe et al., J. Phys. **G23**, 1275 (1997).
   Y. Aritomo et al., Phys.Rev. **C55**, R1011 (1997), and *ibid* **C59**, 796 (1999).
   T. Wada et al., Proc. DANF98, Slovakia, Oct. 1998.
4. W.J. Swiatecki, Nucl. Phys. **A376**, 275 (1982).
   S. Bjornholm and W.J. Swiatecki, Nucl. Phys. **A391**, 471 (1982).
5. T. Wada, Y. Abe and N. Carjan, Phys. Rev. Lett. **70**, 3538 (1993)
   Y. Abe et al., Phys. Reports **C275**, 49 (1996).
6. J. Blocki et al., Ann. Phys. (N.Y.) **105**, 427 (1997).
7. H.A. Kramers, Physica **VII4**, 284 (1940).
8. A.V. Ignatyuk et al., Sov. J. Nucl. Phys. **21**, 255 (1975).
9. Yu. Oganessian et al., preprint JINR, E7-99-53.
10. G.G. Adamian et al., Nucl. Phys. **A627**, 361 (1997).

SIMILARITIES AND DIFFERENCES BETWEEN NUCLEI AND METAL CLUSTERS

C. WALTHER[1], J.V. KRATZ[1], K. LÜTZENKIRCHEN[1], L. SCHWEIKHARD[2]
[1]*Institut für Kernchemie, Universität Mainz, D-55099 Mainz*
[2]*Institut für Physik, Universität Mainz, D-55099 Mainz*

It is not obvious that metal clusters should behave like atomic nuclei - but they do. Despite the quite different energy scales both systems show shell effects, giant resonances and compound states. The similarity is not accidential but rather due to the almost free movement of fermions within a confined space.

At first glance one can hardly think of two systems being more different than atomic nuclei and atomic clusters: Whereas one consists of neutrons and protons the other is built of atoms. Size scales and typical excitation energies differ by six orders of magnitude. The nucleus is bound by the strong nuclear force, which is a two body interaction in first order, whereas the interactions of the cluster-constituents are exclusively of long range electromagnetic nature. However, if compared in detail, both nuclei and clusters exhibit a surprising variety of similarities: Magic numbers known for a long time in nuclear physics are observed for metal clusters, giant resonances (plasmons) play an important role in both systems and long living highly excited cluster states are populated in striking analogy to the phenomenon of compound nuclei.

Early investigations revealed an enhanced abundance of sodium clusters consisting of 8, 20, 40 and 58 atoms as compared to their direct neighbours (fig. 1 left). Although nowadays it almost seemes obvious to conclude on shell closures in analogy to the Nilson model, the credit is due to Knight et al.[1] to have suggested the physical processes. It is the movement of the fermionic constituents inside the systems that gives rise to many of the analogies. Nucleons are delocalized over the whole nucleus being confined in a self generated mean potential. Metal clusters on the other hand can be described as a negativly charged 'fluid' of atomic valence electrons within a positive background potential formed by the ionic cores of the atoms (the so called jellium [2]). Because of the huge mass difference of electrons and ions, the movement of the latter can be neglected on short time scales (Born approximation) and the electrons dominate the structure. Deviations of the magic numbers of clusters as compared to nuclei are due to the weaker spin orbit coupling. It contributes to the total binding energy by less than 1% and can be neglected for most considerations.

In contrast to nuclei metal clusters are electrically neutral (or typically

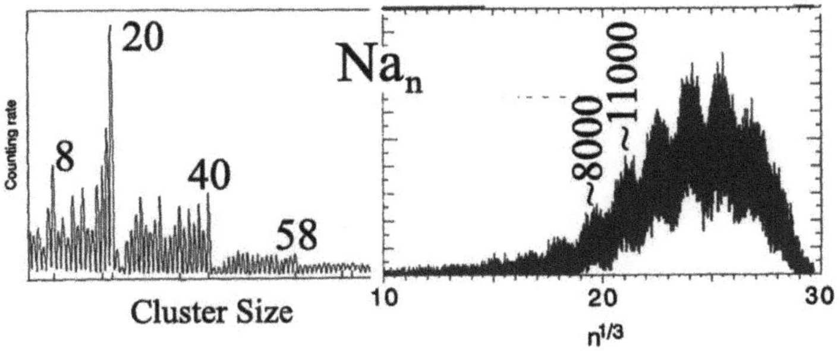

Figure 1. Shell closures of sodium clusters

contain one or two extra charges) and therefore Coulomb instabilities present no size limitation. Sodium clusters of many 10000 atoms can be produced and provide the unique possibiliy to study the evolution of shell effects up to very large numbers [3]. While the magic numbers proceed roughly proportionally to $n^{1/3}$ an additional effect is observed: The stability enhancement in turn is subject to a long periodic change (fig. 1 right) due to the buildup of triangular and rectangular (classical!) orbits of electrons inside the (spherical) cluster. This effect had been predicted earlier for very large nuclei.

Due to the low binding energies of clusters, excitations take place in the eV-domain - easily accessible by VIS and UV lasers. Selby et al. found the sodium clusters Na_{20} to exhibit a broad resonance of photoabsorption cross-section, devoid of any fine structure [4]. Again the analogy to nuclear physics imposed itself: In a 'giant dipole resonance' electrons and ions (instead of neutrons and protons) demix and oscillate against each other. From the width of the resonance a lifetime of some 10^{-15}s was deduced which could be veryfied very recently using fs-lasers. Borggreen et al. investigated this resonance for cluster sizes 3 to 40 and detected the evolution of a resonance splitting with pronounced high energy branch for clusters slightly bigger than a closed shell neighbour [5]. As the clusters increase further the low frequency branch dominates and finally merges into a single resonance again at the next shell closure. As can be shown by use of classical electrodynamics [6] these findings reflect the evolution of cluster geometry from spherical via prolate and oblate back to spherical shape. It is not trivial that such a quantum system can be described as homogeneously charged particle and provides further evidence for the validity of the jellium model.

In order to determine dissociation energies we have developed the method of time-resolved photo dissociation in a time range of about 10 μs to 100 ms.

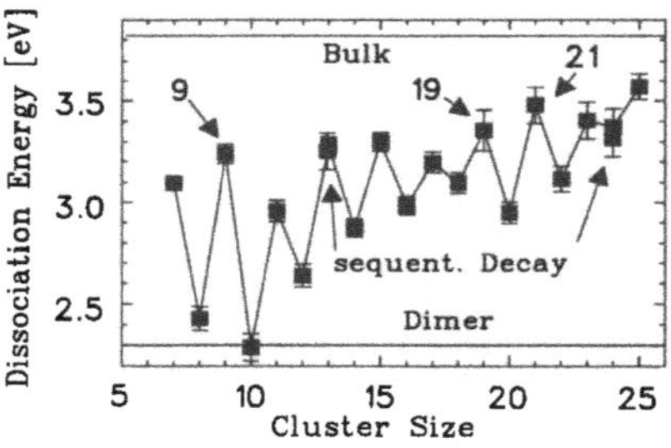

Figure 2. Dissociation energies of Au_n^+

It makes use of the storage capability of a Penning trap [7]. The photon is absorbed by the cluster electronically but due to strong coupling transformed within ps into vibrational excitation. If the energy exceeds the dissociation energy, the system cools by evaporation of a neutral atom and the fragmentation yield is a measure of the photon absorption cross section. By measuring decay times as a function of excitation energy the dissociation energy of the cluster can be determined within the framework of phase space (Weisskopf) theory [8]. Dissociation energies of silver cluster cations were gained this way [9] and currently the investigations are being extended to the homologue element of gold. Fig. 2 shows preliminary results. Clearly visible are the shell closures for Au_9^+ (8 valence electrons), Au_{21}^+ (20 valence electrons) and Au_{19}^+ (closure of the d-subshell) and a pronounced odd-even staggering due to spin coupling. Because of the higher stability as compared to silver clusters gold has to be heated to higher temperatures and radiative cooling has to be taken into account as was shown in the case of vanadium [10].

References

1. W. Knight *et al.*, *Phys. Rev. Lett.* **52**, 2141 (1984).
2. M. Brack, *Rev. Mod. Phys.* **65**, 677 (1993).
3. T.P. Martin, *Phys. Rep.* **273**, 199 (1996).
4. K. Selby *et al.*, *Phys. Rev.* B **40**, 5417 (1989).
5. J. Borggreen *et al*, *Phys. Rev.* B **48**, 17507 (1993).
6. C. Bohren and D.Huffman, *Absorption and scattering of light by small particles*, Wiley - Intersc. Publ. N.Y. (1983).
7. C. Walther *et al*, *Chem Phys. Lett.* **256**, 77 (1996).
8. S. Frauendorf, *Z. Phys.* D **35**, 191 (1995).
9. U. Hild *et al.*, *Phys. Rev.* A **57**, 2786 (1998).
10. C. Walther, Ph.D. thesis Mainz, (1998).

NUCLEAR CLUSTERS ON THE EVE OF THE THIRD MILLENNIUM

HISASHI HORIUCHI

Department of Physics, Kyoto University, Kyoto 606 - 8502, Japan

1 Introduction

When the word of nuclear clustering is understood to express the physics of dynamical assembling and disassembling of nucleons, clustering aspects appear abundantly in many problems in both nuclear structure and nuclear collisions. We need to recall that as is clear in the saturation property of nuclear binding energy, nuclei and nucleons are easy to assemble and disassemble. This implies that the formation of clusters is a fundamental aspect of nuclear many body dynamics together with the formation of mean field [1,2]. The coexistence of the clustering aspect and the mean field aspect (single particle aspect) is a unique feature of nuclear many body system [3].

The omnipresence of the clustering aspect is a basic feature of present day nuclear physics. Talks given in this conference verify this fact. We see that the clustering is playing important roles not only in traditional subjects but also in new phenomena and in new kind of nuclear systems. I think that the knowledge and concept of clustering are being enriched through the studies of new phenomena and new kind of systems. In this summary talk, I would like to emphasize these points. I expect that my talk serves for a talk about prospectives of the study of nuclear clustering toward the third millennium.

I try to classify the talks given in this conference into three categories, namely (i) clustering in nuclear structure, (ii) clustering in nuclear collisions, and (iii) theories and non-nucleonic clusters. My discussions given below follow this classification. I will put some emphasis on the discussion of the formation and dissolution of clusters.

2 Clustering in Nuclear Structure

2.1 Molecular resonances and cluster states

We heard many talks which were devoted to this subject. Betts gave a nice historical survey of the study of molecular resonances starting from the Chalk River experiments on $^{12}C + ^{12}C$ resonant scattering. The search of the possibility of the 6α chain configuration in ^{24}Mg is one of the most exciting issues in recent studies of molecular resonances and has attracted much attention. The experimental techniques used for this search are very elaborate and show the power of the present day experiments. Fulton's talk was the review of the studies of this problem since the last cluster conference at Strasbourg and his conclusion was that the straight chain configuration of 6 alphas is not likely to be formed. Even if the linear chain 6α state is not realized, the experiments tell us that near the 6α breakup threshold there exist clustering states in which the degrees of freedom of 6 alpha clusters are all activated. This is just what we can expect from the Ikeda diagram [4] and hence the experimental studies have given a strong support to the philosophy of the Ikeda diagram. According to the Ikeda diagram for self-conjugate 4N nuclei, the most high-lying alpha cluster states have the structure that is composed of weakly interacting alpha clusters. Those states are expected to be located energetically near the breakup threshold into alpha clusters forming bound or quasi-bound states with respect to this threshold energy, which is because the alpha clusters are interacting weakly one another so as not to fuse into heavier clusters. The Ikeda diagram does not say anything about the configuration of alpha clusters; for example it does not insist the formation of the linear chain configuration of alpha clusters and in fact in ^{12}C the 3α cluster states are known not to have the linear chain configuration. Therefore in order to have more information about the configuration of 6 alpha clusters, we need further experimental and theoretical studies. The talk of Di Pietro about the use of ^{13}N + ^{11}B channel for populating the 6α states is along this direction and the theoretical approach of Sakuragi who uses elaborate coupled channel calculations is very informative for this purpose.

For studying heavier molecular systems of ^{24}Mg + ^{24}Mg and ^{28}Si + ^{28}Si, Uegaki used the strong coupling approach of channels by assuming the formation of geometrical configuration of clusters. He analysed the coupling of vibrational modes of intrinsic configurations with rotational modes. This is in contrast to the weak coupling approach by the use of coupled channels calculations by Sakuragi and Boztosun for lighter molecular systems of ^{12}C + ^{12}C, ^{12}C + ^{16}O, and ^{16}O + ^{16}O. Robson also treated ^{24}Mg + ^{24}Mg and ^{28}Si + ^{28}Si and discussed possibility of hyper deformation. It will be interesting to make close comparison of these approaches with mean field approaches to high spin states such as cranked Nilsson method.

Several talks were presented which aim to explain resonance-like behaviors of heavy-ion reactions without necessarily assuming the formation of the molecular resonance. Ray and Vannucci discussed the orbiting cluster model which has rather long history and Kun and Papa discussed possible reaction mechanisms taking into account memory effects and fluctuation effects.

There were given a few talks which discussed alpha cluster states. Lönnroth reported analyses of alpha scattering on sd-shell nuclei which shows the existence of large number of narrow states at high excitation and Ohkubo discussed alpha clustering in pf-shell region especially about the structure with "core + two alphas". As for the α-clustering in heavy nuclei we heard no talk except a short comment by Lovas at the round-table discussion session. So I here quote only a recent review paper by Lovas and his collaborators [5].

Betts showed one of the recent results by Pandharipande's group which is the intrinsic density distribution of ^{8}Be wave function obtained by the Green's function Monte Carlo method by using realistic nuclear force. This density distribution shows manifestly well-developed two-alpha cluster structure with inter-cluster distance of about 4 fm. This result verifies one of the basic assumptions of the cluster model that the cluster structure has the spatial localization of constituent clusters. Therefore this result unambiguously invalidates the argument that the cluster state is nothing but a special shell model state with a specific spatial symmetry and a specific Elliott's SU(3) symmetry.

In very light nuclei such as ^{7}Li and ^{8}Be, the cluster structure shows up in their ground states [6]. But as the mass number becomes large the mean field is formed in nuclear ground states and the cluster structure appears in excited states. For example in ^{12}C the ground state has a shell-model-like character and alpha cluster states appear in excited states near and above 3α breakup threshold. This is the basic understanding of the cluster model for ^{12}C. Since the *ab initio* calculation by using realistic nuclear force has reached nuclei with mass number A = 8, we would like to expect strongly that such calculation will be made for nuclei with mass number A = 12. When such calculation is realized, the calculation for ^{12}C will show the formation of the mean field in the ground state and the appearance of 3α cluster states as excited states. I think that the *ab initio* calculation by using realistic nuclear force will become a very important subject for studying nuclear clustering.

A kind of approximate *ab initio* calculation by the use of not the realistic nuclear force but the effective nuclear force is now possible by the use of antisymmetrized molecular dynamics (AMD) [1]. By using AMD we can study formation and dissolution of clusters in structure and collision problems. An example is a study of ^{12}C made by Kanada-En'yo [7]. It is shown that the calculation actually reproduces the coexistence of mean field structure of the ground state and the 3α cluster states near and above 3α breakup threshold. Intra and inter band E2 transitions between the ground band and the cluster band are well reproduced and beta decay strengths from the 1^+ ground state of ^{12}N to the ground band and cluster band states are also well reproduced even though these two bands are very different from each other in structure.

The study of formation and dissolution of clusters has been one of the main theoretical subjects of nuclear clustering and many theoretical methods have been used for this study. As we discuss later this study is now becoming more important since the widely extending nuclear chart requires it. Examples of theoretical methods other than AMD are unrestricted Hartree-Fock method, hybrid model of mean field model and cluster model and so on.

2.2 Inter-cluster interaction and molecular states

Long time ago Ali and Bodmer [8] presented a shallow potential which describes α-α scattering very well. But from the microscopic theory using the resonating group method [6] we get deep inter-nucleus potentials [9]. Baye gave a clear review about the relation between deep and shallow inter-nucleus potentials by using his theory of supersymmetric transformation. According to his theory we can say that the deep non-singular potential with no ℓ-dependence is equivalent to the shallow potential with strong ℓ-dependence and singularity. Actually Ali-Bodmer potential is strongly ℓ-dependent and singular. Baye also reported his investigation how much this on-shell equivalence between deep and shallow potentials is modified for systems including off-shell processes such as three-cluster problem and bremsstrahlung.

As for the experimental study of the inter-nucleus potential, von Oertzen gave, in his introductory talk of the conference, a survey starting with the determination of the unique optical potential of alpha particles. In the case of the optical potential of heavy ions it was only about ten years ago that the unique optical potential became to be determined when the discovery of nuclear rainbow phenomena was made in $^{16}O + ^{16}O$ by von Oertzen's group.

Kondō explained us impressively about the determination of the unique deep potential by the aid of the dispersion relation between real and imaginary potentials and also about the study of molecular resonances by the use of deep potentials in the systems of $^{12}C + ^{12}C$, $^{12}C + ^{16}O$, and $^{16}O + ^{16}O$. Nicoli also discussed similar problems based on the analysis of refractive effects.

When we compare the present situation with ten years ago, it is quite impressive to notice the big change that deep potentials are now regarded as being standard potential. Besides the efforts to pin down the unique optical potential for heavy ions, the success of the folded potential model [10] for describing variety of reaction processes was also very influential for this big change.

When two nuclei collides many reaction processes take place beside the elastic and inelastic scattering. If we try to describe many final channels in a unified way, the deep potential is the unique possible choice. When we describe nuclear collision microscopically for example by using time-dependent Hartree-Fock equation, or Boltzmann transport equation [11], or molecular dynamics [12], we clearly see that colliding nuclei pass through each other instead of being reflected by singular repulsive potential at short distance. Yabana and I once showed that the production cross section of high energy gamma ray in heavy ion collisions is largely due to the acceleration of the inter-nucleus relative velocity due to deep inter-nucleus potential [13].

2.3 Fission and cluster decay

Greiner reported, in his introductory talk of the conference, about recent elaborate studies of fission processes by the use of large arrays of Ge detectors, which aim to identify fission fragments by triple γ coincidence. Especially he discussed about the observation of ^{10}Be in the cold ternary spontaneous fission of ^{252}Cf and proposed a possible formation of three-cluster nuclear molecule (^{146}Ba + ^{10}Be + ^{96}Sr) in the exit channel. Poenaru also discussed this ternary fission problem in addition to the extensive nice review of cluster decay. It is surely very much desirable that this kind of elaborate and powerful experiments by the use of large arrays of Ge detectors are further developed in future.

Royer discussed the cluster decay by regarding it as binary fission process and by using proximity potential. La Commara discussed exotic nuclei around ^{100}Sn in relation to cluster emission and Rousseau investigated ^{28}Si + ^{12}C fusion to study possible formation of highly deformed state of ^{40}Ca by detecting cluster emission.

2.4 Superheavy elements

This year we heard big news from Dubna and Berkeley that the new elements with $Z = 114 \sim 118$ are very likely to be synthesized. In this conference we could hear about these hot news from experimental authorities from GSI and Dubna. Münzenberg explained clearly essential points of synthesis experiments of superheavy elements which can exist only by shell effects. His talk covered GSI cold fusion with one-neutron emission and Dubna method with three-neutron emission which led to the synthesis of $Z = 114$. Then he explained recent Berkeley experiment of ^{86}Kr + ^{208}Pb $\rightarrow$ ^{293}118 + n. Itkis gave a nice extensive review of synthesis experiments at Dubna by the use of lighter projectile like ^{48}Ca and explained the recent synthesis of two isotopes of $Z = 114$, ^{287}114 and ^{289}114.

As for the theory, Greiner explained, in his introductory talk, the energy surface by the potential model with various shape parameters, by Skyrme Hartree-Fock, and by relativistic mean field theory. Abe discussed the effect of neutron emission which cools down the system temperature by treating it with Langevin equation. Adamian explained the calculation of the competition between complete fusion and quasifission by using the concept of dinuclear system which evolves to the compound nucleus by the mass asymmetry degree of freedom.

In addition to check experiments of these new discoveries, search for further heavier elements is looked for and also as Reisdorf commented search toward highest neutron number is desirable in view of the nuclear synthesis in the universe.

2.5 Clustering in nuclear matter

Clustering in nuclear matter is not so well studied in spite of its importance especially in dilute nuclear matter. Tohsaki studied this problem by putting clusters at lattice sites of various kinds of lattice by treating antisymmetrization exactly. He found pronounced clustering in the surface region of nuclear matter. It is highly desirable that his study is extended to general situations where nucleons and various clusters are located randomly, since these general situations have been studied only crudely by using quasi-classical molecular dynamics without taking into account antisymmetrization exactly.

Another approach to this problem was presented by Schuck and Röpke by using their theory of cluster mean field approach. Their theory describes for example the change of pairing nature from momentum pairing to spatial pairing as the density becomes dilute. Pairing is expected to be very important in neutron-rich nuclei, especially in neutron skin region with dilute neutron density. We can expect that this theory will contribute much for clarifying the microscopic nature of neutron skin.

The 6α state in ^{24}Mg is also a good example of clustering in dilute nuclear system. As we discuss later clustering in dilute nuclear matter is now experimetally observed explicitly in the Au + Au central collisions.

2.6 Clustering in unstable nuclei

Di-neutron cluster problem in two-neutron halo nuclei such as ^{6}He and ^{11}Li may have been the first subject discussed as a clustering problem in neutron-rich nuclei. Since ^{6}He is a typical two-neutron halo nucleus with simple three-body structure of "$\alpha+n+n$", it was discussed by many talkers in this conference. Suzuki studied this nucleus with his stochastic variational method by using correlated Gaussian basis. He further investigated the distortion effect of the α core by introducing the 4-body model of "$3N + N + n + n$", which was explained in detail by Arai. Wolski reported experimental studies of the two-neutron configuration of ^{6}He by using transfer reactions. He showed the dominance of di-neutron configuration in this Boromean nucleus. He further reported about the study of ^{8}He. Katō also discussed ^{6}He from the viewpoint of $2n$ pairing and by using the method of complex coordinate rotation so as to be able to treat resonance states. He also pointed out the importance of the pairing correlation in ^{11}Li for understanding the inversion of p-shell orbits and sd-shell orbits.

One of the most actively discussed subjects about the clustering in neutron-rich nuclei in recent years is the dimer structure in Be isotopes. Long time ago Okabe, Abe and others discussed molecular orbits around the $\alpha - \alpha$ core [14], and Seya, Kohno, and Nagata extended this model of molecular orbit up to the neutron dripline not only in Be isotopes but also in B isotopes [15]. The *ab initio* model AMD then confirmed theoretically the formation of the $\alpha - \alpha$ core in Be isotopes up to the neutron dripline ^{14}Be nucleus [16,17]. It was von Oertzen who carefully investigated the correspondence between the predictions of the dimer model and large amount of the relevant experimental data. In his introductory talk, von Oertzen explained above-mentioned history, discussed present status, and expressed future prospects. Miljanić reported experimental studies of clustering structure of Be isotopes by using various types of reactions. Itagaki reported theoretical studies of Be isotopes on the basis of molecular orbit model and he suggested strongly that the neutron magic nucleus ^{12}Be does not have closed shell configuration.

Be isotopes can be well understood by the molecular orbit model or the two-center shell model with $\alpha - \alpha$ core and also by AMD whose results are very similar to the former two models. Especially we find that the positive parity orbit with $\Omega^\pi = 1/2^+$ which comes down from the sd-shell and is called the σ-orbit plays a very important role. The famous positive parity ground state of ^{11}Be has the halo structure due to the last halo neutron occupying this σ-orbit. In ^{9}Be the first excited positive parity band has also its last neutron occupying this σ-orbit. What is very impresive is the structure of low lying levels of ^{10}Be. We can classify these levels into three types according to the number of neutrons occupying the σ-orbit. The ground band with $K^\pi = 0^+$ and the side band with $K^\pi = 2^+$ have no neutron in the σ-orbit, while the negative parity excited band with $K^\pi = 1^-$ has one neutron in the σ-orbit and the excited positive parity band with $K^\pi = 0^+$ has two neutrons in the σ-orbit. AMD calculations reproduce reasonably well the intra and inter band E2 and E1 transitions among the ground and excited bands and also reproduce well Gamow Teller strengths from the 3^+ ground state of ^{10}B to ground and side band states [18].

As Itagaki suggested, it is likely that the neutron closed shell structure of ^{12}Be is broken with two valence neutrons occupying the σ-orbit. What is the reason of the breaking of the neutron magic number $N = 8$ in ^{11}Be and ^{12}Be ? Certainly the deformation due to the clustering plays an important role in pulling down the σ-orbit close to the p-shell. But is it the only reason ? In the ^{11}Li ground state it is strongly suggested that here also the neutron closed shell structure is largely broken with two valence neutrons occupying the sd-shell orbits. However at present it is not clear whether any kind of prominent clustering is existent or not in the ^{11}Li ground state. As mentioned before, for this problem of the inversion of the p-shell and sd-shell orbits, Katō et al. insist that the pairing correlation plays an important role. Similarly for ^{11}Be, Sagawa et al. pointed out previously that the pairing correlation plays an important role for the inversion [19]. By using AMD we have also checked that the pairing correlation plays certainly an important role for the inversion in ^{11}Be. In order to solve the problem of the breaking of the neutron magic number $N = 8$, it is surely necessary to have unified understanding of the structure of neutron-rich nuclei at least around ^{11}Be and ^{11}Li.

A strong suggestion of the existence of the ^{6}He + ^{6}He dimer structure in excited states of ^{12}Be was given in the talk by Freer. This suggestion was obtained by the experimental study of the fragmentation of ^{12}Be into ^{6}He + ^{6}He via the inelastic excitation by the proton and carbon targets. Similar experimental indication was reported a few years ago by RIKEN group [20]. The idea of the dimer structure with $\alpha-\alpha$ core leads us naturally to the expectation of the existence of the polymer structure with linear chain configuration of α clusters which are bound by the help of mediating neutrons more stably than without neutrons. At the round-table discussion session, Itagaki discussed the possibility of the trimer state in ^{14}C by showing his molecular orbit model calculations.

An interesting experimental study of two proton correlation in proton-rich ^{17}Ne was shown by Thirolf who investigated the angular correlation of two proton decay, $^{17}\text{Ne} \rightarrow {}^{15}\text{O}$.

Are there any other characteristic phenomena which show an indispensable role of the clustering in unstable nuclei ? We have an attractive possible answer to this inquiry from the AMD study of B isotopes [21]. According to AMD, B isotopes near neutron dripline show prominent di-cluster density distribution. Here five protons are divided spatially into two groups with two and three protons, respectively, which are surrounded by neutrons. Reliability of such AMD results is assured by good reproduction of data for ^{13}B $\sim$

^{19}B which include binding energies, radii, electric quadrupole moments, and magnetic moments. We notice an important fact that the clustering in ^{17}B and the neutron dripline nucleus ^{19}B is entirely different from the clustering feature in stable nuclei. For example, if we regard ^{19}B as having a di-cluster structure of ^{8}He + ^{11}Li or ^{10}He + ^{9}Li, we notice that any of clusters ^{8}He, ^{10}He, ^{9}Li, and ^{11}Li does not satisfy the "cluster condition" that the cluster should be a stable nucleus. Furthermore we notice that the threshold rule of the Ikeda diagram is heavily violated, because ^{19}B is deeply bound by 12.73 MeV and 14.17 MeV from the ^{8}He + ^{11}Li and ^{10}He + ^{9}Li thresholds, respectively. (Also the neutron dripline nucleus ^{14}Be is deeply bound by 9.16 MeV and 11.13 MeV from the ^{8}He + ^{6}He and ^{10}He + α thresholds, respectively.) Thus we are forced to regard that the clustering in ^{17}B and ^{19}B is of a novel type different from the clustering in stable nuclei.

I have a conjecture that in general a novel type of clustering is a candidate of nuclear strucure near neutron dripline. By definition the neutron dripline nucleus has the structure which accommodates maximum number of neutrons for a given number of protons. Such structure is realized by making the volume of the neutron skin region maximum by avoiding the energy loss due to the Pauli principle. When we notice that the thickness of the neutron skin cannot be larger than the nuclear force range because neutrons without protons cannot be bound, we see that deformed shape has bigger volume of neutron skin than spherical shape and that clustering shape with spatially divided protons has even bigger volume of neutron skin.

Since the advent of the radioactive ion beam, we have seen that fundamental concepts of nuclear physics are forced to be changed: Saturation property of density is violated, proportionality of density distributions of protons and neutrons is violated, magic numbers are broken, and so on. We are now at the doorway of renaissance of nuclear physics. In view of this situation, the study of dynamical assembling and disassembling of nucleons in the foreseen widely extended nuclear chart is surely a fundamental task of nuclear physics.

2.7 Clustering in hypernuclei

In this conference, hypernuclei were discussed only in the talk of Suzuki who reported the structure study of single and double Λ hypernuclei with a microscopic cluster model. He proposed that a proper account of the effect of the Pauli-forbidden states for quarks largely resolves the famous underbinding problem of $^{5}_{\Lambda}$He.

Recently gamma transitions in hypernuclei have begun to be detected. For example they were detected in $^{7}_{\Lambda}$Li and in $^{9}_{\Lambda}$Be. Especially in $^{7}_{\Lambda}$Li the

B(E2) strength was measured [22] and it is expected that B(E2) strength in ${}^{9}_{\Lambda}$Be will be also measured soon. While ^{8}Be is unbound sitting above $\alpha + \alpha$ threshold, ${}^{9}_{\Lambda}$Be is bound with its ground and first excited states below α + ${}^{5}_{\Lambda}$He threshold. It is the reason why we can have the B(E2) information in ${}^{9}_{\Lambda}$Be while we cannot get it in ^{8}Be. Although the α - α structure in ${}^{9}_{\Lambda}$Be is slightly modified from that in ^{8}Be due to glue role of the Λ particle which binds two α's more tightly, we can extract good information about the α clustering in ^{8}Be from the B(E2) values in ${}^{9}_{\Lambda}$Be. Of course, besides the B(E2) values, data of the energy spectra of ${}^{9}_{\Lambda}$Be also provide us with very useful information about the α clustering in ^{8}Be. This kind of study was already made in the case of ${}^{7}_{\Lambda}$Li [23].

The original idea that the structure of hypernuclei will be of great help for studying the cluster structure as mentioned above was proposed long time ago by Bandō and his collaborators [24]. These authors also discussed the cluster structure in hypernuclei which has no counterpart in ordinary nuclei. It is easy to imagine that the hypernuclei formed by putting Λ particle(s) to unstable nuclei will provide us with very important information about the structure of unstable nuclei.

3 Clustering in Nuclear Collisions

3.1 Heavy ion collisions and multifragmentation

We heard many talks about heavy ion collisions whose incident energies were very widely distributed up to a few GeV per nucleon. This can be said to be one of the characteristic features of this conference. Reisdorf reported about central collisions of Au + Au above about 100 MeV/nucleon on the basis of FOPI collaboration and pointed out that even at 1 GeV/nucleon about 50 % of protons are contained in clusters. Clusters are regarded as being formed in radially expanding dilute nuclear matter. I think that these experiments present us with the first observation of the clustering in dilute nuclear matter. His reports included also a so-called mixing experiment which gives us important insight into the equilibration of the system and which resulted in partial transparency. Ono reported AMD studies of heavy ion collisions which included central collisions of Au + Au at 150 and 250 MeV/nucleon. He showed that production cross sections of fragments and average fragment kinetic energy as a function of fragment mass in Au + Au collisions are nicely reproduced by taking proper account of coalescence processes. Korolija reported the central Au + Au collisions at 1.06 GeV/nucleon which showed more energy share in collective expansion than in thermal motion. Rami reported the study of system size dependence in cluster production at SIS energies, which revealed the universality in both energy and system size in the case of spectators but the presence of the dependence on both energy and system size in the case of participants. Dželalija reported about the generalized approach to entropy determination for which nucleon occupation function in momentum space were measured for Au + Au. Petrovici discussed cluster flows and their incident energy dependence especially by studying the azimuthal angle dependence.

Trautmann discussed the decay of excited spectator nuclei based on the ALADIN collaboration for Au + Au collisions up to 1 GeV/nucleon, which aims the identification of the predicted liquid-gas phase transition in nuclei. His discussions included the certification of the equlibration of decaying system. This time he put some stress on various ways of extracting temperature which included the new temperature T_{LiBe} in addition to the previously studied T_{HeLi}. By using AMD we recently calculated the caloric curve of the $A = 36$ system, which showed, in spite of its small size, clear phase transition from liquid phase to gas phase with transition phase between the two [25]. We found that the characters of the transition phase including its temperature and fragment mass distribution are very much dependent on the pressure on the system, which implies the necessity of the proper consideration of the system pressure or confining volume.

Neubert reported proton induced fragmentation of light nuclei at 1 GeV putting stress on the derivation of isotopic yield ratio. He discussed the equilibration and comparisons with model calculations.

Cluster production occurs via various mechanisms. In the above we discussed production of clusters in expanding dilute nuclear matter and in relation to liquid-gas phase transition. Rivet gave a nice survey about various kinds of cluster formation in the incident energy region below 100 MeV/nucleon, which was much based on INDRA collaboration. As for the cluster production via the neck formation, she explained detailed understanding of the dynamics such as oriented fission and also the equlibration at the time of rupture of the neck. Ono discussed the neck formation in $^{40}Ca + {}^{40}Ca$ at 35 MeV/nucleon and he showed that the fragment charge distribution is not reproduced with the hard EOS with momentum independent mean field.

The multifragmentation due to the spinordal decomposition of a finite system was discussed by Rivet and especially in detail by Borderie. Borderie explained his microscopic simulation of nuclear collisions which incorporates fluctuations due to stochastic force in the usual Boltzmann transport equation (BNV). He showed good agreement between calculated observables and experimental ones for the head-on collisions of heavy nuclei and concluded that the multifragmentation may be traced back to the spinordal decomposition of a finite system.

Moretto discussed multifragmentation from the statistical exploration of the fragmentation phase space.

Clustering correlation inside the colliding nuclei is also an origin of cluster production. In addition to the direct knock-out of the preformed clusters, the cluster emission via the clustering excited resonance states is also an important process [26]. In Gadioli's talk which discussed α particle emissions in ^{12}C induced reactions there were presented various mechanisms of α particle production for which α particles produced in the ^{12}C breakup play important roles.

Coalescence process is of course an important mechanism of the formation of light clusters. This process is automatically included in microscopic simulation calculations. Běták discussed this process in his exciton model calculations in addition to the introduction of preformed clusters.

As is shown by INDRA collaboration the fusion processes vanish around 30 MeV/nucleon and the binary dissipative collisions with quasiprojectile and quasitarget in the exit channel dominate at higher energies. Eudes reported the study of this change of the reaction process which was made by using the Boltzmann transport equation (LV) adopting Gogny force and appropriate in-medium NN cross section.

3.2 Nuclear astrophysics

In this conference, only Cherubini gave a talk in relation with nuclear astrophysics. He explained one of the indirect methods of measurement of charged particle cross sections in low energy region of astrophysical interest. He discussed the measurement of ^{6}Li + ^{6}Li $\rightarrow 3\alpha$ in extracting the cross section of α + ^{6}Li $\rightarrow 2\alpha$. As was mentioned in Baye's comment at the round-table discussion session, microscopic cluster model calculations are well known to have made large contributions in this field, and furthermore their contributions will be even more necessitated in future.

4 Theories and Non-nucleonic Clusters

4.1 Theories

As symmetries play important roles in any fields of physics, the viewpoint of symmetries in cluster physics has also a long history. In this conference, several talks were devoted to this subject. Cseh and Hess explained their recent symmetry approach which combines microscopically constructed model space with phenomenological interactions and is named semimicroscopic algebraic cluster model. Tumino reported the study of ^{12}C by a three-alpha model which constitutes of two p-bosons for two Jacobi coordinates and one s-boson. Brant reported about extensive studies of medium heavy nuclei by extended versions of interacting boson model which incorporate broken pair.

Understanding of the roles of Pauli principle is a fundamental problem of nuclear cluster physics and has a long history of its investigation. Lovas made what he calls a clarifying talk about clustering and the Pauli principle. Especially he emphasized the importance of the quantity named the amount of clustering. This quantity and the notion of correctly normalized relative wave function by the use of the norm kernel are known to be indispensable for studying the clustering in heavy nuclei. He pointed out the fact that the amount of clustering is by no means small even in lead region and regarded that it implies the generality of clustering mode of motion.

There were several theory talks which were concerned with the relation between clustering and mean field. In Lovas's talk this relation was also discussed. The talks by Schuck and Röpke which I cited before just treated the formation and dissolution of clusters from the side of mean field theory. De la Mota discussed a new formulation which treats the fragment formation in heavy ion collisions by incorporating early cluster recognition algorithm into mean field theory and by treating quantal aspects.

As was mentioned by Borderie and de la Mota, it is now widely recognized that the quantum aspects are important in fragment formation or system disassembling. The AMD-V theory which was explained in Ono's talk is a new theory of molecular dynamics which incorporates a new type of process of the quantum branching of channels in addition to the usual two-nucleon collision process.

4.2 Non-nucleonic clusters

In this conference no talk was devoted to the quark cluster problem except Suzuki's talk where the effect of the Pauli-forbidden states of quarks was discussed in relation to the hypernucleus. It is well known that the resonating group calculations with quark clusters have made big contributions for the study of nuclear force. They are expected to make even more contributions for the extended study of the force between baryons including hyperons.

As for the physics of atomic clusters we heard two illuminating talks. Lutz explained us various interesting properties of ^{4}He clusters including various interactions with other atoms or molecules. An especially impressive new technique was the experimental formation of superfluid ^{4}He-clusters of controllable size by surrounding them with normal ^{3}He. Walther explained similarilties and differences between nuclei and atomic clusters by discussing shell and super-shell structure, giant resonances, fission, and so on. Then he explained a new method to investigate radiative cooling process in the case of V_{13}^{+} clusters. The collision problem of atomic clusters is another interesting subject whose comparison with nuclear processes is proving to be useful [27]. I think it will be valuable to exchange theoretical ideas of describing the quantum branching of channels.

5 Conclusion

The formation of clusters is a fundamental aspect of nuclear many body dynamics together with the formation of mean field. This leads to the omnipresence of the clustering aspect in nuclear phenomena which is a basic feature of present day nuclear physics. I have tried to verify this fact by summarizing the talks given in this conference. We have seen that the clustering is playing important roles not only in traditional subjects but also in new phenomena and in new kind of nuclear systems. Typical examples of the latter are subjects in neutron-rich nuclei and in multifragment production phenomena. We have also seen that the knowledge and concept of clustering are being enriched through the studies of new phenomena and new kind of systems. As examples we can quote new types of clustering in neutron-rich nuclei and various mechanisms of fragment formation.

In order to cope with the present status of nuclear physics where the knowledge and concept of clustering are being enriched, it is indispensable to develop new models, new innovative calculational methods, and new theories. Compared with the physics of mean field, the physics of dynamical assembling and dissasembling of nucleons is more difficult to handle theoretically. It is important future task of ours to further develop more general theory of nuclear clustering.

The concept of clustering is of course not restricted to the many-body system of nucleons. Actually there have been discussions and interchange of ideas and techniques between nuclear cluster physics and quark cluster physics and atomic cluster physics [27]. Developments of present-day nuclear cluster physics in widely extended area will certainly be of great help for the study of these neighboring fields.

References

1. H.Horiuchi, *Clustering in Nuclear Structure and Collisions*, Proc. NATO Advanced Study Institute on Correlations and Clustering Phenomena in Subatomic Physics, Dronten (1996), edited by M. N. Harakeh, J. H. Koch, and O. Scholten, NATO ASI Series B: Physics Vol.**359** (Plenum Press, 1997), pp.29 ~ 51.
2. H.Horiuchi, *Microscopic Study of Clustering Phenomena in Nuclei*, Proc. Int. Symp. in Honor of Akito Arima; Nuclear Physics in 1990's, Santa Fe (1990), edited by D.H.Feng, J.N.Ginocchio, T.Otsuka and D.Strottman, Nucl.Phys. **A522**, 257c (1991).
3. A. Arima, *Summary Talk*, Proc. IV Int. Conf. on Clustering Aspects of Nuclear Structure and Nuclear Reactions, Chester (1984), edited by J.S.Lilley and M.A.Nagarajan, (Reidel; Dordrecht, Boston, Lancaster, 1985), p.399.
4. H.Horiuchi and K.Ikeda, *Cluster Model of the Nucleus*, International Review of Nuclear Physics, Vol.**4** (ed. T.T.S.Kuo and E.Osnes, World Scientific, Singapore, 1986), pp.1 ~ 258.
5. R. G. Lovas, R. J. Liotta, A. Insolia, K. Varga, and D. S. Delion, Phys. Reports **294**, 265 (1998).
6. K. Wildermuth and Y. C. Tang, *Unified Theory of the Nucleus*, (Vieweg, Braunschweig, Germany, 1977)
7. Y. Kanada-En'yo, Phys. Rev. Letters **81**, 5291 (1998).
8. S. Ali and A. R. Bodmer, Nucl. Phys. **80**, 99 (1966).
9. H. Horiuchi, *Microscopic Study of Nucleus-Nucleus Interaction*, Trends in Theoretical Physics, Vol.**2**, Chapt. 13, (ed. P. J. Ellis and Y. C. Tang, Addison-Wesley, Reading MA., 1991), pp.277 ~ 349.
10. M. E. Brandan and G. R. Satchler, Phys. Reports **285**, 143 (1997).
11. G. F. Bertsch and S. Das Gupta, Phys. Reports **160**, 189 (1988).
12. J. Aichelin, Phys. Reports **202**, 233 (1991).
13. K. Yabana and H. Horiuchi, Prog. Theor. Phys. **82**, 86 (1989).
14. Y. Abe, J. Hiura, and H. Tanaka, Prog. Theor. Phys. **49**, 800 (1973); S. Okabe, Y. Abe, and H. Tanaka, *ibid.* **57**, 866 (1977); S. Okabe and Y. Abe, *ibid.* **61**, 1049 (1979).
15. M. Seya, M. Kohno, and S. Nagata, Prog. Theor. Phys. **65**, 204 (1981).
16. H. Horiuchi, T. Maruyama, A. Ohnishi, and S. Yamaguchi, Proc. Int. Conf. on Nuclear and Atomic Clusters, Turku (1991), edited by M. Brenner, T. Lönnroth and F. B. Malik, (Springer; New York, 1992), p.512; Proc. Int. Symp. on Structure and Reactions of Unstable Nuclei, Niigata (1991), edited by K. Ikeda and Y. Suzuki, (World Scientific; Singapore,

1992), p.108.
17. Y. Kanada-En'yo, H. Horiuchi, and A. Ono, Phys. Rev. **C52**, 628 (1995).
18. Y. Kanada-En'yo, H. Horiuchi, and A. Doté, Phys. Rev. **C** to be published.
19. H. Sagawa, B. A. Brown, and H. Esbensen, Phys. Letters **B309**, 1 (1993).
20. A. A. Korsheninnikov, E. Yu. Nikolskii, T. Kobayashi, D. V. Aleksandrov, M. Fujimaki, H. Kumagai, A. A. Ogloblin, A. Ozawa, I. Tanihata, Y. Watanabe, and K. Yoshida, Phys. Letters **B343**, 53 (1995).
21. Y. Kanada-En'yo and H. Horiuchi, Phys. Rev. **C52**, 647 (1995).
22. H. Tamura, private communication and *Proc. Sendai Int. Workshop on the Spectroscopy of Hypernuclei*, Genshikaku Kenkyu Vol. **43**, No.4, 85 (1998).
23. E. Hiyama, M. Kamimura, K. Miyazaki, and T. Motoba, Phys. Rev. **C59**, 2351 (1999).
24. H. Bandō, T. Motoba, and J. Žofka, *Production, Structure and Decay of Hypernuclei*, International Journal of Modern Physics A, Vol.**5**, No.21, (1990), pp.4021 ∼ 4198.
25. Y. Sugawa and H. Horiuchi, Phys. Rev. **C60**, (1999), to be published.
26. H. Takemoto, H. Horiuchi, A. Engel, and A. Ono, Phys. Rev. **C54**, 266 (1996); H. Takemoto, H. Horiuchi, and A. Ono, Phys. Rev. **C57**, 811 (1998).
27. W. Greiner, *Summary Talk*, Proc. Second Int. Conf. on Atomic and Nuclear Clusters '93, Santorini (1993), edited by G.S.Anagnostatos and W.von Oertzen, Z. Physik **A349**, No.3/4, 315 (1994).

Program of the Conference

Monday, June 14

09:00 Conference opening

MORNING SESSION I - *Chairman:* **N. Cindro**, Zagreb

09:10 **W. Greiner**, Frankfurt
From Bochum to Rab 30 years of cluster conferences, a theoretician's point of view

09:55 **W. von Oertzen**, Berlin
30 years of cluster physics, an experimentalist's point of view

10:40 COFFEE BREAK

MORNING SESSION II - *Chairman:* **R.A. Ricci**, Padova/Legnaro

11:10 **R.G. Lovas**, Debrecen
Clusterization and the Pauli principle

11:45 **D. Baye**, Brussels
Discrete ambiguities in the cluster-cluster interaction and supersymmetry

12:20 **A. Tohsaki**, Ueda
Microscopic aspects of alpha-clustering in nucleonic sea

AFTERNOON SESSION - *Chairman:* **R.G. Lovas**, Debrecen

15:30 **J. Cseh**, Debrecen
Clusters, symmetries and predictions

15:55 **S. Brant**, Zagreb
Interacting boson plus broken pair and Interacting boson fermion -fermion model approach to high spin states

16:20 **T. Lönnroth**, Turku
Microscopic Properties of alpha-Cluster States

16:45 **S.Yu. Kun**, Canberra
Spontaneous nuclear clustering

17:10 **A. Ray**, Calcutta
Nuclear orbiting in A=40 and A=100 mass regions

17:35 **M. Papa**, Catania
Fluctuations in the excitation functions of dissipative collisions induced on the ^{27}Al + ^{27}Al system at E_{lab} = 4.4 MeV/nucleon and dinuclear molecules

17:55 **M. La Commara**, Darmstadt
Production of exotic nuclei near ^{100}Sn in fusion reactions followed by cluster emission

18:15 **E. Běták**, Bratislava
Clustering Phenomena and Isotopic Effects in Preequilibrium Reactions

18:35 **A. Tumino**, Catania
The algebraic theory of the three body problem applied to the study of the ^{12}C nucleus

18:50 **M. Rousseau**, Strasbourg
Cluster emission of ^{8}Be in the ^{28}Si+^{12}C fusion reaction at low temperature

20:00 Posting the poster session

Tuesday, June 15

MORNING SESSION I - *Chairman:* **M. Lattuada**, Catania

09:00 **R. Betts**, Argonne/Chicago
Forty years of nuclear molecules

09:35 **B.R. Fulton**, Birmingham
The six-alpha chain story

10:10 **A. Di Pietro**, Edinburgh
6-α particle emission in the reaction ^{13}N+^{11}B

10:30 COFFEE BREAK

MORNING SESSION II - *Chairman:* **Y. Abe**, Kyoto

11:00 **P. Schuck**, Grenoble
Self consistent pair fluctuations in strongly correlated Fermion systems

11:35 **G. Röpke**, Rostock
Cluster mean-field approach for correlations in dense matter

12:10 **Y. Sakuragi**, Osaka
Multi-cluster structure of highly-excited states in ^{24}Mg

12:35 **I. Boztosun**, Oxford
Coupled-channels calculations for deformed light nuclei. Problems and solutions?

AFTERNOON SESSION - *Chairman:* **F. Haas**, Strasbourg

15:00 **M.P. Nicoli**, Birmingham
Refractive effects in the collisions between carbon and oxygen nuclei: from resonances to Airy structures

15:25 **Y. Kondō**, Kyoto
A deep potential description of ^{12}C+^{12}C, ^{16}O+^{12}C and ^{16}O+^{16}O scattering: Gross resonant structures and nuclear rainbows

16:00 **E. Uegaki**, Akita
Spin alignment in heavy-ion reactions

16:25 **B.A. Robson**, Canberra
Hot hyperdeformed coherently rotating clusters in ^{24}Mg+^{24}Mg and ^{28}Si+^{28}Si elastic and inelastic scattering

16:45 Reception at the City Hall (Gradska loža)
Tour of the city of Rab

Wednesday, June 16

MORNING SESSION - *Chairman:* **B.A. Robson**, Canberra

8:30 **H.O. Lutz**, Bielefeld
Size-limited superfluid aggregates: ^{4}He clusters

9:05 **C. Walther**, Karlsruhe
Radiative cooling of V_{13}^+ clusters

9:25 Round table discussion (R. Betts, H. Horiuchi, W. von Oertzen)

10:10 COFFEE BREAK

11:00 Boat excursion to the Islands of Krk and Košljun

Thursday, June 17

MORNING SESSION I - *Chairman:* **G. Münzenberg**, Darmstadt

09:00 **D.N. Poenaru**, Bucharest-Magurele
Multi-cluster nuclear molecules

09:35 **G. Royer**, Nantes
Cluster and alpha emission as an asymmetric fission process

10:00 **P.O. Hess**, Giessen
How to treat clusterization of heavy nuclei: an algebraic attempt

10.105 COFFEE BREAK (Conference photograph)

MORNING SESSION II - *Chairman:* **M. Itkis**, Dubna

11:00 **Y. Suzuki**, Niigata
The utility of a microscopic cluster description in light halo and hypernuclei

11:35 **P.G. Thirolf**, Garching
Particle-unbound clusters far-off stability

12:10 **R. Wolski**, Dubna/Cracow
Cluster structure of ^{6}He studied by means of ^{6}He+^{4}He and ^{6}He+^{1}H reactions at the beam energy of 25 MeV/nucleon

12:45 **W. Neubert**, Dresden
Fragment production in proton induced interactions with light nuclei

AFTERNOON SESSION - *Chairman:* **Y. Suzuki**, Niigata

15:30 **M. Freer**, Birmingham
^{6}He+^{6}He cluster states in ^{12}Be

15:55 **Đ. Miljanić**, Zagreb
Clustering in beryllium nuclei

16:20 **Y. Katō**, Sapporo
Pairing Correlations in Light Neutron-Rich Nuclei

16:45 **L. Vannucci**, Legnaro
Clustering configurations in intermediate medium mass systems

17:10 **S. Ohkubo**, Kochi
^{40}Ca+α+α cluster model of ^{48}Cr

17:30 **K. Arai**, Niigata
Structure of the A=6 nuclei with an extended three-cluster model

17:50 **S. Cherubini**, Louvain-la-Neuve
Nuclear astrophysics with indirect methods

18:10 **N. Itagaki**, Sapporo
^{12}Be is not a magic number nucleus

18:30 - 19:30 POSTER SESSION

Friday, June 18

MORNING SESSION I - *Chairman:* **L.G. Moretto**, Berkeley

09:00 **W. Reisdorf**, Darmstadt
Clusters and nuclear flow

09:35 **A. Ono**, Sendai
Cluster formation dynamics in medium energy nuclear collisions

10:10 **M.F. Rivet**, Orsay
Cluster formation in nuclear dynamics below 100 MeV/nucleon

10:45 COFFEE BREAK

MORNING SESSION II - *Chairman:* **A. Ono**, Sendai

11:15 **W. Trautmann**, Darmstadt
Equilibration and phase transition in spectator reactions

11:50 **V. de la Mota**, Nantes
Describing fragments with LV and DYWAN transport models

12:15 **E. Gadioli**, Milano
Alpha particle emission in the interaction of ^{12}C with nuclei at incident energies of several ten MeV/nucleon

12:40 **Ph. Eudes**, Nantes
Evidence for nuclear matter transparency in central heavy-ion collisions for light systems around Fermi energy

AFTERNOON SESSION I - *Chairman:* **E. Gadioli**, Milan

16:00 **L.G. Moretto**, Berkeley
The role of fragmentation phase space in multifragmentation

16:35 **B. Borderie**, Orsay
Multifragmentation and spinodal decomposition of finite systems

17:00 **M. Petrovici**, Bucharest-Magurele
Clusters as sensitive probes for a deeper understanding of baryon flow observables

17:25 Pause

AFTERNOON SESSION II - *Chairman:* **W. Reisdorf**, Darmstadt

17:45 **M. Dželalija**, Split
Generalized approach to entropy determination in heavy-ion collisions

18:10 **F. Rami**, Strasbourg
System size effects in cluster production at SIS energies

18:35 **M. Korolija**, Zagreb
System expansion and clusters

20:00 Conference dinner

Saturday, June 19

MORNING SESSION - *Chairman:* **D.N. Poenaru**, Bucharest

08:30 **G. Münzenberg**, Darmstadt
Approaching superheavy elements

09:05 **Y. Abe**, Kyoto
Theory of nuclear reactions leading to superheavy elements

09:40 **M. Itkis**, Dubna
Shell effects and the fission of heavy and superheavy nuclei

10:15 **G.G. Adamian**, Giessen
Peculiarities of production of very heavy nuclei

10:35 COFFEE BREAK

11:05 **H. Horiuchi**, Kyoto
Nuclear clusters on the eve of the third millennium

11:50 Conference closing

List of Posters

Fragment Production in Proton Induced Interactions with Light Nuclei
L.N. Andronenko, M.N. Andronenko, W. Neubert, D.M. Seliverstov

The Continuum Level Density in the Microscopic Cluster Model
K. Arai

High Spin Resonances in $^{12}C+^{12}C$ Scattering
C.A. Bremner, S.P.G. Chappell, W.D.M. Rae

^{24}Mg at High Spin
S.P.G. Chappell, C. Bremner, W.D.M. Rae

Calculation of a Nuclear System with the Non-Spherical Core on the Example of $^{10}Li=^{9}Li+n$
G.F. Filippov, S.V. Korennov, and K. Katō

$^{7}Li(d,\alpha\alpha)n$ Reaction and the 2^{+} Isospin-Mixed Doublet in ^{8}Be
N. Godinović, S. Blagus, M. Bogovac, M. Lattuada, M. Milin, Đ. Miljanić, D. Rendić, N. Soić, C. Spitaleri, M. Zadro

Microscopic Coupled-Channel Study of $^{12}C+^{12}C$ Resonance Reactions
M. Ito, Y. Sakuragi, Y. Hirabayashi

Multi-Cluster Structure of Highly-Excited States in ^{24}Mg
M. Ito, Y. Sakuragi, Y. Hirabayashi

$^{6}He+^{6}He$ Molecular States in ^{12}Be
M. Ito, Y. Sakuragi

Heavy-Ion Resonances of the $^{16}O+^{16}O$ System in Multi-Cluster Exit Channels
M. Katsuma, M. Ito, Y. Sakuragi

Study of Molecular States in Boron, Beryllium and Carbon Isotopes
P.J. Leask for the CHARISSA Collaboration

Booklet of Summaries

Retrospective

Fundamentals of nucleon clustering in the nuclear medium

Clustering models

Resonances, nuclear molecules, orbiting

Highly deformed nuclear systems, fission, cluster decay

Clusters in nuclear structure

List of Participants

ABBONDANNO Ugo
Istituto Nazionale di Fisica Nucleare, Sezione di Trieste, 34127 Trieste, Italy
abbondanno@trieste.infn.it

ABE Yasuhisa
Yukawa Institute for Theoretical Physics, Kyoto University, Kyoto 606-8502, Japan
abey@yukawa.kyoto-u.ac.jp

ADAMIAN Gurgen G.
Institut für Theoretische Physik der Justus-Liebig- Universität, Heinrisch - Buff - Ring 16, D-35392 Giessen, Germany
Gurgen.Adamian@physik.uni-giessen.de

ARAI Koji
Department of Physics, Niigata University, Niigata 950-2181, Japan
arai@nt.sc.niigata-u.ac.jp

BASRAK Zoran
Ruđer Bošković Institute, P.O.Box 1016, HR-10001 Zagreb, Croatia
basrak@irb.hr

BAYE Daniel
Physique Nucléaire Théorique et Physique Mathématique C.P. 229, Université Libre de Bruxelles, B 1050 Brussels, Belgium
dbaye@ulb.ac.be

BĚTÁK Emil
Institute of Physics, Slovak Academy of Sciences, SK-84228 Bratislava, Slovakia
betak@savba.sk

BETTS Russell
Argonne National Laboratory, Argonne, USA *and* University of Illinois at Chicago, Chicago, USA
betts@anl.gov

BORDERIE Bernard
Institut de Physique Nucléaire, IN2P3-CNRS, F-91406 Orsay Cedex, France
borderie@ipno.in2p3.fr

BOZTOSUN Ismail
Department of Nuclear Physics, University of Oxford Keble Road, Oxford OX1 3RH, United Kingdom
i.boztosun1@physics.ox.ac.uk

BRANT Slobodan
Department of Physics, Faculty of Science, University of Zagreb, 10 000 Zagreb, Croatia
brant@phy.hr

BREMNER Charles
Department of Nuclear Physics, University of Oxford, Keble Road, Oxford OX1 3RH, United Kingdom
c.bremner1@physics.ox.ac.uk

CHAPPELL Steve
Department of Nuclear Physics, University of Oxford, Keble Road, Oxford OX1 3RH, United Kingdom
s.chappell1@physics.ox.ac.uk

CHERUBINI Silvio
Institut de Physique Nucléaire, Université Catholique de Louvain, 1348 Louvain-la-Neuve, Belgium
cherubini@fynu.ucl.ac.be

CINDRO Nikola
Ruđer Bošković Institute, P.O.Box 1016, HR-10001 Zagreb, Croatia
ncindro@faust.irb.hr

CSEH Jozsef
Institute of Nuclear Research of the Hungarian Academy of Sciences, H-4001 Debrecen, Pf. 51, Hungary
cseh@atomki.hu

ČAPLAR Roman
Ruđer Bošković Institute, P.O.Box 1016, HR-10001 Zagreb, Croatia
caplar@faust.irb.hr

de la MOTA Virginia
Laboratoire SUBATECH, Université de Nantes, École des Mines de Nantes, IN2P3/CNRS, F-44 072 Nantes Cedex 03, France
delamota@subatech.in2p3.fr

DI PIETRO Alessia
Department of Physics and Astronomy University of Edinburgh, Edinburgh, EH9 3JZ, UK
dipietro@ph.ed.ac.uk

DŽELALIJA Mile
Department of Physics, University of Split, 21 000 Split, Croatia
mile@pmfst.hr

EUDES Philippe
Laboratoire SUBATECH
Université de Nantes, École des Mines de Nantes, IN2P3/CNRS, F-44 072 Nantes Cedex 03, France
eudes@subatech.in2p3.fr

FREER Martin
School of Physics and Astronomy, University of Birmingham, Birmingham, B15 2TT, UK
mf@np.ph.bham.ac.uk

FULTON Brian R.
School of Physics and Astronomy University of Birmingham, Birmingham, B15 2TT, England
b.r.fulton@birmingham.ac.uk

GADIOLI Ettore
Dipartimento di Fisica Istituto Nazionale di Fisica Nucleare, Sezione di Milano, Milano, Italy
ettore.gadioli@mi.infn.it

GADIOLI ERBA Enrica
Dipartimento di Fisica
Istituto Nazionale di Fisica Nucleare, Sezione di Milano Milano, Italy
gadioli@mi.infn.it

GODINOVIĆ Nikola
Technical University of Split, Ruđera Boškovića bb, 21 000 Split, Croatia
ngod@fesb.hr

GREINER Walter
Institut für Theoretische Physik der Universität Frankfurt am Main, Frankfurt, Germany
greiner@th.physik.uni-frankfurt.de

HAAS Florent
Institut de Recherches Subatomiques, CNRS-IN2P3-ULP, Strasbourg, France
florent.haas@ires.in2p3.fr

HESS Peter O.
Institut für Theoretische Physik der Justus-Liebig- Universität
Heinrisch-Buff-Ring 16, D-35392 Giessen, Germany
hess@theo.physik.uni-giessen.de

HORIUCHI Hisashi
Department of Physics, Kyoto University, Kyoto 606-8502, Japan
horiuchi@ruby.scphys.kyoto-u.ac.jp

IMANISHI Bunryu
KEK Tanashi branch, 188-8501 Tanashi, Tokyo, Japan
imanishi@tanashi.kek.jp

ITAGAKI Naoyuki
Division of Physics, Graduate School of Science, Hokkaido University
Sapporo 060-0810 *and* RIKEN, Wako, Japan
itagaki@postman.riken.go.jp

ITKIS Mikhail G.
Flerov Laboratory of Nuclear Reactions, JINR, 141 980 Dubna, Russia
itkis@sungraph.jinr.dubna.su

ITO Makoto
Department of Physics, Osaka City University, Osaka 558-8585, Japan
itom@ocunp.hep.osaka-cu.ac.jp

KATŌ Kiyoshi
Division of Physics, Graduate School of Science, Hokkaido University,
Sapporo 060-0810, Japan
kato@nucl.sci.hokudai.ac.jp

KONDŌ Yosio
Department of Natural Sciences, Kyoto Women's University, 35
Kitahiyoshi-cho, Imakumano Higashiyama-ku, Kyoto 605-8501, Japan
kondo@yukawa.kyoto-u.ac.jp

KORENNOV Sergei
Division of Physics, Graduate School of Science, Hokkaido University,
Sapporo 060-0810, Japan
korennov@nucl.sci.hokudai.ac.jp

KOROLIJA Milorad
Ruđer Bošković Institute, P.O.Box 1016 HR-10001 Zagreb, Croatia
korolija@faust.irb.hr

KUN Sergey Yu.
Department of Theoretical Physics, RSPhysSE, IAS, Australian National
University, Canberra ACT 0200, Australia
ksu105@rsphysse.anu.edu.au

LA COMMARA Marco
Gesellschaft für Schwerionenforschung (GSI), D-64291 Darmstadt, Germany
m.lacommara@gsi.de

LATTUADA Marcello
Dipartimento di Fisica, Università di Catania *and* LNS-INFN, Via S. Sofia
44, 95123 Catania, Italy
lattuada@lns.infn.it

LEASK Peter J.
School of Physics and Astronomy, University of Birmingham, Birmingham, B15 2TT, England
P.J.Leask@bham.ac.uk

LOVAS Rezso G.
Institute of Nuclear Research of the Hungarian Academy of Sciences, Debrecen, P.O.Box 51, H-4001, Hungary
rgl@atomki.hu

LÖNNROTH Tom
Department of Physics, Åbo Akademi, FIN-20500 Turku, Finland
tlonnrot@ra.abo.fi

LUTZ Hans O.
Universität Bielefeld, Fakultät für Physik, Universitätsstr. 25, D-33615 Bielefeld, Germany
lutz@physik.uni-bielefeld.de

MARCINKOWSKY Andrzej
Soltan Institute for Nuclear Studies, Hoza street No 69, 00-681 Warszawa, Poland
amar@fuw.edu.pl

MARIANSKI Bohdan
Soltan Institute for Nuclear Studies, Hoza street No 69, 00-681 Warszawa, Poland

MILIN Matko
Ruđer Bošković Institute, P.O.Box 1016, HR-10001 Zagreb, Croatia
milin@vstlnr.irb.hr

MILJANIĆ Đuro
Ruđer Bošković Institute, P.O.Box 1016, HR-10001 Zagreb, Croatia
miljanic@vstlnr.irb.hr

MORETTO Luciano G.
University of California, Berkeley *and* Lawrence Berkeley National Laboratory, MS 88, 94720 Berkeley, California, USA
LGMoretto@lbl.gov

MÜNZENBERG Gottfried
Gesellschaft für Schwerionenforschung (GSI), D-64291 Darmstadt, Germany
G.Muenzenberg@gsi.de

NEUBERT Waldemar
Forschungszentrum Rossendorf, Dresden, Germany
W.Neubert@fz-rossendorf.de

NICOLI Marie-Paule
School of Physics and Astronomy, University of Birmingham, Birmingham, B15 2TT, England
mpn@np.ph.bham.ac.uk

OHKUBO Shigeo
Department of Applied Science and Environment, Kochi Women's University, Kochi 780-8515, Japan
ohkubo@is.kochi-wu.ac.jp

ONO Akira
Department of Physics, Tohoku University, Sendai 980-8578, Japan
ono@nucl.phys.tohoku.ac.jp

PAPA Massimo
INFN, Sezione di Catania, Corso Italia 57, 95129 Catania, Italy
Papa@lns.infn.it

PAVIN Nenad
Department of Physics, Faculty of Science, University of Zagreb, 10 000 Zagreb, Croatia
npavin@phy.hr

PETROVICI Mihai
National Institute for Physics and Nuclear Engineering, P.O.Box MG-6, Bucharest-Magurele, Romania
mpetro@ifin.nipne.ro

POENARU Dorin N.
National Institute of Physics and Nuclear Engineering, P.O.Box MG-6, RO-76900 Bucharest-Magurele, Romania
poenaru@ifin.nipne.ro

RAMI Fouad
IReS, 23,rue du Loess, B.P.28-BAT.20, 67037 Strasbourg Cedex 2, France
fouad.rami@ires.in2p3.fr

RAY Amlan
Variable Energy Cyclotron Center, 1/AF Bidhannagar, Calcutta 700064, India
ray@veccal.ernet.in

REISDORF Willibrord
Gesellschaft für Schwerionenforschung (GSI), D-64291 Darmstadt, Germany
W.Reisdorf@gsi.de

RICCI Renato A.
Dipartimento di Fisica dell'Universita' di Padova, 35131 Padova, Italy *and* INFN, Laboratori Nazionali di Legnaro, 35020 Legnaro, Italy
raricci@lnl.infn.it

RIVET Marie-France
Institut de Physique Nucléaire, IN2P3-CNRS, F-91406 Orsay Cedex, France
rivet@ipno.in2p3.fr

ROBSON B.A.
Department of Theoretical Physics, RSPhysSe, IAS, Australian National University, Canberra ACT 0200, Australia
bar105@rsphysse.anu.edu.au

ROUSSEAU Marc
IReS, CNRS - IN2P3 - ULP, F-67037 Strasbourg Cedex 02, France
marc.rousseau@ires.in2p3.fr

ROYER Guy
Laboratoire SUBATECH, Université de Nantes, École des Mines de Nantes, IN2P3/CNRS, F-44 072 Nantes Cedex 03, France
royer@subatech.in2p3.fr

RÖPKE Gerd
FB Physik, Universität Rostock, D-18051 Rostock, Germany
gerd@darss.mpg.uni-rostock.de

SAKURAGI Yukinori
Department of Physics, Osaka City University, Osaka 558-8585, Japan
sakuragi@ocunp.hep.osaka-cu.ac.jp

SCHUCK Peter
Institut des Sciences Nuclèaires, Universitè Joseph Fourier 53, Avenue des Martyrs, F-38026, Grenoble Cedex, France
schuck@isnhp4.in2p3.fr

SOIĆ Neven
Ruđer Bošković Institute, P.O.Box 1016, HR-10001 Zagreb, Croatia
soic@vstlnr.irb.hr

SUZUKI Yasuyuki
Department of Physics, Niigata University, Niigata 950-2181, Japan
suzuki@nt.sc.niigata-u.ac.jp

SZILNER Suzana
Ruđer Bošković Institute, P.O.Box 1016, HR-10001 Zagreb, Croatia
suzana@faust.irb.hr

TAKASHINA Masaaki
Department of Physics, Osaka City University, Osaka 558-8585, Japan
itom@ocunp.hep.osaka-cu.ac.jp

THIROLF Peter G.
Sektion Physik, Universität München, Am Coulombwall 1, D-85748 Garching, Germany
peter.thirolf@physik.uni-muenchen.de

TOHSAKI Akihiro
Department of Fine Materials Engineering, Shinshu University, Ueda, 386-8567, Japan
asuzuk1@giptc.shinshu-u.ac.jp

TRAUTMANN Wolfgang
Gesellschaft für Schwerionenforschung (GSI), D-64291 Darmstadt, Germany
W.Trautmann@gsi.de

TUMINO Aurora
Dipartimento di Fisica, Università di Catania *and* LNS-INFN, Via S. Sofia 44, 95123 Catania, Italy
tumino@lns.infn.it

UEGAKI Eiji
Department of Physics, Akita University, Akita 010-8502, Japan
ue@uws13.phys.akita-u.ac.jp

VANNUCCI Luigi
INFN, Laboratori Nazionali di Legnaro, 35020 Legnaro, Italy
vannucci@lnl.infn.it

von OERTZEN Wolfram
Hahn-Meitner-Institut, Glienicker Strasse 100, D-14109 Berlin, Germany *and* Fachbereichphysik, Freie Berlin Universität Berlin, Germany
oertzen@hmi.de

WALTHER Clemens
Forschungszentrum Karlsruhe GmbH / INE, Postfach 3640, D-76021 Karlsruhe, Germany
walther@inemail.ine.fzk.de

WOLSKI Roman
Flerov Laboratory of Nuclear Reactions, JINR, Dubna, 141980 Russia *and* Institute of Nuclear Physics, Cracow, Poland
wolski@nrsun.jinr.dubna.su

WOOLLISCROFT Richard
School of Physics and Astronomy, University of Birmingham, Birmingham, B15 2TT, England
r.j.woolliscroft@bham.ac.uk

Author Index

www.ingramcontent.com/pod-product-compliance
Lightning Source LLC
LaVergne TN
LVHW082257241026
850861LV00008B/64
* 9 7 8 9 8 1 0 2 4 2 3 3 6 *